Lecture Notes in Physics

Volume 942

The Lecture Notes in Physics

The series Lecture Notes in Physics (LNP), founded in 1969, reports new developments in physics research and teaching-quickly and informally, but with a high quality and the explicit aim to summarize and communicate current knowledge in an accessible way. Books published in this series are conceived as bridging material between advanced graduate textbooks and the forefront of research and to serve three purposes:

- to be a compact and modern up-to-date source of reference on a well-defined topic
- to serve as an accessible introduction to the field to postgraduate students and nonspecialist researchers from related areas
- to be a source of advanced teaching material for specialized seminars, courses and schools

Both monographs and multi-author volumes will be considered for publication. Edited volumes should, however, consist of a very limited number of contributions only. Proceedings will not be considered for LNP.

Volumes published in LNP are disseminated both in print and in electronic formats, the electronic archive being available at springerlink.com. The series content is indexed, abstracted and referenced by many abstracting and information services, bibliographic networks, subscription agencies, library networks, and consortia.

Proposals should be sent to a member of the Editorial Board, or directly to the managing editor at Springer:

Christian Caron
Springer Heidelberg
Physics Editorial Department I
Tiergartenstrasse 17
69121 Heidelberg/Germany
christian.caron@springer.com

More information about this series at http://www.springer.com/series/5304

Mauro Carfora • Annalisa Marzuoli

Quantum Triangulations

Moduli Space, Quantum Computing, Non-Linear Sigma Models and Ricci Flow

Second Edition

 Springer

Mauro Carfora
Dipartimento di Fisica
Università degli Studi di Pavia
Pavia, Italy

Annalisa Marzuoli
Dipartimento di Matematica
Università degli Studi di Pavia
Pavia, Italy

ISSN 0075-8450   ISSN 1616-6361   (electronic)
Lecture Notes in Physics
ISBN 978-3-319-67936-5   ISBN 978-3-319-67937-2   (eBook)
https://doi.org/10.1007/978-3-319-67937-2

Library of Congress Control Number: 2017955946

Printed on acid-free paper

This Springer imprint is published by Springer Nature
The registered company is Springer International Publishing AG
The registered company address is: Gewerbestrasse 11, 6330 Cham, Switzerland

Preface to the First Edition

The above illustration shows a variant woodcut printer's device on the verso last leaf of a rare sixteenth-century edition of Plato's *Timaeus* (*Divini Platonis Operum a Marsilio Ficino tralatorum, Tomus Quartus. Lugduni, apud Joan Tornaesium M.D.XXXXX*). The printer's device to the colophon shows a medallion with a tetrahedron in the center and the motto round the border: *Nescit Labi Virtus* (Virtue Cannot Fail). This woodcut beautifully illustrates the role of the perfect shape of the tetrahedron in classical culture. The tetrahedron conveys such a profound impression of stability as to be considered an epitome of virtue, unfailingly inspiring us with the depth and elegance of its shape. In the course of history, the geometry of the tetrahedron, of the Platonic solids, and more generally of the highly symmetrical discrete patterns one encounters in nature and art has always been connected with some of the more sophisticated aspects of mathematics and physics of the day. From Plato's *Timaeus* to Piero della Francesca's *Libellus de Quinque Corporibus Regularibus* to Pacioli's *De Divina Proportione* up to Kepler's *Harmonices Mundi*, there have always been attempts to involve the Platonic solids and their many variants to provide mathematical models for the physical universe. What makes these shapes perfectly irresistible to many mathematicians and physicists, both amateur and professional, is culturally related to their long-standing role in natural philosophy but also to the recondite fact that the geometry of these discrete structures often points to unexpected connections between very distinct aspects of mathematics and physics. And modern theoretical physics has drawn even further attention to the latter property.

Indeed, polyhedral manifolds, the natural generalization of Platonic solids, play quite a distinguished role in such settings as Riemann moduli space theory, strings and quantum gravity, topological quantum field theory, condensed matter physics, and critical phenomena. The motivation for such a wide spectrum of applications goes beyond the observation that polyhedral manifolds provide a natural discrete analogue of the smooth manifolds in which a physical theory is framed. Rather, it is often a consequence of an underlying structure, only apparently combinatorial, which naturally calls into play nontrivial aspects of representation theory, complex analysis, and topology in a way that makes manifest the basic geometric structures of the physical interactions involved. In spite of these remarks, one has to admit that, in almost all existing literature, the role of triangulated manifolds remains merely a convenient discretization of the physical theory, a grab bag of techniques that are computationally rather than conceptually apt to disclose the underlying physics and geometry. The restriction to such a computational role may indeed be justified by the physical nature of the problem, as is often the case in critical statistical field theory, but sometimes it is not. This is the discriminating criterion motivating these lecture notes, since, in the broad panorama the theory offers, the relation between polyhedral surfaces, Riemann moduli spaces, noncritical string theory, and quantum computing emerges as a clear path probing the connection between triangulated manifolds and quantum physics to the deepest levels.

Chapter 1 is devoted to a detailed study of the geometry of polyhedral manifolds and in particular triangulated surfaces. This subject, which may be considered a classic, has recently seen a flourishing of many new results of great potential impact in the physical applications of the theory.

Here the focus is on results which are either new or not readily accessible in the standard repertoire. In particular, we discuss from an original perspective the structure of the space of all polyhedral surfaces of a given genus and their stable degenerations. In such a framework, and within the whole landscape of the space of polyhedral surfaces, an important role is played by the conical singularities associated with the Euclidean triangulation of a surface. We provide a detailed analysis of the geometry of these singularities and introduce the associated notion of cotangent cones, circle bundles, and the attendant Euler class on the space of polyhedral surfaces. This is a rather delicate point which appears in many guises in quantum gravity and string theory and which is related to the role that Riemann moduli spaces play in these theories. Not surprisingly, the Witten–Kontsevich model lurks in the background of our analysis, and some of the notions we introduce may well serve to illustrate, from a more elementary point of view, the often very technical definitions that characterize this subject.

We turn in Chap. 2 to the formulation of a powerful dictionary between polyhedral surfaces and complex geometry. It must be noted that, in both the mathematical and the physical applications of the theory, the connection between Riemann surfaces and triangulations typically emphasizes the role of ribbon graphs and the associated metric. The conical geometry of the polyhedral surface is left aside and seems to play no significant role. This attitude can be motivated by Troyanov's basic observation that the conformal structure does not see the conical singularities of a polyhedral surface. However, this gives a rather narrow perspective on the much wider role that the theory has to offer. Thus, we relate a polyhedral surface to a corresponding Riemann surface by fully taking into account its conical geometry. This connection is many-faceted and exploits a vast repertoire of notions, ranging from complex function theory to algebraic geometry. We start by defining the barycentrically dual polytope associated with a polyhedral surface and discuss the geometry of the corresponding ribbon graph.

By adapting to our case an elegant version of Strebel's theorem provided by Mulase, we explicitly construct the Riemann surface associated with the dual polytope. This brings us directly to the analysis of Troyanov's singular Euclidean structures and to the construction of the bijective map between the moduli space $\mathfrak{M}_{g,N_0}$ of Riemann surfaces (M, N_0) with N_0 marked points, decorated with conical angles, and the space of polyhedral structures. In particular, the first Chern class of the line bundles naturally defined over $\mathfrak{M}_{g,N_0}$ by the cotangent space at the i th marked point is related to the corresponding Euler class of the circle bundles over the space of polyhedral surfaces defined by the conical cotangent spaces at the i th vertex of the triangulation. While this is not an unexpected connection, the analogy with Witten–Kontsevich theory being obvious, we stress that the conical geometry adds to this property the possibility of a deep and explicit characterization of the Weil–Petersson form in terms of the edge lengths of the triangulation. This result is obtained by a subtle interplay between the geometry of polyhedral surfaces and 3D hyperbolic geometry, and it will be discussed in detail in Chap. 3, since it explicitly hints at the connection between polyhedral surfaces and quantum geometry in higher dimensions.

Chapter 3 deals with the interplay between polyhedral surfaces and 3D hyperbolic geometry mentioned above and the characterization of the Weil–Petersson form ω_{WP} on the space of polyhedral structures with given conical singularities. An important role in such a setting is played by the interesting recent results by G. Mondello regarding an explicit expression for the Weil–Petersson form for hyperbolic surfaces with geodesic boundaries. In order to construct a combinatorial representative of ω_{WP} for polyhedral surfaces, we exploit this result and the connection between similarity classes of Euclidean triangles and the triangulations of 3-manifolds by ideal tetrahedra. We describe this construction in detail, since it will also characterize a striking mapping between closed polyhedral surfaces and hyperbolic surfaces with geodesic boundaries. Such a mapping has a life of its own, strongly related to the geometry of the moduli space of pointed Riemann surfaces, and it provides a useful framework for discussing such matters as open/closed string dualities.

The content of Chap. 4 (Chap. 5 in this second edition) constitutes an introduction to the basic ideas of 2D quantum field theory and noncritical strings. This is classic material which nevertheless proves useful for illustrating the interplay between quantum field theory, the moduli space of Riemann surfaces, and the properties of polyhedral surfaces which are the *leitmotiv* of this LNP. At the root of this interplay lies 2D quantum gravity. It is well known that such a theory allows for two complementary descriptions: on the one hand, we have a conformal field theory (CFT) living on a 2D world-sheet, a description that emphasizes the geometrical aspects of the Riemann surface associated with the world-sheet, while on the other, the theory can be formulated as a statistical critical field theory over the space of polyhedral surfaces (dynamical triangulations). We show that many properties of such 2D quantum gravity models are related to a geometrical mechanism which allows one to describe a polyhedral surface with N_0 vertices as a Riemann surface with N_0 punctures dressed with a field whose charges describe discretized curvatures (related to the deficit angles of the triangulation). Such a picture calls into play the (compactified) moduli space of genus g Riemann surfaces with N_0 punctures $\mathfrak{M}_{g;N_0}$ and allows one to prove that the partition function of 2D quantum gravity is directly related to computation of the Weil–Petersson volume of $\mathfrak{M}_{g;N_0}$. By exploiting the large N_0 asymptotics of such Weil–Petersson volumes, recently characterized by Manin and Zograf, it is then easy to relate the anomalous scaling properties of pure 2D quantum gravity, the KPZ exponent, to the Weil–Petersson volume of $\mathfrak{M}_{g;N_0}$. This ultimately relates to the difficult problem of constructively characterizing the appropriate functional measures on spaces of Riemannian manifolds, often needed in the study of quantum gravity models and in the statistical mechanics of extended objects. We also address the more general case of the interaction of conformal matter with 2D quantum gravity and in particular the characterization of the associated KPZ exponents. By elaborating on the recent remarkable approach to the spectral theory over polyhedral surfaces due to A. Kokotov, we provide a general framework for analyzing KPZ exponents by discussing the scaling properties of the corresponding discretized Liouville theory.

In a rather general sense, polyhedral surfaces also provide a natural kinematical framework within which we can discuss open/closed string duality. A basic problem in such a setting is to provide an explanation of how open/closed duality is generated dynamically and in particular how a closed surface is related to a corresponding open surface, with gauge-decorated boundaries, in such a way that the quantization of this correspondence leads to an open/closed duality. Typically, the natural candidate for such a mapping is Strebel's theorem, which allows one to reconstruct a closed N-pointed Riemann surface M of genus g from the datum of the quadratic differential associated with a ribbon graph.

Are ribbon graphs, with the attendant BCFT techniques, the only key for addressing the combinatorial aspects of open/closed string duality? The results of Chap. 3 show that from a closed polyhedral surface we naturally get an open hyperbolic surface with geodesic boundaries. This gives a geometrical mechanism describing the transition between closed and open surfaces which, in a dynamical sense, is more interesting than Strebel's construction. Such a correspondence between closed polyhedral surfaces and open hyperbolic surfaces is indeed easily promoted to the corresponding moduli spaces: $\mathfrak{M}_{g;N_0} \times \mathbb{R}_+^N$, the moduli spaces of N_0-pointed closed Riemann surfaces of genus g whose marked points are decorated with the given set of conical angles, and $\mathfrak{M}_{g;N_0}(L) \times \mathbb{R}_+^{N_0}$, the moduli spaces of open Riemann surfaces of genus g with N_0 geodesic boundaries decorated by the corresponding lengths. Such a correspondence provides a nice kinematical setup for establishing an open/closed string duality, by exploiting the recent striking results by M. Mirzakhani on the relation between intersection theory over $\mathfrak{M}(g;N_0)$ and the geometry of hyperbolic surfaces with geodesic boundaries. The results in this chapter connect directly with many deep issues in 3D geometry, ultimately relating to the volume conjecture in hyperbolic geometry and to the role of knot invariants. This eventually brings us to the next topic we will discuss.

Indeed, Chaps. 5 and 6 (Chaps. 6 and 7 in this second edition) deal with the interplay between triangulated manifolds, knots, topological quantum field theory, and quantum computation. As Justin Roberts has aptly emphasized, the standard topological invariants were *created* in order to distinguish between things, and, owing to their intrinsic definitions, it is clear what kind of properties they reflect. For instance, the Euler number χ of a smooth, closed, and oriented surface $\mathscr{S}$ completely determines its topological type and can be defined as $\chi(\mathscr{S}) = 2 - 2g$, where g is the number of handles of $\mathscr{S}$.

On the other hand, quantum invariants of knots and 3-manifolds were *discovered*, but their indirect construction based on quantum group technology often hides information about the purely topological properties they are able to detect. What is lost at the topological level is, however, well paid back by the possibility of connecting this theory with a whole range of issues in pure mathematics and theoretical physics.

To the early connections such as quantum inverse scattering and exactly solvable models, it is worth adding the operator algebra approach used originally by Jones to define his knot polynomial. However, the most profitable development of the theory was the one suggested by Schwarz and formalized by Witten. Indeed, recognizing quantum invariants as partition functions and vacuum expectation values of physical observables in Chern–Simons–Witten topological quantum field theory provides a *physical* explanation for their existence and properties. Even more radically, one could speak of a conceptual explanation, as far as the topological origin of these invariants remains unknown. In this wider sense, quantum topology might be thought of as the mathematical substratum of an SU(2) CSW topological field theory, quantized according to the path integral prescription (the coupling constant $k \geq 1$ is constrained to be an integer related to the deformation parameter q by $q = \exp(\frac{2\pi i}{k+2})$).

The CSW environment provides the physical interpretation for quantum invariants, but in addition, it also includes all the historically distinct definitions. In particular, monodromy representations of the braid group appear in a variety of conformal field theories, since point-like "particles" confined in 2D regions evolve along braided worldlines. As a matter of fact, the natural extension of CSW theory to a 3-manifold $\mathcal{M}^3$ endowed with a nonempty 2D boundary $\partial\mathcal{M}^3$ induces on $\partial\mathcal{M}^3$ a specific quantized boundary conformal field theory, namely, the SU(2) Wess–Zumino–Witten (WZW) theory at level $\ell = k + 2$. The latter provides in turn the framework for dealing with $SU(2)_q$-colored links presented as closures of oriented braids and associated with the Kaul unitary representation of the braid group. A further extension of this representation can be used to construct the quantum 3-manifold invariants explicitly within a purely algebraic setting.

Such quantities are essentially the Reshetikhin–Turaev–Witten invariants evaluated for 3-manifolds presented as complements of knots/links in the 3-sphere S^3, up to an overall normalization. Discretizations of manifolds appear here at a fundamental level, in particular, from $SU(2)$-decorated triangulations of 3D manifolds to triangulated boundary surfaces supporting a (boundary) conformal field theory.

Their use is relevant both in the characterization of the theory and in the actual possibility of computing the topological invariants under discussion. This computational role is a basic property, since the possibility of computing quantities of topological or geometric nature was recognized as a major achievement for quantum information theory by the Fields medalist Michael Freedman and coworkers.

Their *topological quantum computation* setting was designed to comply with the behavior of *modular functors* of 3D Chern–Simons–Witten (CSW) non-abelian topological quantum field theory (TQFT), the gauge group being typically SU(2). In physicists' language, such functors are partition functions and correlators of the quantum theory, and, owing to gauge invariance and invariance under diffeomorphisms, which freeze out local degrees of freedom, they share a global, topological character.

More precisely, the physical observables are associated with topological invariants of knots—the prototype of which is the Jones polynomial—and the generating functional is an invariant of the 3D ambient manifold, the Reshetikhin–Turaev–Witten invariant. We will discuss these matters in detail, with many illustrative examples and diagrams. We think that these case studies provide a good illustration of the richness of the subject, with a repertoire of mathematical techniques and physical concepts that may open up new and exciting territories for research.

Pavia, Italy   Mauro Carfora
Pavia, Italy   Annalisa Marzuoli
April 2011

Preface to the Second Edition

Ever since our LNP appeared 5 years ago, the field has undergone some important developments, among which we list the publication of the remarkable book by Bertrand Eynard *Counting Surfaces* (Progress in Mathematical Physics 70, Springer 2016), where one finds a deep framework for understanding the combinatorial threads among moduli space, matrix models, discretization of surfaces, and integrable systems. This time span has also seen progress in Liouville quantum gravity and in the analysis of the associated scaling laws, turning the elusive KPZ formula into a theorem (B. Duplantier, S. Sheffield, *Liouville Quantum Gravity and KPZ*, Inventiones Mathematicae 185(2), 333–393 (2011), and F. David, A. Kupiainen, R. Rhodes, V. Vargas, *Liouville Quantum Gravity on the Riemann Sphere* Commun. Math. Phys. (2016) 342: 869–907). Adding these topics to this second edition while appropriate in the spirit of the book was not possible for reason of space and the amount of background knowledge needed. We have chosen instead to make a single and focalized major addition to this new edition in the form of a rather vast chapter "The Quantum Geometry of Polyhedral Surfaces: Nonlinear σ Model and Ricci Flow." This includes sections on the geometry of nonlinear σ model, on the relation between Riemannian metric measure spaces and the dilaton, and on renormalization group and its connection with Ricci flow via the background field quantization method. The perturbative embedding of the Ricci flow in the renormalization group flow for NLσM is studied in considerable depth presenting methods and results not treated in any other book. We also discuss new techniques for studying this embedding. The reader can find most of the background knowledge needed carefully explained, with some additional material in an appendix on Riemannian geometry.

It is our pleasure and privilege to express our deep thanks to the many friends and colleagues who positively reacted to the first edition of this book.

Pavia, Italy
Pavia, Italy
July 2017

Mauro Carfora
Annalisa Marzuoli

Acknowledgments

We would like to thank Gaia for her patience with us over the years; our friends and colleagues Jan Ambjørn, Enzo Aquilanti, David Glickenstein, Giorgio Immirzi, Robert Littlejohn, Mario Rasetti, and Tullio Regge; and our young collaborators for so many discussions about the polyhedral side of mathematics and physics.

Tullio Regge would have been not surprised to learn that the subject of triangulations keeps on attracting with wonderful freshness the attention of mathematicians and physicists. The many facets of the theory, such as its beauty, sometimes its elegance, and certainly its unexpected connections with so many fields besides mathematics and physics, are a realm of geometry and the imagination that Tullio loved. This book is dedicated to his memory.

Contents

Chapter 1
Triangulated Surfaces and Polyhedral Structures

In this chapter we introduce the foundational material that will be used in our analysis of triangulated surfaces and of their quantum geometry. We start by recalling the relevant definitions from Piecewise–Linear (PL) geometry, (for which we refer freely to [20, 21]). After these introductory remarks we specialize to the case of Euclidean *polyhedral surfaces* whose geometrical and physical properties will be the subject of the first part of the book. The focus here is on results which are either new or not readily accessible in the standard repertoire. In particular we discuss from an original perspective the structure of the space of all polyhedral surfaces of a given genus and their stable degenerations. This is a rather delicate point which appears in many guises in quantum gravity [6], and string theory, and which is related to the role that Riemann moduli space plays in these theories. Not surprisingly, the Witten–Kontsevich model [10] lurks in the background of our analysis, and some of the notions we introduce may well serve for illustrating, from a more elementary point of view, the often deceptive and very technical definitions that characterize this subject. In such a framework, and in the whole landscaping of the space of polyhedral surfaces an important role is played by the conical singularities associated with the Euclidean triangulation of a surface. We provide, in the final part of the chapter, a detailed analysis of the geometry of these singularities. Their relation with Riemann surfaces theory will be fully developed in Chap. 2.

1.1 Triangulations

A n-simplex $\sigma^n \equiv (x_0, \ldots, x_n)$ with vertices $x_0, \ldots, x_n$ is the following subspace of $\mathbb{R}^d$, (with $d > n$),

$$\sigma^n := \left\{ \sum_{i=0}^{n} \lambda_i x_i \ \middle|\ \lambda_i \geq 0, \ \sum_{i=0}^{n} \lambda_i = 1 , \ x_i \in \mathbb{R}^d \right\} \tag{1.1}$$

© Springer International Publishing AG 2017
M. Carfora, A. Marzuoli, *Quantum Triangulations*, Lecture Notes in Physics 942,
https://doi.org/10.1007/978-3-319-67937-2_1

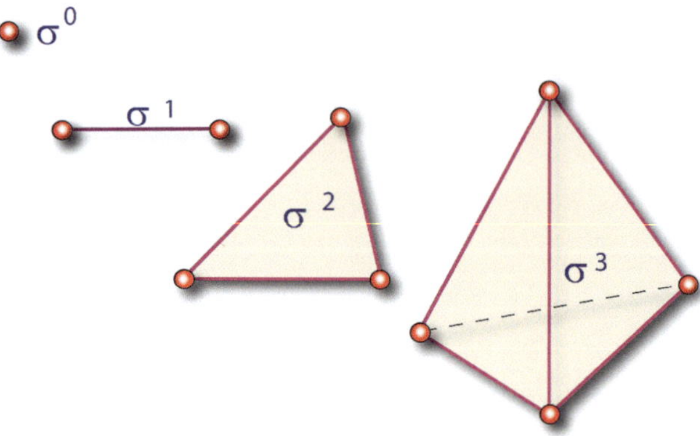

Fig. 1.1 The simplexes σ^n, for $n = 0, 1, 2, 3$

where $x_0, \ldots, x_n$ are $n + 1$ points in general position in $\mathbb{R}^d$. The $\{\lambda_i\}$ provide the barycentric coordinates of the points $x \in \sigma^n$, i.e., $\sigma^n \ni x = \sum_{i=0}^{n} \lambda_i x_i$. In particular, the *barycentre* of σ^n is the point

$$w(\sigma^n) := \sum_{i=0}^{n} \frac{1}{n+1} x_i. \tag{1.2}$$

A *face* of an n-simplex σ^n is any simplex whose vertices are a subset of those of σ^n, and a *simplicial complex* T is a finite collection of simplices in $\mathbb{R}^d$ such that if $\sigma_1^n \in T$ then so are all of its faces, and if $\sigma_1^n, \sigma_2^m \in T$ then $\sigma_1^n \cap \sigma_2^m$ is either a face of σ_1^n or is empty. The *h- skeleton* of T is the subcomplex $K^h \subset T$ consisting of all simplices of T of dimension $\leq h$. Note that every simplicial complex with N vertices admits a canonical embedding into the $(N-1)$-dimensional simplex and consequently into Euclidean space (Fig. 1.1).

Let T be a (finite) simplicial complex. Consider the set theoretic union $|T| \subset \mathbb{R}^d$ of all simplices from T

$$|T| \doteq \cup_{\sigma \in T} \sigma. \tag{1.3}$$

Note that $|T|$ can be also seen as the set of all formal (finite) linear combinations

$$|T| \ni \mu := \left\{ \sum_{\sigma_i^0 \in T} \mu(\sigma_i^0) \, \sigma_i^0 \, \middle| \, 0 \leq \mu_i \leq 1, \, \sum_{\sigma_i^0 \in T} \mu(\sigma_i^0) = 1 \right\}, \tag{1.4}$$

of vertices σ_i^0 of T. The $\{\mu(\sigma_i^0)\}$ are the (barycentric) coordinates of the point $\mu \in |T|$. Introduce on the set $|T|$ a topology that is the strongest of all topologies in which the embedding of each simplex into $|T|$ is continuous, (the set $A \subset |T|$ is closed iff $A \cap \sigma^k$ is closed in σ^k for any $\sigma^k \in T$). The space $|T|$ is the underlying *polyhedron*, geometric carrier of the simplicial complex T, it provides the topological space underlying the simplicial complex. The topology of $|T|$ can be more conveniently described in terms of the *star of a simplex* σ, *star*(σ), the union of all simplices of which σ is a face. The open subset of the underlying polyhedron $|T|$ provided by the interior of the carrier of the star of σ is the *open star* of σ. Notice that the open star is a subset of the polyhedron $|T|$, while the star is a sub-collection of simplices in the simplicial complex T. It is immediate to verify that the open stars can be used to define the topology of $|T|$. The polyhedron $|T|$ is said to be triangulated by the simplicial complex T. More generally, we adopt the following

Definition 1.1 (Triangulation of a topological space) A topological space is triangulable if it is homeomorphic to the polyhedron of a simplicial complex T, i.e. if $M \simeq |T|$. A triangulation (T_l, M) of a topological space M is a simplicial complex T together with a map $f : T \to M$ which is a homeomorphism of the simplicial polyhedron $|T|$ onto M.

Let

$$V(T) := \{\sigma^0(1), \ldots, \sigma^0(N_0(T))\} \in K^0 \tag{1.5}$$

denote the set of $N_0(T)$ vertices of a triangulation (M, T), and by

$$E(T) := \{\sigma^1(1), \ldots, \sigma^1(N_1(T))\} \in K^1 \tag{1.6}$$

the corresponding set of edges. Then, we have

Definition 1.2 The incidence (or boundary) relation of the triangulation (T_l, M) is the map

$$\partial(T) : E(T) \longrightarrow \frac{V(T) \times V(T)}{\mathfrak{G}_2}, \tag{1.7}$$

$$(\sigma^1(1), \ldots, \sigma^1(N_1(T))) \longmapsto \left[\left(\sigma^0(a_1), \sigma^0(b_1)\right) / \sim, \ldots, \left(\sigma^0(a_{N_1}), \sigma^0(b_{N_1})\right) / \sim \right],$$

which to each edge $\sigma^1(k)$ of (T_l, M) associates a corresponding *unordered* pair $\left(\sigma^0(a_k), \sigma^0(b_k)\right) / \sim$ of distinct vertices of (T_l, M), where $\sim$ is the natural relabeling action $\mathfrak{G}_2 \left(\sigma^0(a_k), \sigma^0(b_k)\right) := \left(\sigma^0(b_k), \sigma^0(a_k)\right)$.

Since two vertices of (T_l, M) are connected at most by one edge, we have that the cardinality of $\left|\partial^{-1} \left(\sigma^0(a_k), \sigma^0(b_k)\right)\right|$ can be 0 or 1. The symmetric $(0, 1)$ matrix

$$I_{jk} := \left|\partial^{-1} \left(\sigma^0(a_j), \sigma^0(b_k)\right)\right|, \tag{1.8}$$

is the *incidence matrix* of (T_l, M). It follows that the number of edges incident on the vertex $\sigma^0(k)$ is provided by

$$q(k) = \sum_{h \neq k} I_{hk}, \tag{1.9}$$

(for a 2-dimensional triangulation this is also the number of triangles incident on a vertex). An automorphism $\Phi(T) := (\Phi_{V(T)}, \Phi_{E(T)})$ of (T_l, M) is a pair of bijective maps

$$\Phi_{V(T)} : V(T) \longrightarrow V(T) \qquad \Phi_{E(T)} : E(T) \longrightarrow E(T), \qquad (1.10)$$

that preserve the incidence relation $\partial(T) : E(T) \rightarrow \frac{V(T) \times V(T)}{\mathfrak{S}_2}$ defining (M, T). We have

Definition 1.3 Let (T_l, M) be a triangulation with $N_0(T)$ vertices and $N_1(T)$ edges. The group of automorphism $Aut(T)$ of (T_l, M) is the group generated by the set of automorphisms $\{\Phi(T)\}$.

A simplicial map $f : T \rightarrow L$ between two simplicial complexes T and L is a continuous map $f : |T| \rightarrow |L|$ between the corresponding underlying polyhedrons which takes n-simplices to n- simplices for all n, (piecewise straight-line segments are mapped to piecewise straight-line segments). The map f is a simplicial isomorphism if $f^{-1} : L \rightarrow T$ is also a simplicial map. Such maps preserve the natural combinatorial structure of $\mathbb{R}^n$. Note that a simplicial map is determined by its values on vertices. In other words, if $f : K^0 \rightarrow L^0$ carries the vertices of each simplex of T into some simplex of L, then f is the restriction of a unique simplicial map.

Sometimes, one refers to a simplicial complex T as a simplicial division of $|T|$. A subdivision T' of T is a simplicial complex such that $|T'| = |T|$ and each n-simplex of T' is contained in an n- simplex of T, for every n. A property of simplicial complex T which is invariant under subdivision is a combinatorial property or *Piecewise-Linear* (PL) property of T, and a Piecewise-Linear homeomorphism $f : T \rightarrow L$ between two simplicial complexes is a map which is a simplicial isomorphism for some subdivisions T' and L' of T and L. A well–known subdivision is provided by the *barycentric subdivision*. This is almost a graphical notion, simple to draw (in low dimensions), but a little bit annoying to define. It can be characterized inductively: the barycenter of a vertex σ^0 is the vertex itself, and assume that we have defined the barycentric subdivision for the generic $k-1$ simplex $\in T$, for $k \geq 2$. The construction is extended to the k-dimensional simplex $\sigma^k \in T$ by introducing a new vertex σ'^0 which is the barycentric average of the vertices of σ^k, and by taking the cone, (see the next paragraph), with respect to σ'^0 of the barycentric subdivision (Fig. 1.2) of the $k - 1$-dimensional boundary of σ^k.

1.2   Piecewise-Linear Manifolds

A PL manifold of dimension n is a polyhedron $M \simeq |T|$ each point of which has a neighborhood PL homeomorphic to an open set in R^n. PL manifolds are realized by simplicial manifolds under the equivalence relation generated by PL homeomorphism. Any piecewise linear manifold can be triangulated, and viceversa a triangulated space can be characterized as a PL-manifold according to the [21]

Fig. 1.2 Barycentric
subdivision of a
2-dimensional simplicial
complex

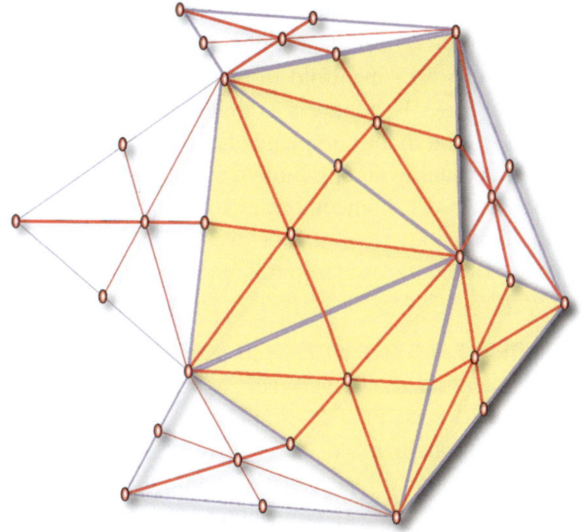

Theorem 1.1 *A simplicial complex T is a simplicial manifold of dimension n if for all r-simplices $\sigma^r \in T$, the link of σ^r, $link(\sigma^r)$ has the topology of the boundary of the standard $(n-r)$-simplex, viz. if $link(\sigma^r) \simeq \mathbb{S}^{n-r-1}$.*

Recall that the link of a simplex σ in a simplicial complex T is the union of all faces σ_f of all simplices in $star(\sigma)$ satisfying $\sigma_f \cap \sigma = \emptyset$, also recall that the *Cone* on the link $link(\sigma^r)$, $C(link(\sigma^r))$, is the product $link(\sigma^r) \times [0,1]$ with $link(\sigma^r) \times \{1\}$ identified to a point. The above theorem follows [21] by noticing that a point in the interior of an r-simplex σ^r has a neighborhood homeomorphic to $B^r \times C(link(\sigma^r))$, where B^r denotes the ball in $\mathbb{R}^n$. Since $link(\sigma^r) \simeq \mathbb{S}^{n-r-1}$, and $C(\mathbb{S}^{n-r-1}) \simeq B^{n-r}$, we get that $|T|$ is covered by neighborhoods homeomorphic to $B^r \times B^{n-r} \simeq B^n$ and thus it is a manifold. Note that the theorem holds whenever the links of vertices are $(n-1)$-spheres. As long as the dimension $n \leq 4$, the converse of this theorem is also true. But this is not the case in larger dimensions, and there are examples of triangulated manifolds where the link of a simplex is not a sphere. In general, necessary and sufficient conditions for having a manifolds out of a simplicial complex require that the link of each cell has the homology of a sphere, (see e.g. [21] for further details).

Often one generates an n-dimensional PL-manifold by glueing a finite set of n-simplices $\{\sigma^n\}$. A rather detailed analysis of such glueing procedures is given in Thurston's notes [21], and here we simply recall the most relevant facts.

Definition 1.4 Given a finite set of simplices and the associated collection of faces, a glueing is a choice of pairs of faces together with simplicial identifications maps between faces such that each face appears in exactly one of the pairs.

The identification space resulting from the quotient of the union of the simplices by the equivalence relation generated by the identification maps, is homeomorphic to the polyhedron of a simplicial complex. In particular, the glueing maps are linear, and the simplicial complex T obtained by glueing face-by-face n-simplices has the structure of a PL- manifold in the complement of the $(n-2)$-skeleton. Since the link of an $(n-2)$-simplex is a circle, it is not difficult to prove that the PL-structure actually extend to the complement of the $(n-3)$- skeleton, and that the identification space of a glueing among finite n-simplices is a PL-manifold if and only if the link of each cell is PL-homeomorphic to the standard PL-sphere.

In dimension $n > 2$, not every simplicial complex T obtained by glueing simplices along faces is a simplicial manifold, and in general one speaks of *pseudo-manifolds*.

Definition 1.5 The map $T \to M$ generates (the triangulation of) an n-dimensional pseudomanifold if: (i) every simplex of T is either an n-simplex or a face of an n-simplex; (ii) each $(n-1)$-simplex is a face of at most two n- simplices; (iii) for any two simplices σ^n, τ^n of T, there exists a finite sequence of n-simplices $\sigma^n = \sigma_0^n, \sigma_1^n, \ldots, \sigma_j^n = \tau^n$ such that σ_i^n and σ_{i+1}^n have an $(n-1)$-face in common, (i.e., there is a simplicial path connecting σ^n and τ^n).

Recall that a regular point p of a polyhedron $|T|$ is a point having a neighborhood in $|T|$ homeomorphic to an n-dimensional simplex, otherwise p is called a *singular point*. Absence of singular points in a pseudo-manifolds characterizes triangulated manifolds. Moreover, an n-dimensional polyhedron $|T|$ is a pseudo- manifold if and only if the set of regular point in $|T|$ is dense and connected and the set of all singular points is of dimension less than $n-1$.

In order to construct a simplicial manifold T by glueing simplices σ^n, through their $n-1$-dimensional faces, the following constraints must be satisfied

$$\sum_{i=0}^{n}(-1)^i N_i(T) = \chi(T),\tag{1.11}$$

$$\sum_{i=2k-1}^{n}(-1)^i \frac{(i+1)!}{(i-2k+2)!(2k-1)!}N_i(T) = 0,\tag{1.12}$$

if n is even, and $1 \leq k \leq n/2$. Whereas if n is odd

$$\sum_{i=2k}^{n}(-1)^i \frac{(i+1)!}{(i-2k+1)!2k!}N_i(T) = 0,\tag{1.13}$$

with $1 \leq k \leq (n-1)/2$. These relations are known as the Dehn-Sommerville equations. The first, (1.11), is just the Euler-Poincaré equation for the triangulation T of which $N_i(T)$ denotes the number of i-dimensional simplices, the *f-vector* of the triangulation T. The conditions (1.12) or (1.13), are a consequence of the fact that in a simplicial manifold, constructed by glueings, the link of every $(2k-1)$-simplex (if n is odd) or $2k$-simplex (if n is even), is an odd-dimensional sphere, and hence it has Euler number zero. Note that in order to generate an n-dimensional polyhedron

Fig. 1.3 The triangular
pillow $(T^{pill}, \mathbb{S}^2)$ is a
semi–simplicial triangulated
surface generated by gluing
the boundaries of two distinct
triangular faces

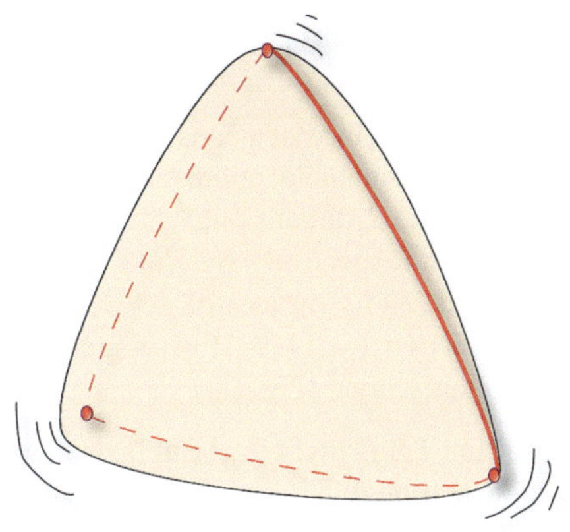

there is a *minimum number* $\hat{q}(n)$ of simplices σ^n that must join together at any
$n - 2$–simplex. For instance, in dimension 2, in order to have a simplicial complex
we must have at least three triangles incident on the generic vertex $\sigma^0(j)$. If this
condition is not met we generally speak, with a slight abuse of language, of a *a semi–
simplicial complex* . Note that this does not necessarily imply a failure of the Dehn–
Sommerville relations (1.12) and (1.13). For instance, in what follows an important
role is played by the following elementary semi–simplicial object (Fig. 1.3)

Definition 1.6 The *triangular pillow* $(T^{pill}, \mathbb{S}^2)$ is the semi–simplicial triangulation
of the sphere $\mathbb{S}^2$ obtained by gluing the edges of two distinct triangles without
identifying their interior.

Is elementary to check that in this way we get a semi–simplicial complex with f–
vector $N_0(T) = 3$, $N_1(T) = 3$, $N_2(T) = 2$, and where the star of each vertex $\sigma^0(j)$
contains just two triangles. Nonetheless, the relations (1.11) and (1.12) still yield
$\sum_{i=0}^{n}(-1)^i N_i(T) = \chi(T) = 2$ and $2N_1(T) = 3N_2(T)$, as for a regular triangulation.
The paper of W. Thurston [22] discusses nice examples of generalized triangulations
of the sphere where two edges of a triangle are folded together to form a vertex
incident to a single triangle.

By the very definition of PL manifolds, it follows that there exist distinct
triangulations, $T^{(i)}$, of the some PL manifold M. For later convenience, it is better
to formalize this remark, and recall the following standard characterization by W.
Tutte [25]:

Definition 1.7 Two triangulations, $T^{(1)}$ and $T^{(2)}$ of the some underlying PL manifold M are identified if there is a one-to-one mapping of vertices, edges, faces,
and higher dimensional simplices of $T^{(1)}$ onto vertices, edges, faces, and higher
dimensional simplices of $T^{(2)}$ which preserves incidence relations. If no such
mapping exists the corresponding triangulations are said to be *distinct*.

Sometimes, (e.g., W. Thurston [21], p.105), such triangulations are said to be
combinatorially equivalent. However, we shall avoid this terminology since often
combinatorial equivalence is used as synonymous of PL- equivalence.

1.3   Polyhedral Surfaces

By a closed surface of genus g we mean a 2-dimensional smooth (C^∞) manifold which is an orientable, compact, connected and without boundary. Let T denote an oriented finite 2-dimensional semi-simplicial complex with underlying polyhedron $|T|$. Denote respectively by $F(T)$, $E(T)$, and $V(T)$ the set of $N_2(T)$ faces, $N_1(T)$ edges, and $N_0(T)$ vertices of T, where $N_i(T) \in \mathbb{N}$ is the number of i-dimensional subsimplices $\sigma^i(\ldots) \in T$. If we assume that $|T|$ is homeomorphic to a closed surface M of genus g, then (Fig. 1.4)

Definition 1.8 (Polyhedral Surfaces) A polyhedral surface (T_l, M) of genus g is a Euclidean triangulation of the closed surface M such that

$$(T_l, M) := f : T \longrightarrow M \tag{1.14}$$

is a realization of the homeomorphism $|T_l| \to M$ where each edge $\sigma^1(h, j)$ of T is a rectilinear simplex of variable length $l(h, j)$. In simpler terms, T is generated by Euclidean triangles glued together by isometric identification of adjacent edges.

This is a good place to mention that in physics there is a well–established use of the term Regge surfaces for indicating Euclidean triangulations with fixed connectivity. However, when we speak of a polyhedral surfaces we shall always mean a triangulation (T_l, M) of M whose connectivity is not fixed a priori. For this reason it is worthwhile stressing that

Remark 1.1 (Regge vs. Polyhedral surfaces) In these lecture notes we shall avoid the term *Regge surface*, and explicitly refer to (T_l, M) as a Euclidean triangulation of M, or simply as a *polyhedral surface*. We also note that in such a general setting a

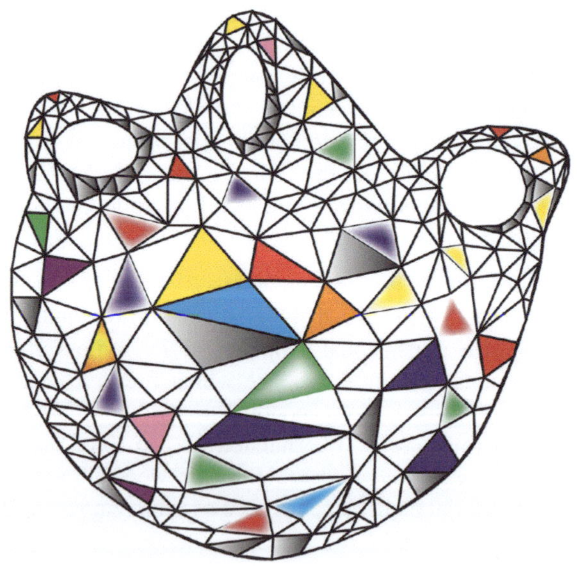

Fig. 1.4 A polyhedral surface of genus $g = 3$

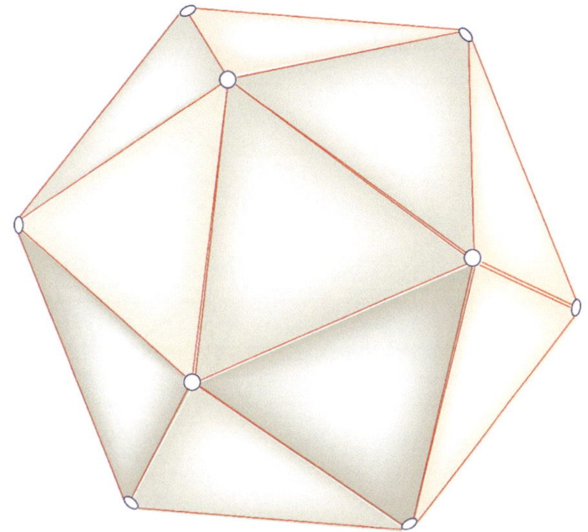

Fig. 1.5 The open
polyhedral surface obtained
by removing vertices

dynamical triangulation $(T_{l=a}, M)$ is a particular case [1, 2] of a polyhedral surface realized by rectilinear and equilateral simplices of a fixed edge-length $l = a$.

Remark 1.2 We can naturally associate with a Euclidean triangulation (T_l, M) the topologically open surface $M' := M \setminus K^0(T)$ obtained by removing the 0–skeleton, (i.e. the collection of vertices $\{\sigma^0(1), \ldots, \sigma^0(N_0(T))\}$), from M (Fig. 1.5). In terms of M' we can characterize admissible paths (Fig. 1.6) and path homotopy on (T_l, M) according to [24]

Definition 1.9 A path $c : [0, 1] \rightarrow M$ on a polyhedral surface (T_l, M) is admissible if it has finitely many intersections with the edges $\{\sigma^1(i, k)\}$ of the triangulation and if $c(s) \in M'$ for any $0 < s < 1$. A family of paths c_t, $t \in [0, 1]$, is an admissible homotopy if $s \rightarrow c_t(s) \in M$ is an admissible path for any $t \in [0, 1]$.

Note that whereas the path $c(s)$ is required to evolve in M' for $0 < s < 1$, it can start, for $s = 0$, at a vertex $\sigma^0(i)$ and end, for $s = 1$, at a vertex $\sigma^0(k)$ of (T_l, M). By slightly perturbing $c(s = 0)$ and $c(s = 1)$, we can always assume, if necessary, that both $c(s = 0)$ and $c(s = 1)$ are interior points of some simplex σ^j, with $j = 1, 2$. In this connection we have

Definition 1.10 Let x_0 be a chosen base point in M. The set of all admissible homotopy classes in (T_l, M) starting and ending at the base point x_0 form a group, with respect to path–concatenation, which can be identified with the fundamental group $\pi_1(M', x_0)$ of $(M', x_0) := (M, x_0) \setminus K^0(T)$.

For a chosen fixed triangle $\sigma_0^2 \in T$, let us consider the pair $(\sigma^j, [c])$ where σ^j is a j–simplex of T, $j = 0, 1, 2$, and $[c]$ is an admissible homotopy class of paths joining a point in σ_0^2 to a point in σ^j. The collection of such pairs $\{(\sigma^j, [c])\}$ together with the projection map $P : (\sigma^j, [c]) \mapsto \sigma^j$ define the j–simplices $\widehat{\sigma}^j := (\sigma^j, [c])$

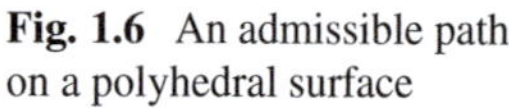

Fig. 1.6 An admissible path on a polyhedral surface

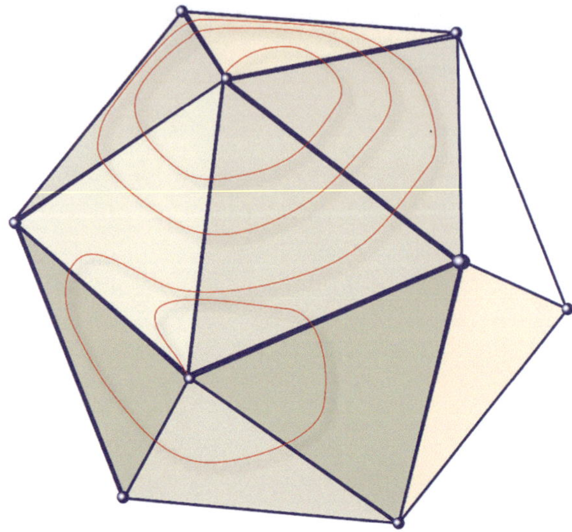

of a non locally finite simplicial complex whose geometrical realization $(\widehat{M}, \widehat{T}, P)$ characterizes the universal branched cover of (T_l, M) according to [24]

Definition 1.11 The universal branched cover $(\widehat{M}, \widehat{T}, P)$ of the piecewise flat surface (T_l, M) is the geometric realization of the simplicial complex $\widehat{T}$ obtained by requiring $P : \widehat{\sigma}^j \mapsto \sigma^j$ to be an isometry on each simplex.

There is a natural simplicial action of $\pi_1(M', x_0)$ on $\widehat{T}$ given by

$$\pi_1(M', x_0) \times \widehat{T} \longrightarrow \widehat{T} \tag{1.15}$$

$$\left([a], (\sigma^j, [c])\right) \longmapsto (\sigma^j, [c\,a]).$$

This extends to an isometric action of $\pi_1(M', x_0)$ on $(\widehat{M}, \widehat{T}, P)$ and one has the identification $(T_l, M) = (\widehat{M}, \widehat{T}, P)/\pi_1(M', x_0)$. These elementary topological properties play an important role in the characterization of the geometry of triangulated Euclidean surfaces.

1.4 The Metric Geometry of Polyhedral Surfaces

The metric geometry of a Euclidean triangulation is defined by the distribution of edge-lengths $\sigma^1(m, n) \to l(m, n) \in \mathbb{R}_+$ satisfying the appropriate triangle inequalities $l(h, j) \leq l(j, k) + l(k, h)$, whenever $\sigma^2(k, h, j) \in F(T)$. Such an assignment uniquely characterizes the Euclidean geometry of the triangles $\sigma^2(k, h, j) \in T$ and in particular, via the cosine law, the associated vertex angles $\theta_{jkh} \doteq \angle[l(j, k), l(k, h)]$, $\theta_{khj} \doteq \angle[l(k, h), l(h, j)]$, $\theta_{hjk} \doteq \angle[l(h, j), l(j, k)]$; e.g.

$$\cos \theta_{jkh} = \frac{l^2(j,k) + l^2(k,h) - l^2(h,j)}{2l(j,k)l(k,h)}. \tag{1.16}$$

If we note that the area $\Delta(j,k,h)$ of $\sigma^2(j,k,h)$ is provided, as a function of θ_{jkh}, by

$$\Delta(j,k,h) = \frac{1}{2} l(j,k)l(k,h) \sin \theta_{jkh}, \tag{1.17}$$

then the angles θ_{jkh}, θ_{khj}, and θ_{hjk} can be equivalently characterized by the formula

$$\cot \theta_{jkh} = \frac{l^2(j,k) + l^2(k,h) - l^2(h,j)}{4\,\Delta(j,k,h)}, \tag{1.18}$$

which will be useful later on. The cosine law (1.16) shows that datum of the edge–lengths $\sigma^1(m,n) \to l(m,n) \in \mathbb{R}_+$ is equivalent to the assignment of a metric, in each triangle of (T_l, M), that can be extended to a piecewise flat metric on the triangulated surface (T_l, M) in an obvious way. Explicitly [24], the generic triangle $\sigma^2(j,k,h) \in (T_l, M)$ has a Euclidean length structure, (naturally invariant under the automorphisms corresponding to permutations of its vertices), which allows to define the Euclidean length $l(c)$ of a curve $c : [0,1] \to \sigma^2(j,k,h)$. Since the length structure on $\sigma^2(h,j,k)$ is defined by the distribution of edge-lengths $\sigma^1(h,j), \sigma^1(j,k), \sigma^1(k,h) \mapsto l(h,j), l(j,k), l(k,h) \in \mathbb{R}_+$, it can be coherently extended to all adjacent triangles, say to $\sigma^2(m,j,h)$. In particular, for a curve $c : [0,1] \to \sigma^2(h,j,k) \cup \sigma^2(m,j,h)$ which is the concatenation, $c = c_1 c_2$ of $c_1 : [0,\frac{1}{2}] \to \sigma^2(h,j,k)$ and $c_2 : [\frac{1}{2},1] \to \sigma^2(m,j,h)$, we have $l(c) = l(c_1) + l(c_2)$. By iterating such a procedure, we can evaluate the length of any piecewise–smooth curve $[0,1] \ni t \mapsto c(t) \in (T_l, M)$ and define the distance $d(p,q)$ between any two points p, q of (T_l, M) as the infimum of the length of curves joining the two given points. It follows that $(T, M; d)$ is a length space, (in particular a geodesic space since $(T, M; d)$ is complete). Note that the distance function so defined is intrinsic.

As we shall see momentarily, the $3N_2(T)$ angles $\{\theta_{jkh}\}$ carry the curvature structure of the polyhedral surface (T_l, M), and, as in standard Riemannian geometry, a natural question is to what extent this curvature information characterizes (T_l, M). The natural starting point for answering to this basic question is provided by the following characterization [19]

Definition 1.12 (Rivin's local Euclidean structure) The assignment

$$\mathscr{E}(T) : \left\{ \sigma^2(k,h,j) \right\}_{F(T)} \longrightarrow \mathbb{R}_+^{3\,N_2(T)} \tag{1.19}$$

$$\sigma^2(k,h,j) \longmapsto (\theta_{jkh}, \theta_{khj}, \theta_{hjk})$$

of the angles θ_{jkh}, θ_{khj}, and θ_{hjk} (with the obvious constraints $\theta_{jkh} > 0$, $\theta_{khj} > 0$, $\theta_{hjk} > 0$, and $\theta_{jkh} + \theta_{khj} + \theta_{hjk} = \pi$) to each $\sigma^2(k,h,j) \in T$ defines the local Euclidean structure $\mathscr{E}(T)$ of (T_l, M).

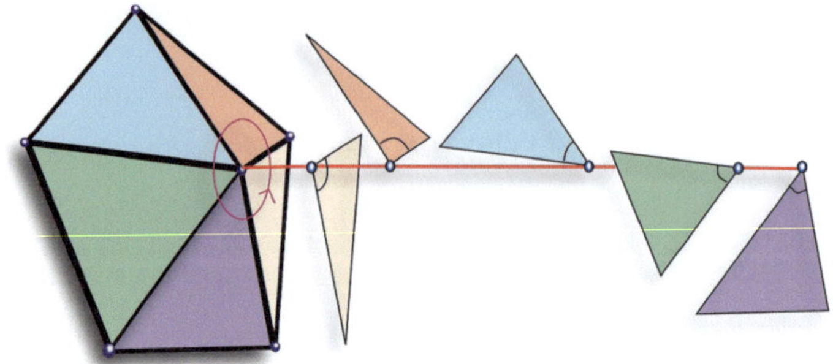

Fig. 1.7 $GL_2(\mathbb{R})$ holonomy around a vertex

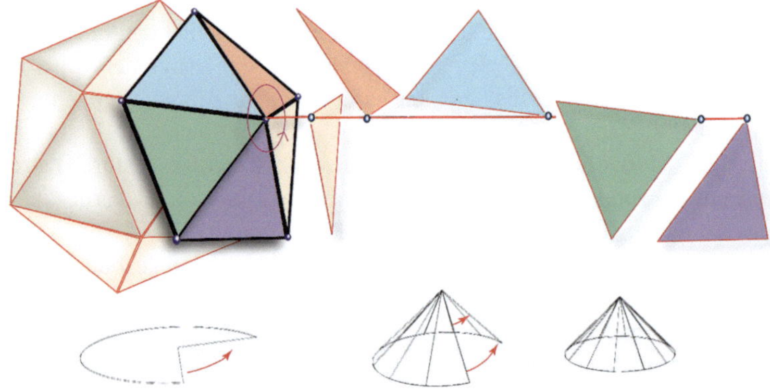

Fig. 1.8 Euclidean holonomy and the corresponding deficit angle around a vertex

It must be stressed that the datum of $\mathscr{E}(T)$ does not allow to reconstruct the metric geometry of a triangulated surface. It only characterizes the similarity classes of the realization of each $\sigma^2(k, h, j)$ as an Euclidean triangle; in simpler words, their shape and not their actual size.

As emphasized by Rivin [19], the knowledge of the locally Euclidean structure on $|T| \to M$ corresponds to the holonomy representation (Fig. 1.7)

$$H(T) : \pi_1(M \setminus K^0(T)) \longrightarrow GL_2(\mathbb{R}) \tag{1.20}$$

of the fundamental group of the punctured surface $M \setminus K^0(T)$ into the general linear group $GL_2(\mathbb{R})$, and the action of $GL_2(\mathbb{R})$ is not rigid enough for defining a coherent Euclidean glueing of the corresponding triangles $\sigma^2(k, h, j) \in T$. A few subtle properties of the geometry of Euclidean triangulations are at work here, and to put them to the fore let us consider $q(k)$ triangles $\sigma^2(k, h_\alpha, h_{\alpha+1})$ incident on the generic vertex $\sigma^0(k) \in T_l$ and generating the star

$$Star[\sigma^0(k)] \doteq \cup_{\alpha=1}^{q(k)} \sigma^2(k, h_\alpha, h_{\alpha+1}), \quad h_{q(k)+1} \equiv h_1. \tag{1.21}$$

We have the following characterization of the geometry of $Star[\sigma^0(k)]$ (Fig. 1.8)

Proposition 1.1 *To any given locally Euclidean structure*

$$\mathscr{E}\left(Star[\sigma^0(k)]\right) \doteq \{(\theta_{\alpha+1,k,\alpha},\ \theta_{k,\alpha,\alpha+1},\ \theta_{\alpha,\alpha+1,k})\} \tag{1.22}$$

on $Star[\sigma^0(k)]$ *there corresponds a conical defect*

$$\Theta(k) \doteq \sum_{\alpha=1}^{q(k)} \theta_{\alpha+1,k,\alpha}, \tag{1.23}$$

supported at $\sigma^0(k)$, *and a logarithmic dilatation* [19], *with respect to the vertex* $\sigma^0(k)$, *of the generic triangle* $\sigma^2(k, h_\alpha, h_{\alpha+1}) \in Star[\sigma^0(k)]$, *given by*

$$D(k, h_\alpha, h_{\alpha+1}) \doteq \ln \sin \theta_{k,\alpha,\alpha+1} - \ln \sin \theta_{\alpha,\alpha+1,k}. \tag{1.24}$$

Proof If $\{l(m,n)\}$ is a distribution of edge-lengths to the triangles $\sigma^2(k, h_\alpha, h_{\alpha+1})$ of $Star[\sigma^0(k)]$ compatible with $\mathscr{E}(T)$, then by identifying

$$\theta_{k,\alpha,\alpha+1} \equiv \angle[l(k, h_\alpha),\ l(h_\alpha, h_{\alpha+1})] \tag{1.25}$$

$$\theta_{\alpha,\alpha+1,k} \equiv \angle[l(h_\alpha, h_{\alpha+1}),\ l(h_{\alpha+1}, k)],$$

and by exploiting the law of sines, we can write

$$D(k, h_\alpha, h_{\alpha+1}) = \ln \frac{l(h_{\alpha+1}, k)}{l(k, h_\alpha)},. \tag{1.26}$$

In terms of this parameter, we can define ([19]) the dilatation holonomy of $Star[\sigma^0(k)]$ according to

$$H(Star[\sigma^0(k)]) \doteq \sum_{\alpha=1}^{q(k)} D(k, h_\alpha, h_{\alpha+1}). \tag{1.27}$$

The vanishing of $H(Star[\sigma^0(k)])$ implies that if we circle around the vertex $\sigma^0(k)$, then the lengths $l(h_{\alpha+1}, k)$ and $l(k, h_{\alpha+1})$ of the pairwise adjacent oriented edges $\sigma^1(h_{\alpha+1}, k)$ and $\sigma^2(k, h_{\alpha+1})$ match up for each $\alpha = 1, \ldots, q(k)$, with $h_\alpha = h_\beta$ if $\beta = \alpha \bmod q(k)$, and $Star[\sigma^0(k)]$ is endowed with a *conically complete* singular Euclidean structure. $\diamond$

Definition 1.13 A local Euclidean structure $\mathscr{E}(T)$ such that the dilatation holonomy $H(Star[\sigma^0(k)])$ vanishes for each choice of star $Star[\sigma^0(k)] \subset T$ is called *conically complete*.

Later on we shall see that under a natural topological condition this notion of completeness implies the following basic result

Theorem 1.2 (I. Rivin [19], M. Troyanov [23]) *There exist a Euclidean metric on* (T_l, M) *with preassigned conical singularities at the set of vertices* $V(T)$ *if and only if the local Euclidean structure* $\mathscr{E}(T)$ *is conically complete.*

In other words, the triangles in T can be coherently glued into a Euclidean triangulation of the surface, with the preassigned deficit angles

$$\varepsilon(k) \doteq 2\pi - \Theta(k) \tag{1.28}$$

generated by the given $\mathscr{E}(T)$, if and only if $\mathscr{E}(T)$ is complete in the above sense. Thus, (T_l, M) is locally Euclidean everywhere except at the vertices $\{\sigma^0(k)\}$ where the sum $\Theta(k)$ of the angles $\theta_{\alpha+1,k,\alpha}$, associated with the triangles incident on $\sigma^0(k)$, is in excess (negative curvature) or in defect (positive curvature) with respect to the 2π flatness constraint. In particular, if the deficit angles $\{\varepsilon(k)\}_{V(T)}$ all vanish, we end up in the familiar notion of holonomy associated with the completeness of the Euclidean structure associated with (T_l, M) and described by a developing map whose rotational holonomy around any vertex is trivial.

1.5   Complex–Valued Holonomy

There is a natural way to keep track of both dilation factors and conical defects by introducing a complex-valued holonomy. This is a first indication of the important role that function theory has in capturing the geometry of polyhedral surfaces. Let (τ_k, τ_h, τ_j) a ordered triple of complex numbers describing the vertices of a realization, in the complex plane $\mathbb{C}$, of the oriented triangle $\sigma^2(k, h, j)$, with edge lengths $l(k, h)$, $l(h, j)$, $l(j, k)$. By using Euclidean similarities we can always map (τ_k, τ_h, τ_j) to $(0, 1, \tau_{jkh})$, with

$$\tau_{jkh} \doteq \frac{\tau_j - \tau_k}{\tau_h - \tau_k} = \frac{l(j, k)}{l(k, h)} \, e^{\sqrt{-1}\,\theta_{jkh}}, \tag{1.29}$$

where

$$\arg \tau_{jkh} = \arg(\tau_j - \tau_k) - \arg(\tau_h - \tau_k) = \theta_{jkh} \in [0, 2\pi), \tag{1.30}$$

(thus $Im\,\tau_{jkh} > 0$). The triangle $(0, 1, \tau_{jkh})$ is in the same similarity class $\mathscr{E}(\sigma^2(k, h, j))$ of $\sigma^2(k, h, j)$, and the vector τ_{jkh}, the *complex modulus* (Fig. 1.9) of the triangle $\sigma^2(k, h, j)$ with respect to $\sigma^0(k)$, parametrizes $\mathscr{E}(\sigma^2(k, h, j))$. The same similarity class is obtained by cyclically permuting the vertex which is mapped to 0, i.e., $(\tau_h, \tau_j, \tau_k) \to (0, 1, \tau_{khj})$ and $(\tau_j, \tau_k, \tau_h) \to (0, 1, \tau_{hjk})$, where

$$\tau_{khj} \doteq \frac{\tau_k - \tau_h}{\tau_j - \tau_h} = \frac{l(k, h)}{l(h, j)} \, e^{\sqrt{-1}\,\theta_{khj}}, \tag{1.31}$$

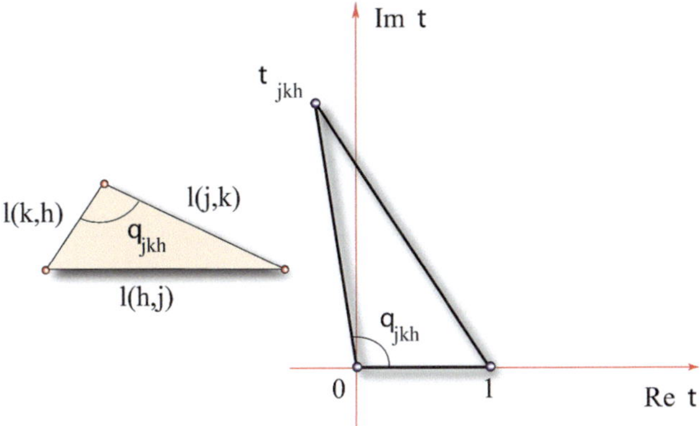

Fig. 1.9 The complex modulus τ describing the similarity class of a triangle

$$\tau_{hjk} \doteq \frac{\tau_h - \tau_j}{\tau_k - \tau_j} = \frac{l(h,j)}{l(j,k)}\, e^{\sqrt{-1}\,\theta_{hjk}}, \tag{1.32}$$

are the moduli of $\mathscr{E}(\sigma^2(k,h,j))$ with respect to the vertex $\sigma^0(h)$ and $\sigma^0(j)$, respectively.

Elementary geometrical considerations imply that the triangles

$$(\tau_{jkh}\,\tau_{hjk}\,\tau_{khj},\, 0,\, \tau_{hjk}\,\tau_{jkh}), \tag{1.33}$$

$$(0,\, 1,\, \tau_{jkh}),$$

$$(\tau_{jkh},\, \tau_{hjk}\,\tau_{jkh},\, 0),$$

are congruent. This yields the relations

$$\tau_{jkh}\,\tau_{hjk}\,\tau_{khj} = -1, \tag{1.34}$$

$$\tau_{jkh}\,\tau_{hjk} = \tau_{jkh} - 1,$$

according to which a choice of a moduli with respect a particular vertex specifies also the remaining two moduli. For instance, if we describe $\mathscr{E}(\sigma^2(k,h,j))$ by the modulus $\tau_{jkh} \doteq \tau$, $(Im\,\tau > 0)$, with respect to $\sigma^0(k)$ then we get

$$\tau_{jkh} \doteq \tau, \tag{1.35}$$

$$\tau_{khj} = \frac{1}{1-\tau},$$

$$\tau_{hjk} = 1 - \frac{1}{\tau}.$$

By selecting the standard branch on $\mathbb{C}_{(-\infty,0]} := \mathbb{C} - (-\infty, 0]$ of the natural logarithm, we also get

$$\ln \tau_{jkh} = \ln \tau, \tag{1.36}$$

$$\ln \tau_{khj} = -\ln (1 - \tau),$$

$$\ln \tau_{hjk} = \ln (1 - \tau) - \ln \tau + \pi \sqrt{-1}.$$

These remarks directly imply the

Proposition 1.2 *The logarithmic dilation of the generic triangle $\sigma^2(k, h_\alpha, h_{\alpha+1}) \in Star[\sigma^0(k)]$, can be extended to its complexified form according to*

$$D^{\mathbb{C}}(k, h_\alpha, h_{\alpha+1}) \doteq \ln \tau_{h_{\alpha+1},k,h_\alpha} = \ln \frac{l(h_{\alpha+1}, k)}{l(k, h_\alpha)} + \sqrt{-1} \; \theta_{h_{\alpha+1},k,h_\alpha}, \tag{1.37}$$

where $\tau_{h_{\alpha+1},k,h_\alpha}$ is the complex modulus of the triangle $\sigma^2(k, h_\alpha, h_{\alpha+1})$ with respect to the vertex $\sigma^0(k)$. We can define the associated dilatation holonomy as

$$H^{\mathbb{C}}(Star[\sigma^0(k)]) \doteq \sum_{\alpha=1}^{q(k)} D^{\mathbb{C}}(k, h_\alpha, h_{\alpha+1}) = \tag{1.38}$$

$$= \sum_{\alpha=1}^{q(k)} D(k, h_\alpha, h_{\alpha+1}) + \sqrt{-1} \sum_{\alpha=1}^{q(k)} \theta_{h_{\alpha+1},k,h_\alpha} =$$

$$= H(Star[\sigma^0(k)]) + \sqrt{-1} \; \Theta(k),$$

where $\Theta(k)$ is the conical defect supported at the vertex $\sigma^0(k)$.
It follows that the triangulation $|T_l| \to M$ will be conicaly complete if and only if $Re\, H^{\mathbb{C}}(Star[\sigma^0(k)]) = 0$ for every vertex star, and its conical defects are provided by $\{Im\, H^{\mathbb{C}}(Star[\sigma^0(k)])\}$. In other words, a necessary and sufficient condition on the locally Euclidean structure $\{\theta_{jkh}, \theta_{khj}, \theta_{hjk}\}_{F(T)}$ in order to define a glueing and hence a Euclidean triangulation (T_l, M) is the requirement that

$$\prod_{k=1}^{q(k)} \tau_{h_{\alpha+1},k,h_\alpha} \in U(1), \tag{1.39}$$

for each $Star[\sigma^0(k)]$, i.e. that the image of $H^{\mathbb{C}}(T)$ lies in the group $U(1)$. Complete conical structures on polyhedral surfaces have a number of subtle properties which somehow contrast with their ancillary role of geometrical objects approximating smooth surfaces. Actually they have a life of their own that can be quite more

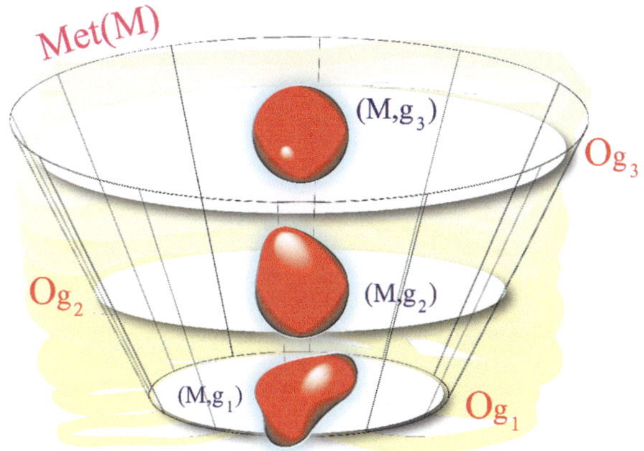

Fig. 1.10 The orbit space describing distinct Riemannian structures

fascinating than expected. This is particularly manifest when we fix our attention not just on a particular metrically triangulated surface but rather on the collection of all possible such objects.

1.6   The Space of Polyhedral Structures $POL_{g,N_0}(M)$

To begin with, it is worthwhile to recall a few facts from the smooth framework.

The space of smooth Riemannian metrics $\mathcal{M}et(M)$, on a compact surface M, is the open subset, in the Fréchet space of symmetric 2–tensor fields $C^\infty(M, \otimes_S^2 T^*M)$, defined by the positive definite bilinear forms (see Appendix A for details). Since $a\,g_{(1)} + b\,g_{(2)}$ is in $\mathcal{M}et(M)$, for any $g_{(1)}, g_{(2)} \in \mathcal{M}et(M)$ and $a, b \in \mathbb{R}_+$, it follows that $\mathcal{M}et(M)$ is an ∞-dimensional open convex cone in the compact open C^∞ topology. Let $\mathcal{D}iff(M)$ denote the ∞-dimensional (topological) group of smooth diffeomorphisms of M. If $\varphi^* g$, $\phi \in \mathcal{D}iff(M)$ denotes the pull–back of a metric $g \in \mathcal{M}et(M)$, then the orbit

$$\mathcal{O}_g := \left\{ g' \in \mathcal{M}et(M) \,\middle|\, g' = \phi^* g \,,\ \phi \in \mathcal{D}iff(M) \right\}, \tag{1.40}$$

describes all metrics $g' \in \mathcal{M}et(M)$ which are geometrically equivalent to a given $g \in \mathcal{M}et(M)$ (Fig. 1.10). Since the action of $\mathcal{D}iff(M)$ on $\mathcal{M}et(M)$ has fixed points corresponding to the isometries $\mathcal{I}_g := \{\varphi \in \mathcal{D}iff(M) \,|\, \phi^*g = g ,\}$ of the given $g \in \mathcal{M}et(M)$, each orbit $\mathcal{O}_g$ is labelled by the corresponding isometry group $\mathcal{I}_g$, (the closed Lie subgroup of $\mathcal{D}iff(M)$, whose Lie algebra is isomorphic to the algebra generated by the Killing vector fields of (M, g)). Note that the set of metrics for which $\mathcal{I}_g \simeq \{id_M\}$, (i.e. no symmetries), is an open set in $\mathcal{M}et(M)$, and geometries of larger symmetry are contained in the boundary of geometries with smaller isometry groups. In general, if one denotes by $\mathcal{M}et(M)_G := \{g \in \mathcal{M}et(M) ,\ \ \mathcal{I}_g \approx G\}$, where G denotes a (conjugacy class of) isometry Lie subgroup of $\mathcal{D}iff(M)$, then we

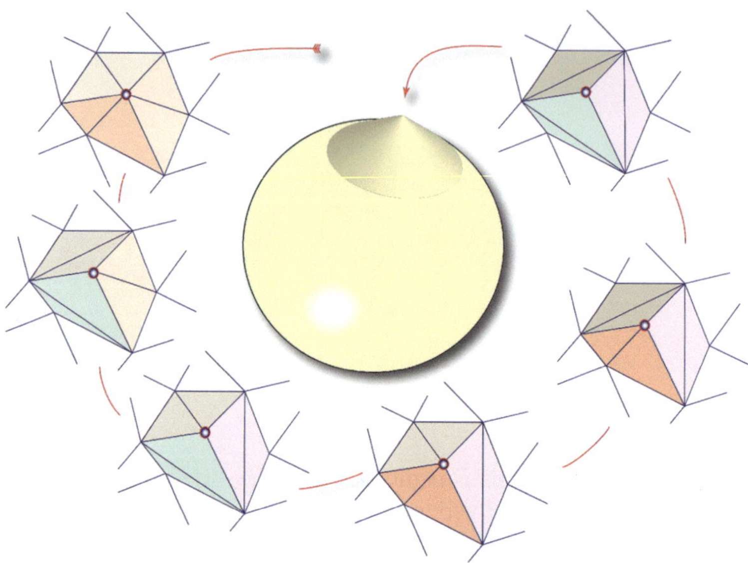

Fig. 1.11 Portions of distinct polyhedral surfaces (T, M), (T', M), (T'', M) ... yielding for the some Polyhedral structure around the given vertex

can write the orbit space $\mathscr{M}et(M)/\mathscr{D}iff(M)$, the *space of Riemannian structures*, as a stratified ∞-dimensional space, i.e.

$$\frac{\mathscr{M}et(M)}{\mathscr{D}iff(M)} \simeq \cup_G \mathscr{S}_G , \qquad \mathscr{S}_G := \mathscr{M}et(M)_G/\mathscr{D}iff(M) , \tag{1.41}$$

where the strata, $\mathscr{S}_G$, are the (∞-dimensional) orbifold of Riemannian structures with the given isometry group $\mathscr{I}_g \simeq G$, partially ordered by the (countable) conjugacy classes of (compact) subgroups $G \in \mathscr{D}iff(M)$, i.e. $\mathscr{S}_G \subset \overline{\mathscr{S}_H}$ whenever $\mathscr{S}_G \cap \overline{\mathscr{S}_H} \neq \emptyset$ and $H \subset G$, $H \neq G$, ($\overline{\mathscr{S}_H}$ denoting the closure of $\mathscr{S}_H$).

As recalled above, the distribution of edge-lengths $\sigma^1(m, n) \to l(m, n) \in \mathbb{R}_+$ of the polyhedral surface (T_l, M), gives rise to an intrinsic distance function d turning (T_l, M) into a metric space. This suggests the following geometrical characterization (Fig. 1.11)

Definition 1.14 (Polyhedral structures) Any two polyhedral surfaces (T_l, M) and $(T'_{l'}, M)$, of genus g and with N_0 labelled vertices, are said to define the same *Polyhedral structure* if the associated distribution of edge lengths induces the same metric geometry on the underlying surface M. We shall denote by $POL_{g, N_0}(M)$ the set of all Polyhedral structures generated by polyhedral surfaces (T_l, M) of genus g, with N_0 labelled vertices.

Note that Polyhedral structures are singular Euclidean structures (in the sense of Troyanov [23]), where the adjective *singular* refers here to the fact that polyhedral surfaces (T_l, M) in general exhibit conical singularities on their vertex set $V(T)$. The geometrical implications of the existence of such a singular set will be discussed at length in Chap. 2. For now, we have the following

Remark 1.3 $POL_{g,N_0}(M)$ can be characterized as a space homeomorphic to an open set of $\mathbb{R}^{N_1}_+$, with $N_1 = 3N_0 + 6g - 6$, topologized by the standard ϵ-neighborhoods $U_\epsilon \subset \mathbb{R}^{N_1}_+$. Explicitly, let (T_l, M) be a polyhedral surface with

$$l_T := \min \left\{ l(h,j) \mid \sigma^1(h,j) \in E(T) \right\}, \tag{1.42}$$

and let $\left\{ d_T(h,k) := d_T\left(\sigma^0_T(h), \sigma^0_T(k) \right) \right\}$ be the corresponding distances among the N_0 labelled vertices $\{\sigma^0_T(j)\} \in (T_l, M)$. If ϵ is a positive number such that $0 < \epsilon < l_T/2$, then the ϵ–neighborhood $U_\epsilon(T_l, M)$ of (T_l, M) in $POL_{g,N_0}(M)$ is the set of all metric triangulations $(T'_{l'}, M)$ whose vertex distances, $\{d_{T'}(h,k)\}$, are such that

$$d_T(h,k) - \epsilon < d_{T'}(h,k) < d_T(h,k) + \epsilon. \tag{1.43}$$

(Note that, since any such $(T'_{l'}, M)$ necessarily has $N_1 = 3N_0 + 6g - 6$ edges, the relation (1.43) yields a corresponding bound on the edge lengths $\{l_{T'}(h,j)\}$ of $(T'_{l'}, M)$).

1.7 The Space of Polyhedral Surfaces $\mathcal{T}^{met}_{g,N_0}(M;\{\Theta(k)\})$

Can we characterize $POL_{g,N_0}(M)$ as the set of all polyhedral surfaces (T_l, M) modulo the action of a mapping factoring out distinct isometry classes of (T_l, M)? This question is naturally suggested by the obvious analogy between $POL_{g,N_0}(M)$ and the space of Riemannian structures on surfaces, and its answer amounts finding out what role plays here the diffeomorphism group. The discretization involved in triangulating the surface M strongly reduces $\mathscr{D}iff(M)$, and the residual invariance is provided by the group of diffeomorphisms, $\mathscr{D}iff(M, N_0)$, which leave the vertex set $V(T)$ of (T_l, M) invariant. In the case we are considering, the action of $\mathscr{D}iff(M, N_0)$ is taken over implicitly by a natural combinatorial mapping among polyhedral surfaces that will be discussed momentarily. Also note that, under a rather obvious analogy, the group of automorphisms $Aut(T)$ of (T_l, M), will play a role analogous to the one played by the isometry groups $\mathscr{I}_g$ for the space of Riemannian structures. In this connection, it is worthwhile to recall the following, (see Chap. 13 of [21]), [3].

Definition 1.15 (Thurston Orbifold) A smooth (metrizable) topological space X is an orbifold in the sense of Thurston if it possess an atlas defined by a collection $\left\{ (U_k, \widetilde{U}_k, \varphi_k, \Gamma_k) \right\}_{k \in I}$, where for each k varying in the index set I, U_k is an open subset of X with $\cup U_k = X$, each $\widetilde{U}_k$ is an open subset of $\mathbb{R}^{n-1} \times [0, \infty)$, the local chart $\varphi_k : \widetilde{U}_k \to U_k$ is a continuous map, and Γ_k is a finite group of orientation preserving diffeomorphisms of $\widetilde{U}_k$ such that each φ_k factors through a homeomorphism between $\widetilde{U}_k/\Gamma_k$ and U_k. Moreover, for every $x \in \widetilde{U}_k$ and $y \in \widetilde{U}_h$ with $\varphi_k(x) = \varphi_h(y)$, there is a diffeomorphism ψ between a neighborhood V of x and a neighborhood W of y such that $\varphi_h(\psi(z)) = \psi_k(z)$ for all $z \in V$.

Thus, a n-dimensional orbifold X is basically a space endowed with an atlas of coordinate neighborhoods which are diffeomorphic to quotients of $\mathbb{R}^n$ by a finite diffeomorphism group preserving the orientation, i.e., a manifold modulo the action of a finite group. In our setting we have.

Proposition 1.3 *The group of automorphims $Aut(T)$ of a polyhedral surface (T_l, M) acts naturally on $\mathbb{R}_+^{N_1(T)}$ via the homomorphism $Aut(T) \to \mathfrak{G}_{N_1(T)}$, where $\mathfrak{G}_{N_1(T)}$ denotes the symmetric group over $N_1(T)$ elements. The resulting quotient space $\mathbb{R}_+^{N_1(T)}/Aut(T)$ is a differentiable orbifold in the sense of Thurston.*

Proof (We are modelling part of our analysis on the very readable and detailed presentation [15]). Let us remark that $\mathfrak{G}_{N_1(T)}$ acts on the $N_1(T)$-dimensional Euclidean space $\mathbb{R}^{N_1(T)}$ by permutations of each $\mathbb{R}$ factor, (i.e., by permutation of axes). This permutation is a particular case of an orthogonal transformation of $\mathbb{R}^{N_1(T)}$, and since the positive quadrant $\mathbb{R}_+^{N_1(T)}$ is naturally embedded into $\mathbb{R}^{N_1(T)}$, we can think of $Aut(T)$ as acting on $\mathbb{R}_+^{N_1(T)}$ through a finite subgroup of the orthogonal group $O(N_1(T))$, with respect to the Euclidean structure of $\mathbb{R}_+^{N_1(T)}$. It follows that the resulting quotient space $\mathbb{R}_+^{N_1(T)}/Aut(T)$ is a differentiable orbifold in the sense of Thurston [21]. $\qquad\qquad\square$

Definition 1.16 Let M be a closed orientable surface of genus g. We define the following sets of triangulations of M:

(i) $\mathcal{T}_{g,N_0}(M)$: The set of all *distinct* triangulations (T, M), (in the sense of Definition 1.7), with $N_0(T) = N_0$ labelled vertices satisfying the topological constraints $N_0 - N_1(T) + N_2(T) = 2 - 2g$, $2N_1(T) = 3N_2(T)$;

(ii) $\mathcal{T}_{g,N_0}^{met}(M)$: The set of all polyhedral surfaces $\{(T_l, M)\}$ obtained by attributing edge–lengths to the triangulated surfaces $(T, M) \in \mathcal{T}_{g,N_0}(M)$,

(iii) $\mathcal{T}_{g,N_0}^{met}(M; \{\Theta(k)\})$: The set of all polyhedral surfaces $\{(T_l, M)|\{\Theta(k)\}\}$ in $\mathcal{T}_{g,N_0}^{met}(M)$ with a prescribed set of conical angles $\{\Theta(k)\}_{k=1}^{N_0}$ over the labelled vertices $\{\sigma^0(k)\} \in (T, M)$.

Note that the triangulations in $\mathcal{T}_{g,N_0}(M)$ are not metrical. In particular any two triangulations $(T^{(1)}, M)$ and $(T^{(2)}, M)$ in $\mathcal{T}_{g,N_0}(M)$ are considered equivalent iff they have the same incidence relations.

We have the following explicit characterization of the space $POL_{g,N_0}(M)$ of polyhedral structures associated with polyhedral surfaces

Proposition 1.4 *The space $POL_{g,N_0}(M)$ of polyhedral structures is a differentiable orbifold*

$$POL_{g,N_0}(M) := \coprod_{[T]\in\mathscr{T}_{g,N_0}(M)} \frac{\mathbb{R}_+^{N_1}}{Aut(T)}, \tag{1.44}$$

of dimension

$$\dim\left[POL_{g,N_0}(M)\right] = N_1 = 3N_0 + 6g - 6, \tag{1.45}$$

locally modelled by

$$\frac{\mathscr{T}_{g,N_0}^{met}(M)}{Aut(T)}. \tag{1.46}$$

Note that the union $\coprod_{[T]\in\mathscr{T}_{g,N_0}(M)}$ runs over the finite set of equivalence classes $[T]$ of *distinct* triangulations (T,M) in $\mathscr{T}_{g,N_0}(M)$ labelling the distinct metrical orbicells $\mathbb{R}_+^{N_1}/Aut(T)$. Also, recall that a polyhedral structure M_{pol} on a surface M is the equivalence class of distinct polyhedral surfaces $\{(T_l,M)\}$ whose edge–lengths distribution induce the same metric geometry on (M,N_0). Let $U_\epsilon(M_{pol}) \subset POL_{g,N_0}(M)$ denote a sufficiently small neighborhood of a given $M_{pol} \in POL_{g,N_0}(M)$. What the above result says is that the polyhedral structures in $U_\epsilon(M_{pol})$ are locally parametrized by the polyhedral surfaces (T_l,M) in $T_{g,N_0}^{met}(M)$ obtained by edge–length assignments $E(T) \mapsto \mathbb{R}_+^{N_1}$ which are in an ϵ–neighborhood of those defining M_{pol}, (see (1.43)). This parametrization is defined up to the natural action of the automorphism group of (T_l,M), (as we shall see, there is a natural sense under which $Aut(T)$ can be associated with M_{pol}).

Proof The open cells of $POL_{g,N_0}(M)$, being associated with triangulations (T_l,M) with a given number of metrized edges, have dimension given by $N_1 = 3N_0+6g-6$ which is constant on the set, $\mathscr{T}_{g,N_0}(M)$, of distinct triangulations (T_l,M) of M with N_0 labeled vertices, (see Definition 1.7). This proves (1.45).

In order to explicitly construct the local parametrization of (1.44) in terms of (1.46), let us recall that by Pachner theorem [17, 18] any triangulation (T',M) in $\mathscr{T}_{g,N_0}(M)$ can be reached from a given (T_l,M) by applying a finite number of *flip–moves*. To define this move compatibly with the topology of $\mathbb{R}_+^{N_1}$ let $\sigma^2(h,j,k)$ and $\sigma^2(k,j,m)$ two adjacent triangles in (T_l,M) sharing the edge $\sigma^1(j,k)$. Denote by $d(h,m)$ the distance of the two vertices $\sigma^0(h)$ and $\sigma^0(m)$ opposite to the shared edge $\sigma^1(j,k)$. The *isometric flip move* (Fig. 1.12)

$$Fl[T,T'] : \mathscr{T}_{g,N_0}^{met}(M) \longrightarrow \mathscr{T}_{g,N_0}^{met}(M), \tag{1.47}$$

$$(T_l,M) \longmapsto Fl[T,T'](T_l,M) = (T',M),$$

can be defined by a finite sequence of local mappings (see Fig. 1.12) of the form

$$Fl_{hm}^{jk}[T,T'] : \sigma^2(h,j,k) \cup \sigma^2(k,j,m) \longmapsto \sigma^2(h,j,m) \cup \sigma^2(m,k,h), \tag{1.48}$$

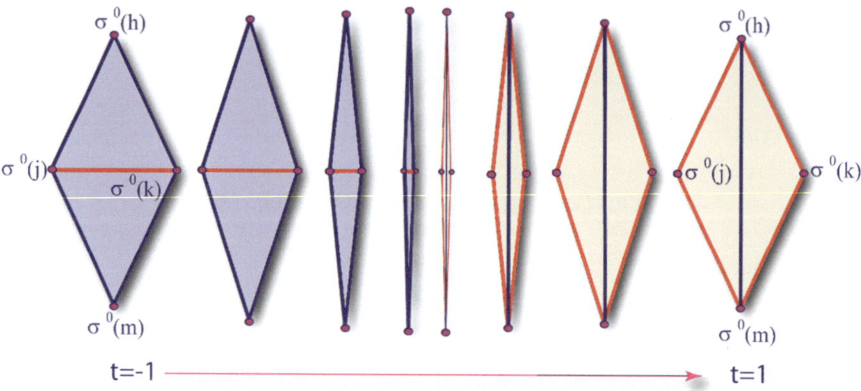

Fig. 1.12 The smooth edge–collapse and edge–expansion used to describe the isometric flip move used in the proof of Proposition 1.4

provided by the 1–parameter deformation which collapses the edge $\sigma_t^1(j,k)$ in (T_l, M) and generates a new edge $\sigma_t^1(h,m)$, (and a new triangulation (T', M) of M), according to

$$\partial \sigma_t^1(j,k) = \left(\sigma^0(j), \sigma^0(k)\right) , \quad \sigma_t^1(h,m) = \emptyset , \quad t \in [-1,0), \qquad (1.49)$$

$$\partial \sigma_t^1(h,m) = \left(\sigma^0(h), \sigma^0(m)\right) , \quad \sigma_t^1(j,k) = \emptyset , \quad t \in [0,1], \qquad (1.50)$$

where ∂ denotes the (incidence) boundary operator, and where the corresponding edge lengths are give by

$$\sigma_t^1(j,k) \longmapsto l_t(j,k) := |t|\, l(j,k) , \qquad t \in [-1,0], \qquad (1.51)$$

$$\sigma_t^1(h,m) \longmapsto l_t(h,m) := (1-t)[l(h,k) + l(k,m)] + t\, d(h,m) , \quad t \in [0,1].$$

This collapse–expansion deformation smoothly interpolates between the pair of triangles $\sigma^2(h,j,k) \cup \sigma^2(k,j,m)$ in the given triangulation (T_l, M) and the new pair

$$Fl_{hm}^{jk}[T, T'] \left(\sigma^2(h,j,k) \cup \sigma^2(k,j,m)\right) := \sigma^2(h,j,m) \cup \sigma^2(m,k,h), \qquad (1.52)$$

which necessarily belongs to a distinct triangulation (T', M) of the surface M. Note that the Euclidean distances between corresponding points in the quadrilaterals $(\sigma^0(h), \sigma^0(j), \sigma^0(m), \sigma^0(k))_{t=-1}$ and $(\sigma^0(h), \sigma^0(j), \sigma^0(m), \sigma^0(k))_{t=1}$ are preserved. This implies that the flip–move (1.51) is well defined in the $\mathbb{R}_+^{N_1}$ topology, and it is an isometry between (T_l, M) and (T', M).

Remark 1.4 (Equilateral triangulations) To avoid misunderstandings, it is perhaps worthwhile to stress that the flip move is not an isometry when constrained to the subset of $\mathscr{T}_{g,N_0}^{met}(M)$ defined by the equilateral triangulations $(T_{l=a}, M)$ of the surface

M. In such a case the elementary flip move (1.51) must be tuned in such a way that the final flipped pair $\sigma^2(h,j,m) \cup \sigma^2(m,k,h)$ remains equilateral. This forces a t–dependent change of the distance function $d(h,m)$. Indeed, the deformation $[-1,1] \ni t \mapsto d_t(h,m)$ must interpolate from the original, $(t=-1)$, value $d(h,m) = \sqrt{3}\,a$ to the final $d_t(h,m)|_{t=1} = a$, (a being the side length of the equilateral triangles $\in (T_{l=a},M)$). This entails a deformation of the metric geometry of $(T_{l=a},M)$, actually mapping $(T_{l=a},M)$ into a metrically distinct equilateral surface $(T'_{l=a},M)$. This is a basic property that lies at the heart of dynamical triangulation theory, since it allows to use flip moves to explore (a subset of the) inequivalent metric structures on the space of polyhedral surfaces. It must be stressed that in this way one probes, (in the large N_0–limit), at most a dense subset of the possible metric structures, and that the stratification of the resulting space does not have the orbifold property, (this directly follows from the analysis of the set of dual ribbon graphs associated with equilateral triangulations [15, 16]).

In our more general setting, equilateral triangulations do not play such a distingushed role since an equilateral triangulation $(T_{l=a},M)$ is mapped, by a sequence of elementary flip moves (1.51), into an isometric polyhedral surface $(T'_{l'\neq a},M)$ which is no longer equilateral.

Coming back to the proof of Proposition 1.4, let us consider the mapping

$$\widetilde{\pi} \;:\; \mathscr{T}^{met}_{g,N_0}(M) \longrightarrow POL_{g,N_0}(M), \tag{1.53}$$

$$(T_l,M) \longmapsto Fl_{(T_l,M)},$$

that associates to (T_l,M) its orbit under a generic finite sequence, $Fl[T,T'](T_l,M)$, of isometric flip moves

$$Fl_{(T_l,M)} := \left\{ (T'_{l'},M) \in \mathscr{T}^{met}_{g,N_0}(M) \,\middle|\, (T'_{l'},M) = Fl[T,T'](T_l,M) \right\}. \tag{1.54}$$

$Fl_{(T_l,M)}$ is the *singular Euclidean structure* M_{pol} associated with the polyhedral surface (T_l,M) (Fig. 1.13). Since automorphisms $\in Aut(T)$ generate equivalent polyhedral surfaces in $\mathscr{T}^{met}_{g,N_0}(M)$, the orbit map (1.53) descend to the quotient space

$$\pi \;:\; \frac{\mathscr{T}^{met}_{g,N_0}(M)}{Aut(T)} \longrightarrow POL_{g,N_0}(M), \tag{1.55}$$

$$(T_l,M) \;\sim_{Aut(T)}\; (T'_{l'},M) \longmapsto Fl_{(T_l,M)} \simeq Fl_{(T'_{l'},M)}.$$

Conversely, let $M_{pol} \in POL_{g,N_0}(M)$ a singular Euclidean structure and let (T_l,M) and $(T'_{l'},M)$ two distinct polyhedral surfaces in $\pi^{-1}(M_{pol}) \subset \mathscr{T}^{met}_{g,N_0}(M)$. Denote by $Aut(T)$ the automorphisms group of (T_l,M). Let $Fl[T,T']$ be the (sequence of) flip moves such that $Fl[T,T'](T_l,M) = (T'_{l'},M)$, and let $\beta \in Aut(T)$. We have

$$\left(Fl[T,T'] \circ \beta \circ Fl[T,T']^{-1}\right)(T'_{l'},M) = \left(Fl[T,T'] \circ \beta\right)(T_l,M) \tag{1.56}$$

$$= Fl[T,T'](T_l,M) = (T'_{l'},M),$$

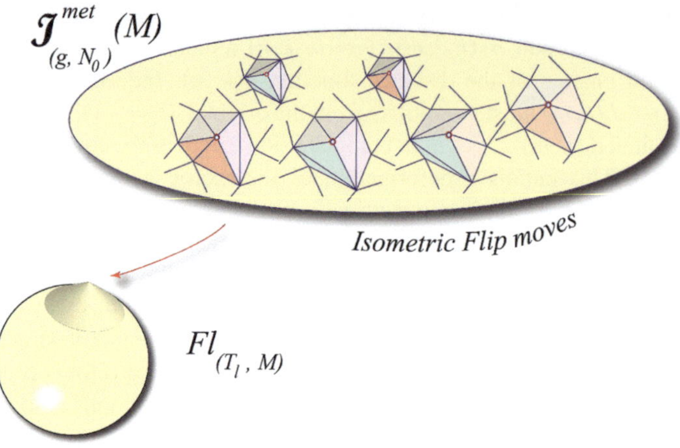

Fig. 1.13 The flip move orbit defining a singular Euclidean structure

thus $\beta' := Fl[T,T'] \circ \beta \circ Fl[T,T']^{-1} \in Aut(T')$. In other words, the action of $Aut(T)$ on $POL_{g,N_0}(M)$ is defined modulo a conjugacy class of flip moves on any polyhedral surface (T_l, M) representative of M_{pol}.

For ϵ small enough, let $U_\epsilon(M_{pol})$ denote an ϵ–neighborhood of a given $M_{pol} \in POL_{g,N_0}(M)$. Its inverse image, $\widetilde{\pi}^{-1}\left(U_\epsilon(M_{pol})\right)$, under the map $\widetilde{\pi}$ is a subset of $\mathcal{J}^{met}_{g,N_0}(M)$. Let (T_l, M) be a polyhedral surface generating, (under flip moves), the given M_{pol}, i.e., $M_{pol} = Fl_{(T_l,M)}$, (see (1.54)). There is an obvious continuous bijection between $U_\epsilon(M_{pol})$ and $\widetilde{\pi}^{-1}\left(U_\epsilon(M_{pol})\right)$ defined by representing the generic $\widehat{M}_{pol} \in U_\epsilon(M_{pol})$ via the corresponding orbit map $\widehat{M}_{pol} = Fl_{(T_{l\pm\epsilon},M)}$. Moreover, according to the above remarks, the action of the automorphism group $Aut(T)$ on this map is well–defined under flip conjugation. It follows that

$$\pi \; : \; \frac{\widetilde{\pi}^{-1}\left(U_\epsilon(M_{pol})\right)}{Aut(T)} \longrightarrow U_\epsilon(M_{pol}) \subset POL_{g,N_0}(M), \tag{1.57}$$

is a homeomorphism providing a local coordinate chart

$$(\mathcal{U}_T,\, \pi_T) := \left(\frac{\widetilde{\pi}^{-1}\left(U_\epsilon(M_{pol})\right)}{Aut(T)} ,\, \pi \right), \tag{1.58}$$

for $POL_{g,N_0}(M)$. If a singular Euclidean structure M^*_{pol} belongs to the intersection $(\mathcal{U}_{T(1)},\, \pi_{T(1)}) \cap (\mathcal{U}_{T(2)},\, \pi_{T(2)})$ of any two such a local coordinate chart, then there are two isometric polyhedral surfaces, say $(T^{(1)}_{l_1}, M)$ and $(T^{(2)}_{l_2}, M)$, representing locally M^*_{pol}. According to the above remarks, we can always find a finite sequence of isometric flip moves such that we can transform $(T^{(1)}_{l_1}, M)$ into $(T^{(2)}_{l_2}, M)$, modulo a conjugated action of $Aut(T^{(1)})$. Since the flip move is defined by an interpolation which is smooth in the $\mathbb{R}^{N_1}$ topology, the two local coordinate charts $(\mathscr{U}_{T(1)}, \pi_{T(1)})$ and $(\mathscr{U}_{T(2)}, \pi_{T(2)})$ have diffeomorphic intersection, and consequently $POL_{g,N_0}(M)$ is a differentiable orbifold locally modelled on $(\mathscr{U}_T, \pi_T) \subset \mathscr{T}^{met}_{g,N_0}(M)/Aut(T)$ as stated.

A direct consequence of Proposition 1.4 is the following

Theorem 1.3 *The space of Polyhedral structures $POL_{g,N_0}(M)$ generated by closed polyhedral surfaces (T_l, M) of genus g with N_0 vertices has*

$$\mathscr{C}ard \left\{ \mathscr{T}_{g,N_0}(M) \right\}, \tag{1.59}$$

distinct orbifold strata, where $\mathscr{C}ard$ denotes the cardinality of the set of distinct triangulations considered.

Proof Since, according to Proposition 1.4, the strata $\dfrac{\mathbb{R}^{N_1}_{+}}{Aut(T)}$ of $POL_{g,N_0}(M)$ are labelled by the finite set of distinct triangulations $[T] \in \mathscr{T}_{g,N_0}(M)$, the theorem immediately follows. ∎

In order to explore the basic properties of the space of polyhedral structures $POL_{g,N_0}(M)$ we need to discuss the structure of $\mathscr{T}^{met}_{g,N_0}(M)$ in the neighborhood of a given polyhedral surface (T_l, M). This will provide the appropriate notion of local deformation of polyhedral surfaces. A thorough analysis of the deformation theory of polyhedral surface, (in the general case of Euclidean, Hyperbolic, and Spherical triangulated surfaces), is presented in a series of remarkable papers by Feng Luo [7, 12, 13], who, building on the influential work of Thurston [22], Rivin [19], and Leibon [11], provides a very detailed and illustrative geometrical framework for the theory. Here we take a slightly different point of view, more apt to the physical applications, emphasizing the natural connections with the Witten–Kontsevich theory.

1.8 Cotangent Cones and Circle Bundles $\mathscr{Q}_{(k)}$ Over $\mathscr{T}^{met}_{g,N_0}(M)$

A basic question in discussing the deformation theory of $\mathscr{T}^{met}_{g,N_0}(M)$ is to characterize the space of the infinitesimal deformations of a given polyhedral surface (T_l, M), i.e. a suitable notion of tangent space to $\mathscr{T}^{met}_{g,N_0}(M)$ at (T_l, M). The idea is to formalize the intuitive picture of what a deformation of a neighborhood of a vertex $\sigma^0(k)$ in a polyhedral surface looks like. To this end, we inject the given vertex $\sigma^0(k)$ in the origin O of $(\mathbb{R}^3; O, x, y, z)$, where we identify $\mathbb{R}^3$, endowed with the vector product $\times$, with the Lie algebra $\mathfrak{so}(3)$ of the rotation group $SO(3)$. Since $\sigma^0(k)$ is a conical point whose generators are edges of Euclidean triangles, we can associate with $\sigma^0(k)$ the polyhedral cone with apex O, whose generators have lengths normalized to 1 and whose directrix is a piecewise–geodesic curve on the surface of a unit sphere $\mathbb{S}^2$ centered at the origin. This curve is a spherical polygon whose side–lengths are provided by the original vertex angles $\{\theta_\alpha\}$ generating the conical geometry around $\sigma^0(k)$. This construction has a natural $SO(3)$ symmetry which, when reduced, generates a set of spherical polygons which parametrizes all possible distinct polyhedral cones in $(\mathbb{R}^3; O, x, y, z)$ having the same intrinsic metric geometry of the local Euclidean structure $\mathscr{E}\left(Star[\sigma^0(k)]\right)$ near $\sigma^0(k)$. This set characterizes the *polyhedral cotangent cone* to $\mathscr{E}\left(Star[\sigma^0(k)]\right)$ at $\sigma^0(k)$. There is a natural bundle associated with this construction which, to some extent, provides a suitable notion of (co)tangent space to $\mathscr{T}_{g,N_0}^{met}(M)$ at (T_l, M).

We begin with a few notational remarks. Recall that if $q(k) \geq 2$ triangles $\sigma^2(h_\alpha, k, h_{\alpha+1})$ are incident on $\sigma^0(k)$, then

$$Star[\sigma^0(k)] \doteq \cup_{\alpha=1}^{q(k)} \sigma^2(h_\alpha, k, h_{\alpha+1}), \quad h_{q(k)+1} \equiv h_1, \tag{1.60}$$

where the index α is taken modulo $q(k)$. We denote by

$$\mathscr{E}\left(Star[\sigma^0(k)]\right) \doteq \{(\theta_{\alpha,k,\alpha+1}, \theta_{k,\alpha+1,\alpha}, \theta_{\alpha+1,\alpha,k})\}, \tag{1.61}$$

the generic conically complete locally Euclidean structure on $Star[\sigma^0(k)]$, and consider the ordered sequence of vertex angles at $\sigma^0(k)$

$$\{\theta_{\alpha,k,\alpha+1}\}_{\alpha=1}^{\alpha=q(k)}, \quad \theta_{q(k),k,q(k)+1} \equiv \theta_{q(k),k,1}, \tag{1.62}$$

as a point of $[0, \pi]^{q(k)}$, modulo cyclic permutations in $\mathbb{Z}_{q(k)}$. The boundary points 0 and π corresponds to the degeneration of those triangles $\sigma^2(h_\alpha, k, h_{\alpha+1})$ whose corresponding vertex angle $\theta_{\alpha,k,\alpha+1}$ either goes to 0, (a *thinning* triangle), or to π, (a *fattening* triangle). Note that in the former case we have that the edge length $l(h_\alpha, h_{\alpha+1}) \to 0$ with $l(l(k, h_\alpha)$ and $l(k, h_{\alpha+1})$ both > 0; whereas, in the latter case we have $l(h_\alpha, h_{\alpha+1}) = l(k, h_\alpha) + l(k, h_{\alpha+1})$. According to Proposition 1.1 to any $\mathscr{E}\left(Star[\sigma^0(k)]\right)$ we can associate a conical defect

$$\Theta(k) \doteq \sum_{\alpha=1}^{q(k)} \theta_{\alpha,k,\alpha+1}, \tag{1.63}$$

supported at $\sigma^0(k)$, which induces a conical geometry on $Star[\sigma^0(k)]$. Since (1.63) depends only on the $\sigma^0(k)$ angles $\{\theta_{\alpha,k,\alpha+1}\} \in \mathscr{E}\left(Star[\sigma^0(k)]\right)$, it is natural to define the set of *vertex angle structures* at $\sigma^0(k)$ associated with $\mathscr{E}\left(Star[\sigma^0(k)]\right)$ according to

$$\mathscr{V}ert_{q(k)} := \left\{\{\theta_\alpha\}_k \in [0, \pi]^{q(k)}/\mathbb{Z}^{q(k)} \,\middle|\, \exists \mathscr{E}\left(Star[\sigma^0(k)]\right) \ s.t. \ \{\theta_\alpha\}_k = \{\theta_{\alpha,k,\alpha+1}\}\right\}, \tag{1.64}$$

(we shall often drop the index k in $\{\theta_\alpha\}_k$ if it is clear from the geometrical context which vertex we are considering). Similarly, for a given $\Theta(k) \in \mathbb{R}^+$, we define the set of *vertex angle structures* associated with the conical angle $\Theta(k)$ as

$$\mathscr{V}ert_{q(k)}[\Theta(k)] := \left\{ \{\theta_\alpha\}_k \in \mathscr{V}ert_{q(k)} \Big| \sum_{\alpha=1}^{q(k)} \theta_\alpha = \Theta(k) \right\}. \tag{1.65}$$

Remark 1.5 Since the automorphism group $Aut(T)$ of a given a triangulation (T_l, M) in $\mathscr{T}^{met}_{g,N_0}(M)$ fixes every vertex, we can consistently label the vertices of (T_l, M) by the index set $V := (1,\ldots,k,\ldots,N_0)$. This also induces a corresponding labeling on the vertex angles $\{\theta(h,k,j)\}$ of (T_l,M) which is coherent with the vertex star–grouping (1.62). Explicitly, if we define the ordering

$$\ldots \prec \{\theta_\alpha\}_k \prec \{\theta_\beta\}_{k+1} \prec \ldots, \tag{1.66}$$

i.e.,

$$\ldots \prec \{\theta_{\alpha,k,\alpha+1}\}_{\alpha=1}^{\alpha=q(k)} \prec \{\theta_{\beta,k+1,\beta+1}\}_{\beta=1}^{\alpha=q(k+1)} \prec \ldots, \tag{1.67}$$

we can label the vertex angles $\{\theta(h,k,j)\}$ by an index set Φ with cardinality $3N_2$. In particular we can associate to the polyhedral surface (T_l,M) the vertex angles vector

$$\mathscr{T}^{met}_{g,N_0}(M) \longrightarrow (0,\pi)^{3N_2} = \oplus_{k=1}^{N_0} \mathscr{V}ert_{q(k)} \tag{1.68}$$

$$(T_l, M) \longmapsto \vec{\vartheta}(T_l) := \left(\{\theta_\alpha\}_1, \ldots, \{\theta_\beta\}_{N_0} \right),$$

and the *angle–vertex* incidence matrix

$$A_{T_l} := [a_{kJ}]_{k \in V,\, J \in \Phi}, \tag{1.69}$$

where $a_{kJ} \equiv 1$ if the vertex angle $\angle_J$ is incident on the vertex $\sigma^0(k)$, $a_{kJ} \equiv 0$ otherwise. Namely, $a_{kJ} \equiv 1$ iff $\angle_J = \theta_{\alpha,k,\alpha+1}$ for some $\alpha \in (1,\ldots,q(k))$. Note in particular that the subset of triangulations $\mathscr{T}^{met}_{g,N_0}(M; \{\Theta(k)\})$ with a prescribed set of conical angles $\{\Theta(k)\}$ can be equivalently characterized by the condition

$$\mathscr{T}^{met}_{g,N_0}(M; \{\Theta(k)\}) = \left\{ (T_l,M) \in \mathscr{T}^{met}_{g,N_0}(M) \big| A_{T_l} \vec{\vartheta}(T_l)^t = \vec{\Theta}(T_l)^t \right\}, \tag{1.70}$$

where $A^{T_l} \vec{\vartheta}(T_l)^t$ denotes the matrix product between A_{T_l} and the transposed (column) vector $\vec{\vartheta}(T_l)$, and where $\vec{\Theta}(T_l)$ denotes the given conical angles vector associated with (T_l,M), i.e.

$$\vec{\Theta}(T_l) := (\Theta(1),\ldots,\Theta(k),\ldots,\Theta(N_0)). \tag{1.71}$$

$\square$

The general idea behind the introduction of $\mathscr{V}ert_{q(k)}$ and $\mathscr{V}ert_{q(k)}[\Theta(k)]$ is that these sets may serve as coordinate spaces for the conical angles $\{\Theta(k)\}$ associated with polyhedral surfaces. However this fails short of being a significant parametrization since the incidence order $q(k)$ may vary with the vertex $\sigma^0(k)$ and with the polyhedral surface $(T_l, M) \in \mathscr{T}^{met}_{g,N_0}(M)$ considered. To bypass this difficulty, it is convenient to *telescope* [4] the set $\{\theta_\alpha\}_{\alpha=1}^{q(k)}$ in $\mathscr{V}ert_{q(k)}$ as $q(k) \to \infty$, i.e. to inject $\{\theta_\alpha\}_{\alpha=1}^{q(k)}$ not in a particular $\mathscr{V}ert_{q(k)}$ but rather in the inductive limit of $\mathscr{V}ert_{q(k)}$ as $q(k) \to \infty$. This is easily realized by observing that there is a natural sequence of inclusions

$$\left(\theta_1, \ldots, \theta_{q(k)}\right) \hookrightarrow \mathscr{V}ert_{q(k)} \hookrightarrow \ldots \mathscr{V}ert_{q(k)+h} \hookrightarrow \mathscr{V}ert_{q(k)+h+1} \hookrightarrow \ldots, \qquad (1.72)$$

defined, (e.g., in the $\mathscr{V}ert_{q(k)} \hookrightarrow \mathscr{V}ert_{q(k)+1}$ case), by the $q(k) + 1$ distinct injections

$$\left(\theta_1, \ldots, \theta_{q(k)}\right) \overset{j_\beta^{q,q+1}}{\longrightarrow} \left(\theta_1, \ldots, \theta_{\beta-1}, 0, \theta_\beta, \ldots, \theta_{q(k)}\right), \qquad (1.73)$$

where $\beta = 1, \ldots, q(k) + 1$. Thus, independently from the particular vertex $\sigma^0(k)$ and triangulation (T_l, M) considered, we introduce the following characterization

Definition 1.17 The inductive limit of the family $\left\{\mathscr{V}ert_q, j_\beta^{q,q+1}\right\}$, as $q \to \infty$, defines the space of *vertex angle structures*

$$\mathscr{V}ert := \lim_{q \to \infty} \left\{\mathscr{V}ert_q \hookrightarrow \mathscr{V}ert_{q+1}, j_\beta^{q,q+1}\right\} \simeq [0, \pi]^\infty, \qquad (1.74)$$

associated with the set of Euclidean vertex stars $\left\{\mathscr{E}\left(Star[\sigma^0(k)]\right)\right\}$ of polyhedral surfaces in $\mathscr{T}^{met}_{g,N_0}(M)$.

Explicitly, we may think of a sequences $\{\theta_\alpha\}_{\alpha=1}^{q(k)} \in \mathscr{V}ert_{q(k)}$, associated with $\mathscr{E}\left(Star[\sigma^0(k)]\right)$, as the coordinates of the corresponding conical structure in the local chart $\mathscr{V}ert_{q(k)}$ of $\mathscr{V}ert$, i.e.

$$\oplus_{k=1}^{N_0} \mathscr{V}ert_{q(k)} \hookrightarrow \mathscr{V}ert. \qquad (1.75)$$

We can locally model the geometrical realizations of the local chart $\mathscr{V}ert_{q(k)}$ in terms of spherical polygons (Fig. 1.14). This will characterize a natural polygonal bundle over $\mathscr{V}ert$ which, as described in the introductory remarks, will play the role of the cotangent bundle to $\mathscr{T}^{met}_{g,N_0}(M)$. To carry out such a construction explicitly, let us consider the link of $\sigma^0(k)$ in $Star[\sigma^0(k)] \in (T_l, M) \in \mathscr{T}^{met}_{g,N_0}(M)$,

$$link(k) := link_{(T_l,M)}\left(\sigma^0(k)\right) \simeq \mathbb{S}^1, \qquad (1.76)$$

viz. the union of all edges $\sigma^1(h_\alpha, h_{\alpha+1})$ in $Star[\sigma^0(k)]$ satisfying $\sigma^1(h_\alpha, h_{\alpha+1}) \cap \sigma^0(k) = \emptyset$, (see Theorem 1.1). We can naturally map $link(k)$ to a closed rectifiable piecewise–smooth Jordan curve $c_{(k)}$ on the unit sphere

$$c_{(k)} : [0, 1] \longrightarrow link(k) \hookrightarrow \mathbb{S}^2, \qquad (1.77)$$

$$t \longmapsto x(t) \hookrightarrow c_{(k)}(t),$$

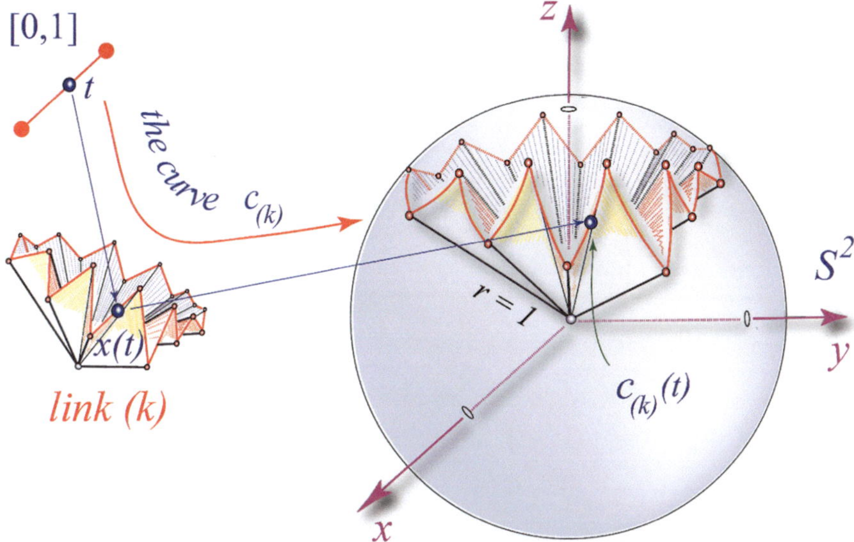

Fig. 1.14 The image of the curve (1.77) is characterized by the intersection, in $(\mathbb{R}^3; O, x, y, z)$, of a geometric realization of $\mathscr{E}\left(Star[\sigma^0(k)]\right)$ with a unit sphere $\mathbb{S}^2$ centered in the origin $O \equiv \sigma^0(k)$. This intersection is well-defined up to a natural $SO(3)$ action

where $x(t) \in link(k)$, and where $c_{(k)}(0) = c_{(k)}(1)$. The image $c_{(k)}([0,1])$ of the curve $c_{(k)}$ can be explicitly constructed by intersecting, in $\left((\mathbb{R}^3, \times) \cong \mathfrak{so}(3); O, x, y, z\right)$, a geometrical realization of the given Euclidean structure $\mathscr{E}\left(Star[\sigma^0(k)]\right)$ with a unit sphere $\mathbb{S}^2$ centered in the origin O, (see Fig. 1.14). This generates a cone whose intrinsic metric depends only on the length $|c_{(k)}|$ of the curve $c_{(k)}$, (see Fig. 1.15), and we have the characterization, (see also [5])

Lemma 1.1 *The conical angle $\Theta(k)$ at the vertex $\sigma^0(k)$ is the length $|c_{(k)}|$ of the Jordan curve* (1.77). $\square$

The mapping (1.77) also induces on the curve $c_{(k)}$ the decoration defined by the $q(k)$ points $\{p_\alpha\}$ images of the vertices $\{\sigma^0(h_\alpha)\} \in link(k)$. In particular the vertex angle $\theta_\alpha(k)$ associated with the triangle $\sigma^2(h_\alpha, k, h_\alpha)$ provides the length

$$\left|c_{(k)}(\alpha, \alpha + 1)\right| := \theta_\alpha(k), \tag{1.78}$$

of the arc of geodesic $c_{(k)}(\alpha, \alpha + 1)$ between the point p_α and $p_{\alpha+1}$.

Remark 1.6 The $\{c_{(k)}(\alpha, \alpha + 1)\}$ are geodesic arcs since they are defined by the intersection between $\mathbb{S}^2$ and planes passing through the center of $\mathbb{S}^2$.

M. Kapovich and J. Millson have studied in depth the properties of the set of spherical polygons and of the associated moduli spaces, (see [9] and references therein). Here we exploit a few elementary aspects of their analysis playing an important role in our framework. We start by recalling the

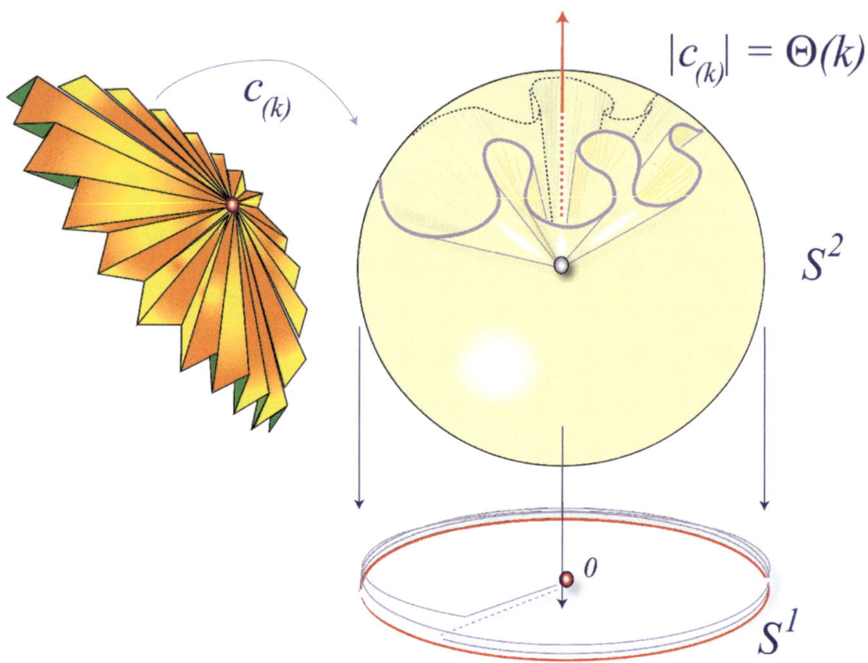

Fig. 1.15 The cone associated with the curve $c_{(k)}$ on the unit sphere S^2. The intrinsic metric of such a cone depends only on the length $|c_{(k)}|$ of the curve $c_{(k)}$, (to emphasize this we have rounded the image of the curve). This length is the conical angle $\Theta(k)$ at the vertex. If the cone generated by $c_{(k)}$ is cut open along a generator and unfolded into the (complex) plane, the curve $c_{(k)}$ covers the circle S^1 and $\Theta(k)$ gives the length (with multiplicity) of the circuitation around S^1

Definition 1.18 (Labeled Spherical Polygons [9]) A $q(k)$–tuple of oriented geodesic arcs $c_{(k)}(\alpha, \alpha + 1) \in \mathbb{S}^2$ of lengths

$$0 < \left| c_{(k)}(\alpha, \alpha + 1) \right| < \pi, \tag{1.79}$$

such that the end–point of the geodesic segment $c_{(k)}(\alpha, \alpha + 1)$ coincides with the initial point of the geodesic segment $c_{(k)}(\alpha + 1, \alpha + 2)$ defines a closed spherical $q(k)$–gon

$$P_{(k)}(c_{(k)}) := \{c_{(k)}(\alpha, \alpha + 1)\}. \tag{1.80}$$

$\square$

Let us remark that the labeling here refers to the fact that the geodesic segments $c_{(k)}(\alpha, \alpha + 1)$ defining the edges of the polygon are path–ordered. Moreover, as long as $0 < \left| c_{(k)}(\alpha, \alpha + 1) \right| < \pi$, the geodesic arcs $c_{(k)}(\alpha, \alpha + 1)$ are equivalently characterized by the ordered sequence of $q(k)$ points $\{p_\alpha\}$, and we can identify $P_{(k)}(c_{(k)}) = \{p_\alpha\}$. The more general case $\left| c_{(k)}(\alpha, \alpha + 1) \right| \in [0, \pi]$, which corresponds to spherical polygons associated with degenerating triangles in $\mathscr{E}(Star[\sigma^0(k)])$, can be handled with a detailed analysis of the possible singular polygonal configurations associated with the triangle degenerations. In order to avoid the complications due to these singular configurations, we shall consider uniquely the generic case under the restriction (1.79).

There is a natural $SO(3)$ diagonal action on the spherical polygons $P_{(k)}(c_{(k)})$ defined by [9]

$$g \cdot P_{(k)}(c_{(k)}) := \left(g \cdot \overrightarrow{Op}_1, \dots, g \cdot \overrightarrow{Op}_{q(k)} \right), \tag{1.81}$$

where $g \in SO(3)$ and where $\overrightarrow{Op}_\alpha$ is the vector in $\mathbb{R}^3$ connecting the center O of $\mathbb{S}^2$ with the point $p_\alpha \in \mathbb{S}^2$. Denote by $\mathscr{P}_{q(k)}(\mathbb{S}^2)$ the space of spherical polygons [9]. Since, according to the above remarks, there is a one–to–one correspondence between $SO(3)$ equivalence classes of spherical polygons and polyhedral cones modelling a neighborhood of the vertex $\sigma^0(k)$ it is natural to introduce the

Definition 1.19 The quotient space of $\mathscr{P}_{q(k)}(\mathbb{S}^2)$ by the $SO(3)$ action described above,

$$\mathscr{Q}_{q(k)}(\mathbb{S}^2) := \mathscr{P}_{q(k)}(\mathbb{S}^2) \, / \, SO(3), \tag{1.82}$$

defines the space of *polyhedral cones* over the vertex $\sigma^0(k) \in (T_l, M)$.

In order to fix the $SO(3)$ gauge freedom in $\mathscr{P}_{q(k)}(\mathbb{S}^2)$ in (1.82) we start by remarking that the configuration space of spherical polygons with given side-lengths $\{|c_{(k)}(\alpha, \alpha + 1)|\}$ can be characterized as the set

$$\mathscr{P}_{(k)}(\mathbb{S}^2; \{|c_{(k)}(\alpha, \alpha + 1)|\}) := \tag{1.83}$$

$$\left\{ \{p_\alpha\} \in \mathbb{S}^{2\,q(k)} \,\middle|\, \overrightarrow{Op}_\alpha \cdot \overrightarrow{Op}_{\alpha+1} = \cos \,|c_{(k)}(\alpha, \alpha + 1)| \right\},$$

where $\cdot$ denotes the scalar product in $\mathbb{R}^3$. In particular, we can always choose the geodesic arc $c_{(k)}(q(k), 1)$ of the spherical polygon $P_{(k)}(c_{(k)})$ to lie in the (O, x, y)–plane $\subset \mathbb{R}^3$. Let $(\overrightarrow{\varepsilon}_1, \overrightarrow{\varepsilon}_2)$ be orthonormal basis vectors in (O, x, y), and let us denote by p_0 the midpoint of the geodesic arc $c_{(k)}(q(k), 1)$. We take p_0 to be the origin of the piecewise geodesic path $c_{(k)}$. Rotate the polygon $P_{(k)}(c_{(k)})$ around the origin in the (O, x, y)–plane so that $\overrightarrow{Op}_0 \equiv \overrightarrow{\varepsilon}_1$. This fixes the position of the vector $\overrightarrow{Op}_1$ according to (Fig. 1.16)

$$\overrightarrow{Op}_1 = \cos \frac{|c_{(k)}(q(k), 1)|}{2} \, \overrightarrow{\varepsilon}_1 + \sin \frac{|c_{(k)}(q(k), 1)|}{2} \, \overrightarrow{\varepsilon}_2. \tag{1.84}$$

In such a representation there is a natural map defined by

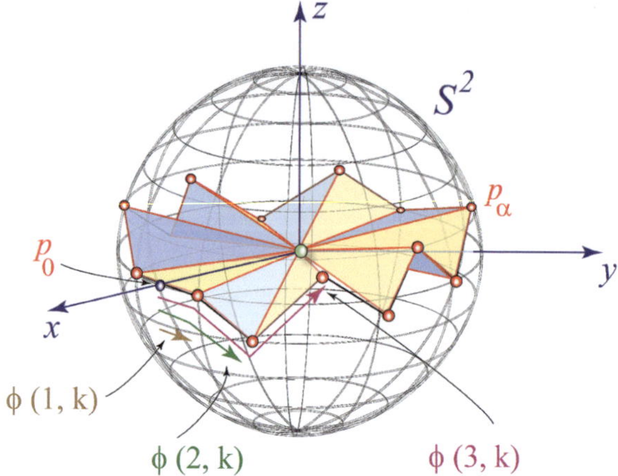

Fig. 1.16 The decoration defined by the $q(k)$ points $\{p_\alpha\}$ images of the vertices $\{\sigma^0(h_\alpha)\} \in link(k)$. The arc–length computed from the origin p_0, defined by choosing the geodesic arc $c_{(k)}(q(k), 1)$ of the spherical polygon $P_{(k)}(c_{(k)})$ to lie in the (O, x, y)–plane $\subset \mathbb{R}^3$, characterizes the map $\phi(\alpha, k)$ associated with the point p_α

$$\pi_{\phi(k)} \; : \; \mathscr{P}_{q(k)}(\mathbb{S}^2) \; \longrightarrow \; \left[\mathbb{S}^1/2\pi\right]^{q(k)} \tag{1.85}$$

$$\{p_\alpha\} \; \longmapsto \; \{\phi(\alpha, k)\}_{\alpha=1}^{q(k)} \; := \; \left\{\frac{\left|c_{(k)}(0, \alpha)\right|}{\Theta(k)}\right\}_{\alpha=1}^{q(k)},$$

where $\left|c_{(k)}(0, \alpha)\right|$ denotes the length of the piecewise geodesic arc of curve $c_{(k)}(0, \alpha)$ between the midpoint $p_0 \in c_{(k)}(q(k), 1)$, (the origin of the piecewise geodesic path $c_{(k)}$), and the polygon vertex p_α. Note that the normalization to the total conical angle $\Theta(k) = \sum_{\alpha=1}^{q(k)} \theta_\alpha$ is well–defined since, according to Lemma 1.1,

$$\Theta(k) = \sum_{\alpha=1}^{q(k)} \left|c_{(k)}(\alpha, \alpha + 1)\right|. \tag{1.86}$$

Moreover, we have

$$\phi(\alpha + 1, k) \; = \; \phi(\alpha, k) + \frac{\left|c_{(k)}(\alpha, \alpha + 1)\right|}{\Theta(k)}, \qquad \alpha = 1, \ldots, q(k) - 1, \tag{1.87}$$

which directly follows from the definition of $\phi(\alpha, k)$, and

$$\phi(1, k) \; = \; \phi(q(k), k) + \frac{\left|c_{(k)}(q(k), 1)\right|}{\Theta(k)} - 1, \qquad \alpha = q(k), \tag{1.88}$$

which is a consequence of the identity

$$\frac{\left|c_{(k)}(q(k), 1)\right|}{\Theta(k)} = 1 - \sum_{\alpha=1}^{q(k)-1} \left(\phi(\alpha + 1, k) - \phi(\alpha, k)\right). \tag{1.89}$$

Conversely,

$$\left|c_{(k)}(\alpha, \alpha + 1)\right| = \Theta(k)\left[\phi(\alpha + 1, k) - \phi(\alpha, k)\right], \quad \alpha = 1, \ldots, q(k) - 1, \tag{1.90}$$

and

$$\left|c_{(k)}(q(k), 1)\right| = \Theta(k)\left[\phi(1, k) - \phi(q(k), k) + 1\right]. \tag{1.91}$$

It follows that, relative to the framing (1.84), we can uniquely represent any spherical polygon, with given edge lengths $\{\left|c_{(k)}(\alpha, \alpha + 1)\right| = \theta_\alpha(k)\}$, in terms of the set

$$\mathcal{Z}_{\{\phi(\alpha,k)\}} := \left\{ \{\phi(\alpha + 1, k) - \phi(\alpha, k)\} \mid \{\phi(\alpha, k)\} \in \left[\mathbb{S}^1/2\pi\right]^{q(k)} \right\}. \tag{1.92}$$

The quotient map

$$\mathscr{P}_{(k)}(\mathbb{S}^2; \{\left|c_{(k)}(\alpha, \alpha + 1)\right|\}) \longrightarrow \frac{\mathscr{P}_{(k)}(\mathbb{S}^2; \{\left|c_{(k)}(\alpha, \alpha + 1)\right|\})}{SO(3)}, \tag{1.93}$$

associating to a polygon $P_{(k)}(c_{(k)})$ its representative in the framing (1.84), shows that (1.90) and (1.91) provide a homeomorphism

$$\mathcal{Z}_{q(k)}(\{\phi(\alpha, k)\}) \longrightarrow \frac{\mathscr{P}_{(k)}(\mathbb{S}^2; \{\left|c_{(k)}(\alpha, \alpha + 1)\right|\})}{SO(3)}, \tag{1.94}$$

which characterizes $\mathcal{Z}_{q(k)}(\{\phi(\alpha, k)\})$ as a slice for the $SO(3)$ action on the spaces of spherical polygons with given edge–lengths. It follows that we can represent the space of polyhedral cones over the vertex $\sigma^0(k)$, $\mathcal{Q}_{q(k)}(\mathbb{S}^2)$, according to

$$\mathcal{Q}_{q(k)}(\mathbb{S}^2) \simeq \mathcal{Z}_{q(k)}(\{\phi(\alpha, k)\}). \tag{1.95}$$

In particular, if $V(T)$ denotes the vertex set of (T_l, M), then the map

$$\pi_k : V(T) \times \mathcal{V}ert \longrightarrow \sqcup_{k=1}^{N_0} \mathcal{Q}_{q(k)}(\mathbb{S}^2) \tag{1.96}$$

$$(\sigma^0(k), \{\theta_\alpha\}_k) \longmapsto (\{\phi(\alpha, k)\}, \{\phi(\alpha + 1, k) - \phi(\alpha, k)\} = \{\theta_\alpha\}_k),$$

is a choice of a polyhedral cone (Fig. 1.17) having, near the given vertex $\sigma^0(k)$, the same intrinsic Euclidean structure $\mathscr{E}\left(Star[\sigma^0(k)]\right)$ defined by the vertex angles $\{\theta_\alpha\}_k$. Thus, it is natural to introduce the following

Definition 1.20 The mapping

$$(T_l, M)\big|_{\sigma^0(k)} \longmapsto \mathcal{Q}_{q(k)}(\mathbb{S}^2), \tag{1.97}$$

defines the polyhedral cotangent cone to (T_l, M) at the vertex $\sigma^0(k)$.

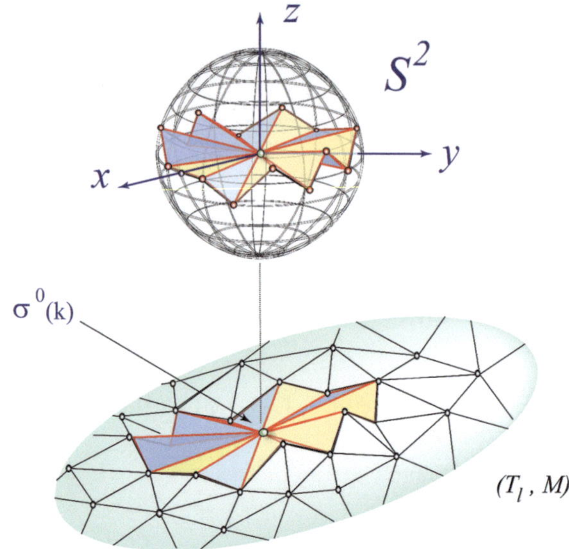

Fig. 1.17 $\mathcal{Q}_{q(k)}(\mathbb{S}^2)$ characterizes the polyhedral cotangent cone to (T_l, M) at the vertex $\sigma^0(k)$

Roughly speaking, the identification of $\mathcal{Q}_{q(k)}(\mathbb{S}^2)$ with the polyhedral cotangent space at $\sigma^0(k)$ is also justified by noticing that the polyhedra in $\mathcal{Q}_{q(k)}(\mathbb{S}^2)$ are side–wise dual to the tangent space [21] over the corresponding open triangle $\breve{\sigma}^2(\alpha, k, \alpha + 1)$.

In a well–defined sense, the pair $\left(\sigma^0(k), \mathcal{Q}_{q(k)}(\mathbb{S}^2)\right)$ comes decorated with *coordinates*: $(\{\phi(\alpha, k)\})$ associated with the vertex $\sigma^0(k)$, and $(\{\phi(\alpha + 1, k) - \phi(\alpha, k)\})$ associated with the choice of a polyhedral cone in $\mathcal{Q}_{q(k)}(\mathbb{S}^2)$ over $\sigma^0(k)$. For later use it is useful to formalize this remark into the

Definition 1.21 The map

$$(\sigma^0(k), \mathcal{Q}_{q(k)}(\mathbb{S}^2)) \longmapsto (\{\phi(\alpha, k)\}, \{\phi(\alpha + 1, k) - \phi(\alpha, k)\}), \tag{1.98}$$

defines the polyhedral cotangent coordinates associated with the vertex $\sigma^0(k) \in (T_l, M)$. In particular the assignment of a 1-form in $(\sigma^0(k), \mathcal{Q}_{q(k)}(\mathbb{S}^2))$,

$$v_k := \sum_{\alpha=1}^{q(k)} (\phi(\alpha, k) - \phi(\alpha + 1, k)) \, d(\phi(\alpha, k)), \tag{1.99}$$

(recall that $\phi(\alpha + 1, k) := \phi(1, k)$ for $\alpha = q(k)$), characterizes the choice of a polyhedral cone in $(\sigma^0(k), \mathcal{Q}_{q(k)}(\mathbb{S}^2))$.

Note that the 1-form v_k is intrinsically defined over $\mathcal{Q}_{(k)}(\mathbb{S}^2)$, in the sense that it does not depend on the chosen origin point p_0 used for characterizing the polygonal arc-length map $\phi(\alpha, k)$. The above characterization of the polyhedral cotangent cone to (T_l, M) at the vertex $\sigma^0(k)$ can be extended (Fig. 1.18) to $\mathcal{T}^{met}_{g,N_0}(M)$ according to the

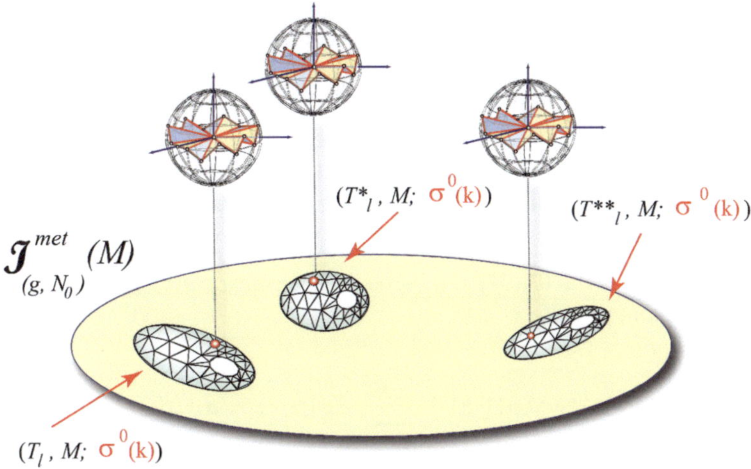

Fig. 1.18 The polyhedral cotangent cones $\mathcal{Q}_{q(k)}(\mathbb{S}^2)$ to the polyhedral surface (T_l, M) at the vertex $\sigma^0(k)$, characterizes as $(T_l, M; \sigma^0(k))$ varies in $\mathcal{T}^{met}_{g,N_0}(M)$, a (polygonal) circle bundle over $\mathcal{T}^{met}_{g,N_0}(M)$. Since there are N_0 labeled vertices for each (T_l, M), there are N_0 such a bundle over $\mathcal{T}^{met}_{g,N_0}(M)$

Lemma 1.2 *Let* $C_k := (\{\phi(\alpha, k)\}, \{\phi(\alpha + 1, k) - \phi(\alpha, k)\})$ *denote the generic polyhedral cone in* $\mathcal{Q}_{q(k)}(\mathbb{S}^2)$, *then the mapping*

$$\mathcal{T}^{met}_{g,N_0}(M) \longrightarrow \otimes^{N_0}_{k=1} \mathcal{Q}_{q(k)}(\mathbb{S}^2) \tag{1.100}$$

$$(T_l, M) \longmapsto (C_1 \ldots, C_k, \ldots, C_{N_0})$$

defined by the N_0 polyhedral cotangent cones at the (labelled vertices of the) the polyhedral surfaces $(T_l, M) \in \mathcal{T}^{met}_{g,N_0}(M)$, *defines N_0 circle bundles* $\mathcal{Q}_{(k)}$ *over* $\mathcal{T}^{met}_{g,N_0}(M)$.

Proof The $\mathbb{S}^1$ fiber of $\mathcal{Q}_{(k)}$ is naturally associated with the spherical polygon defining the polyhedral cones in $\mathcal{Q}_{(k)}$. Moreover, the definition of $\phi(k, \alpha)$ easily implies that the bundle projection

$$pr_k : \mathcal{Q}_{(k)} \longrightarrow \mathcal{T}^{met}_{g,N_0}(M) \tag{1.101}$$

$$C_k \longmapsto C_k|_{\sigma^0(k)\in(T_l,M)}\,,$$

is continuous when restricted to open sets of (T_l, M) in $\mathcal{T}^{met}_{g,N_0}(M)$. By considering the inductive limit $\varXi_\infty$ of the family of polyhedral spaces $\{\varXi_{\{\phi(\alpha,k)\}}\}$ as $q(k) \to \infty$, and by considering the polyhedral cotangent cone $\mathcal{Q}_{(k)}$ as a point in $\varXi_\infty$, it is also easy to see that pr_k is well–defined under the *flip move* (1.51), (we need to use $\varXi_\infty$ because the incidence number $q(k)$ changes under flip moves). ∎

1.9   The Conical Symplectic Form on $\mathscr{T}_{g,N_0}^{met}(M,\{\Theta(k)\})$

Let us turn our attention to the space $\mathscr{T}_{g,N_0}^{met}(M,\{\Theta(k)\})$ of polyhedral surfaces with a given set of conical angles $\{\Theta(k)\}_{k=0}^{N_0}$ over their N_0 labelled vertices $\{\sigma^0(k)\}$, (see Definition 1.16). Consider, at the generic point vertex $\sigma^0(k)$ of (T_l,M) the 1-form v_k (1.99) describing the choice of a polyhedral cone in $\mathscr{Q}_{q(k)}(\mathbb{S}^2)$. For $q(k)=2$, v_k is obviously closed. However, as soon as $q(k)\geq 3$ we have

Lemma 1.3 *If $\sigma^0(k)\in(T_l,M)$ is a vertex with incidence $q(k)\geq 3$, then v_k is not closed and its differential*

$$d\,v_k = d\,\phi(q(k),k)\wedge d\,\phi(1,k) + \sum_{\beta=1}^{q(k)-1} d\,\phi(\beta,k)\wedge d\,\phi(\beta+1,k) \qquad (1.102)$$

can be pulled–back under the map (1.96) π_k *to the two form* $\omega_k = \pi_k^*(dv_k)$ *defined on* $\mathscr{V}ert_{q(k)}[\Theta(k)]$ *by*

$$\omega_k = \sum_{1\leq\alpha<\beta\leq q(k)-1} d\left(\frac{\theta_\alpha}{\Theta(k)}\right)\wedge d\left(\frac{\theta_\beta}{\Theta(k)}\right). \qquad (1.103)$$

Proof We provide a proof by induction. As a preliminary step, notice that, since we work at fixed conical angle $\Theta(k)$, we can introduce the $\Theta(k)$–normalized angles

$$\widehat{\theta}_\alpha(k) := \frac{\theta_\alpha(k)}{\Theta(k)}. \qquad (1.104)$$

Moreover, from $d\widehat{\theta}_\alpha(k) = \pi_k^*(d(\phi(\alpha+1,k)-\phi(\alpha,k)))$ it follows that we can write

$$\sum_{1\leq\alpha<\beta\leq q(k)-1} d\widehat{\theta}_\alpha\wedge d\widehat{\theta}_\beta \qquad (1.105)$$

$$= \pi_k^*\left(\sum_{1\leq\alpha<\beta\leq q(k)-1} d(\phi(\alpha+1,k)-\phi(\alpha,k))\wedge d(\phi(\beta+1,k)-\phi(\beta,k))\right).$$

Exploiting this representation it is easily checked that $\omega_k = \pi_k^*(dv_k)$ holds for $q(k)=3$. We consider this as the first inductive step and complete the induction by showing that if $\omega_k = \pi_k^*(dv_k)$ holds for $q(k)=n>3$, then it also holds for $q(k)=n+1$. We start with the identity

$$\sum_{1\leq\alpha<\beta\leq q(k)-1} d(\phi(\alpha+1,k)-\phi(\alpha,k))\wedge d(\phi(\beta+1,k)-\phi(\beta,k)) \qquad (1.106)$$

$$= \sum_{\alpha=1}^{q(k)-2} d\left(\phi(\alpha+1,k)-\phi(\alpha,k)\right) \wedge \sum_{\beta=\alpha+1}^{q(k)-1} \left(\phi(\beta+1,k)-\phi(\beta,k)\right),$$

from which it easily follows that, in passing from $q(k)$ to $q(k)+1$, we have the recurrence relation (for notational ease, we drop the explicit k-dependence in $\phi(\alpha,k)$),

$$\sum_{1\leq\alpha<\beta\leq q(k)} d\left(\phi(\alpha+1)-\phi(\alpha)\right) \wedge d\left(\phi(\beta+1)-\phi(\beta)\right) \tag{1.107}$$

$$= \sum_{1\leq\alpha<\beta\leq q(k)-1} d\left(\phi(\alpha+1)-\phi(\alpha)\right) \wedge d\left(\phi(\beta+1)-\phi(\beta)\right)$$

$$+ \sum_{\alpha=1}^{q(k)-1} d\left(\phi(\alpha+1)-\phi(\alpha)\right) \wedge d\left(\phi(q(k)+1)-\phi(q(k))\right).$$

Since

$$\sum_{\alpha=1}^{q(k)-1} d\left(\phi(\alpha+1)-\phi(\alpha)\right) \wedge d\left(\phi(q(k)+1)-\phi(q(k))\right) \tag{1.108}$$

$$= d\phi(1) \wedge d\phi(q(k)) + d\phi(q(k)) \wedge d\phi(q(k)+1) + d\phi(q(k)+1) \wedge d\phi(1),$$

we can explicitly rewrite (1.107) as

$$\sum_{1\leq\alpha<\beta\leq q(k)} d\left(\phi(\alpha+1)-\phi(\alpha)\right) \wedge d\left(\phi(\beta+1)-\phi(\beta)\right) \tag{1.109}$$

$$= \sum_{1\leq\alpha<\beta\leq q(k)-1} d\left(\phi(\alpha+1)-\phi(\alpha)\right) \wedge d\left(\phi(\beta+1)-\phi(\beta)\right)$$

$$+ d\phi(1) \wedge d\phi(q(k)) + d\phi(q(k)) \wedge d\phi(q(k)+1) + d\phi(q(k)+1) \wedge d\phi(1).$$

This recurrence relation immediately yields the required equality $\omega_k = \pi_k^*\left(dv_k\right)$ at the induction step $q(k) \mapsto q(k)+1$. $\blacksquare$

By exploiting the forms ω_k as $\sigma^0(k)$ varies in $(T_l,M) \in \mathscr{T}_{g,N_0}^{met}\left(M,\{\Theta(k)\}\right)$ we can naturally introduce a globally defined 2-form $\omega_T(\{\Theta\})$ on each $(T_l,M) \in \mathscr{T}_{g,N_0}^{met}\left(M,\{\Theta(k)\}\right)$ according to

$$\omega_T(\{\Theta\}) := \sum_{k=1}^{N_0} \Theta^2(k)\, \omega_k = \sum_{k=1}^{N_0} \sum_{1 \le \alpha < \beta \le q(k)-1} d\,\theta_\alpha(k) \wedge d\,\theta_\beta(k). \qquad (1.110)$$

We have

Theorem 1.4 *Let $(T_l, M) \in \mathscr{T}_{g,N_0}^{met}(M, \{\Theta(k)\})$ be a polyhedral surface decorated with a given assignment of conical angles $\{\Theta(k)\}_{k=1}^{N_0}$ over its vertices, then the mapping*

$$(T_l, M) \longmapsto \omega_T(\{\Theta\}) \qquad (1.111)$$

defines a symplectic form over $\mathscr{T}_{g,N_0}^{met}(M, \{\Theta(k)\})$.

Proof From the relation $3\,N_2 = 2\,N_1$ it follows that the face number N_2 of a polyhedral surface (T_l, M) is always even. Without loss in generality we can assume that $N_2 \ge 4$, otherwise we would end up in the case $N_2 = 2$ corresponding to the triangular pillow which generates a trivially closed dv_k for each vertex, ($q(k) = 2$, $k = 1, 2, 3$). Let us note that the 2-form $\omega_T(\{\Theta\})$ is actually defined over the space

$$\mathscr{V}ert\,[(T_l, M)] := \times_{k=1}^{N_0} \mathscr{V}ert\,[\Theta(k)] \simeq \mathbb{R}^{\sum_{k=1}^{N_0}(q(k)-1)} \qquad (1.112)$$

of vertex angles associated with the given triangulation $(T_l, M) \in \mathscr{T}_{g,N_0}^{met}(M, \{\Theta(k)\})$. Since $\sum_{k=1}^{N_0} q(k) = 3\,N_2$, from $3\,N_2 = 2\,N_1$ and Euler relation (1.11), we easily compute

$$\sum_{k=1}^{N_0}(q(k) - 1) = \frac{5}{2} N_2 + 2g - 2, \qquad (1.113)$$

where g denotes the genus of the surface M. It follows, (since $N_2 \ge 4$), that $\sum_{k=1}^{N_0}(q(k) - 1)$ is an even number. Thus, $\omega_T(\{\Theta\})$ is defined over the even dimensional manifold $\mathscr{V}ert\,[(T_l, M)]$. Since we have the natural map

$$(T_l, M) \longrightarrow \mathscr{V}ert\,[(T_l, M)] \ , \qquad (1.114)$$

by abusing notation we can think of $\omega_T(\{\Theta\})$ as defined (by pull-back) over $\mathscr{T}_{g,N_0}^{met}(M, \{\Theta(k)\})$.

In order to show that $\omega_T(\{\Theta\})$ is non–degenerate we need to introduce *tangent vectors* to $\mathscr{V}ert\,[(T_l, M)]$, namely, infinitesimal variations of the vertex structures of (T_l, M) (Fig. 1.19). In particular, we shall be interested in those vector fields over $\mathscr{V}ert\,[(T_l, M)]$ that generates metrical deformations of the polyhedral surface (T_l, M). To this end let us denote by $l(k, \alpha)$ the length of the edges $\sigma^1(k, h_\alpha) \in Star\,[\sigma^0(k)]$ and by $l(\alpha, \alpha + 1)$ the length of the edges $\sigma^1(h_\alpha, h_{\alpha+1}) \in link(k)$, (see (1.76)). Let us consider the directed edges of the oriented triangle

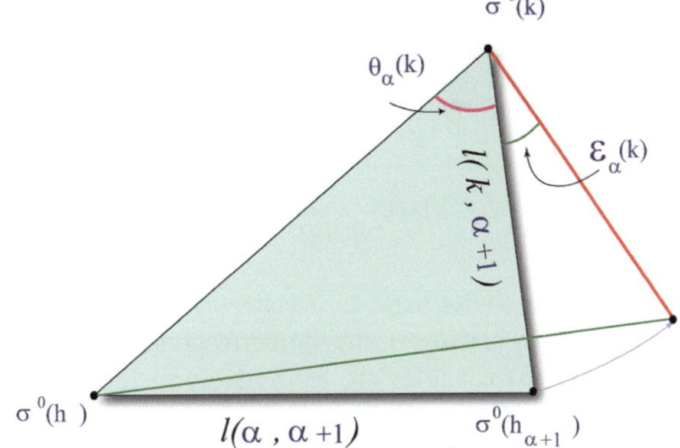

Fig. 1.19 The geometrical set up for introducing tangent vectors to $\mathscr{V}ert\,[(T_l,M)]$ describing infinitesimal variations of the vertex structures of (T_l,M)

$\sigma^2(h_\alpha,k,h_{\alpha+1}) \in Star\,[\sigma^0(k)]$ realized by the vectors $\vec{l}(k,\alpha)$, $\vec{l}(\alpha,\alpha+1)$, and $\vec{l}(\alpha+1,k)$ in the Euclidean plane $(\mathbb{R}^2,\delta)$. As we vary the vertex angle $\theta_\alpha(k)$,

$$\theta_\alpha(k) \longmapsto \theta_\alpha(k) + t\,\epsilon_\alpha(k)\,, \qquad 0 \le t \le 1\,, \tag{1.115}$$

the directed edge $\vec{l}(\alpha,\alpha+1)$ varies according to

$$\vec{l}(\alpha,\alpha+1) \longmapsto \vec{l}(\alpha,\alpha+1) + \epsilon_\alpha(k)\,l(k,\alpha+1)\,\frac{\partial}{\partial\theta_\alpha(k)}, \tag{1.116}$$

where $\frac{\partial}{\partial\theta_\alpha(k)}$ denotes the tangent vector to the unit circle $\mathbb{S}^1$ centered at the vertex $\sigma^0(k)$. Thus, the metric variations of $link(k)$ induced by variations in $\mathscr{V}ert[\Theta(k)]$ are described by

$$link(k) \longmapsto \left\{ \epsilon_\alpha(k)\,l(k,\alpha+1)\,\frac{\partial}{\partial\theta_\alpha(k)} \right\}_{\alpha=1}^{q(k)-1} \in T\mathscr{V}ert[\Theta(k)]. \tag{1.117}$$

By letting the vertex $\sigma^0(k)$ run over $(T_l,M) \in \mathscr{T}^{met}_{g,N_0}(M,\{\Theta(k)\})$, we can extend (1.117) to the map

$$v_{(T)} : \mathscr{T}^{met}_{g,N_0}(M,\{\Theta(k)\}) \longrightarrow \qquad T_{(T_l,M)}\mathscr{V}ert \tag{1.118}$$

$$(T_l,M) \longmapsto v_{(T)}(T_l,M) := \left\{ \epsilon_\alpha(k)\,l(k,\alpha+1)\,\frac{\partial}{\partial\theta_\alpha(k)} \right\}_{\alpha=1}^{q(k)-1}, \quad k=1,\ldots,N_0,$$

where $T_{(T_l,M)}\mathscr{V}ert := \times_{k=1}^{N_0} T\mathscr{V}ert[\Theta(k)]$. We may formally interpret such a mapping as defining a *tangent vector* to $\mathscr{T}_{g,N_0}^{met}(M, \{\Theta(k)\})$ at (T_l, M). To show that $\omega_T(\{\Theta\})$ is non–degenerate, let us localize $v_{(T)}$ to the generic oriented triangle $\sigma^2(h,k,j)$ of (T_l, M) by considering the vector field in $T_{(T_l,M)}\mathscr{V}ert$ defined by

$$v_{(T)} := \left(l(k,h)\epsilon(k)\frac{\partial}{\partial\theta(k)},\ l(j,k)\epsilon(j)\frac{\partial}{\partial\theta(j)},\ l(h,j)\epsilon(h)\frac{\partial}{\partial\theta(h)} \right), \qquad (1.119)$$

where $(\theta(k), \theta(j), \theta(h)) \in \mathscr{V}ert[\Theta(k)] \times [\Theta(j)] \times [\Theta(h)]$ are the vertex angles of $\sigma^2(h,k,j)$, and $l(\cdot,\cdot)$ are the corresponding edge–lengths. Assuming that $q(k)$, $q(j)$, and $q(h)$ triangles are respectively incident on the vertices of $\sigma^2(h,k,j)$, we order $\mathscr{V}ert[\Theta(k)] \times [\Theta(j)] \times [\Theta(h)]$ by identifying $(\theta(k), \theta(j), \theta(h))$ with the triple $(\theta_1(k), \theta_1(j), \theta_1(h))$. Let us consider the interior product between $\omega_T(\{\Theta\})$ and $v_{(T)}$

$$i_{v_{(T)}}\, \omega_T(\{\Theta\}) = \sum_{a=k,j,h}\ \sum_{1\leq\alpha<\beta\leq q(a)-1} i_{v_{(T)}}\left(d\theta_\alpha(a) \wedge d\theta_\beta(a) \right). \qquad (1.120)$$

From

$$i_{v_{(T)}}\left(d\theta_\alpha(a) \wedge d\theta_\beta(a) \right) = i_{v_{(T)}}\left(d\theta_\alpha(a) \right) d\theta_\beta(a) - i_{v_{(T)}}\left(d\theta_\beta(a) \right) d\theta_\alpha(a), \qquad (1.121)$$

and $i_{v_{(T)}}\left(d\theta_\alpha(a) \right) = d\theta_\alpha(a)\left(v_{(T)} \right) = v_{(T)}^\alpha$, we easily compute

$$i_{v_{(T)}}\, \omega_T(\{\Theta\}) = l(k,h)\epsilon(k) \sum_{\alpha=2}^{q(k)-1} d\theta_\alpha(k) + l(j,k)\epsilon(j) \sum_{\alpha=2}^{q(j)-1} d\theta_\alpha(j) \qquad (1.122)$$

$$+\, l(h,j)\epsilon(h) \sum_{\alpha=2}^{q(h)-1} d\theta_\alpha(h).$$

Since $i_{v_{(T)}}\, \omega_T(\{\Theta\}) = 0$ iff $\epsilon(k) = \epsilon(j) = \epsilon(h) = 0$, there are no non–trivial (infinitesimal) deformations of Euclidean triangles $\sigma^2(h,k,j)$ in ker $\omega_T(\{\Theta\})$. Thus $\omega_T(\{\Theta\})$ is non–degenerate on $T_{(T_l,M)}\mathscr{V}ert$ and provides a symplectic form on $\mathscr{V}ert$. ∎

Remark 1.7 We can consider more general infinitesimal deformations of the Euclidean triangle $\sigma^2(h,k,j)$ by setting

$$d\,l(j,h) = \sum_{a=k,j,h} \frac{\partial l(j,h)}{\partial\theta(a)}\, d\theta(a), \qquad (1.123)$$

and similarly for the remaining edge–lengths $l(h,k)$ and $l(k,j)$. Note that such deformations are *always* such that the one–form

$$\theta(k)\, d \ln l(j,h) + \theta(j)\, d \ln l(h,k) + \theta(h)\, d \ln l(k,j) \tag{1.124}$$

is closed [19], i.e.,

$$d\theta(k) \wedge d \ln l(j,h) + d\theta(j) \wedge d \ln l(h,k) + \theta(h) \wedge d \ln l(k,j) = 0. \tag{1.125}$$

Such a relation can be considered as the two-dimensional counterpart of the Schläfli formula (see [14] for a thorough analysis of the geometrical meaning of these forms).

Not surprisingly, there are many points of contact between the above introduction of a symplectic structure on the space of polyhedral surfaces $\mathcal{T}_{g,N_0}^{met}(M, \{\Theta(k)\})$ and the seminal analysis by M. Kontsevich of the interplay between the geometry of the moduli space of Riemann surfaces and space of ribbon graphs with given boundary lengths [10]. The duality between ribbon graphs and polyhedral surface, and their connection to Riemann moduli theory, is the most obvious explanation. However, as we shall see, the singular metric geometry of polyhedral surfaces adds some surprising zest to such matters.

The decoration of $\mathcal{T}_{g,N_0}^{met}(M)$ with the N_0 polyhedral cotangent cones $\mathcal{Q}_{(k)}$ corresponds to modelling a polyhedral manifold (T_l, M) as a surface (M, N_0) with N_0 marked points (the vertices) decorated with a field of polyhedral cones (*curvature!*). As we have seen, the analysis of such a field provides, via the symplectic form $\omega_T(\{\Theta\})$, information on the local deformation of $\mathcal{T}_{g,N_0}^{met}(M)$ in a neighborhood of the given (T_l, M). Does $\omega_T(\{\Theta\})$ also capture global properties of the space of polyhedral surfaces? The hinted connection with Kontsevich theory allows an obvious affirmative answer, and in line with these remarks we explicitly connect the local forms $\{\omega_k\}$ to the Euler class of the N_o combinatorial bundles $\mathcal{Q}_{(k)}$ over $\mathcal{T}_{g,N_0}^{met}(M, \{\Theta(k)\})$.

1.10   The Euler Class of the Circle Bundle $\mathcal{Q}_{(k)}$

Let $U_\phi(T_l, M)$ and $U_{\widetilde{\phi}}(T_l, M)$, with $U_\phi(T_l, M) \cap U_{\widetilde{\phi}}(T_l, M) \neq \emptyset$, denote two open neighborhoods in $\mathcal{T}_{g,N_0}^{met}(M, \{\Theta(k)\})$ of a polyhedral surface (T_l, M). Assume that over $U_\phi(T_l, M)$ the combinatorial bundle $\mathcal{Q}_{(k)}$ is trivialized by the coordinates $(\{\phi(\alpha, k)\}, \{\phi(\alpha + 1, k) - \phi(\alpha, k)\})$ whereas over $U_{\widetilde{\phi}}(T_l, M)$ the trivialization is provided by $(\{\widetilde{\phi}(\alpha, k)\}, \{\widetilde{\phi}(\alpha + 1, k) - \widetilde{\phi}(\alpha, k)\})$. The relation between these two trivializations is provided by

$$\widetilde{\phi}(\alpha, k) = \phi(\alpha, k) + f\left(\{\phi(\mu, k)\}_{\mu=1}^{q(k)}\right), \tag{1.126}$$

corresponding to a $\{\phi(\mu,k)\}_{\mu=1}^{q(k)}$–dependent rotation $e^{2\pi\sqrt{-1}f}$ of the circle $\mathbb{S}^1/2\pi$ over which the polygonal arc lengths $\{\phi(\alpha,k)\}_{\alpha=1}^{q(k)}$ are defined. We can associate with $\mathcal{Q}_{(k)}|_{U_\phi}$ the $\mathfrak{u}(1)$ valued 1-form

$$A_{(k)}|_{U_\phi} := \tag{1.127}$$

$$-\sum_{\alpha=1}^{q(k)}\left(\phi(\alpha+1,k)-\phi(\alpha,k)-\frac{1}{q(k)}\right)e^{-2\pi\sqrt{-1}\phi(\alpha,k)}\,d\,e^{2\pi\sqrt{-1}\phi(\alpha,k)}.$$

We have

Lemma 1.4 *The 1-form $A_{(k)}|_{U_\phi}$ defines a $U(1)$-connection over the circle bundle* $(\mathcal{Q}_{(k)},\ \mathcal{T}_{g,N_0}^{met}(M,\{\Theta(k)\}))$.

Proof If for notational ease we drop the explicit k-dependence then, under the change of trivialization (1.126), we get

$$\widetilde{A}|_{U_{\widetilde{\phi}}} := -\sum_{\alpha=1}^{q}\left(\widetilde{\phi}(\alpha+1)-\widetilde{\phi}(\alpha)-\frac{1}{q}\right)e^{-2\pi\sqrt{-1}\widetilde{\phi}(\alpha)}\,d\,e^{2\pi\sqrt{-1}\widetilde{\phi}(\alpha)} \tag{1.128}$$

$$= A|_{U_\phi} - \left(\sum_{\alpha=1}^{q}\left(\widetilde{\phi}(\alpha+1)-\widetilde{\phi}(\alpha)-\frac{1}{q}\right)\right)e^{-2\pi\sqrt{-1}f}\,d\,e^{2\pi\sqrt{-1}f}$$

$$= A|_{U_\phi} + e^{-2\pi\sqrt{-1}f}\,d\,e^{2\pi\sqrt{-1}f},$$

where we have exploited the fact that (1.126) leaves $(\phi(\alpha+1)-\phi(\alpha))$ (form) invariant and that $\sum_{\alpha=1}^{q}\left(\widetilde{\phi}(\alpha+1)-\widetilde{\phi}(\alpha)-\frac{1}{q}\right)=-1.$ ∎

From the Definition (1.127) of the connection form $A_{(k)}$ we have

$$A_{(k)} = 2\pi\sqrt{-1}\sum_{\alpha=1}^{q0}\left(\phi(\alpha,k)-\phi(\alpha+1,k)+\frac{1}{q(k)}\right)d\phi(\alpha,k) \tag{1.129}$$

$$= 2\pi\sqrt{-1}\,v_k + \frac{2\pi\sqrt{-1}}{q(k)}\sum_{\alpha=1}^{q(k)}d\phi(\alpha,k),$$

where v_k denotes the 1-form associated with the choice of a polyhedral cone in $\mathcal{Q}_{q(k)}(\mathbb{S}^2)$, (see (1.99)). Since $d\left(\sum_{\alpha=1}^{q(k)}d\phi(\alpha,k)\right)\equiv 0$, we can compute the curvature of $A_{(k)}$ according to

$$\Omega_{(k)} := dA_{(k)} = 2\pi\sqrt{-1}\,dv_k, \tag{1.130}$$

and we get that the Euler class of the circle bundle $(\mathcal{Q}_{(k)}, \mathcal{T}^{met}_{g,N_0}(M, \{\Theta(k)\}))$ is given by the 2-form

$$\frac{\sqrt{-1}}{2\pi}\, \Omega_{(k)} = -\, d\, v_k. \tag{1.131}$$

According to Lemma 1.3 we have

Theorem 1.5 (The Euler class of $\mathcal{Q}_{(k)}$) *The 2-form ω_k is the pull–back, under the polyhedral map π_k defined by (1.96), of the Euler class $\frac{\sqrt{-1}}{2\pi}\, \Omega_{(k)}$ of the circle bundle $(\mathcal{Q}_{(k)}, \mathcal{T}^{met}_{g,N_0}(M, \{\Theta(k)\}))$,*

$$\omega_k = \sum_{1 \leq \alpha < \beta \leq q(k)-1} d\left(\frac{\theta_\alpha(k)}{\Theta(k)}\right) \wedge d\left(\frac{\theta_\beta(k)}{\Theta(k)}\right) = -\pi_k^*\left(\frac{\sqrt{-1}}{2\pi}\, \Omega_{(k)}\right).$$
$$\tag{1.132}$$

■

This result is the counterpart, directly expressed in terms of the geometry of polyhedral surfaces in $\mathcal{T}^{met}_{g,N_0}(M, \{\Theta(k)\}))$, of the celebrated Kontsevich analysis of the Ribbon graph complex associated with the combinatorial description of the moduli space of pointed Riemann surfaces. (It is worthwhile to remark that, in such a Riemann moduli setting, the explicit use of a connection for computing the Euler class of a family of cyclically ordered set is discussed in [8]). We shall explicitly exploit the characterization (1.132) in Chap. 4, when evaluating the symplectic volume of $\mathcal{T}^{met}_{g,N_0}(M, \{\Theta(k)\})$. We conclude this introductory chapter by examining the possible degenerations of polyhedral surfaces. Since polyhedral surfaces turn out to be Riemann surfaces in disguise, we shall model the possible degenerations of a $(T_l, M) \in \mathcal{T}^{met}_{g,N_0}(M))$ along the standard degenerations relevant in Riemann moduli space theory.

1.11 Degenerations and Stable Polyhedral Surfaces

The space of polyhedral structures $POL_{g,N_0}(M)$ is not a compact orbifold since triangulated surfaces can degenerate. For instance, let $\sigma^1(h,j), \sigma^1(j,k)$, and $\sigma^1(k,h)$ three adjacent edges not bounding a triangle $\sigma^2(h,j,k)$ in (T_l, M). Assume that $l(h,j) = l(j,k) = l(k,h) := (1-t)l, \quad t \in [0,1]$. Then, as $t \searrow 1$, the edge–path $\sigma^1(h,j) \cup \sigma^1(j,k) \cup \sigma^1(k,h)$ collapses and either disconnects a component of (T_l, M) or pinches a handle of (T_l, M) (Fig. 1.20).

If we allow for semi–simplicial complexes, (as we do in this lecture notes), then a limiting and important case of degeneration is obtained as follows.

Definition 1.22 (Slitting and replicating edges) Let $(T_l, M)_{(hj)} \in POL_{g,N_0}(M)$ be a given Euclidean triangulated surface with a marked edge $\sigma^1(h,j)$. The *slit–open transformation* on $(T_l, M)_{(hj)}$ is defined by removing (slitting) the interior of the

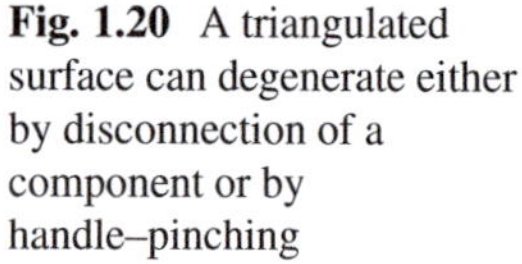

Fig. 1.20 A triangulated surface can degenerate either by disconnection of a component or by handle–pinching

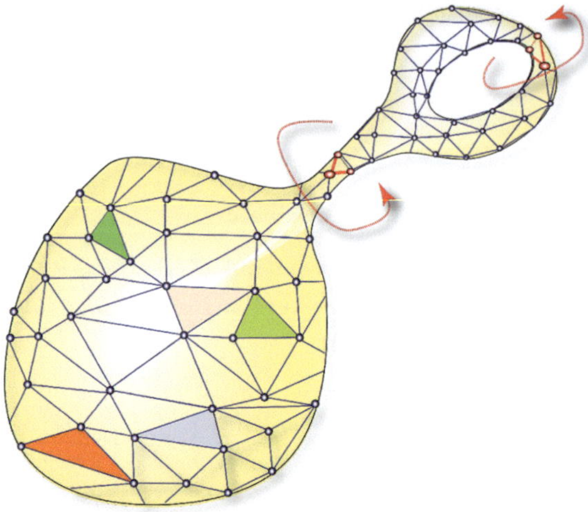

marked edge $\sigma^1(h, j)$ and by connecting (replicating) the vertices $\sigma^0(h)$ and $\sigma^0(j)$ with two distinct oriented edges $\sigma^1_{(-)}(h, j)$ and $\sigma^1_{(+)}(j, h)$,

$$
(T_l, M)_{(hj)} \vdash \quad \begin{array}{c} edge - slit \\ \& \\ edge - replicate \end{array} \quad \longrightarrow (\widetilde{T}_l, \widetilde{M}_{(h,j)}) \ . \tag{1.133}
$$

This procedure generates a (semi–simplicial) triangulated surface $(\widetilde{T}_l, \widetilde{M}_{(h,j)})$ with boundary $\partial \widetilde{M}_{(h,j)} = \sigma^1_{(-)}(h, j) \cup \sigma^1_{(+)}(h, j) \approx \mathbb{S}^1$. If $l(h, j)$ is the length of the marked edge $\sigma^1(h, j)$, then we stipulate that the length of the boundary $\partial \widetilde{M}_{(h,j)}$ is provided by $2\, l(h, j)$ (Fig. 1.21).

Consider the surface with boundary $\Sigma_{(h,j)}$ obtained from the surface $\partial \sigma^3(j, h, b, a)$ of a tetrahedron by slitting open and replicating the edge $\sigma^1(h, j)$. Similarly, let $(\widetilde{T}_l, \widetilde{M}_{(h,j)})$ be the triangulated surface obtained by the action of the slit–replica deformation on a given Euclidean triangulated surface with a marked edge $(T_l, M)_{(hj)} \in POL_{g,N_0}(M)$. Let us glue, via an orientation reversing simplicial map, $\Sigma_{(h,j)}$ to $(\widetilde{T}_l, \widetilde{M}_{(h,j)})$. In this way we obtain a closed (semi–simplicially) triangulated surface $(T_l^*, M) := (\widetilde{T}_l, \widetilde{M}_{(h,j)}) \cup \Sigma_{(h,j)} \in POL_{g,N_0+2}(M)$, (with the same topology of $(T_l, M)_{(hj)}$, since we are gluing back a disk). Let us parametrize the length of the neck $\sigma^1_{(-)}(h, j) \cup \sigma^1_{(+)}(h, j)$ in (T_l^*, M) according to $t \mapsto L_{neck} := 2(1 - t)\, l(h, j), t \in [0, 1]$. Then as $t \to 1$, we obtain a family of triangulated surfaces $(T_l^*, M)_t$ which degenerate by pinching–off a *pillow* tail $(T^{pill}, \mathbb{S}^2)$, (see Definition 1.6), into a surface with N_0 ordinary vertices and a *pinching node* σ^0_*, i.e.

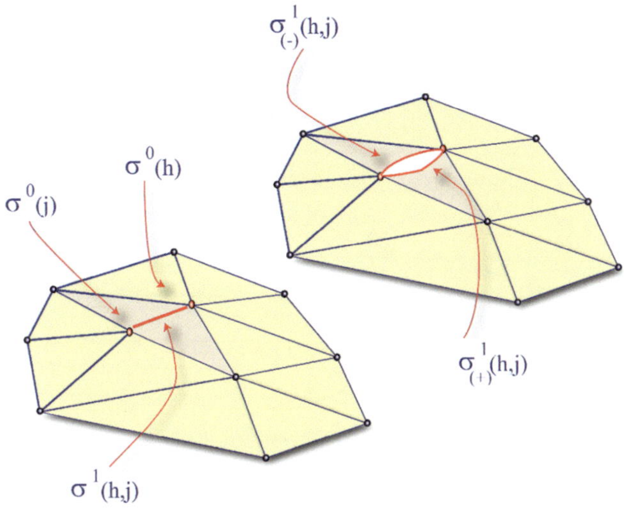

Fig. 1.21 The slit-open transformation in a portion of a Euclidean triangulated surface

$$(T_l^*, M)_t \vdash \underset{goes\ to\ 1}{\overset{as\ t}{\downarrow}} \longrightarrow (\widehat{T}_l, M) := (T_l^\times, M) \underset{\sigma_*^0}{\cup} (T^{pill}, \mathbb{S}^2) , \qquad (1.134)$$

where $(T_l^\times, M)$ is a polyhedral surface with $N_0 - 2$ vertices connected to the pillow tail $(T^{pill}, \mathbb{S}^2)$ through σ_*^0.

Remark 1.8 Note that the *pinching node* σ_*^0 is not counted for as a vertex of $(\widehat{T}_l, M)$ (Fig. 1.22).

Remark 1.9 (Euclidean and Hyperbolic pillow tail) Here the polyhedral pillow tail $(T^{pill}, \mathbb{S}^2)$ is geometrically realized by collapsing the neck $\sigma_{(-)}^1(h,j) \cup \sigma_{(+)}^1(h,j)$ of the slit–open surface of a tetrahedron $\sigma^3(j, h, b, a)$. This is equivalent to gluing the edges of the two distinct oriented Euclidean triangles $\sigma^2(a, j, b)$ and $\sigma^2(a, h, b)$. The resulting "triangulated" surface, even if hard to visualize, (it cannot be rendered as the boundary of a convex 3-dimensional region in the Euclidean space $\mathbb{E}^3$), is a perfectly sound polyhedral surface topologically equivalent to the sphere. Its three vertices $\sigma^0(\alpha)$, $\alpha = 1, 2, 3$, support positive deficit angles $\epsilon(\alpha) > 0$, and the (discretized form of the) Gauss–Bonnet theorem provides indeed

$$\sum_{\alpha=1}^{3} \frac{\epsilon(\alpha)}{2\pi} = 2 = \chi(\mathbb{S}^2) , \qquad (1.135)$$

where $\chi(\mathbb{S}^2)$ is the Euler characteristic of $\mathbb{S}^2$. If we slightly round off these conical points by smearing vertex curvature, and add some positive curvature

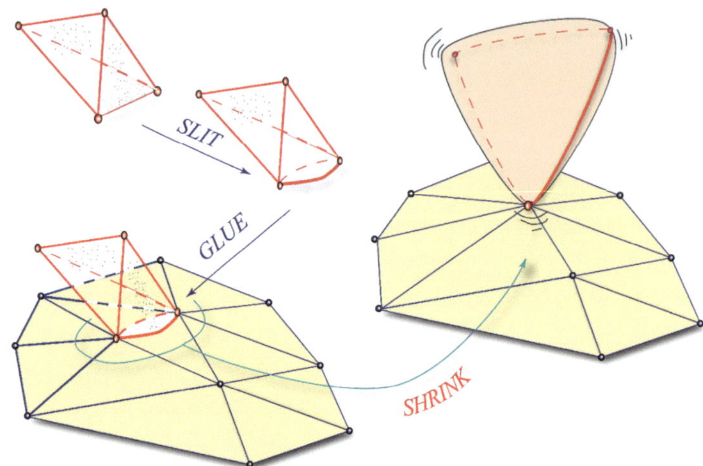

Fig. 1.22 The slitting and glueing procedure giving rise to a family of triangulated surfaces which degenerate by forming a pillow surface tail

(negative is even better! see below) to the two triangular faces, then we get the (incorrect) visualization of $(T^{pill}, \mathbb{S}^2)$ adopted, for illustrative purposes, in the figures. Since the thrice-punctured sphere is the largest subdomain of $\mathbb{S}^2$ supporting a hyperbolic metric, if we replace the Euclidean triangles $\sigma^2(a, j, b)$ and $\sigma^2(a, h, b)$ with corresponding hyperbolic ideal triangles $\sigma^2_{hyp}(a, j, b)$ and $\sigma^2_{hyp}(a, h, b)$, (with all removed vertices pushed on the circle at infinity), then the above gluing pattern among the two (ideal) triangles, (defined up to translations, since the edges have infinite lengths), generates the well–known hyperbolic structure of finite area on the thrice-punctured sphere $(T^{pill}_{hyp}, \mathbb{S}^2/\{\sigma^0(\alpha)\})$:   the *hyperbolic pillow tail*, (this is nicely described in [21]). The (finite) automorphism group of $(T^{pill}_{hyp}, \mathbb{S}^2/\{\sigma^0(\alpha)\})$ allows to map the punctures (vertices) at the points $(0, 1, \infty)$ of the extended complex plane $\mathbb{C} \cup \{\infty\}$ and one can also denote the (hyperbolic) pillow tail as $\left(T^{pill}_{hyp}, \mathbb{CP}^1_{(0,1,\infty)}\right)$. The pillow tail has a particular role in analyzing the degenerations of both polyhedral and smooth surfaces. The underlying rationale is related to the fact that *Geometric Invariant Theory* requires at least three marked points for having a form of stability in the degenerations of geometrical structures. This is a well-known phenomenon for Riemann surfaces when discussing moduli space compactifications. It may appear rather obvious for polyhedral surfaces since one cannot form semi–simplicial triangulated surfaces with less than 3 vertices. However, at a deeper level, the stability of the pillow tail can be considered as a direct consequence of the fact that to a Euclidean triangulated surface one can naturally associate [23] a corresponding Riemann surface. Explicitly, we have the following result that will be proved in Chap. 2.

Theorem 1.6 *The pillow tail* $(T^{pill}, \mathbb{S}^2)$ *is conformally equivalent to the thrice-punctured sphere* $\mathbb{CP}^1_{(0,1,\infty)}$ *and thus it is stable in the Riemann moduli sense.*
These remarks suggest the following natural characterization.

Theorem 1.7 (Stable Polyhedral Surfaces) *Let* $(T_l, M) \in POL_{g,N_0}(M)$ *be a polyhedral surface of genus* g *with* N_0 *vertices. Consider, according to Definition 1.9, a finite collection* $\{S_i\}$ *of admissible paths which are embedded circles in* $M' :=$ $M \backslash K^0(T)$, *where* $K^0(T) := \{\sigma^0(1), \ldots, \sigma^0(N_0)\}$ *is the 0–skeleton of* (T_l, M), *and where each circle* S_i *is in a distinct isotopy class relative to* M'. *Also assume that none of these circles bound a disk in* M *containing at most a vertex of* (T_l, M). *Then by contracting such circles to zero length, via an area preserving length–contraction isotopy,* (T_l, M) *can degenerate only by:*

(i) Pinching into separate components,
(ii) Handle–pinching,
(iii) Pillow-tail pinching, and gives rise to a stable polyhedral surface with N_0 *vertices* $(\widehat{T}_l, M)$.

Proof The somewhat delicate point of the above statement is to show that the degeneration of $(T_l, M) \in POL_{g,N_0}(M)$, generated by contracting the circles S_i withing a given isotopy class, yields a (semi–simplicial) polyhedral surface with N_0 vertices. Note that *(iii)* can be considered as a particular case of *(i)*, and that handle–pinching can be reduced to *(i)* by cutting out the handle, (in two non–pinching regions), and gluing it back after pinching. Thus, it is sufficient to examine in detail only case *(i)*, i.e. the degeneration associated with pinching (T_l, M) into separate components.

To discuss case *(i)*, fix the attention on a S_i which is a disconnecting circle in M, i.e. if D_i is the disk bounded by S_i, then D_i and $M \backslash D_i$ are disconnected subsets of M. Since any of the embedded circle $S_i : [0, 1] \to M'$, $s \mapsto S_i(s)$, is an admissible path, it has finitely many intersections with the edges of $(T_l, M) \in POL_{g,N_0}(M)$. Thus, without loss in generality, and up to an isotopy, we may assume that a given S_i intersects a finite number of edges $\{\sigma^1(j)\}_{j=1}^{J_i}$, and each edge at most once, say at the edge–length midpoint. Explicitly, let $\{\sigma^0(k)\}_{k=1}^{K_i}$, $K_i \geq 2$, denote the vertices contained in the disk D_i bounded by the embedded circle S_i. Note that D_i may be topologically non–trivial. Let $\{\sigma^1(j)\}_{j=1}^{J_i}$ be the set of edges in (T_l, M) connecting the vertices $\{\sigma^0(k)\}_{k=1}^{K_i} \in D_i$ with the vertices in $M \backslash D_i$. Denote by $\{\widehat{\sigma}_j^0(S_i)\}$, with $j = 1, \ldots, j_i$, the mid–point intersections $S_i \cap \{\sigma^1(j)\}_{j=1}^{J_i}$ and, as the notation suggests, characterize them as auxiliary vertices generating, by adding corresponding edges, a new triangulation (T_l', M) (Fig. 1.23). This new auxiliary triangulation (with $N_0 + J_i$ vertices) describes the some polyhedral structure associated with (T_l, M), since the added auxiliary vertices are not conical. Let L_i denote the length of the circle S_i as induced by the metric geometry of (T_l, M). Consider a length–contraction isotopy (relative to M') of the given S_i defined by the family of paths $S_i^{(t)}$, $t \in [0, 1]$, $s \mapsto S_i^{(t)}(s) := (1 - t)S_i(s)$, with the requirement that the (Euclidean) area of the corresponding disk $D_i^{(t)}$, bounded by $S_i^{(t)}$, remains fixed for all $t \in [0, 1]$. Since we

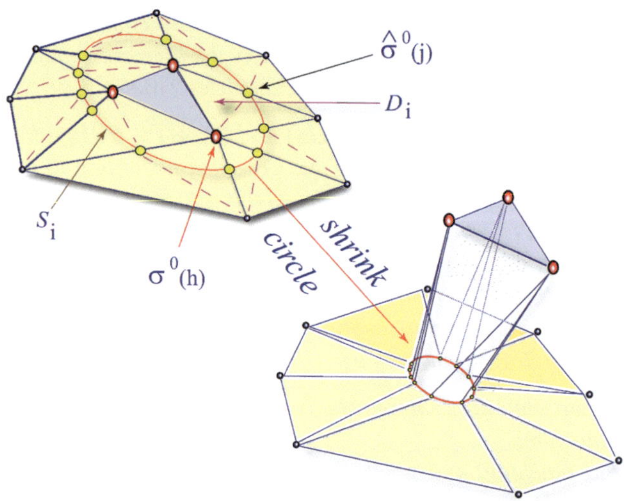

Fig. 1.23 The geometrical set up for proving Theorem 1.7. Dashed lines represent the added edges which, together with the mid–point vertices $\widehat{\sigma}_j^0(S_i)$ intercepted by the shrinking circle S_i, characterize the auxiliary triangulation (T_l', M)

are considering a length contraction in a given isotopy class relative to M', $D_i^{(t)} \cup S_i^{(t)}$ contains $J_i(t) + K_i$ vertices, *viz.* the original set $\{\sigma^0(k)\}_{k=1}^{K_i}$ which by isotopy must be always contained in $D_i^{(t)}$, plus the auxiliary $\{\widehat{\sigma}_j^0(S_i^{(t)})\}$ intercepted by the boundary shrinking circle $S_i^{(t)}$. Similarly, the open surface $M \backslash D_i^{(t)}$ contains $N_0 - (J_i(t) + K_i)$ vertices. As $t \nearrow 1$, $S_i^{(t)}$ shrinks to a point and the $J_i(t)$ auxiliary vertices $\{\widehat{\sigma}_k^0(S_i)\}$ coalesce into a nodal point $\sigma_*^0(S_i)$. Correspondingly, the triangulation $(T_l', M)_t$ pinches off into two distinct components: a closed triangulated surface $D_i^* \cup \sigma_*^0(S_i)$ with $K_i + 1$ vertices joined, via the node $\sigma_*^0(S_i)$, to an open triangulated surface $(T_l^*, M \backslash \sigma_*^0(S_i))$ with $N_0 - (K_i + 1)$ vertices. Thus, the union

$$\left(\widehat{T}_l, M\right)_{S_i} := \left(T_l^*, M \backslash \sigma_*^0(S_i)\right) \cup \left(D_i^* \cup \sigma_*^0(S_i)\right), \tag{1.136}$$

is a polyhedral surface with N_0 vertices, (recall that $\sigma_*^0(S_i)$ is not counted as a vertex), and describes the pinching–off degeneration of (T_l, M) along the embedded circle S_i (Figs. 1.24, 1.25 and 1.26).

Remark 1.10 Note that we can arrange the pinching in such a way that the conical angle $\Theta(S_i)$ at $\sigma_*^0(S_i)$, that may result from the S_i–collapsing, is the same on both side of the pinching node. Also, by a finite number of flip moves, we can always make the incidence on the node $\sigma_*^0(S_i)$ the same as seen from both sides of the resulting stable triangulation.

Since the number of isotopy classes of embedded circles relative to M' is finite, there is at most a finite number of such degenerations. In particular, if S_i contains just two

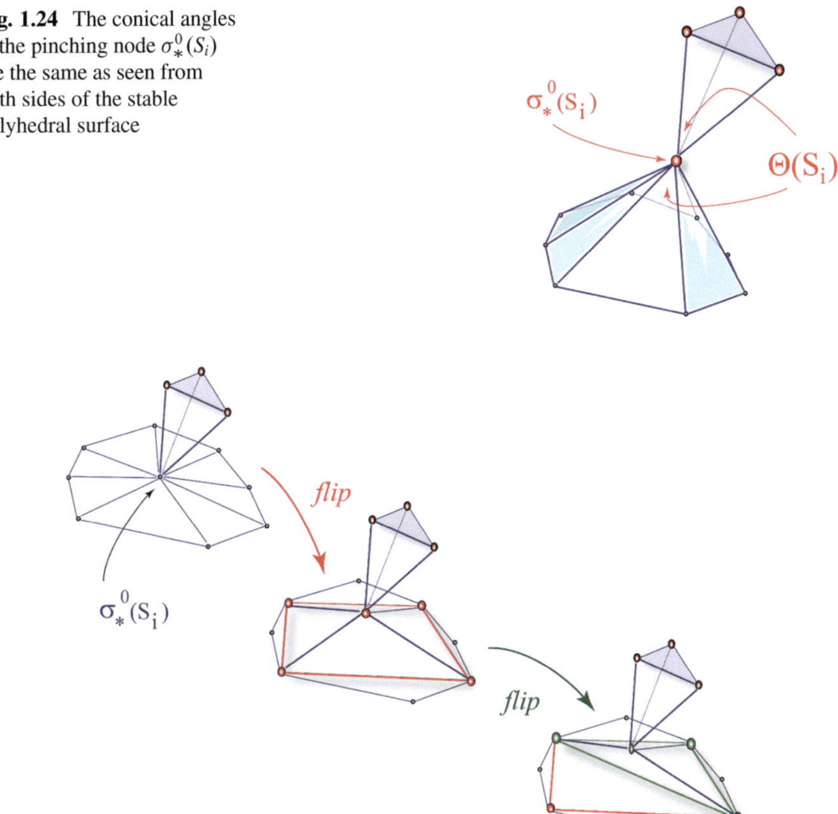

Fig. 1.24 The conical angles at the pinching node $\sigma_*^0(S_i)$ are the same as seen from both sides of the stable polyhedral surface

Fig. 1.25 By a finite sequence of flip moves we can always arrange the pinching in such a way that the incidence on the node $\sigma_*^0(S_i)$ is the same as seen from both sides of the resulting stable triangulation, (here $q[\sigma_*^0(S_i)] = 3$). In order to illustrate the flips, the figure does not correctly render the fact that also the conical angles on both sides are the same

vertices, (and hence their slit open connecting edge), then it is easy to see that the resulting $D_i^* \cup \sigma_*^0(S_i)$ is a triangular pillow $(T^{pill}, \mathbb{S}^2)$, and $(\widehat{T}_l, M)_{S_i}$ represents a stable pillow tail degeneration of (T_l, M). Finally, if S_i is a nonseparating circle, then S_i must be a representative, in the given isotopy class relative to M', of one of the g disjoint simple closed curves which can be cut from an orientable surface of genus g without disconnecting it. In such a case, $(\widehat{T}_l, M)_{S_i}$ is associated with a handle–pinching degeneration.

Such a characterization of the possible degenerations of a polyhedral surface is clearly modelled on what happens in Riemann surface theory, an analogy which will be fully justified in Chap. 2, and that suggests to compactify $POL_{g,N_0}(M)$ by completing it with the stable polyhedral surfaces $\{(\widehat{T}_l, M)_{S_i}\}$. Since pillow–tail degeneration is stable, we define

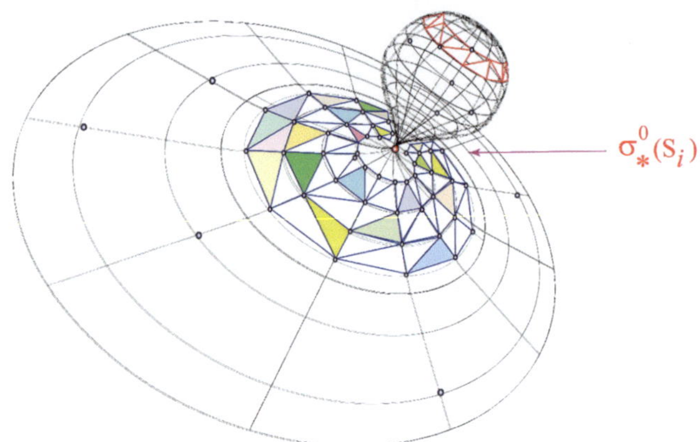

Fig. 1.26 The pinching point $\sigma^0_*(S_i)$ associated with a shrinking circle S_i does not count as an ordinary vertex. Rather, it is a nodal point in the stable polyhedral surface $(\widehat{T}_l, M)_{S_i}$

Fig. 1.27 Since the point p is disjoint from the vertex set $\{\sigma^0(k)\}_{k=1}^{N_0}$ of (T_l, M), it is promoted to be a new vertex $p := \sigma^0(N_0 + 1)$

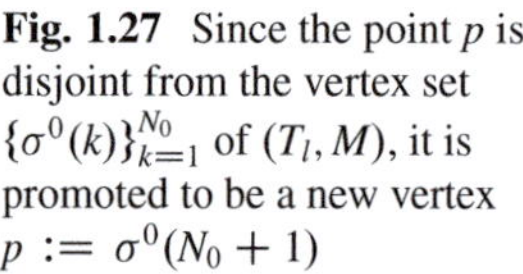

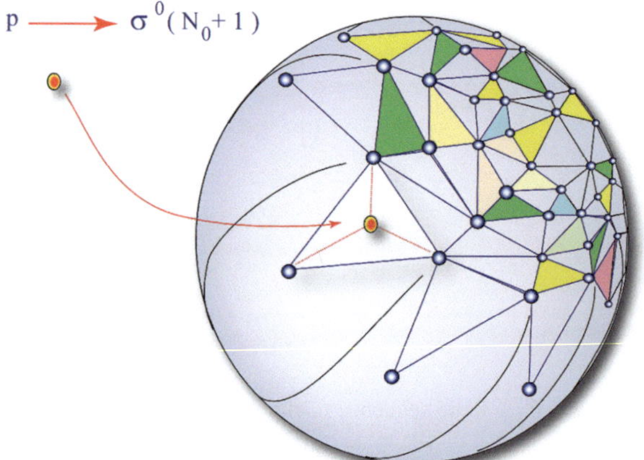

Definition 1.23 The space $\overline{POL}_{g,N_0}(M)$ of stable polyhedral surfaces of genus g with N_0 vertices is defined by completing $POL_{g,N_0}(M)$ in such a way that the closure $\partial POL_{g,N_0}(M)$ of $POL_{g,N_0}(M)$ in $\overline{POL}_{g,N_0}(M)$ consists of stable polyhedral surfaces $\{(\widehat{T}_l, M)_{S_i}\}$ with N_0 vertices and with a finite set of pillow–tail pinching degenerations.

Pushing further the analogy with standard moduli space theory, let us observe that any point p on a stable polyhedral surface $(T_l, M) \in \overline{POL}_{g,N_0}(M)$ defines a natural mapping

$$(T_l, M) \longrightarrow \overline{POL}_{g,N_0+1}(M) \tag{1.137}$$

that determines a stable polyhedral surface $(\widehat{T}_l, M) \in \overline{POL}_{g,N_0+1}(M)$. Explicitly, we can distinguish three possible cases (Figs. 1.27, 1.28 and 1.29):

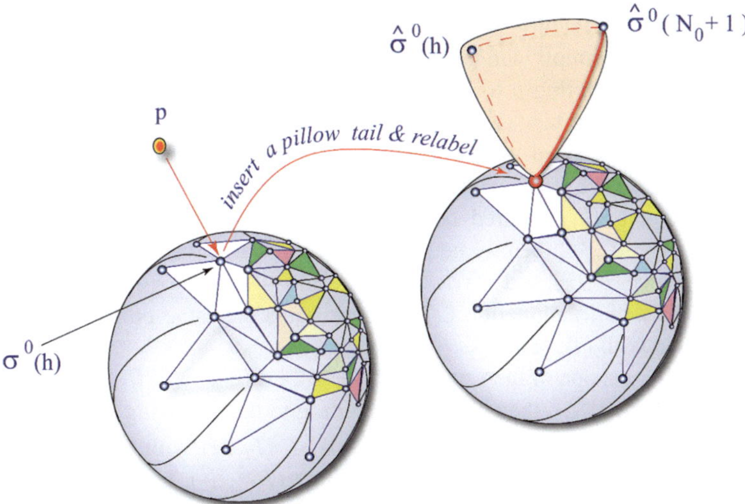

Fig. 1.28 If the point p is a vertex, say $\sigma^0(h)$, of (T_l, M), we introduce a pillow tail $(T^{pill}, \mathbb{S}^2)$, this *removes* the vertex $\sigma^0(h)$ transforming it in a pinching node. The operation is completed by relabelling the two new vertices of $(T^{pill}, \mathbb{S}^2)$ as $\widehat{\sigma}^0(h)$ and $\widehat{\sigma}^0(N_0 + 1)$

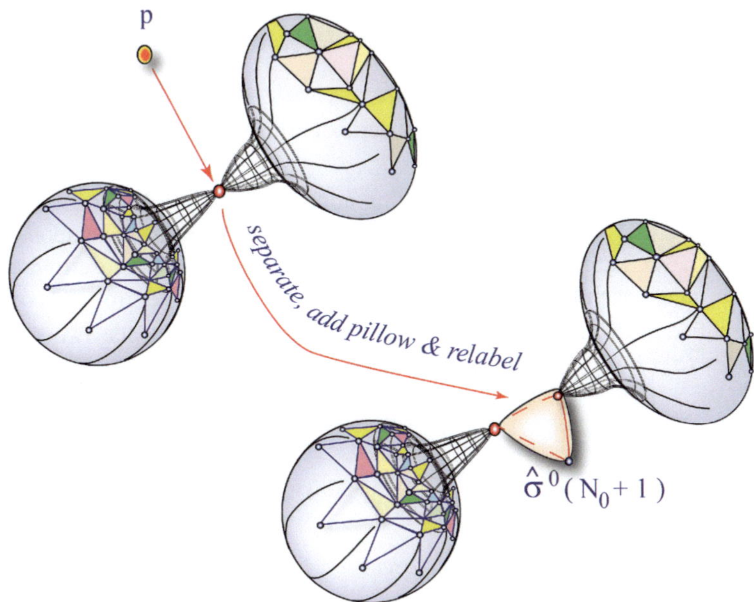

Fig. 1.29 If p coincides with a pinching node $\sigma^0_*(S_i)$ of (T_l, M), then separate the pinched components and insert a pillow tail $(T^{pill}, \mathbb{S}^2)$. This generates two pinching nodes and a new vertex in $(T^{pill}, \mathbb{S}^2)$ which gets the label $\widehat{\sigma}^0(N_0 + 1)$

(A) As long as the point $p \in (T_l, M)$ is disjoint from the vertex set $\{\sigma^0(k)\}_{k=1}^{N_0}$ of (T_l, M) one simply defines $(\widehat{T}_l, M)$ to be the new triangulation we get from (T_l, M) by promoting p to a be a vertex, i.e., $p := \sigma^0(N_0 + 1)$, and adding the corresponding new edges, (if p is in the interior of a triangle $\sigma^2(k, h, j) \in (T_l, M)$ then we need to add three new edges issuing from $\sigma^0(N_0 + 1)$ and connecting to the vertices $\sigma^0(k)$, $\sigma^0(h)$, and $\sigma^0(j)$ of $\sigma^2(k, h, j)$; If p belongs to an edge $\sigma^1(h, j)$ shared between two adjacent triangles $\sigma^2(k, h, j)$, and $\sigma^2(h, i, j) \in (T_l, M)$ then we need to add two new edges connecting $\sigma^0(N_0 + 1)$ to the vertices $\sigma^0(k)$, and $\sigma^0(i)$). Since the added vertex does not carry a conical angle, the new triangulation $(\widehat{T}_l, M)$ represents the same polyhedral structure associated with (T_l, M).

(B) If the point p is one of the vertices of (T_l, M), i.e., if $p = \sigma^0(h)$ for some vertex $\sigma^0(h) \in \{\sigma^0(k)\}_{k=1}^{N_0}$, then proceed according to the following steps: *(i)* for any $1 \leq i \leq N_0$, with $i \neq h$, identify $\widehat{\sigma}^0(i) \in (\widehat{T}_l, M)$ with the corresponding $\sigma^0(i) \in (T_l, M)$; *(ii)* take a triangular pillow $(T^{pill}, \mathbb{S}^2)$, label with a the index h one of its three vertices $(\sigma_{pill}^0(1), \sigma_{pill}^0(2), \sigma_{pill}^0(3))$, say $\sigma_{pill}^0(1) \mapsto \sigma_{pill}^0(h)$, and attach it to the given $p = \sigma^0(h) \in (T_l, M)$ so as to generate a pinching node σ_*^0, (and remove it from the N_0 count of the vertices); *(iii)* relabel the remaining two vertices $(\sigma_{pill}^0(2), \sigma_{pill}^0(3)) \in (T^{pill}, \mathbb{S}^2)$ as $\widehat{\sigma}^0(h)$ and $\widehat{\sigma}^0(N_0 + 1)$. In this way, we get a genus g stable polyhedral surface

$$s_h\left[(T_l, M)\right] = (\widehat{T}_l, M) := (T_l, M)_h \underset{p=\sigma_*^0}{\cup} (T^{pill}, \mathbb{S}^2) \in \overline{POL}_{g, N_0 + 1}(M)$$

$$(1.138)$$

with a pillow tail and with a double vertex corresponding to the original vertex $p = \sigma^0(h)$.

(C) Finally, if p coincides with a pinching node $\sigma_*^0(S_i)$ of (T_l, M), then $(\widehat{T}_l, M) \in \overline{POL}_{g, N_0 + 1}(M)$ results from: *i)* separating the two pinched components of (T_l, M); *ii)* Setting $\widehat{\sigma}^0(j) := \sigma^0(j)$ for any $1 \leq j \leq N_0$; *iii)* Inserting a copy of a triangular pillow $(T^{pill}, \mathbb{S}^2)$ with two of its vertices, say $(\sigma_{pill}^0(2), \sigma_{pill}^0(3))$ identified with the preimage of $p = \sigma^0(h)$ so as to generate two pinching nodes $(\sigma_*^0(2), \sigma_*^0(3))$; *(iv)* identify the third vertex $\sigma_{pill}^0(1) \in (T^{pill}, \mathbb{S}^2)$ with $\widehat{\sigma}^0(N_0 + 1)$.

Conversely, let

$$\pi : \overline{POL}_{g, N_0 + 1}(M) \longrightarrow \overline{POL}_{g, N_0}(M) \qquad (1.139)$$

$$(\widehat{T}_l, M) \vdash \quad \overset{\textit{forget}}{\underset{\textit{collapse}}{\&}} \quad \longrightarrow (T_l, M)$$

the projection which forgets the $(N_0 + 1)^{st}$ vertex $\widehat{\sigma}^0(N_0 + 1)$ of $(\widehat{T}_l, M)$ and, if $\widehat{\sigma}^0(N_0 + 1)$ is in a pillow tail component $(T^{pill}, \mathbb{S}^2) \in (\widehat{T}_l, M)$, collapses $(T^{pill}, \mathbb{S}^2)$ and replaces the corresponding pinching node σ_*^0 with the remaining vertex of $(T^{pill}, \mathbb{S}^2)$.

The fiber of π over (T_l, M) is parametrized by the map (1.137), and if (T_l, M) has a trivial automorphism group $Aut(T)$ then $\pi^{-1}(T_l, M)$ is by definition the polyhedral surface (T_l, M), otherwise it is identified with the quotient $(T_l, M)/Aut(T)$. Thus, under the action of π, we can consider $\overline{POL}_{g,N_0+1}(M)$ as a family (in the orbifold sense) of polyhedral surfaces over $\overline{POL}_{g,N_0}(M)$. Note that, by construction, $\overline{POL}_{g,N_0+1}(M)$ comes endowed with the N_0 natural sections $s_1, \ldots, s_{N_0}$

$$s_h : \overline{POL}_{g,N_0}(M) \longrightarrow \overline{POL}_{g,N_0+1}(M), \tag{1.140}$$

defined by (1.138).

It is important to realize that the N_0 combinatorial bundles defined by the polyhedral cotangent cones at the (labelled vertices of the) (T_l, M) in $\mathscr{T}_{g,N_0}^{met}(M)$, (see Lemma 1.2), naturally extend to the stable polyhedral surfaces in $\overline{POL}_{g,N_0}(M)$. Again this is in a rather obvious analogy with the line bundles on the moduli space of N_0–pointed Riemann surfaces, (see Chap. 2), and Kontsevich's characterization of combinatorial classes over the moduli space of ribbon graphs. However, here the situation is geometrically simpler.

References

1. Ambjørn, J., Carfora, M., Marzuoli, A.: The Geometry of Dynamical Triangulations. Lecture Notes in Physics, vol. m50. Springer, Berlin/Heidelberg (1997)
2. Ambjörn, J., Durhuus, B., Jonsson, T.: Quantum Geometry. Cambridge Monograph on Mathematical Physics. Cambridge University Press, Cambridge (1997)
3. Boileau, M., Leeb, B., Porti, J.: Geometrization of 3-dimensional Orbifolds. Ann. Math. **162**, 195290 (2005)
4. Bott, R., Tu, L.W.: Differential Forms in Algebraic Topology. GTM, vol. 82. Springer, New York (1982)
5. Busemann, H.: Convex Surfaces. Interscience Tracts in Pure and Applied Mathematics, vol. 6. Interscience Publisher, New York (1958)
6. Carfora, M., Marzuoli, A.: Conformal modes in simplicial quantum gravity and the Weil-Petersson volume of moduli space. Adv. Math. Theor. Phys. **6**, 357 (2002). [arXiv:math-ph/0107028]
7. Guo, R., Luo, F.: Rigidity of polyhedral surfaces, II. Geom. Topol. **13**, 1265–1312 (2009). arXiv:0711.0766
8. Igusa, K.: Combinatorial Miller-Morita-Mumford classes and Witten cycles. Algebraic Geom Topol **4**, 473–520 (2004)
9. Kapovich, M., Millson, J.J.: On the moduli space of a spherical polygonal linkage. Can. Math. Bull. **42**(3), 307–320 (1999)
10. Kontsevitch, M.: Intersection theory on the moduli space of curves and the matrix airy functions. Commun. Math. Phys. **147**, 1–23 (1992)
11. Leibon, G.: Characterizing the Delaunay decompositions of compact hyperbolic surfaces. Geom. Topol. **6**, 361–391 (2002)
12. Luo, F.: Rigidity of polyhedral surfaces. arXiv:math/0612714
13. Luo, F.: Rigidity of polyhedral surfaces, III. arXiv:1010.3284
14. Luo, F.: Variational principles on triangulated surfaces. In: Handbook of Geometric Analysis. arXiv:0803.4232 (to appear)

15. Mulase, M., Penkava, M.: Ribbon graphs, quadratic differentials on Riemann surfaces, and algebraic curves defined over $\overline{\mathbb{Q}}$. Asian J. Math. **2**(4), 875–920 (1998). math-ph/9811024 v2
16. Mulase, M., Penkava, M.: Periods of strebel differentials and algebraic curves defined over the field of algebraic numbers. Asian J. Math. **6**(4), 743–748 (2002)
17. Pachner, U.: Konstruktionsmethoden und das kombinatorische Homöomorphieproblem für Triangulationen kompakter semilinearer Mannigfaltigkeiten. Abh. Math. Sem. Univ. Hamburg **57**, 69 (1986)
18. Pachner, U.: P.L. Homeomorphic manifolds are equivalent by elementary shellings. Eur. J. Comb. **12**, 129–145 (1991)
19. Rivin, T.: Euclidean structures on simplicial surfaces and hyperbolic volume. Ann. Math. **139**, 553–580 (1994)
20. Rourke, C.P., Sanderson, B.J.: Introduction to Piecewise-Linear Topology. Springer, New York (1982)
21. Thurston, W.P.: Three-Dimensional Geometry and Topology, vol. 1 (edited by S. Levy). Princeton University Press, Princeton (1997). See also the full set of Lecture Notes (December 1991 Version), electronically available at the Mathematical Sciences Research Institute, Berkeley
22. Thurston, W.P.: Shapes of polyhedra and triangulations of the sphere. Geom. Topol. Monogr. **1**, 511 (1998)
23. Troyanov, M.: Prescribing curvature on compact surfaces with conical singularities. Trans. Am. Math. Soc. **324**, 793 (1991); see also Troyanov, M.: Les surfaces euclidiennes a' singularites coniques. L'Enseignment Mathematique **32**, 79 (1986)
24. Troyanov, M.: On the moduli space of singular Euclidean surfaces. arXiv:math/0702666v2 [math.DG]
25. Tutte, W.J.: A census of planar triangulations. Can. J. Math. **14**, 21–38 (1962)

Chapter 2
Singular Euclidean Structures and Riemann Surfaces

As we have seen in Chap. 1, a Euclidean triangulated surface (T_l, M) characterizes a polyhedral metric with conical singularities associated with the vertices of the triangulation. In this chapter we show that around any such a vertex we can introduce complex coordinates in terms of which we can write down the conformal conical metric, locally parametrizing the singular structure of (T_l, M). This makes available a powerful dictionary between 2–dimensional triangulations and complex geometry. It must be noted that, both in the mathematical and in the physical applications of the theory, the connection between Riemann surfaces and triangulations typically emphasizes only the role of ribbon graphs and of the associated metric. The conical geometry of the polyhedral surface (T_l, M) is left aside and seems to play no significant a role. Whereas this attitude can be motivated by the observation (due to M. Troyanov [8]) that the conformal structure does not see the conical singularities (see below, for details), it gives a narrow perspective of the much wider role that the theory has to offer. Thus, it is more appropriate to connect a polyhedral surface (T_l, M) to a corresponding Riemann surface by taking fully into account the conical geometry of (T_l, M). This connection is many–faceted and exploits a vast repertoire of notion ranging from complex function theory to algebraic geometry. We start by defining the barycentrically dual polytope (P_T, M) associated with a polyhedral surface (T_l, M) and discuss the geometry of the corresponding ribbon graph. Then, by adapting to our case the elegant approach in [4], we explicitly construct the Riemann surface associated with (P_T, M). This directly bring us to the analysis of Troyanov's singular Euclidean structures and to the construction of the bijective map between the moduli space $\mathfrak{M}_{g,N_0}$ of Riemann surfaces (M, N_0) with N_0 marked points, decorated with conical angles, and the space of polyhedral structures. In particular the first Chern class of the line bundles naturally defined over $\mathfrak{M}_{g,N_0}$ by the cotangent space at the i–th marked point is related with the corresponding Euler class of the circle bundles over the space of polyhedral surfaces defined by the conical cotangent spaces at the i–th vertex of (T_l, M). Whereas this is not an unexpected connection, the analogy with Witten–Kontsevich theory being obvious,

© Springer International Publishing AG 2017

M. Carfora, A. Marzuoli, *Quantum Triangulations*, Lecture Notes in Physics 942,
https://doi.org/10.1007/978-3-319-67937-2_2

we stress that the conical geometry adds to this property the possibility of a deep and explicit characterization of the Weil–Petersson form in terms of the edge–lengths of (T_l, M). This result [1] is obtained by a subtle interplay between the geometry of (T_l, M) and 3–dimensional hyperbolic geometry, and it will be discussed in detail in Chap. 3 since it explicitly hints to the connection between polyhedral surfaces and quantum geometry in higher dimensions.

2.1 The Barycentrically Dual Polytope of a Polyhedral Surface

Many subtle aspects of the geometry of the set $POL_{g,N_0}(M)$ of polyhedral surfaces (T_l, M) are captured by the properties of the barycentrically dual polytope associated with (T_l, M). In this section we study in full detail this polytope, eventually making contact with ribbon graph theory.

Denote by $(T^{(1)}, M) := |T_l^{(1)}| \to M$ the (first) barycentric subdivision of $(T_l, M) = |T_l| \to M$, realized by the medians of the triangles $\sigma^2(i, jk) \in T$, (recall that by Ceva's theorem the medians all intersect at the barycenter of $\sigma^2(i, jk)$). This procedure divides each $\sigma^2(i, jk)$ into six new triangles and we have (Fig. 2.1).

Definition 2.1 The closed stars in $(T^{(1)}, M)$ of the vertices of the original triangulation (T_l, M) form a collection of 2-cells $\{\rho^2(i)\}_{i=1}^{N_0(T)}$ characterizing the *conical* polytope $(M, P_T) := |P_{T_l}| \to M$ barycentrically dual to (T_l, M).
The adjective conical emphasizes that here we are considering a geometrical presentation $|P_{T_l}| \to M$ of P where the 2-cells $\{\rho^2(i)\}_{i=1}^{N_0(T)}$ retain the conical geometry induced on the barycentric subdivision by the original conical metric structure of (T_l, M).

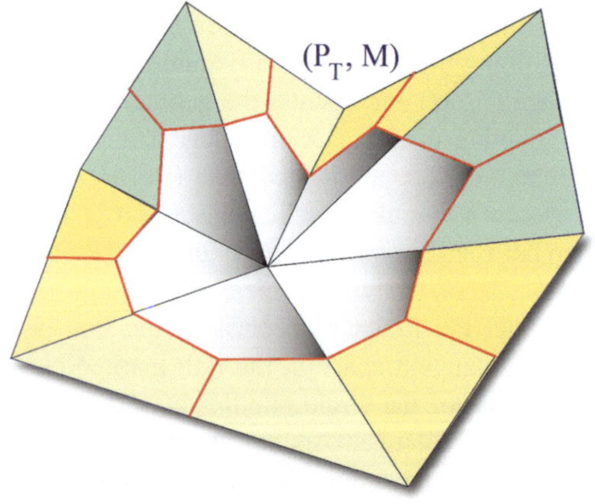

Fig. 2.1 The conical dual polytope

In order to describe explicitly the geometry of (P_T, M) let us consider the generic oriented (counter clockwise) triangle $\sigma^2(h, j, k) \in (T_l, M)$ of sides $\sigma^1(h, j)$, $\sigma^1(j, k)$, and $\sigma^1(k, h)$. Note that each edge $\sigma^1(h, j)$ has two orientations, $\sigma^1(h, j) = -\sigma^1(j, h)$. Denote by

$$\sigma^0(h) \searrow \sigma^1(h, j), \tag{2.1}$$

$$\sigma^1(h, j) \nearrow \sigma^0(j) \tag{2.2}$$

the *source* and the *target* of the edge $\sigma^1(h, j)$, respectively. These define the incidence (or boundary) map among the generic edge $\sigma^1(h, j)$ and the corresponding source $\sigma^0(h)$, and target $\sigma^0(j)$ vertices.

Let $W(h, j)$, $W(j, k)$, and $W(k, h)$ respectively denote the barycenters of the edges $\sigma^1(h, j)$, $\sigma^1(j, k)$, and $\sigma^1(k, h)$. We have

Definition 2.2 The oriented *half–edges* $\rho^1(\bullet, \bullet)^{\pm}$, connecting $W(h, j)$, $W(j, k)$, and $W(k, h)$ to the vertex $\rho^0(h, j, k)$ of the polytope (M, P_T), are defined by the incidence relations

$$W(j, h) \searrow \rho^1(h, j)^+ \nearrow \rho^0(j, k, h) \searrow \rho^1(h, k)^- \nearrow W(k, h), \tag{2.3}$$

$$W(k, j) \searrow \rho^1(j, k)^+ \nearrow \rho^0(k, h, j) \searrow \rho^1(j, h)^- \nearrow W(h, j), \tag{2.4}$$

$$W(h, k) \searrow \rho^1(k, h)^+ \nearrow \rho^0(h, j, k) \searrow \rho^1(k, j)^- \nearrow W(j, k). \tag{2.5}$$

Note that

$$\rho^1(h, j) = \rho^1(h, j)^- \cup W(j, h) \cup \rho^1(h, j)^+, \tag{2.6}$$

and $\rho^1(h, j)^+ \equiv -\rho^1(j, h)^-$. We denote by

$$\widehat{L}(\bullet, \bullet)^{\pm} := \left| \rho^1(\bullet, \bullet)^{\pm} \right|. \tag{2.7}$$

the length of the given half edge (Fig. 2.2).

Remark 2.1 If we fix our attention to a given triangle $\sigma^2(h, j, k) \in (T_l, M)$ we can exploit a simpler notation for the lengths of the half edges incident on the vertex, $\rho^0(h, j, k) \in (M, P_T)$, dual to $\sigma^2(h, j, k)$, i.e.,

$$\widehat{L}(k) := |\rho^1(h, j)^+| = |\rho^1(j, h)^-|, \tag{2.8}$$

$$\widehat{L}(h) := |\rho^1(j, k)^+| = |\rho^1(k, j)^-|, \tag{2.9}$$

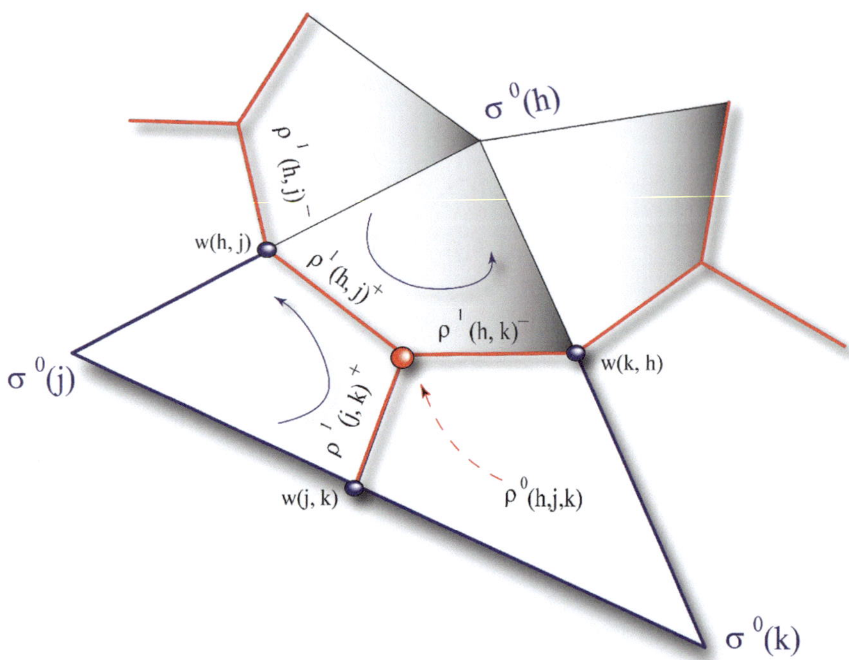

Fig. 2.2 The geometry of half edges

$$\widehat{L}(j) := |\rho^1(k,h)^+| = |\rho^1(h,k)^-|. \tag{2.10}$$

It must be stressed that this notation is ambiguous as soon as we consider two adjacent triangles in (T_l, M) and the corresponding trivalent vertices in (P_T, M). In this latter case we have to use the more explicit labeling $\widehat{L}(\bullet, \bullet)^\pm := |\rho^1(\bullet, \bullet)^\pm|$.

Since the medians of an arbitrary triangle $\sigma^2(h,j,k) \in (T_l, M)$ divide one another in the ratio 2:1, the lengths of the medians $W(j,k)\rho^2(h)$, $W(k,h)\rho^2(j)$, $W(h,j)\rho^2(k)$ of the triangle $\sigma^2(h,j,k)$ are given by $3\widehat{L}(h)$, $3\widehat{L}(j)$, and $3\widehat{L}(k)$, respectively, (we can use here the simpler notation described in Remark 2.1). Moreover, since one can always construct a triangle from the medians of an arbitrary triangle we have that $\widehat{L}(h)$, $\widehat{L}(j)$, and $\widehat{L}(k)$ satisfy the triangle inequality (Fig. 2.3)

$$\widehat{L}(k) \le \widehat{L}(h) + \widehat{L}(j) \,, \qquad \circlearrowleft (k,h,j), \tag{2.11}$$

where $\circlearrowleft (k,h,j)$ stands for cyclic permutation. An elementary computation provides the relations connecting the side lengths $l(\bullet, \bullet)$ to the $\widehat{L}(\bullet)$'s

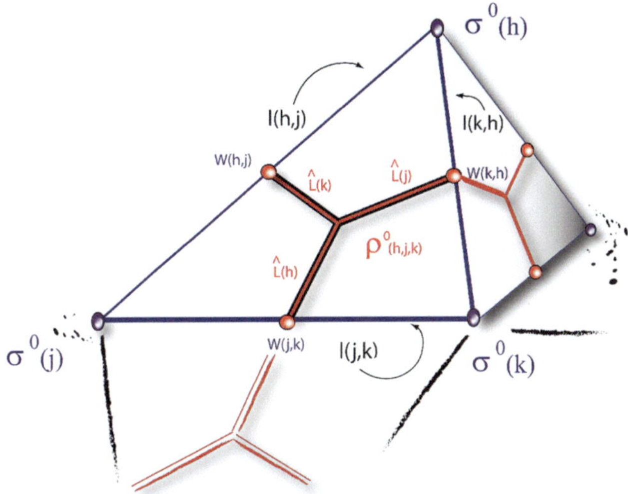

Fig. 2.3 The relation between the edge-lengths of the conical polytope and the edge-lengths of the triangulation

$$
\begin{aligned}
\widehat{L}^2(j) &= \tfrac{1}{18}l^2(j,k) + \tfrac{1}{18}l^2(h,j) - \tfrac{1}{36}l^2(k,h)\\
\widehat{L}^2(k) &= \tfrac{1}{18}l^2(k,h) + \tfrac{1}{18}l^2(j,k) - \tfrac{1}{36}l^2(h,j)\\
\widehat{L}^2(h) &= \tfrac{1}{18}l^2(h,j) + \tfrac{1}{18}l^2(k,h) - \tfrac{1}{36}l^2(j,k),
\end{aligned}
\tag{2.12}
$$

$$
\begin{aligned}
l^2(k,h) &= 8\widehat{L}^2(h) + 8\widehat{L}^2(k) - 4\widehat{L}^2(j)\\
l^2(h,j) &= 8\widehat{L}^2(j) + 8\widehat{L}^2(h) - 4\widehat{L}^2(k)\\
l^2(j,k) &= 8\widehat{L}^2(k) + 8\widehat{L}^2(j) - 4\widehat{L}^2(h).
\end{aligned}
$$

Let $\theta_{khj}, \theta_{hjk}, \theta_{jkh}$ respectively denote the angles associated with the vertices $\sigma^0(h)$, $\sigma^0(j)$, $\sigma^0(k)$ of the triangle $\sigma^2(h,j,k)$. Denote by

$$
\phi_{jk} := \angle[\widehat{L}(j), \widehat{L}(k)],
\tag{2.13}
$$

$$
\phi_{kh} := \angle[\widehat{L}(k), \widehat{L}(h)],
$$

$$
\phi_{hj} := \angle[\widehat{L}(h), \widehat{L}(j)],
$$

the angles formed by the half edges incident on the 3–valent vertex $\rho^0(h,j,k)$. The medians of $\sigma^2(h,j,k)$ induce a splitting of the vertex angles $\theta_{khj}, \theta_{hjk}, \theta_{jkh}$ according to (Fig. 2.4)

$$
\theta_{khj} = \theta_{khj}^- + \theta_{khj}^+ , \qquad \circlearrowleft (k,h,j),
\tag{2.14}
$$

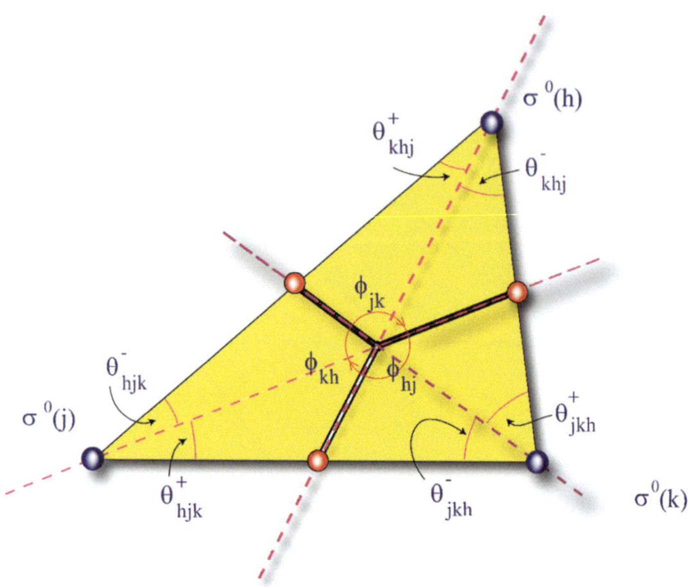

Fig. 2.4 The relation between the median splitting of the angles at the vertices of the triangle $\sigma^2(h, j, k)$ and the angles among half edges

and the similitude between the triangles $\sigma^2(h, j, k)$ and $(W(h, j), W(j, k), W(k, h))$ yields the relations

$$\phi_{jk} = \pi - \theta^-_{jkh} - \theta^+_{hjk}, \tag{2.15}$$

$$\phi_{kh} = \pi - \theta^-_{khj} - \theta^+_{jkh},$$

$$\phi_{hj} = \pi - \theta^-_{hjk} - \theta^+_{khj}.$$

Finally, let us denote by $\alpha^\pm(khj)$ the angles between the median issued from the vertex $\sigma^0(h)$ and the oriented edge $\sigma^1(j, k)$. We have

$$\alpha^-_{khj} = \theta_{jkh} + \theta^-_{khj}, \tag{2.16}$$

$$\alpha^+_{khj} = \theta_{hjk} + \theta^+_{khj}.$$

Similarly

$$\alpha^-_{hjk} = \theta_{khj} + \theta^-_{hjk}, \tag{2.17}$$

$$\alpha^+_{hjk} = \theta_{jkh} + \theta^+_{hjk}, \tag{2.18}$$

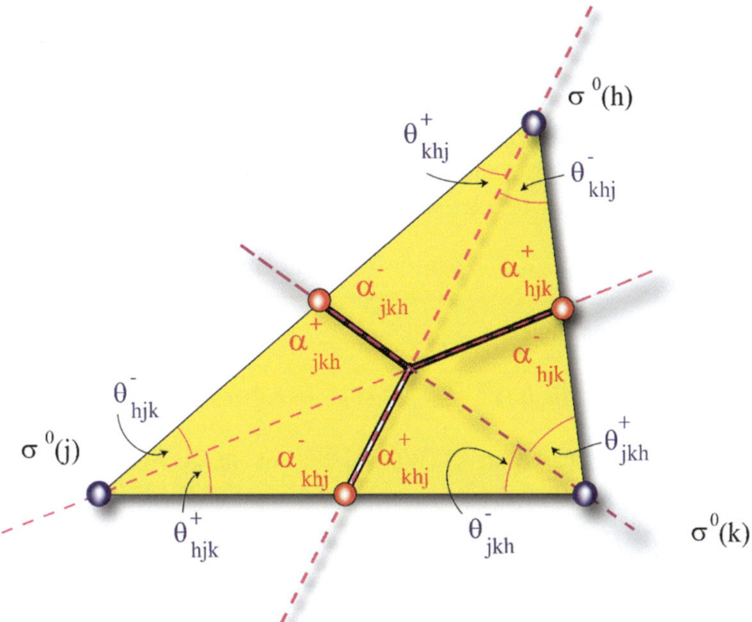

Fig. 2.5 The six median angles $\alpha^{\pm}_{\circlearrowleft(k,h,j)}$

and

$$\alpha^-_{jkh} = \theta_{hjk} + \theta^-_{jkh}, \tag{2.19}$$

$$\alpha^+_{jkh} = \theta_{khj} + \theta^+_{jkh}, \tag{2.20}$$

for the angles between the oriented edges $\sigma^1(k,h), \sigma^1(h,j)$ and the medians issued from $\sigma^0(j)$ and $\sigma^0(k)$, respectively (Fig. 2.5).

The set of these elementary relations allow to recover, as the indices (h,j,k) vary, the metric geometry of the dual polytope (P_T, M) from the geometry of the original singular Euclidean triangulation (T_l, M).

2.2 Polytope Automorphisms and Ribbon Graphs

The geometrical realization of the polytope (P_T, M) directly implies that we can associate with (P_T, M) the graph

$$\Gamma_P = \left(\{\rho^0(h,j,k)\} \overset{N_1(T)}{\bigsqcup} \{W(h,j)\}, \{\rho^1(h,j)^-\} \overset{N_1(T)}{\bigsqcup} \{\rho^1(h,j)^+\} \right). \tag{2.21}$$

whose vertex set

$$v(\Gamma_P) := \{\rho^0(h,j,k)\}^{N_2(T)} \cup \{W(h,j)\}^{N_1(T)} \tag{2.22}$$

is identified with the barycenters of the triangles $\{\sigma^o(h,j,k)\}^{N_2(T)}$, and with the barycenters $\{W(h,j)\}^{N_1(T)}$ of the edges $\{\sigma^1(h,j)\}$ belonging to the original triangulation (T_l, M), whereas the edge set

$$e(\Gamma_P) := \{\rho^1(h,j)\}^{N_1(T)} = e_-(\Gamma_P) \cup e_+(\Gamma_P) \tag{2.23}$$

is generated by the half-edges

$$e_-(\Gamma_P) := \{\rho^1(h,j)^-\}^{N_1(T)} \,, \quad e_+(\Gamma_P) := \{\rho^1(h,j)^+\}^{N_1(T)}, \tag{2.24}$$

joined through the barycenters $\{W(h,j)\}^{N_1(T)}$ of the corresponding $\sigma^1(h,j)$ (Fig. 2.6).

Proposition 2.1 *The graph Γ_P is the edge refinement of the three-valent graph*

$$K^1(P) := \left(\{\rho^0(h,j,k)\}^{N_2(T)}, \{\rho^1(h,j)\}^{N_1(T)}, \mathscr{I}_P\right) \tag{2.25}$$

defined by the incidence relation $\mathscr{I}_P$ of the 1–skeleton of the polytope (P_T, M).

Proof Let G_e denote of a (directed graph) $G := (v, e, \mathscr{I})$, defined by a set of vertices v of valence ≥ 3, a set of directed edges e, and an incidence relation $\mathscr{I} : e \to v \times v$. Recall [4] that the *edge refinement* G_e of G is the graph obtained by adding a midpoint W_e on each edge $\rho^1 \in e$. We have $G_e = (v \cup v_e, e_- \cup e_+, \mathscr{I}_e)$ where $v_e := \{W_e\}$ is the set of added 2–valent vertices, $e_- \cup e_+$ is the set of edges splitted in half–edges by the midpoints $\{W_e\}$, and where the induced incidence

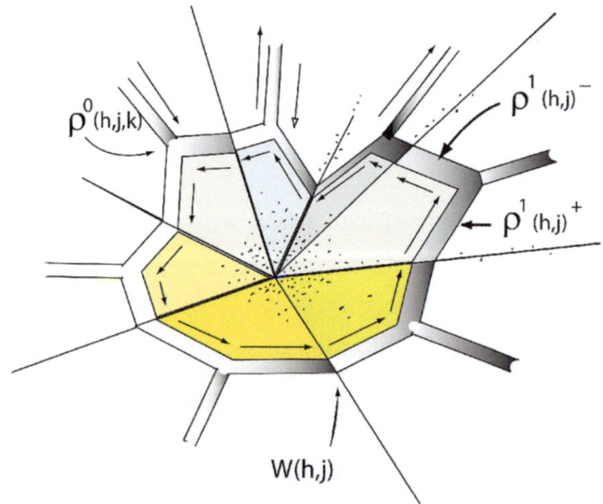

Fig. 2.6 The ribbon graph associated with the barycentrically dual polytope

relation $\mathscr{I}_e$ is consistent with $\mathscr{I}$. The (counterclockwise) orientation in the 2-cells $\{\rho^2(k)\}$ of (P_T, M) gives rise to a cyclic ordering on the set of edges $\{\rho^1\}$ of the 3–valent graph $K^1(P, M)$ defined by the 1–skeleton of (P_T, M). Thus, the edges $\{\rho^1\}^{N_1(T)}$ of $K^1(P, M)$ are directed and the barycenters $\{W(h, j)\}^{N_1(T)}$ generate a mid–point splitting which gives rise to the graph Γ_P. The incidence relations of Γ_P are clearly consistent with those of $K^1(P)$. In particular, since $K^1(P)$ is a trivalent graph, by eliminating all 2–valent vertices from Γ_P we can immediately recover the 1–skeleton of (P_T, M) from its edge refinement Γ_P. $\diamond$

The relevance of the edge refinement Γ_P stems from the observation that the natural automorphism group $Aut(P)$ of (P, M) , (i.e., the set of bijective maps preserving the incidence relations defining the geometrical realization of (P, M)), is not the standard automorphism group $Aut\,(K^1_{abs}(P))$ of $K^1(P)$ thought of as the underlying abstract graph, but rather the automorphism group of its edge refinement Γ_P, i.e.,

$$Aut\,(P) \; := \; Aut(\Gamma_P). \tag{2.26}$$

Explicitly, by adapting the definition of automorphism of the edge–refinement of a graph [4], we have

Definition 2.3 A graph automorphism of Γ_P is a triple of bijections

$$\alpha_v : \{\rho^0(h, j, k)\} \longrightarrow \{\rho^0(h, j, k)\}, \tag{2.27}$$

$$\alpha_e : \{W(h, j)\} \longrightarrow \{W(h, j)\},$$

$$\beta : \{\rho^1(h, j)^+\} \cup \{\rho^1(h, j)^-\} \longrightarrow \{\rho^1(h, j)^+\} \cup \{\rho^1(h, j)^-\},$$

which are compatible with the incidence relation of Γ_P.

Note that the orientation in the 2-cells $\{\rho^2(k)\}$ of (P_T, M) gives rise also to a cyclic ordering on the set of half-edges $\{\rho^1(h, j)^{\pm}\}^{N_1(T)}$ incident on the vertices $\{\rho^0(h, j, k)\}^{N_2(T)}$. Thus,

$$Aut(\Gamma_P) \subset Aut\,(K^1_{abs}(P)), \tag{2.28}$$

since the elements in $Aut(\Gamma_P)$ are the automorphisms in $Aut\,(K^1_{abs}(P))$ that preserve the cyclic ordering at each vertex.

Recall that we can associate with the polyhedral surface (T_l, M) the topologically open surface $M' := M \setminus K^0$ obtained by removing the vertices $V(T)$ from (T_l, M). Then, M' admits a canonical deformation retraction onto the *spine* of M', defined by the union of the closed simplices of the barycentric subdivision of (T_l, M) that lie in M'. Namely, the inclusion $K^1(P) \hookrightarrow M'$ is a homotopy equivalence.

We have the following characterization

Proposition 2.2 *The edge-refinement Γ_P of the 1-skeleton $K^1(P)$ of the barycentrically dual polytope (P_T, M) of a polyhedral surface of genus g has the following properties:*

(i) Γ_P is a ribbon graph [4], i.e., the edge refinement of an abstract graph Γ endowed with a cyclic ordering on the set of the half-edges,

$$\{\rho^1(h,j)^+, \rho^1(j,k)^+, \rho^1(k,h)^+\}, \tag{2.29}$$

incident on each 3–valent vertex $\rho^0(h,j,k) \in \Gamma$. The ribbon graph Γ_P has $N_0(T)$ boundary components $\{\partial_{(k)}\Gamma_P\}$, $N_1(T)$ edges, and $N_2(T)$ 3–valent vertices with

$$2g - 2 + N_0(T) = \frac{1}{2}N_2(T). \tag{2.30}$$

(ii) Γ_P is metrized by the function

$$\widehat{L} : e_\pm(\Gamma) \longrightarrow \mathbb{R}_+^{3N_2(T)} \tag{2.31}$$

$$\{(\rho^1(h,j)^\pm, \rho^1(j,k)^\pm, \rho^1(k,h)^\pm)\} \longmapsto \left\{\left(\widehat{L}(k), \widehat{L}(h), \widehat{L}(j)\right)\right\}, \tag{2.32}$$

that assigns to each (unoriented) half–edge, incident on $\rho^0(h,j,k)$, a positive real number.

Conversely, any metrized edge refinement of a ribbon graph Γ, with $n_0(\Gamma)$ boundary components, $n_1(\Gamma)$ edges, and $n_2(\Gamma)$ 3–valent vertices such that

$$2g - 2 + n_0(\Gamma) = n_1(\Gamma) - n_2(\Gamma), \tag{2.33}$$

characterizes an oriented surface $\widetilde{M}(\Gamma)$ with $n_0(\Gamma)$ boundary components $\partial_{(k)}\widetilde{M}$, possessing Γ as a spine. By capping the boundary components $\partial_{(k)}\widetilde{M}$ with $n_0(\Gamma)$ topological disks we get a surface M of genus g endowed with a Euclidean triangulation $|T| \to M$, with f–vector $N_0(T) = n_0(\Gamma)$, $N_1(T) = n_1(\Gamma)$, $N_2(T) = n_2(\Gamma)$, admitting Γ as the edge–refinement of the 1–skeleton of the barycentrically dual polytope (P_T, M) associated with (T_l, M).

Proof The first part of the proposition directly follows from the connection, established above, between the ribbon graph Γ_P and the edge refinement of 1–skeleton $K^1(P)$ associated with the polyhedral surface (T_l, M).

For the second part, let us recall that an abstract, 3–valent, ribbon graph $\Gamma := (v, e, \mathscr{I})$ with $n_0(\Gamma)$ boundary components, $n_1(\Gamma)$ edges, and $n_2(\Gamma)$ 3–valent vertices such that $2g - 2 + n_0(\Gamma) = n_1(\Gamma) - n_2(\Gamma)$, characterizes the notion of *map of a surface*. This implies that geometrically one can realize Γ by using a double index notation to represent (half–)edges on an oriented two–plane, and connect any two edges, according to the incidence $\mathscr{I}$, in such a way that the double index notation (and hence the orientation) is consistent. This immediately implies that Γ is topologically an oriented topological surface with boundary which retracts on Γ. According to the Euler relation, by capping the boundaries with disks components we get a closed surface of genus g, as stated.[1] Let $\Gamma_e := (v \cup v_e,\ e_- \cup e_+,\ \mathcal{I}_e)$ be the edge refinement of Γ. Denote by $v(h,j,k) \in v(\Gamma)$ the generic 3–valent vertex, and let $\big(e^{\pm}(h,j),\ e^{\pm}(j,k),\ e^{\pm}(k,h)^{\pm}\big) \in e(\Gamma_e)$ the three half–edges incident on it. Γ_e is metrized by assigning, for each $v(h,j,k) \in v(\Gamma)$, a length to the corresponding half–edges, according to

$$\big(e^{\pm}(h,j),\ e^{\pm}(j,k),\ e^{\pm}(k,h)^{\pm}\big) \longmapsto \big(\widehat{L}(k), \widehat{L}(h), \widehat{L}(j)\big), \tag{2.34}$$

where each triple $\big(\widehat{L}(k), \widehat{L}(h), \widehat{L}(j)\big) \in \mathbb{R}^3_+$, satisfies the triangle inequality

$$\widehat{L}(k) \le \widehat{L}(h) + \widehat{L}(j), \qquad \circlearrowleft (k,h,j). \tag{2.35}$$

This assignment can be used to characterize a Euclidean triangulation of the $n_0(\Gamma)$ disks capping the boundary components of Γ according to the following steps (Fig. 2.7):

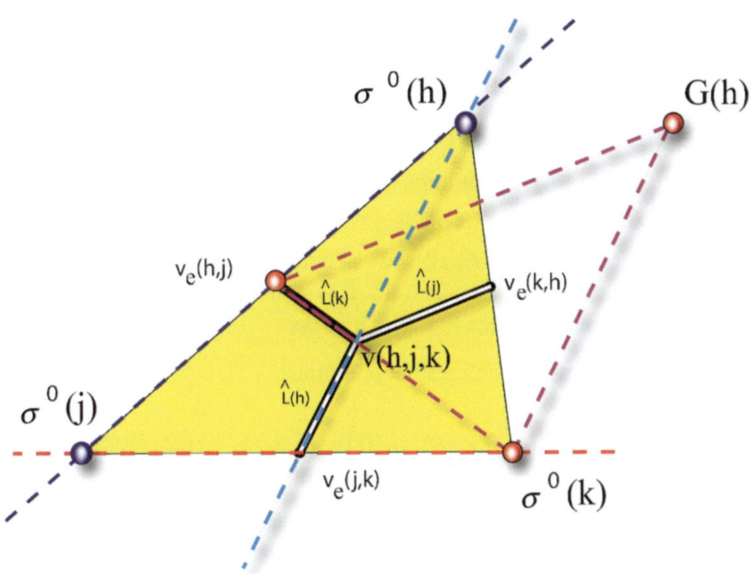

Fig. 2.7 The construction of the triangle $\sigma^2(h,j,k)$ out of its median triangle $\Delta[v_e(h,j), \sigma^0(k), G(h)]$

---

[1] See [2–4] for an in depth analysis of the various aspects of the connection between ribbon graphs and surface theory. The name ribbon graph which is indeed quite evocative, as confronted to equivalent denominations as fat graphs, or maps of surfaces, apparently first appears in [2].

(i) place the 3–valent vertex $v(h, j, k) \in v(\Gamma)$ in the origin of an oriented (counter clockwise) Euclidean 2–plane $(\mathbb{R}^2, \delta)$, (δ the Euclidean metric). Each incident edge $(e(h, j), e(j, k), e(k, h))$ and the associated 2–valent vertices

$$(v_e(h, j),\, v_e(j, k),\, v_e(k, h)), \tag{2.36}$$

defining the edge refinement, are injectively mapped into $(\mathbb{R}^2, \delta)$ modulo a rotation, i.e., $v_e(h, j)$ can be on any point of the circumference C_k of radius $\widehat{L}(k)$, centered at the origin of $(\mathbb{R}^2, \delta)$. Similarly $v_e(j, k)$, and $v_e(k, h)$ can be anywhere on the associated circumferences, with $v_e(h, j) \neq v_e(j, k) \neq v_e(k, h)$.
(ii) Select and fix in $(\mathbb{R}^2, \delta)$ an incident half–edge, say $e^+(h, j)$, entering in $v(h, j, k)$, and consider, in $(\mathbb{R}^2, \delta)$, the directed segment of length $3\widehat{L}(k)$ generated by $e^+(h, j)$. This segment will connect the point $v_e(h, j) \in (\mathbb{R}^2, \delta)$ to another point that we denote $\sigma^0(k)$. Over the segment $(v_e(h, j), \sigma^0(k))$ construct the unique oriented triangle $\Delta[v_e(h, j), \sigma^0(k), G(h)]$, with vertex $G(h) \in (\mathbb{R}^2, \delta)$, and sides of lengths $3\widehat{L}(h)$ and $3\widehat{L}(j)$, respectively.
(iii) Draw the directed parallel to the side $(\sigma^0(k), G(h))$ passing through $v(h, j, k)$. Along this oriented line fix a new vertex $\sigma^0(h)$ by considering the directed segment $(v(h, j, k), \sigma^0(h))$ of length $2\widehat{L}(h)$. We can now fix the action of the rotation group on $(\mathbb{R}^2, \delta)$ by declaring that this oriented line contains the mid–vertex $v_e(j, k)$. The corresponding segment $(v_e(j, k), \sigma^0(h))$ has length $3\widehat{L}(h)$. Finally, the two oriented lines $[\sigma^0(h), v_e(h, j)]$ and $[\sigma^0(k), v_e(j, k)]$ meet at a point that we denote $\sigma^0(j)$. The triangle $\sigma^2(h, j, k)$, the vertices of which are defined by the points $\sigma^0(h)$, $\sigma^0(j)$, $\sigma^0(k)$, has medians of lengths $\left(3\widehat{L}(k), 3\widehat{L}(h), 3\widehat{L}(j)\right)$ and its median triangle is $\Delta[v_e(h, j), \sigma^0(k), G(h)]$.
(iv) By exploiting the relations (2.12) we can compute the lengths $l(h, j), l(j, k)$, and $l(k, h)$ of the sides of the triangle $\sigma^2(h, j, k)$ so defined. Finally, from (1.16) also the associated vertex angles θ_{hjk}, θ_{jkh}, and θ_{khj}.

This elementary construction defines the mapping

$$(\Gamma_e, \widehat{L}(\bullet)) \ni \left(v(h, j, k),\, \widehat{L}(k), \widehat{L}(h), \widehat{L}(j) \right) \longmapsto \sigma^2(h, j, k) \in (\mathbb{R}^2, \delta), \tag{2.37}$$

which associates to each 3–valent vertex $v(h, j, k)$ of the metrized ribbon graph Γ a well–defined Euclidean triangle $\sigma^2(h, j, k)$.

Let us denote by $\{v(h, j_\alpha, j_{\alpha+1})\}_{(h)}$, $j_{q(h)+1} \equiv j_1$ the collection of 3–valent vertices in a given boundary component $\partial_{(h)}\Gamma$ of the ribbon graph Γ, containing $q(h)$ vertices. According to the above construction, we can associate to any such component a corresponding finite simplicial complex

$$\partial_{(h)}\Gamma \longmapsto T_{(h)} := \left\{ \sigma^2(h, j_\alpha, j_{\alpha+1}) \right\}_{(h)}, \tag{2.38}$$

defined by $q(h)$ triangles $\{\sigma^2(h, h, j_\alpha, j_{\alpha+1})\}$ all incident on the same vertex $\sigma^0(h)$, and characterizing a Euclidean triangulation $f_{(h)} : T_{(h)} \longrightarrow D_{(h)} \subset M$, of the disk

$D_{(h)}$ capping $\partial_{(h)}\Gamma$. The triangulated capping disk $(T_{(h)}, D_{(h)})$ has a natural conical geometry associated with the conical angle

$$\Theta(h) := \sum_{\alpha=1}^{q(h)} \theta_{\alpha+1, h\alpha} \tag{2.39}$$

supported at $\sigma^0(h)$. If $(\lambda(h), \beta(h))$ denote polar coordinates, (based at $\sigma^0(h)$), of a point $p \in D_{(h)}$, then $D_{(h)}$ can be geometrically realized as the space

$$\{(\lambda(h), \beta(h)) : \lambda(h) \geq 0; \beta(h) \in \mathbb{R}/(2\pi - \varepsilon(h))\mathbb{Z}\}/ (0, \beta(h)) \sim (0, \beta'(h)) \tag{2.40}$$

endowed with the conical metric

$$d\lambda(h)^2 + \lambda(h)^2 d\beta(h)^2. \tag{2.41}$$

This construction can be coherently extended to all boundary components $\partial_{(h)}\Gamma$ of Γ and characterize (T_l, M) as the polyhedral surface associated with Γ, correspondingly identifying the given ribbon graph with the (edge refinement of the) 1–skeleton of the barycentrically dual polytope of (T_l, M). ■

2.3 Remarks on Metric Ribbon Graphs

Let us denote by RG_{g,N_0} the set of all connected ribbon graphs Γ of genus g, with given edge-set $e(\Gamma)$ and N_0 labeled boundary components, with $2 - 2g - N_0(T) < 0$, and such that every vertex has valency ≥ 3. The set RG_{g,N_0}^{met} of metric ribbon graphs obtained from RG_{g,N_0} by assigning a length to each edge $\in e(\Gamma)$ can be characterized [3, 4] as a space homeomorphic to $\mathbb{R}_+^{|e(\Gamma)|}$, ($|e(\Gamma)|$ denoting the number of edges in $e(\Gamma)$), topologized by the standard ϵ-neighborhoods $U_\epsilon \subset \mathbb{R}_+^{|e(\Gamma)|}$, (this is strictly true only for the top-dimensional component associated with trivalent ribbon graph. The general case is discussed in great detail in [4]). The automorphism group $Aut(\Gamma)$ acts naturally on such a space via the homomorphism $Aut(\Gamma) \to \mathfrak{G}_{e(\Gamma)}$, where $\mathfrak{G}_{e(\Gamma)}$ denotes the symmetric group over $|e(\Gamma)|$ elements. The topology on the set of ribbon graphs is characterized by a (Whitehead) expansion $\succeq$ and collapse $\preceq$ procedure which amounts to expanding or collapsing edges and separating or coalescing vertices. Explicitly, if $l(t) = tl$ is the length of an edge $\rho^1(j)$ of a ribbon graph $\Gamma_{l(t)} \in RG_{g,N_0}^{met}$, then, as $t \to 0$, we get the metric ribbon graph $\widehat{\Gamma} := \preceq (\Gamma)$ which is obtained from $\Gamma_{l(t)}$ by collapsing the edge $\rho^1(j)$. The expansion is realized by reversing this construction. Denote [4] by $\mathscr{X}_{\succeq\Gamma}^{met}$ the set of all metric ribbon graphs that can be obtained by Γ through all possible expansions of Γ, modulo identifications under ribbon graph isomorphisms preserving the half–edges structure. In full analogy with the case of polyhedral surfaces we have discussed in detail in Sect. 1.6, the quotient space $\mathbb{R}_+^{|e(\Gamma)|}/Aut(\Gamma)$ is a differentiable orbifold in terms of which we can write

$$RG_{g,N_0}^{met} = \bigsqcup_{\Gamma \in RG_{g,N_0}} \frac{\mathbb{R}_+^{|e(\Gamma)|}}{Aut(\Gamma)}. \tag{2.42}$$

The set RG_{g,N_0}^{met} is [3, 4] a locally compact Hausdorff space modelled on strata (rational orbicells) $\frac{\mathscr{X}_{\succeq\Gamma}^{met}}{Aut(\Gamma)}$ of codimension $\sum_{\rho^0\in\Gamma}\left(deg(\rho^0)-3\right)$.

Let us denote by $K_1 RP_{g,N_0}^{met}$ the top-dimensional, $\left(\dim\left[K_1 RP_{g,N_0}^{met}\right] = 3N_0 + 6g - 6\right)$, open orbicell in RG_{g,N_0}^{met}, associated with the 3–valent ribbon graphs in RG_{g,N_0} barycentrically dual to metrical triangulations in $\mathscr{T}_{g,N_0}^{met}$. Note that $K_1 RP_{g,N_0}^{met}$ is modelled on a subset of metric ribbon graphs $\mathscr{X}_{\succeq\Gamma}^{met}$ which is acted upon by a collapse–expansion move, dual of the flip move acting on the metrical triangulations $\in \mathscr{T}_{g,N_0}^{met}$. According to Proposition 2.2 there is a natural combinatorial isomorphism

$$POL_{g,N_0}(M) \simeq K_1 RP_{g,N_0}^{met} \subset RG_{g,N_0}^{met}, \tag{2.43}$$

between the space of Polyhedral structures on surfaces of genus g with N_0 vertices and the space of (isomorphism classes of) 3–valent metrical ribbon graphs $K_1 RP_{g,N_0}^{met}$ which are barycentrically dual to polyhedral surfaces (T_l, M).

2.4 The Riemann Surface Associated with (P_T, M)

In this section we explicitly construct the Riemann surface associated with the conical polytope (P_T, M). To this end, let Γ denote the ribbon graph defined by the edge–refinement of the 1–skeleton $K^1 P_T$ of the conical polytope (P_T, M). Let $\rho^2(h)$, $\rho^2(j)$, and $\rho^2(k)$ respectively be the two-cells $\in (P_T, M)$ barycentrically dual to the vertices $\sigma^0(h)$, $\sigma^0(j)$, and $\sigma^0(k)$ of a triangle $\sigma^2(h, j, k) \in (T_l, M)$. Denote by $\rho^0(h, j, k)$ the 3-valent, cyclically ordered, vertex of Γ defined by (Fig. 2.8)

$$\rho^0(h,j,k) \doteq \partial\rho^2(h) \underset{\Gamma}{\bigcap} \partial\rho^2(j) \underset{\Gamma}{\bigcap} \partial\rho^2(k). \tag{2.44}$$

Similarly, if $\rho^1(h,j)^+$ and $\rho^1(j,h)^-$, respectively are the oriented half–edges in $\rho^2(h)$ and $\rho^2(j)$ incident on $\rho^0(h,j,k)$, we can formally write

$$\rho^1(h,j)^+ \bigsqcup \rho^1(j,h)^- \doteq \partial\rho^2(h) \underset{\Gamma}{\bigcap} \partial\rho^2(j). \tag{2.45}$$

A notation, this latter that emphasizes the distinct orientation of the same half–edge of Γ as seen by the two adjacent oriented cells $\rho^2(h)$ and $\rho^2(j)$. Recall

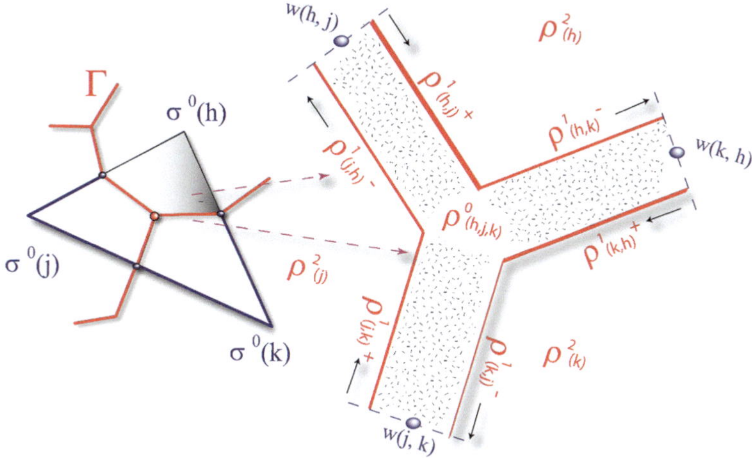

Fig. 2.8 The oriented half–edges of the polytopal ribbon graph Γ around the vertex $\rho^0(h,j,k)$

that the full oriented edge $\rho^1(h,j)$ of Γ shared by $\rho^2(h)$ and $\rho^2(j)$ is provided by $\rho^1(h,j)^- \cup \rho^1(h,j)^+$ as seen from $\rho^2(h)$ and by $\rho^1(j,h)^- \cup \rho^1(j,h)^+$ as seen from $\rho^2(j)$. The complementary half–edge $\rho^1(h,j)^-$, (and its reversed orientation counterpart $\rho^1(j,h)^+$), is incident on some other vertex of Γ and should not be confused with $\rho^1(j,h)^-$! Such a notation may seem unwieldy, but it has the advantage of being algorithmic and easily allows to propagate data along the ribbon graph Γ once we have chosen a labeling for the $N_0(T)$ 2–cells of (P_T, M).

Let $\partial(\rho^2(h))$ be the oriented boundary of $\rho^2(k) \subset (P_T, M)$. The open disk

$$B^2(h) \doteq \{p \in \rho^2(h)\backslash\partial(\rho^2(h))\}, \tag{2.46}$$

centered on the vertex $\sigma^0(k) \in (T_l, M)$, is contained in the star $st(\sigma^0(h))$. Note that any two such balls, say $B^2(h)$ and $B^2(j)$, $h \neq j$, are pairwise disjoint, and that the complex

$$(T_l, M) \Big/ \bigcup_{k=1}^{N_0(T)} B^2(k) \tag{2.47}$$

retracts on the 1-skeleton $K^1 P_T$ of (P_T, M). Let us denote by

$$L(h) = \sum_{\alpha=1}^{q(h)} \left(\widehat{L}(h, \alpha)^- + \widehat{L}(h, \alpha)^+\right) \tag{2.48}$$

the length of the boundary $\partial(\rho^2(h))$ of $\rho^2(h)$, where $\widehat{L}(h, \alpha)^\mp$ are the lengths of the $2\,q(k)$ ordered half–edges $\{\rho^1(h, \alpha)^\mp\} \in \partial(\rho^2(h))$.

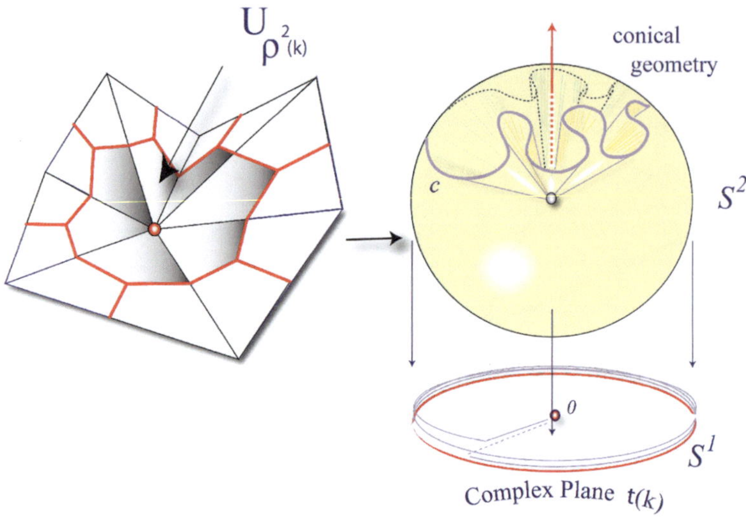

Fig. 2.9 The conical geometry around a vertex $\sigma^0(k)$ can be represented in terms of a complex variable $t(k)$

With these preliminary remarks along the way, we associate to any vertex $\sigma^0(h) \in (T_l, M)$ the complex coordinate $t(h) \in \mathbb{C}$, defined in the open disk $U_{\rho^2(h)} \subset \mathbb{C}$ of unit radius (Fig. 2.9),

$$
B^2(h) \xrightarrow{t(h)} U_{\rho^2(h)} \doteq \{t(h) \in \mathbb{C} \mid |t(h)| < 1\} \tag{2.49}
$$

$$
p \longmapsto t(h)[p] , \quad t(h)[\sigma^0(h)] \equiv 0.
$$

In terms of $t(h)$ we can explicitly write down the singular Euclidean metric locally characterizing the conical Euclidean structure in $B^2(h)$, according to

$$
ds^2_{(h)} := \frac{L^2(h)}{4\pi^2} |t(h)|^{-2\left(\frac{\varepsilon(h)}{2\pi}\right)} |dt(h)|^2 , \tag{2.50}
$$

where $L(h)$ is the length of $\partial(\rho^2(h)) = \partial \overline{B}^2(h)$ and $\varepsilon(h)$ is the corresponding deficit angle in $B^2(h)$. One easily recognizes in (2.50) the metric of a Euclidean cone of total angle $\Theta(h) = 2\pi - \varepsilon(h)$.

If $v_h : B^2 \to \mathbb{R}$ is a continuous function (C^2 on $B^2 - \{\sigma^0(h)\}$) such that, for $t(h) \to 0$, $|t(h)|\frac{\partial v_h}{\partial t(h)} \to 0$ and $|t(h)|\frac{\partial v_h}{\partial \overline{t}(h)} \to 0$, then we can also consider the *conformal class* $\mathscr{C}_{(h)}$ of all metrics possessing the same conical structure of $ds^2_{(h)}$

$$
\mathscr{C}_{(h)} \equiv \left[ds^2_{(h)} \right]. \tag{2.51}
$$

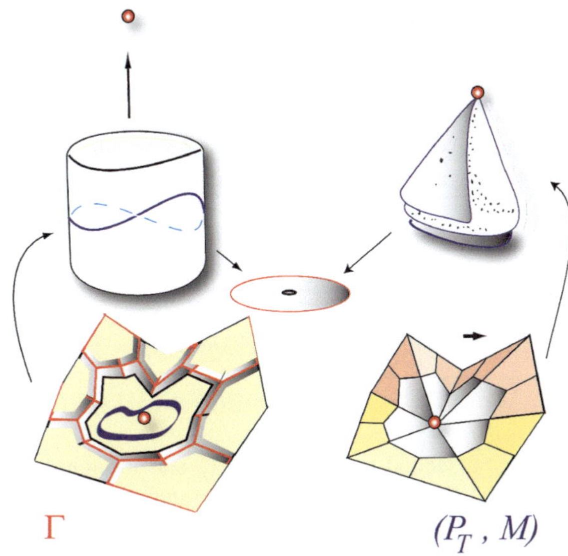

Fig. 2.10 The cylindrical and the conical metric over a polytopal cell are conformally related

In particular we have that $\left(M, \mathscr{C}_{sg}\right)$ can be (locally) represented as

$$ds^2_{(h)} = e^{2v_h} \frac{L^2(h)}{4\pi^2 \, |t(h)|^2} \, |dt(h)|^2, \tag{2.52}$$

where the conformal factor v_h is given by

$$v_h \doteq \left(\frac{\Theta(h)}{2\pi} \right) \ln |t(h)| . \tag{2.53}$$

This explicitly shows that the conical metric is conformal to a smooth metric, (for later convenience, here we have chosen the cylindrical metric on the punctured disk, but we could have chosen the metric $|dt(h)|^2$ as well), and thus establishes a basic observation due to Troyanov [6, 7] (Fig. 2.10).

Proposition 2.3 *The conical singularities are invisible from the conformal viewpoint.*

A somewhat direct consequence of such a remark is that we can, without giving particular analytical attention to the conical singularities, glue together the uniformizations $\{U_{\rho^2(h)}\}_{h=1}^{N_0(T)}$ along the pattern defined by the 1-skeleton of (P_T, M) and generate on M the quasi-conformal structure

$$\left((M; N_0), \mathscr{C}_{sg}\right) \doteq \bigcup_{(P_T, M)} \{U_{\rho^2(h)};\, ds^2_{(h)}\}_{h=1}^{N_0(T)} \tag{2.54}$$

giving (T_l, M) the structure of a Riemann surface with N_0 marked points. As we shall see, the conical singularities will appear, in such a picture, as a decoration of the marked points.

In order to carry out explicitly such a construction, we exploit the basic observation that the conical metric (2.52) is conformal, (see (2.53)), to the cylindrical metric canonically associated with the quadratic differential

$$\phi\big|_{\rho_h^2} := \frac{L^2(h)}{4\pi^2 \, |t(h)|^2} \, |dt(h)|^2, \tag{2.55}$$

on the punctured disk $\Delta_h^* \subset U_{\rho^2(h)}$,

$$\Delta_h^* := \{t(h) \in \mathbb{C} | \, 0 < |t(h)| < 1\}. \tag{2.56}$$

$\phi\big|_{\rho_h^2}$ has a second order pole in Δ_h^* and it naturally extends as a Jenkins–Strebel quadratic differential to the vertex and half–edge uniformizations of the whole ribbon graph Γ. To see this explicitly, let us associate to the oriented half–edge $\rho^1(h,j)^+$ of $\Gamma \cap \rho^2(h)$ a complex coordinate $z(h,j)^+$ defined in the strip

$$U_{\rho^1(h,j)+} \doteq \{z(h,j)^+ \in C | 0 < Re\, z(h,j) < \widehat{L}(h,j)^+\}, \tag{2.57}$$

$\widehat{L}(h,j)^+$ being the length of the half–edge considered. The coordinate $\zeta(h,j,k)$, corresponding to the 3-valent vertex $\rho^0(h,j,k) \in \Gamma \cap \rho^2(h)$, is defined in the open set

$$U_{\rho^0(h,j,k)} \doteq \{\zeta(h,j,k) \in C | \, |\zeta(h,j,k)| < \delta, \ \zeta(h,j,k)[\rho^0(h,j,k)] = 0\}, \tag{2.58}$$

where $\delta > 0$ is a suitably small constant. Finally, the generic two-cell $\rho^2(k)$ is, according to the remarks above, parametrized in the unit disk

$$U_{\rho^2(k)} \doteq \{t(k) \in C | \, |t(k)| < 1, \ t(k)[\sigma^0(k)] = 0\}, \tag{2.59}$$

where $\sigma^0(k)$ is the vertex $\in (T_l, M)$ corresponding to the given two-cell.

The coordinate neighborhoods $\{U_{\rho^0(h,j,k)}\}_{(h,j,k)}^{N_2(T)}$, $\{U_{\rho^1(h,j)+}\}_{(h,j)}^{N_1(T)}$, and $\{U_{\rho^2(k)}\}_{(k)}^{N_0(T)}$ can be coherently glued, along the pattern associated with the ribbon graph Γ, by noticing that to each oriented half–edge $\rho^1(h,j)^+$ we can associate the standard quadratic differential on $U_{\rho^1(h,j)+}$ given by

$$\phi\big|_{\rho^1(h,j)+} = dz(h,j)^+ \otimes dz(h,j)^+. \tag{2.60}$$

Such $\phi\big|_{\rho^1(h,j)+}$ can be extended to the remaining local uniformizations $U_{\rho^2(k)}$, and $U_{\rho^0(h,j,k)}$, by exploiting a classic result in Riemann surface theory according to which a quadratic differential ϕ has a finite number of zeros $n_{zeros}(\phi)$ with orders k_i and a finite number of poles $n_{poles}(\phi)$ of order s_i such that

$$\sum_{i=1}^{n_{zero}(\phi)} k_i - \sum_{i=1}^{n_{pole}(\phi)} s_i = 4g - 4. \tag{2.61}$$

In our case we must have $n_{zeros}(\phi) = N_2(T)$ with $k_i = 1$, (corresponding to the fact that the 1-skeleton of (P_T, M) is a trivalent ribbon graph Γ), and $n_{poles}(\phi) = N_0(T)$ with $s_i = s$ $\forall i$, for a suitable positive integer s. According to such remarks (2.61) reduces to

$$N_2(T) - sN_0(T) = 4g - 4. \tag{2.62}$$

From the Euler relation $N_0(T) - N_1(T) + N_2(T) = 2 - 2g$, and $2N_1(T) = 3N_2(T)$ we get $N_2(T) - 2N_0(T) = 4g - 4$. This is consistent with (2.62) if and only if $s = 2$. Thus the extension ϕ of $\phi(h)|_{\rho^1(h)}$ along the 1-skeleton Γ of (P_T, M) must have $N_2(T)$ zeros of order 1 corresponding to the trivalent vertices $\{\rho^0(h, j, k)\}$ of Γ and $N_0(T)$ quadratic poles corresponding to the polygonal cells $\{\rho^2(k)\}$ of perimeter lengths $\{L(k)\}$, (see (2.55)). Around a zero of order one and a pole of order two, every (Jenkins-Strebel) quadratic differential ϕ has a canonical local structure which (along with (2.60)) is given by [4]

$$(|P_{T_l}| \to M) \to \phi \doteq \begin{cases} \phi|_{\rho^1(h,j)+} = dz(h,j)^+ \otimes dz(h,j)^+, \\[2mm] \phi|_{\rho^0(h,j,k)} = \tfrac{9}{4}\, \zeta(h,j,k)\, d\zeta(h,j,k) \otimes d\zeta(h,j,k), \\[2mm] \phi|_{\rho^2(k)} = -\dfrac{[L(k)]^2}{4\pi^2 t^2(k)}\, dt(k) \otimes dt(k), \end{cases} \tag{2.63}$$

where $\{\rho^0(h, j, k), \rho^1(h, j)^+, \rho^2(k)\}$ runs over the set of vertices, half–edges, and 2-cells of (P_T, M). Since $\phi|_{\rho^1(h,j)+}$, $\phi|_{\rho^0(h,j,k)}$, and $\phi|_{\rho^2(k)}$ must be identified on the non-empty pairwise intersections $U_{\rho^0(h,j,k)} \cap U_{\rho^1(h,j)+}$, $U_{\rho^1(h,j)+} \cap U_{\rho^2(k)}$ we can associate to the polytope (P_T, M) a complex structure $((M; N_0), \mathscr{C}_{sg})$ by coherently gluing, along the pattern associated with the ribbon graph Γ, the local uniformizations $\{U_{\rho^0(h,j,k)}\}^{N_2(T)}$, $\{U_{\rho^1(h,j)+}\}^{2N_1(T)}$, and $\{U_{\rho^2(k)}\}^{N_0(T)}$. Explicitly, let $\{U_{\rho^1(h,j)+}\}$, $\{U_{\rho^1(j,k)+}\}$, $\{U_{\rho^1(k,h)+}\}$ be the three generic open strips associated with the three cyclically oriented half–edges $\rho^1(h, j)^+$, $\rho^1(j, k)^+$, $\rho^1(k, h)^+$ incident on the vertex $\rho^0(h, j, k)$. Then the corresponding coordinates $z(h, j)^+$, $z(j, k)^+$, and $z(k, h)^+$ are related to $\zeta(h, j, k)$ by the transition functions[2] (Fig. 2.11)

---

[2]If we glue directly the flat stripes $z(h, j)^+$, $z(j, k)^+$, and $z(k, h)^+$ by suitably identifying their extremities then we would get the familiar conical singularity of 3π associated with a zero of order 1 of a J–S quadratic differential. Recall that a zero of order k generates a conical singularity given by $\pi(k + 2)$. This conical singularity is rather annoying since it is not directly related with the conical singularities of the polyhedral surface (T_l, M). It can be eliminated by introducing, at the vertex $\rho^0(h, j, k)$, the uniformizing coordinate $\zeta(h, j, k)$ with $\zeta(h, j, k) = [z(k, h)^+]^{2/3}$. This

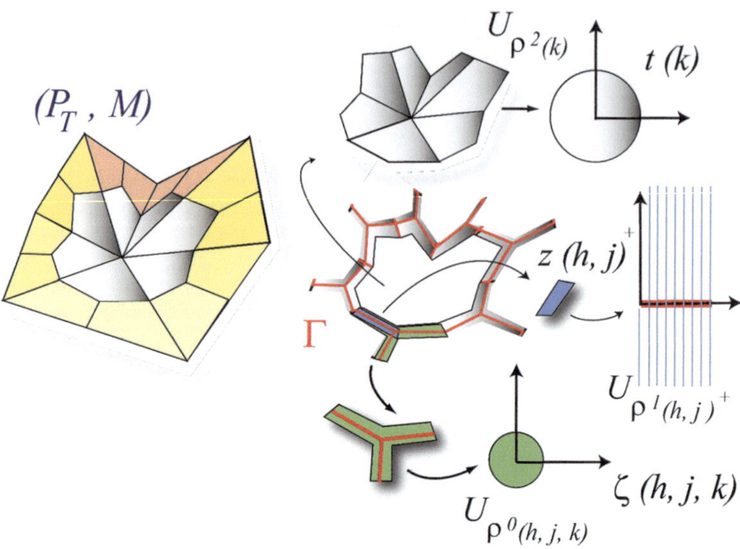

Fig. 2.11 The complex coordinate neighborhoods associated with the polytope (P_T, M)

$$\zeta(h,j,k) = \begin{cases} \left[z(k,h)^+\right]^{\frac{2}{3}}, \\[2ex] e^{\frac{2\pi}{3}\sqrt{-1}}\left[z(h,j)^+\right]^{\frac{2}{3}}, \\[2ex] e^{\frac{4\pi}{3}\sqrt{-1}}\left[z(j,k)^+\right]^{\frac{2}{3}}, \end{cases} \tag{2.64}$$

Similarly, if $\{U_{\rho^1(h,j_\beta)^\pm}\}$, $\beta = 1, 2, \ldots, q(k)$ are the open strips associated with the $q(k)$ (oriented) half edges $\{\rho^1(h,j_\beta)^\pm\}$ boundary of the generic polygonal cell $\rho^2(h)$, then the transition functions between the corresponding coordinate $t(h)$ and the $\{z(h,j_\beta)^\pm\}$ are given by

$$t(h) = \exp\left(\frac{2\pi\sqrt{-1}}{L(h)}\left(\sum_{\beta=1}^{\nu-1}\widehat{L}(h,j_\beta) + z(h,j_\nu)^\pm\right)\right), \quad \nu = 1, \ldots, q(h),$$

$$\tag{2.65}$$

with $\sum_{\beta=1}^{\nu-1} \cdot \doteq 0$, for $\nu = 1$, and where $L(h)$ denotes the perimeter of $\partial(\rho^2(h))$. By iterating such a construction for each vertex $\{\rho^0(h,j,k)\}$ in the polytope (P_T, M) we get a very explicit characterization of a complex structure that we denote by $((M; N_0), \mathscr{C})$.

---

can be seen as another manifestation of Troyanov's basic observation that conical singularities are invisible from the conformal viewpoint.

As for the metrical properties of this complex structure note that for any closed curve $c : \mathbb{S}^1 \to U_{\rho^2(k)}$, homotopic to the boundary of $\overline{U}_{\rho^2(k)}$, we get

$$\oint_c \sqrt{\phi|_{\rho^2(k)}} = L(\partial(\rho^2(k))) := L(k). \tag{2.66}$$

In general, let

$$\left|\phi|_{\rho^2(k)}\right| = \frac{[L(k)]^2}{4\pi^2 |t(k)|^2}\, |d\,t(k)|^2, \tag{2.67}$$

denote the cylindrical metric canonically associated with a quadratic differential with a second order pole. If we define $\Delta_k^r := \{t(k) \in \mathbb{C}|\ r \le |t(k)| \le 1\}$, then in terms of the area element associated with the flat metric $\left|\phi|_{\rho^2(k)}\right|$ we get

$$\int_{\Delta_k^r} \frac{i}{2} \frac{[L(k)]^2}{4\pi^2 |t(k)|^2}\, dt(k) \wedge\ \overline{dt}(k) = \frac{[L(k)]^2}{2\pi} \ln\left(\frac{1}{r}\right), \tag{2.68}$$

which, as $r \to 0^+$, diverges logarithmically. Thus, the (punctured) coordinate neighborhood $U_{\rho^2(k)}$, endowed with the flat metric (2.67), $\left|\phi|_{\rho^2(k)}\right|$, is isometric to a flat semi-infinite cylinder. Even if this latter geometry is perfectly consistent with the metric ribbon graph structure, it is not the natural metric to use since we wish to explicitly keep track of the polyhedral surface (T, M) which generates the given ribbon graph Γ. According to Proposition 2.3, given a normalized deficit angle $\frac{\varepsilon(k)}{2\pi} \doteq 1 - \frac{\Theta(k)}{2\pi}$, the conical geometry (2.52) and the cylindrical geometry (2.67) on the punctured disk $\Delta_k^* \subset U_{\rho^2(k)}$ can be conformally related by choosing the conformal factor e^{2v_k} in (2.52) according to, (see (2.53)),

$$e^{2v_k} = |t(k)|^{2\left(\frac{\Theta(k)}{2\pi}\right)}. \tag{2.69}$$

It follows that the conical metric

$$ds_{(k)}^2 := e^{2v_k}\ \left|\phi|_{\rho^2(k)}\right| \tag{2.70}$$

$$= \frac{[L(k)]^2}{4\pi^2\ |t(k)|^2}\ |t(k)|^{2\left(\frac{\Theta(k)}{2\pi}\right)}\ |d\,t(k)|^2$$

$$= \frac{[L(k)]^2}{4\pi^2}\ |t(k)|^{-2\left(\frac{\varepsilon(k)}{2\pi}\right)}\ |d\,t(k)|^2,$$

can be consistently propagated along the ribbon graph associated with the polytope (P_T, M). We have thus established the

Theorem 2.1 *Let (T_l, M) be a polyhedral surface of genus g with N_0 vertices, and let Γ be the (edge–refinement of the) ribbon graph associated with the corresponding barycentrically dual polytope (P_T, M), then the map*

$$(P_T, M) \longrightarrow \left((M; N_0), \mathscr{C}_{sg}\right) = \tag{2.71}$$

$$= \bigcup_{\{\rho^0(h,j,k)\}}^{N_2(T)} U_{\rho^0(h,j,k)} \bigcup_{\{\rho^1(h,j)^+\}}^{2N_1(T)} U_{\rho^1(h,j)} + \bigcup_{\{\rho^2(k)\}}^{N_0(T)} \left(U_{\rho^2(k)}, \, ds^2_{(k)} := e^{2\nu_k} \left|\phi|_{\rho^2(k)}\right|\right),$$

defines the decorated Riemann surface $((M; N_0), \mathscr{C}_{sg})$, *with N_0 marked points, associated with the singular Euclidean structure defined by the polyhedral surface* (T_l, M).

2.5 Troyanov's Singular Euclidean Structures

Note that $((M; N_0), \mathscr{C}_{sg})$ is a N_0–pointed Riemann surface decorated with the deficit angles $\epsilon(h)$ and the cell–perimeter $L(h)$, associated with the marked point $\sigma^0(h)$ and with the uniformizing 2–cell $\rho^2(h)$, respectively. Since we are dealing with Euclidean triangles, we have $\sum_{k=1}^{N_0(T)} \Theta(k) = \pi N_2(T)$. From the Euler and Dehn–Sommerville relations (1.11) and (1.12), $N_0(T) - N_1(T) + N_2(T) = \chi(T)$ and $2N_1(T) = 3N_2(T)$, we immediately get that the decoration of the marked points of $((M; N_0), \mathscr{C}_{sg})$, by the deficit angles is constrained by

$$\sum_{k=1}^{N_0(T)} \frac{\epsilon(k)}{2\pi} = 2 - 2g = \xi(M), \tag{2.72}$$

which appears here as a version of the Gauss–Bonnet theorem. One can do better than this, by following the strategy so nicely described by M. Troyanov [7]. Let us start by extending the normalized angular deficit function $\frac{\epsilon(k)}{2\pi}$ from the 0–skeleton $K^0(T)$ to the whole triangulated surface (T_l, M) by setting

$$\frac{\epsilon(x)}{2\pi} := - \left(\frac{\Theta(x)}{2\pi} - 1\right), \quad \forall x \in K^0(T)$$

$$\tag{2.73}$$

$$\frac{\epsilon(x)}{2\pi} := 0, \quad \forall x \notin K^0(T).$$

Correspondingly we can define a flat surface (M, ds^2) with a discrete set of isolated conical singularities of deficit angle $\frac{\epsilon(x)}{2\pi}$ to be a surface of genus g, M, endowed with a Riemannian metric ds^2 which is isometric to the conical metric (2.50) in the neighborhood of every point $x \in M$, where the deficit function $\frac{\epsilon(x)}{2\pi} \in (-\infty, 1)$ is provided by (2.73). The deficit angle decoration of $((M; N_0), \mathscr{C}_{sg})$ can be naturally summarized [6, 7] in a formal linear combination of the points $\{\sigma^0(k)\}$ with coefficients given by the corresponding deficit angles (normalized to 2π).

Definition 2.4 ([7]) The *real divisor* associated with the Riemann surface $((M; N_0), \mathscr{C}_{sg})$ defined by (T_l, M) is the formal linear combination

$$Div(T) := \sum_{k=1}^{N_0(T)} \left(-\frac{\varepsilon(k)}{2\pi} \right) \sigma^0(k) = \sum_{k=1}^{N_0(T)} \left(\frac{\Theta(k)}{2\pi} - 1 \right) \sigma^0(k) \tag{2.74}$$

supported on the set of vertices $\{\sigma^0(i)\}_{i=1}^{N_0(T)}$.
According to (2.72), the degree $|Div(T)|$ of the divisor is provided by

$$|Div(T)| := \sum_{k=1}^{N_0(T)} \left(\frac{\Theta(k)}{2\pi} - 1 \right) = -\chi(M) . \tag{2.75}$$

The real divisor $|Div(T)|$ characterizes the Euler class of the Riemann surface with conical singularities $((M; N_0), \mathscr{C}_{sg})$, (hence of (T_l, M)), and yields for a corresponding Gauss-Bonnet formula. Explicitly, the Euler number associated with $((M; N_0), \mathscr{C}_{sg})$ is defined, [6, 7], by

$$e(T_l, M) := \chi(M) + |Div(T)|. \tag{2.76}$$

and the Gauss-Bonnet formula reads [6, 7]:

Theorem 2.2 (Troyanov) (Gauss-Bonnet for Riemannian surfaces with conical singularities) *Let $((M; N_0), \mathscr{C}_{sg})$ be a Riemann surface with conical singularities described by the divisor*

$$Div(T) \doteq \sum_{k=1}^{N_0(T)} \left(\frac{\Theta(k)}{2\pi} - 1 \right) \sigma^0(k), \tag{2.77}$$

associated with the marked points (vertices) $\{\sigma^0(k)\}_{k=1}^{N_0(T)}$. Let ds^2 be the conformal metric (2.50) representing the divisor $Div(T)$, and denote by K the Gaussian curvature function defined on the complement of the support of the divisor $Div(T)$, (i.e. on $(T_l, M) - \{\sigma^0(k)\}_{k=1}^{N_0(T)}$). Then

$$\frac{1}{2\pi} \int_M K dA = e(T_l, M), \tag{2.78}$$

where dA is the area element of the local metric ds^2.
According to (2.75) $e(T_l, M) = 0$, and the Gauss-Bonnet formula implies

$$\frac{1}{2\pi} \int_M K dA = 0. \tag{2.79}$$

Thus, the Riemann surface $((M; N_0), \mathscr{C}_{sg})$ associated with a Euclidean triangulation (T_l, M) naturally carries a conformally flat structure. Clearly this is a rather obvious result, (since the metric in $(T_l, M) - \{\sigma^0(i)\}_{i=1}^{N_0(T)}$ is flat). However, it admits a very deep converse proved by M. Troyanov [7, 8], (see also the prescient paper by E. Picard [5]):

Theorem 2.3 (Troyanov) *Let $((M, \mathscr{C}), Div)$ be a Riemann surface $(M, \mathscr{C})$, with a divisor such that $e(M, Div) = 0$. Then there exists on M a unique (up to homothety) conformally flat metric ds^2 with $ds^2 \in \mathscr{C}$ representing the divisor Div, and depending smoothly from the conformal structure $\mathscr{C}$ and the divisor Div. Moreover, given a compact oriented surface M, there are natural bijections between the following structures:*

(i)   The set of geometric equivalence classes of Euclidean triangulations (T_l, M) on M, defined up to homothety;

(ii) The set of flat metrics ds^2 on M with conical singularities up to homothety;

(iii) The set of conformal structures $\mathscr{C}$ on M decorated by a finite real divisor Div, associated with normalized deficit angles $\{\frac{\epsilon(k)}{2\pi}\} < 1$, with vanishing Euler class $e(T_l, M)$.

Even if we are not going to prove directly Troyanov's theorem here, a few comments are in order.

Let $((M; N_0), \mathscr{C}_{sg})$ be the pointed Riemann surface associated with a polyhedral surface (T, M). Denote by $\{p_k\}_{k=1}^{N_0} \in (M; N_0)$ the marked points associated with the vertices $\{\sigma^0(k)\} \in (T, M)$. The conformal class $[ds^2] := [e^{2v} |\phi||]$ of the conical metric representing $((M; N_0), \mathscr{C}_{sg})$ can be written, around the generic marked point p_k, as

$$\left[ds^2_{(k)}\right] = \left[e^{2v_k} \left|\phi\right|_{\rho^2(k)}\right|\right] = e^{2V_k} ds^2_0, \tag{2.80}$$

where ds^2_0 is a smooth metric on M, and where the conformal factor V_k is provided by

$$V\big|_{U_{\rho^2(k)}} = -\frac{\varepsilon(k)}{2\pi} \ln |t(k)| + u. \tag{2.81}$$

The function u is assumed continuous and C^2 on $U_{\rho^2(k)} - p_k$ and is such that, for $t(k) \to 0$, $|t(k)| \frac{\partial u}{\partial t(k)}$, and $|t(k)| \frac{\partial u}{\partial \bar{t}(k)}$ both $\to 0$, (see (2.52)). Note that over the Riemann surface $((M; N_0), \mathscr{C}_{sg})$ associated with a polyhedral surface (T_l, M), we can always (locally!) take the flat metric $|dt(k)|^2$ as the reference ds^2_0. However, for later use, is more profitable to consider the general case, not assuming a priori the local flatness of ds^2_0. The Gaussian curvature K of ds^2, defined on $(T_l, M) - \{\sigma^0(k)\}_{k=1}^{N_0(T)}$, is related to the Gaussian curvature K_0 of the smooth metric ds^2_0 by the relation

$$K dA = K_0 dA_0 - d * dV, \tag{2.82}$$

where dA_0 and $dA = e^{2V} dA_0$ are the area elements of ds_0^2 and ds^2, respectively. By interpreting $d * dV$ as a $(1, 1)$ current on $((M; N_0), \mathscr{C}_{sg})$, a direct computation yields

$$d * dV = \pi \sum_{k=1}^{N_0} \left(-\frac{\varepsilon(k)}{2\pi} \right) \left[\bar{\partial} \left(\frac{dt(k)}{2\pi \sqrt{-1}t(k)} \right) - \partial \left(\frac{d\bar{t}(k)}{2\pi \sqrt{-1}\bar{t}(k)} \right) \right] + 2\sqrt{-1}\partial\bar{\partial}u,$$

$$(2.83)$$

$$= -2\pi \sum_{k=1}^{N_0} \left(-\frac{\varepsilon(k)}{2\pi} \right) \frac{\sqrt{-1}}{2} \delta_{p_k} dt(k) \wedge d\bar{t}(k) + 2\sqrt{-1}\partial\bar{\partial}u.$$

From which we get the (inhomogeneous) Liouville equation associated with the conformal metric ds^2 with conical singularities $\{-\frac{\varepsilon(k)}{2\pi}\}$, i.e.

$$-4\partial_t\partial_{\bar{t}}u = e^{2V}K - K_0 - 2\pi \sum_{k=1}^{N_0} \left(-\frac{\varepsilon(k)}{2\pi} \right) \delta_{p_k}, \qquad (2.84)$$

where δ_{p_k} is the Dirac distribution supported on the marked points p_k. Note that from $\frac{1}{2\pi} \int_M K_0 dA_0 = \chi(M)$ and (2.72), we get, upon integrating (2.84) over M,

$$\frac{1}{2\pi} \int_M K \, dA = \frac{1}{2\pi} \int_M e^{2V} K dA_0 = \frac{1}{2\pi} \int_M K_0 dA_0 + \sum_{k=1}^{N_0} \left(-\frac{\varepsilon(k)}{2\pi} \right) = 0, \qquad (2.85)$$

in agreement with the vanishing of the Euler class (2.78) of $((M; N_0), \mathscr{C})$.

According to Theorem 2.1 we can associate with a polyhedral surface (T_l, M) a decorated Riemann surface $((M; N_0), \mathscr{C}_{sg})$. This suggests that there may be a direct correspondence between the space of polyhedral surfaces $POL_{g, N_0}(M)$ and the moduli space $\mathfrak{M}_{g, N_0}$ suitably decorated. This is indeed the case, and we have the following

Theorem 2.4 *There is a bijective map*

$$\Upsilon \; : \; POL_{g, N_0}(M) \longrightarrow \mathfrak{M}_{g, N_0} \times \mathbb{R}_+ \times \mathbb{R}_{\geq 0}^{N_0 - 1}, \qquad (2.86)$$

$$(T_l, M) \longmapsto \left[((M; N_0), \mathscr{C}) \, , a \, , \{1 - \frac{\epsilon(k)}{2\pi} \}^{N_0} \right],$$

where $\mathfrak{M}_{g, N_0}$ is the moduli space of Riemann surfaces of genus g with N_0 marked points, $a := A(T_l, M) > 0$ is the (Euclidean) area of (T_l, M), and $\{\epsilon(k)\}_{k=1}^{N_0}$ is the sequence (satisfying the constraint $\sum_{k=1}^{N_0} \epsilon(k) = 2 - 2g$), of the deficit angles of (T_l, M).

Proof The existence of a correspondence, defined up to a homothety, between $\mathfrak{M}_{g, N_0} \times \mathbb{R}_{\geq 0}^{N_0 - 1}$ and $POL_{g, N_0}(M)$ follows from Troyanov's Theorem 2.3.

Conversely, to a polyhedral surface $(T_l, M) \in POL_{g,N_0}(M)$ of given area $A(T_l, M) \equiv a$ there corresponds a well–defined Riemann surface $((M; N_0), \mathscr{C}_{sg})$ with area $A\left[((M; N_0), \mathscr{C}_{sg})\right] = a$, decorated with the real divisor $Div(T) := \sum_{k=1}^{N_0} \left(\frac{\epsilon(k)}{2\pi}\right) \sigma^0(k)$ defined by the deficit angles $\{\epsilon(k)\}$. In order to prove that the map Υ so defined is a bijection we observe that $(T_l, M) \in POL_{g,N_0}(M)$ characterizes a unique 3–valent metric ribbon graph Γ decorated with the cell–perimeters $\{L(k)\}^{N_0}$ associated with the dual polytope (P_T, M). Viceversa, to $((M, \mathscr{C}), Div, |(M, \mathscr{C})| = a)$ there corresponds a unique polyhedral surface $\in POL_{g,N_0}(M)$ decorated with a corresponding 3–valent metric ribbon graph Γ with N_0 labelled boundary components of given lengths $\{L(k)\}^{N_0}$, and with the conical structure defined by the divisor Div. We can identify the pair (Γ, Div) with a conical polytope (P_T, M). Since 3–valent metric ribbon graphs are dense in the space of all metric ribbon graphs RG_{g,N_0}^{met}, the map Υ naturally descends to the map $\psi : \mathfrak{M}_{g,N_0} \times \mathbb{R}_+^{N_0} \to RG_{g,N_0}^{met}$, between the $\{L(k)\}^{N_0}$–decorated moduli space $\mathfrak{M}_{g,N_0} \times \mathbb{R}_+^{N_0}$ and the space of all metric ribbon graphs Γ with N_0 labeled boundary components, defined by Strebel theorem. It is well–known that this latter map is a bijection between the two orbifold spaces $\mathfrak{M}_{g,N_0} \times \mathbb{R}_+^{N_0}$ and RG_{g,N_0}^{met}, (the paper [4] provides an in depth analysis of this result).

2.6 Chern and Euler Classes Over $POL_{g,N_0}(M)$

The above result characterizes the spaces of polyhedral surfaces $POL_{g,N_0}(M)$ as a local covering for $\mathfrak{M}_{g,N_0}$. This is strongly related with the familiar covering of $\mathfrak{M}_{g,N_0}$ generated by ribbon graphs via Strebel theory, (see Theorem B.2), and which plays a basic role in Witten–Kontsevich theory [2]. In full analogy with this latter case, we have a natural correspondence between the circle bundles $\{\mathscr{Q}_{(k)}\}$ decorating $POL_{g,N_0}(M)$ and the line bundles $\{\mathscr{L}_i\}$ naturally defined over $\mathfrak{M}_{g,N_0}$ by the cotangent space $T_i^* M$ at the i-th marked point of $((M; N_0), \mathscr{C}_{sg})$, (see Appendix B).

In order to make this correspondence more explicit, let us consider in $POL_{g,N_0}(M)$ the subset $POL_{g,N_0}(M, \{\Theta(k)\}, A(M))$ of polyhedral structures of given area $A(M)$, and with a given sequence of conical angles $\{\Theta(k)\}_{k=1}^{N_0}$. According to the representation provided in Proposition 1.4, this is an orbifold

$$POL_{g,N_0}(M, \{\Theta(k)\}, A(M)) := \left. \coprod_{[T] \in \mathscr{T}_{g,N_0}(M)} \frac{\mathbb{R}_+^{N_1}}{Aut(T)} \right|_{(\{\Theta(k)\}, A(M))}, \tag{2.87}$$

of dimension

$$\dim\left[POL_{g,N_0}(M)\right]\Big|_{(\{\Theta(k)\}, A(M))} = 2N_0 + 6g - 6, \tag{2.88}$$

locally modeled by the polyhedral surfaces in

$$\frac{\mathscr{T}_{g,N_0}^{met}\left(M,\{\Theta(k)\}\right)}{Aut(T)},\tag{2.89}$$

with given area $A(T_l, M) = A(M)$.

As a consequence of Lemma 1.2, the space $POL_{g,N_0}\left(M, \{\Theta(k)\}, A(M)\right)$ comes with the natural decoration provided by the N_0 circle bundles $\{\mathscr{Q}_{(k)}\}$ associated with the sequence of polyhedral cotangent cones $\{\mathscr{Q}_{q(k)}(\mathscr{S}^2)\}_{k=1}^{N_0}$ defined over the N_0 vertices of the polyhedral surface (T_l, M). Under the bijective map Υ defined by Theorem 2.4, these circle bundles naturally correspond to the N_0 line bundles $\{\mathscr{L}_i\}$ over $\mathfrak{M}_{g,N_0}$ defined by the cotangent space $T_i M$ at the i-th marked point of $((M; N_0), \mathscr{C}_{sg})$, i.e., for the generic $k = 1, \ldots, N_0$, we can write

$$\mathscr{Q}_{(k)} = \Upsilon^*\left(\mathscr{L}_k\right).\tag{2.90}$$

Where the bundle pull-back action simply corresponds to decorate $\mathscr{L}_k$ with the conical metric induced by vertex structure around the given $\sigma^0(k) \in (T_l, M)$. Similarly, the first Chern class $c_1(\mathscr{L}_k)$ of $\mathscr{L}_k$, (see Appendix B), is naturally mapped into the Euler class of $\mathscr{Q}_{(k)}$. We have

Theorem 2.5 *The Euler class of the circle bundle $\mathscr{Q}_{(k)}$ is the pull–back, under the map Υ, of the first Chern class $c_1(\mathscr{L}_k)$ of the line bundle $\mathscr{L}_k$ over $\mathfrak{M}_{g,N_0}$,*

$$\frac{\sqrt{-1}}{2\pi}\,\Omega_{(k)} = \Upsilon^*\left(c_1(\mathscr{L}_k)\right),\tag{2.91}$$

and we can write

$$\sum_{1\leq\alpha<\beta\leq q(k)-1} d\left(\frac{\theta_\alpha(k)}{\Theta(k)}\right) \wedge d\left(\frac{\theta_\beta(k)}{\Theta(k)}\right) = -\pi_k^*\left[\Upsilon^*\left(c_1(\mathscr{L}_k)\right)\right],\tag{2.92}$$

where π_k is the polyhedral map defined by (1.96).

Proof The first part of the theorem follows from (2.90), whereas the explicit expression (2.92) is an obvious consequence of Theorem 1.5. ∎

References

1. Carfora, M., Dappiaggi, C., Gili, V.L.: Triangulated surfaces in twistor space: a kinematical set up for open/closed string duality. JHEP **0612**, 017 (2006). arXiv:hep-th/0607146
2. Kontsevich, M.: Intersection theory on the moduli space of curves and the matrix Airy functions. Commun. Math. Phys. **147**, 1–23 (1992)

3. Looijenga, E.: Intersection theory on Deligne-Mumford compactifications. Séminaire Bourbaki 768 (1992–1993)
4. Mulase, M., Penkava, M.: Ribbon graphs, quadratic differentials on Riemann surfaces, and algebraic curves defined over $\overline{\mathbb{Q}}$. Asian J. Math. **2**(4), 875–920 (1998). math-ph/9811024 v2
5. Picard, E.: De l'integration de l'equation $\Delta u = e^u$ sur une surface de Riemann fermee'. Crelle's J. **130**, 243 (1905)
6. Troyanov, M.: Les surfaces euclidiennes a' singularites coniques. L'Enseignment Mathematique **32**, 79 (1986)
7. Troyanov, M.: Prescribing curvature on compact surfaces with conical singularities. Trans. Am. Math. Soc. **324**, 793 (1991)
8. Troyanov, M.: On the moduli space of singular Euclidean surfaces. In: Papadopoulos, A. (ed.) Handbook of Teichmuller Theory, Volume I. IRMA Lectures in Mathematics and Theoretical Physics, vol. 11, pp. 507–540. European Mathematical Society (EMS), Zurich (2007). arXiv:math/0702666v2 [math.DG]

Chapter 3
Polyhedral Surfaces and the Weil–Petersson Form

Let $\overline{\mathfrak{M}}_{g,N_0}$ denote the Deligne–Mumford compactification of the moduli space $\mathfrak{M}_{g,N_0}$ of N_0–pointed Riemann surfaces of genus g, (see Appendix B). It is well–known that the Chern classes $\{c_1(\mathscr{L}_k)\}$ introduced in the previous chapter can be used to define the Witten–Kontsevich intersection theory over $\overline{\mathfrak{M}}_{g,N_0}$. In such a setting it is also possible [9, 20] to characterize various relevant properties of the Weil–Petersson volume of $\overline{\mathfrak{M}}_{g,N_0}$. Such a connection is rather involved and deeply related to the algebraic-geometrical subtleties of Witten–Kontsevich theory. Thus, it comes as a pleasant surprise that the conical geometry of polyhedral surface allows to explicitly construct a representative of the Weil-Petersson form ω_{W-P} on the space of polyhedral structures with given conical singularities $POL_{g,N_0}(M, \{\Theta(k)\}, A(M))$, (to our knowledge this connection first appeared in [4]; a similar property has been proved for ribbon graphs by G. Mondello in the remarkable papers [11, 12], and recently by other authors, see e.g. [6]). In order to construct such a combinatorial representative of ω_{W-P} we exploit the connection between similarity classes of Euclidean triangles and the triangulations of 3–manifolds by ideal tetrahedra. This is a well–known property in hyperbolic geometry, (see e.g. [3]), that we are going to describe in some detail since it will play a basic role in connecting the quantum geometry of polyhedral surfaces to 3–dimensional manifolds.

3.1 Horospheres in $\mathbb{H}^3$

To set the stage, let $\mathbb{H}^3$ denote the 3-dimensional hyperbolic space thought of as the subspace of Minkowski spacetime $(M^4, \langle \cdot, \cdot \rangle)$ defined by [3]

$$\mathbb{H}^3 = \left\{ \vec{x} \doteq (x^0, x^1, x^2, x^3) \mid \langle \vec{x}, \vec{x} \rangle = -1, \ x^0 > 0 \right\}, \tag{3.1}$$

© Springer International Publishing AG 2017
M. Carfora, A. Marzuoli, *Quantum Triangulations*, Lecture Notes in Physics 942,
https://doi.org/10.1007/978-3-319-67937-2_3

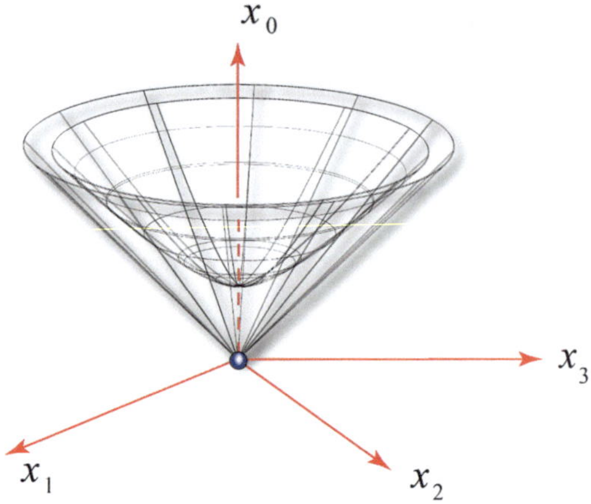

Fig. 3.1 A two-dimensional rendering of the 3-dimensional hyperbolic space $\mathbb{H}^3$ realized as a subspace of Minkowski spacetime

and equipped with the induced Riemannian metric defined by the restriction to the tangent spaces $T_x\mathbb{H}^3$ of the standard Minkowski inner product (Fig. 3.1)

$$\langle \vec{x}, \vec{y} \rangle \doteq -x^0 y^0 + x^1 y^1 + x^2 y^2 + x^3 y^3. \tag{3.2}$$

Recall that the group of orientation preserving isometries of $\mathbb{H}^3$ can be identified with the group $PSL(2,\mathbb{C})$ which acts transitively on $\mathbb{H}^3$ with point stabilizer provided by $SU(2)$. Let $x \in \mathbb{H}^3$ and $\vec{y} \in T_x\mathbb{H}^3$ with $\langle \vec{y}, \vec{y} \rangle = 1$, then the geodesic in $\mathbb{H}^3$ starting at x with velocity $\vec{y}$ is traced by the intersection of $\mathbb{H}^3$ with the two-dimensional hyperplane of M^4 generated by the position vector $\vec{x}$ and the velocity $\vec{y}$, and is described by the mapping

$$\mathbb{R} \ni t \longmapsto \gamma(t) = \cosh(t)\,\vec{x} + \sinh(t)\,\vec{y}. \tag{3.3}$$

Let $\gamma(\infty)$ denote the endpoint of γ on the sphere at infinity $\partial\mathbb{H}^3 \simeq \mathbb{S}^2$, a closed horosphere centered at $\gamma(\infty)$ is a closed surface $\Sigma \subset \mathbb{H}^3$ which is orthogonal to all geodesic lines in $\mathbb{H}^3$ with endpoint $\gamma(\infty)$. In particular, two horospheres with centre at the same point at infinity are at a constant distance. Note that, as a set, the horospheres can be parametrized by future-pointing null vectors belonging to the future light-cone

$$\mathbb{L}^+ \doteq \{\vec{x} \doteq (x^0, x^1, x^2, x^3) \mid \langle \vec{x}, \vec{x} \rangle = 0,\ x^0 > 0\}, \tag{3.4}$$

by identifying the generic horosphere Σ_w with the intersection between $\mathbb{H}^3$ and the null (affine) hyperplane $\langle \vec{y}, \vec{w} \rangle = -1/\sqrt{2}$ defined by the null vector $\vec{w}$ (Fig. 3.2), i.e.,

$$\vec{w} \longmapsto \Sigma_w \doteq \left\{ y \in \mathbb{H}^3 \mid \langle \vec{y}, \vec{w} \rangle = -\frac{1}{\sqrt{2}},\ \langle \vec{w}, \vec{w} \rangle = 0 \right\}. \tag{3.5}$$

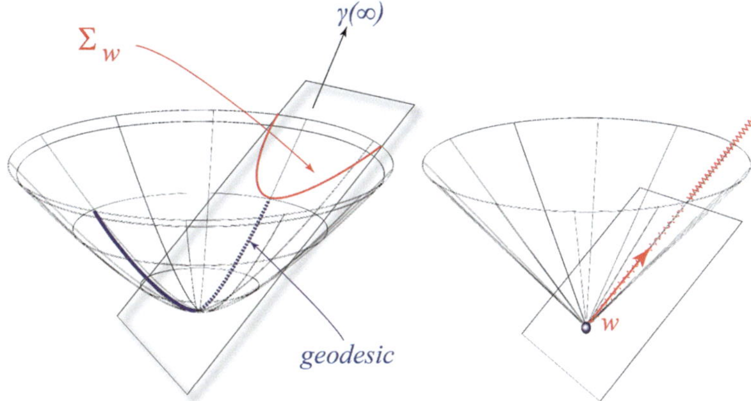

Fig. 3.2 A horosphere Σ_w can be realized as the non–void section of $\mathbb{H}^3$ with a null plane parallel to a corresponding tangent plane to the future light–cone $\mathbb{L}^+$. It follows that $\mathbb{L}^+$ can be used to parametrize the set of all horospheres. The figure also shows the geodesic $t \mapsto \gamma(t)$ orthogonal to Σ_w and whose endpoint $\gamma(\infty)$ provides the center of Σ_w

More explicitly, (see e.g., [7]), given a null (position) vector $\vec{w} := \overrightarrow{OP} \in \mathbb{L}^+$, the affine plane

$$\left\{ y \in M^4 \mid \langle \vec{y}, \vec{w} \rangle = -\frac{1}{\sqrt{2}}, \ \langle \vec{w}, \vec{w} \rangle = 0 \right\}, \tag{3.6}$$

is parallel to the plane $\left\{ y \in M^4 \mid \langle \vec{y}, \vec{w} \rangle = 0 \right\}$ providing the tangent space $T_w \mathbb{L}^+$ of $\mathbb{L}^+$ at P, and it has Σ_w as a non void intersection with $\mathbb{H}^3$. Conversely, given a horosphere Σ in $\mathbb{H}^3$, this can be always characterized as the intersection of $\mathbb{H}^3$ with an affine hyperplane parallel to a null–hyperplane, tangent to $\mathbb{L}^+$ along a future–pointing light ray issuing from the origin. This light ray characterizes a unique null (position) vector $\vec{w}$ such that (3.6) holds, (note that we are here following the normalization in [17]; this provides the correct relation between λ and H lengths. This normalization differs from the standard one in [15] and adopted in [4], giving results which are off by factors of $\sqrt{2}$).

The one–to–one correspondence between null (position) vectors and horospheres shows that future–pointing light rays in $\mathbb{L}^+$ characterize the set of parallel horospheres and allows to associate a natural functional with any pair of horospheres Σ_u and Σ_v according to

$$\lambda\left(\Sigma_u, \Sigma_v\right) \doteq \sqrt{-\langle \vec{u}, \vec{v} \rangle}. \tag{3.7}$$

the quantity $\lambda\left(\Sigma_u, \Sigma_v\right)$ defines the *lambda length* [15] between Σ_u and Σ_v. If $\gamma(p, q)$ denotes the unique geodesic in $\mathbb{H}^3$ connecting the respective centers p and q of Σ_u and Σ_v, then $\lambda\left(\Sigma_u, \Sigma_v\right)$ can be related do the signed geodesic distance

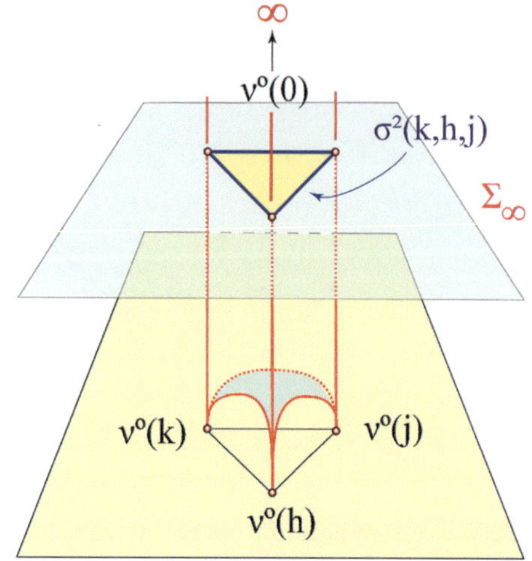

Fig. 3.3 The ideal tetrahedron σ_{hyp}^3 in $\mathbb{H}_{up}^{3,+}$. The intersection between σ_{hyp}^3 and a horosphere Σ_∞, centered at $v^0(0) \doteq \infty$, is a Euclidean triangle $T_\infty(\sigma_{hyp}^3) \equiv \sigma^2(k,h,j)$ in the plane Σ_∞

$\delta(u,v)$ between the intersection points $\gamma(p,q) \cap \Sigma_u$ and $\gamma(p,q) \cap \Sigma_v$, according to

$$\lambda\left(\Sigma_u, \Sigma_v\right) = \sqrt{e^{\delta(u,v)}}, \tag{3.8}$$

($\delta(u,v)$ is by convention < 0 if Σ_u and Σ_v cross each other).

To discuss the connection between polyhedral surfaces and hyperbolic geometry, it will be convenient to represent $\mathbb{H}^3$ by the upper half-space model $\mathbb{H}_{up}^{3,+}$, i.e. as the open upper half space $\{(X,Y,Z) \in \mathbb{R}^3 \mid Z > 0\}$ endowed with the Poincaré metric $Z^{-2}(dX^2 + dY^2 + dZ^2)$. The boundary of $\mathbb{H}^3$ is here provided by $\partial\mathbb{H}_{up}^{3,+} = (\mathbb{R}^2 \times \{0\}) \cup \{\infty\}$, and, up to isometries, we can always map a given point p to ∞. Geodesics in the half-space model are obtained by parametrization of vertical lines $\{x\} \times \mathbb{R}_+$ and circles orthogonal to $\mathbb{R}^2 \times \{0\}$. In particular, since geodesics with end point ∞ are vertical lines, it easily follows that in $\mathbb{H}_{up}^{3,+}$ the horospheres (centered at ∞) are horizontal hyperplanes. It is also worthwhile recalling that the hyperbolic distance between two points p, and $q \in \mathbb{H}^3$ is explicitly provided in $\mathbb{H}_{up}^{3,+}$ by, (see e.g., [3], Corollary A.5.8),

$$d_{\mathbb{H}^3}(p,q) = 2\tanh^{-1}\left[\frac{(X_p - X_q)^2 + (Y_p - Y_q)^2 + (Z_p - Z_q)^2}{(X_p - X_q)^2 + (Y_p - Y_q)^2 + (Z_p + Z_q)^2}\right]^{\frac{1}{2}}. \tag{3.9}$$

In particular, if we take any two geodesics l_1 and l_2 with end-point ∞ and evaluate their hyperbolic distance $d_{\mathbb{H}^3}(l_1, l_2)$ along the horospheres $\Sigma_1 \doteq \{z = t_1\}$ and $\Sigma_2 \doteq \{z = t_2\}$, with $t_2 > t_1$, separated by a distance $d_{\mathbb{H}^3}(\Sigma_1, \Sigma_2)$, then we get the useful relation, (see e.g., section 3.8 of the electronic version of [18]),

$$d_{\mathbb{H}^3}(l_1, l_2)|_{\Sigma_2} = d_{\mathbb{H}^3}(l_1, l_2)|_{\Sigma_1} e^{d_{\mathbb{H}^3}(\Sigma_1, \Sigma_2)}. \tag{3.10}$$

3.2 Ideal Tetrahedra in $\mathbb{H}^{3,+}_{up}$

Let $\sigma^3_{hyp} \doteq (v^0(0), v^0(k), v^0(h), v^0(j))$ be an ideal simplex in $\mathbb{H}^{3,+}_{up}$, i.e., a simplex whose faces are hyperbolic triangles, edges are geodesics, and with vertices lying on $\partial\mathbb{H}^{3,+}_{up}$ (Fig. 3.3). In order to describe the basic properties of σ^3_{hyp} recall that, up to isometries of $\mathbb{H}^{3,+}_{up}$, we can always assume that one of its four vertices, say $v^0(0)$, is at the point ∞ whereas the remaining three $v^0(k)$, $v^0(h)$, and $v^0(j)$ lie on the circumference intersection of $\mathbb{R}^2 \times \{0\}$ with a Euclidean half-sphere $\mathbb{D}^2_r$ of radius r and centre $c \in \{(X, Y, Z) \in \mathbb{R}^3 \,|\, Z = 0\}$. Note that $\mathbb{D}^2_r$ inherits from $\mathbb{H}^{3,+}_{up}$ the structure of a two-dimensional hyperbolic space and that, consequently the simplex $\sigma^2_{hyp} \doteq (v^0(k), v^0(h), v^0(j))$, providing the two-dimensional face of $\sigma^3_{hyp} \doteq (v^0(0), v^0(k), v^0(h), v^0(j))$ opposite to the vertex $v^0(0) \simeq \infty$, is itself an ideal simplex in $\mathbb{D}^2_r$. Denote by $\Delta_\infty(v^0(0))$ the intersection between σ^3_{hyp} and a horosphere Σ_∞ centered at $v^0(0) \doteq \infty$ and sufficiently near to $v^0(0)$. Since all horospheres are congruent, Σ_∞ can be mapped onto a horizontal plane $z = t$ $\subset \mathbb{H}^{3,+}_{up}$ by a conformal mapping fixing ∞, to the effect that $\Delta_\infty(v^0(0))$ is a Euclidean triangle $T_\infty(\sigma^3_{hyp}) \equiv \sigma^2(k, h, j)$ in the plane of the horosphere. This latter remark implies that the vertex angles $(\theta_{jkh}, \theta_{khj}, \theta_{hjk})$ of $T_\infty(\sigma^3_{hyp})$ can be identified with the inner dihedral angles at the three edges $v^1(\infty, k)$, $v^1(\infty, h)$, and $v^1(\infty, j)$ of σ^3_{hyp}, i.e.,

$$\theta_{jkh} \longmapsto \phi_{\infty k} \doteq \angle\left[v^2(0, j, k), v^2(0, k, h)\right], \tag{3.11}$$

$$\theta_{khj} \longmapsto \phi_{\infty h} \doteq \angle\left[v^2(0, k, h), v^2(0, h, j)\right],$$

$$\theta_{hjk} \longmapsto \phi_{\infty j} \doteq \angle\left[v^2(0, h, j), v^2(0, j, k)\right],$$

where $v^2(., ., .)$ denote the faces of σ^3_{hyp}. It is easy to prove, again by intersecting σ^3_{hyp} with horospheres Σ_k, Σ_h, Σ_j sufficiently near to the respective vertices $v^0(k)$, $v^0(h)$, $v^0(j)$, that dihedral angles along opposite edges in σ^3_{hyp} are pairwise equal $\phi_{\infty k} = \phi_{hj}$, $\phi_{\infty h} = \phi_{jk}$, $\phi_{\infty j} = \phi_{kh}$. This implies that the (Euclidean) triangles cut by the horospheres Σ_k, Σ_h, Σ_j are all similar to $T_\infty(\sigma^3_{hyp})$ (Fig. 3.4). In particular, note that the geometrical realizations of the simplices

$$\sigma^2_{hyp}(k, h, j) \doteq v^2(k, h, j), \tag{3.12}$$

$$\sigma^2_{hyp}(\infty, k, h) \doteq v^2(0, k, h),$$

$$\sigma^2_{hyp}(\infty, j, k) \doteq v^2(0, j, k),$$

are ideal triangles in $\mathbb{H}^3$. It follows that the above construction is independent from the choice of which of the four vertices of σ^3_{hyp} is mapped to ∞ and we can parametrize the ideal tetrahedra σ^3_{hyp} in $\mathbb{H}^{3,+}_{up}$ in terms of the similarity class

$[\sigma^2(k,h,j)]$ of the associated Euclidean triangle $T(\sigma^3_{hyp})$: any two ideal tetrahedra σ^3_{hyp} in $\mathbb{H}^{3,+}_{up}$ are congruent iff the associated triangles $T(\sigma^3_{hyp})$ are similar. This is in line with the basic property of $\mathbb{H}^3$ according to which if a diffeomorphism of $\mathbb{H}^3$ preserves angles then it also preserves lengths.

3.3 A Sky–Mapping for Polyhedral Surfaces

The above remarks concerning the connection between Euclidean triangles and ideal tetrahedra σ^3_{hyp} suggest that it may be worthwhile to consider the decoration of each triangle $\sigma^2(k,h,j)$ and of each vertex $\sigma^0(i)$ of a polyhedral surface $(T_l, M) \in \mathscr{T}^{met}_{g,N_0}(M)$ with a corresponding null vector, i.e.,

$$\sigma^2(k,h,j) \longmapsto \overrightarrow{\xi}(k,h,j) \in \mathbb{L}^+ , \tag{3.13}$$

$$\sigma^0(i) \longmapsto \overrightarrow{\xi}(i) \in \mathbb{L}^+ . \tag{3.14}$$

The rationale underlying this mapping is that the null vector $\overrightarrow{\xi}(k,h,j)$ characterizes a *visual* horosphere $\Sigma_\infty(k,h,j)$ where an observer $\mathcal{O}_\infty$ in a neighborhood of $\{\infty\} \in \mathbb{H}^{3,+}_{up}$ sees a Euclidean triangle $\sigma^2(k,h,j) \in (T_l, M)$ resulting from the projection of a hyperbolic triangle $\sigma^2_{hyp}(k,h,j)$ living in the $\mathbb{R}^2 \times \{0\}$ portion of the boundary of $\mathbb{H}^{3,+}_{up}$. The projection takes place along the $\mathbb{H}^{3,+}_{up}$ geodesics defined by the vertex null

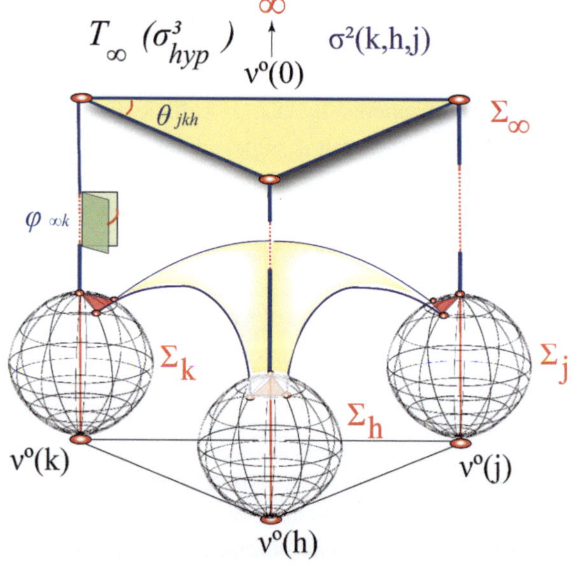

Fig. 3.4 The Euclidean triangles cut by the horospheres Σ_k, Σ_h, Σ_j are all similar to the Euclidean triangle $T_\infty(\sigma^3_{hyp}) \equiv \sigma^2(k,h,j)$ in the plane Σ_∞

Fig. 3.5 The sky mapping between the horosphere Σ_∞, acting as the observer screen, and the $\mathbb{R}^2 \times \{0\}$ portion of the boundary of $\mathbb{H}^{3,+}_{up}$: a hyperbolic triangle $\sigma^2_{hyp}(k,h,j)$ with vertices on $\mathbb{R}^2 \times \{0\}$ is seen as a Euclidean triangle $\sigma^2(k,h,j)$ on Σ_∞

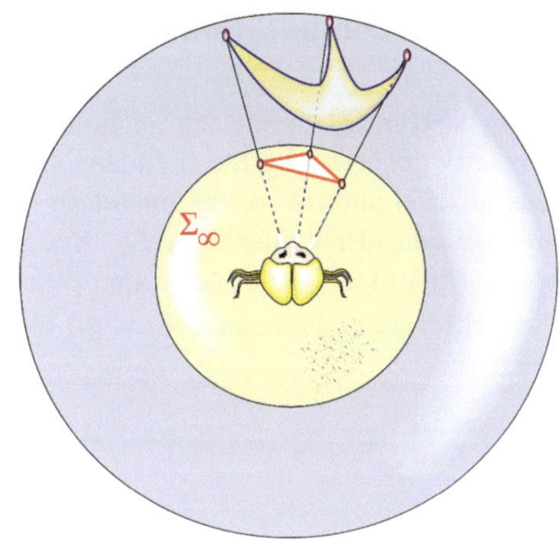

vectors $\{\overrightarrow{\xi}(k)\}$. Thus, the horosphere $\Sigma_\infty(k,h,j)$ represents a (local) *screen* and the pair $(\Sigma_\infty(k,h,j), \Sigma_k)$ characterizes the *visual incoming direction* from which the observer sees the vertex $\sigma^0_{hyp}(k) := v^0(k)$. Note that the geometrical information encoded in each pair of null vectors $(\overrightarrow{\xi}(k,h,j), \overrightarrow{\xi}(k))$ can be equivalently provided by a twistor field attached to the vertex $\sigma^0(k)$, (see [4] for details on this alternative point of view).

As the triangles $\{\sigma^2(k,h,j)\}$ and the vertices $\{\sigma^0(i)\}$ vary in (T_l, M), this visual decoration characterizes a mapping from the space of triangulated surfaces $\mathscr{T}^{met}_{g,N_0}(M)$ to (N_2 copies of) the future light cone $\mathbb{L}^+$, defined by

$$\mathscr{T}^{met}_{g,N_0}(M) \longrightarrow \times^{N_2}_{k=1} \mathbb{L}^+ \tag{3.15}$$

$$(T_l, M) \longmapsto \left\{ \left(\overrightarrow{\xi}(k), \overrightarrow{\xi}(h), \overrightarrow{\xi}(j) \right); \ \overrightarrow{\xi}(k,h,j) \right\}.$$

If we let the observer $\mathscr{O}_\infty$ sit exactly at the point $\infty \in \mathbb{H}^{3,+}_{up}$, then the above construction corresponds to take the cone in $\mathbb{H}^{3,+}_{up}$, from the point $\infty \in \mathbb{H}^{3,+}_{up}$, over the set of hyperbolic triangles $\{\sigma^2_{hyp}(k,h,j)\}$ in the hyperbolic subspace $\mathbb{R}^2 \times \{0\}$. A horosphere Σ_∞ in a neighborhood of ∞, (such as $\Sigma_\infty(k,h,j)$), is thus a *screen* surrounding the observer $\mathscr{O}_\infty \equiv \infty \in \mathbb{H}^{3,+}_{up}$ and (3.15) characterizes what this observer sees under her *sky mapping* (Fig. 3.5).

A basic issue concerning this visual correspondence is what kind of triangulation in the hyperbolic space $\mathbb{R}^2 \times \{0\}$ gives rise to the polyhedral surface the observer at $\infty \in \mathbb{H}^{3,+}_{up}$ sees in her sky. In particular, is there a hyperbolyc surface whose image in a neighborhood of $\{\infty\} \in \mathbb{H}^{3,+}_{up}$ is a polyhedral surface (T_l, M) with conical singularities? Quite surprisingly answering to such a question provides a deep connection between the geometry of $\mathscr{T}^{met}_{g,N_0}(M)$, the Weil–Petersson form, and the moduli space of hyperbolic surfaces with boundaries.

3.4 The Computation of Lambda-Lengths

A key step in discussing the relation among polyhedral surfaces and hyperbolic geometry generated by the sky mapping involves the computation of the lambda-lengths (3.7) in terms of the Euclidean lengths of the edges of $\sigma^2(k,h,j)$. To this end, we consider horospheres Σ_k, Σ_h, Σ_j sufficiently near to the vertices $v^0(k)$, $v^0(h)$, $v^0(j)$ of $\sigma^2_{hyp}(k,h,j)$. We start by evaluating the lambda-lengths (3.7) along the vertical geodesics connecting $v^0(0) \simeq \infty$ with the triangle $\sigma^2_{hyp}(k,h,j)$ (Fig. 3.4). We have

Lemma 3.1

$$\lambda\left(\Sigma_\infty, \Sigma_k\right) = \sqrt{\frac{t}{z_k}}, \tag{3.16}$$

$$\lambda\left(\Sigma_\infty, \Sigma_h\right) = \sqrt{\frac{t}{z_h}},$$

$$\lambda\left(\Sigma_\infty, \Sigma_j\right) = \sqrt{\frac{t}{z_j}},$$

where $z = z_k$, $z = z_h$, and $z = z_j$ respectively define the z coordinates of the intersection points between the horospheres Σ_k, Σ_h, Σ_j and the corresponding vertical geodesics.

Proof We compute explicitly $\lambda\left(\Sigma_\infty, \Sigma_k\right)$, the remaining cases being completely similar. We can assume that $t > z_k$, and evaluate the distance between the points $(x(v^0(k)), y(v^0(k)), z = z_k) := \Sigma_k \cap \gamma(v^0(k), v^0(0))$ and $(x(v^0(k)), y(v^0(k)), z = t) := \Sigma_\infty \cap \gamma(v^0(k), v^0(0))$, along the vertical geodesics $\gamma(v^0(k), v^0(0))$ connecting $v^0(0) \simeq \infty$ with $v^0(k)$. From (3.9) we get

$$\tanh \frac{\delta(\Sigma_\infty, \Sigma_k)}{2} = \frac{t - z_k}{t + z_k}, \tag{3.17}$$

or, more explicitly,

$$\frac{e^{\delta(\Sigma_\infty, \Sigma_k)} - 1}{e^{\delta(\Sigma_\infty, \Sigma_k)} + 1} = \frac{t - z_k}{t + z_k}. \tag{3.18}$$

This immediately yields $\exp\left(\delta(\Sigma_\infty, \Sigma_k)\right) = \frac{t}{z_k}$ and (3.8) provides the required lambda–length according to

$$\lambda\left(\Sigma_\infty, \Sigma_k\right) = \sqrt{\frac{t}{z_k}}. \tag{3.19}$$

∎

The computation of the remaining λ–lengths $\lambda\left(\Sigma_k, \Sigma_h\right)$, $\lambda\left(\Sigma_h, \Sigma_j\right)$, and $\lambda\left(\Sigma_j, \Sigma_k\right)$ is provided by

Lemma 3.2

$$\lambda\left(\Sigma_k, \Sigma_h\right) = \frac{l(k,h)}{\sqrt{z_k z_h}}, \tag{3.20}$$

$$\lambda\left(\Sigma_h, \Sigma_j\right) = \frac{l(h,j)}{\sqrt{z_h z_j}}, \tag{3.21}$$

$$\lambda\left(\Sigma_j, \Sigma_k\right) = \frac{l(j,k)}{\sqrt{z_j z_k}}, \tag{3.22}$$

where $l(k,h)$, $l(h,j)$, and $l(j,k)$ are the Euclidean lengths of the triangle $\sigma^2(k,h,j)$ in Σ_∞.

Proof Consider the intersection of the ideal triangle $\sigma^2_{hyp}(\infty, k, h)$ with the horospheres Σ_∞, Σ_k, and Σ_h. Each such an intersection characterizes a corresponding horocyclic segment F_∞, F_k, F_h whose hyperbolic length defines the *h-length* of the horocyclic segment. In particular, the horocyclic segment traced by $\sigma^2_{hyp}(\infty, k, h) \cap \Sigma_\infty$ is the side $\sigma^1(k, h)$ of the Euclidean triangle $\sigma^2(k, h, j)$. According to (3.9), its h-length is provided by

$$H\left(\sigma^0(k), \sigma^0(h)\right) := d_{\mathbb{H}^3}\left(\sigma^0(k), \sigma^0(h)\right) = 2 \tanh^{-1}\sqrt{\frac{l^2(k,h)}{l^2(k,h) + 4t^2}}. \tag{3.23}$$

On the other hand, the horocyclic segment $\sigma^1(k, h)$ is opposite to the geodesic segment intercepted by the horospheres Σ_k, and Σ_h along the hyperbolic edge $\sigma^1_{hyp}(k, h)$. The lambda-length of this segment is $\lambda\left(\Sigma_k, \Sigma_h\right)$, and according to a result by R. Penner [15, 17], (respectively, Proposition 2.8 and Lemma 4.4), these quantities are related by[1]

$$H\left(\sigma^0(k), \sigma^0(h)\right) = \frac{\lambda\left(\Sigma_k, \Sigma_h\right)}{\lambda\left(\Sigma_\infty, \Sigma_k\right)\lambda\left(\Sigma_\infty, \Sigma_h\right)}, \tag{3.24}$$

from which, by taking into account (3.16), we get

---

[1]The reader is cautioned that the whole subject of λ–lengths and H–lengths is often plagued by factors of 2, (and related $\sqrt{2}$'s), appearing and disappearing from the relevant formulae. This largely depends from the normalization chosen. In a recent review paper [17] R. Penner has clarified this normalization issue. Here we adopt the conventions stipulated in [17].

$$\lambda\left(\Sigma_k, \Sigma_h\right) = \frac{2t}{\sqrt{z_k z_h}}\, \tanh^{-1} \sqrt{\frac{l^2(k,h)}{l^2(k,h) + 4t^2}}. \tag{3.25}$$

Similarly, we compute

$$\lambda\left(\Sigma_h, \Sigma_j\right) = \frac{2t}{\sqrt{z_h z_j}}\, \tanh^{-1} \sqrt{\frac{l^2(h,j)}{l^2(h,j) + 4t^2}}, \tag{3.26}$$

$$\lambda\left(\Sigma_j, \Sigma_k\right) = \frac{2t}{\sqrt{z_j z_k}}\, \tanh^{-1} \sqrt{\frac{l^2(j,k)}{l^2(j,k) + 4t^2}}. \tag{3.27}$$

Note that these relations must hold for any Σ_∞ approaching the ideal vertex $v^0(0) \simeq \infty$. In particular, if we take the limit $\Sigma_\infty \to v^0(0) \simeq \infty$, corresponding to $t \nearrow +\infty$, we easily find (Fig. 3.6)

$$\lambda\left(\Sigma_k, \Sigma_h\right) = \frac{l(k,h)}{\sqrt{z_k z_h}}, \tag{3.28}$$

$$\lambda\left(\Sigma_h, \Sigma_j\right) = \frac{l(h,j)}{\sqrt{z_h z_j}}, \tag{3.29}$$

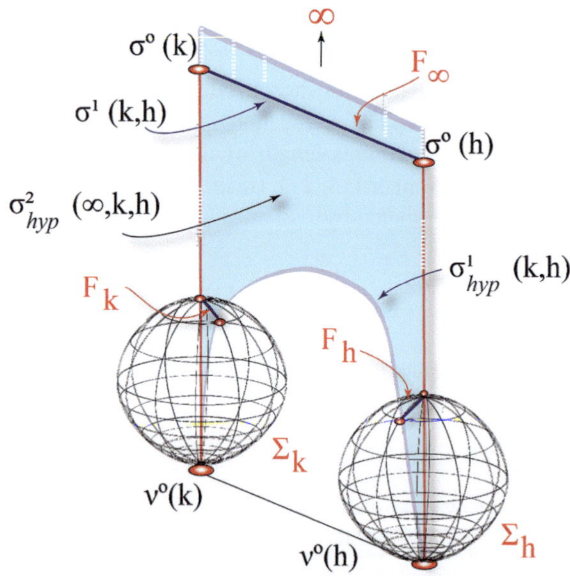

Fig. 3.6 The intersection of the ideal triangle $\sigma^2_{hyp}(\infty, k, h)$ with the horospheres Σ_∞, Σ_k, and Σ_h defines the corresponding horocyclic segments F_∞, F_k, F_h

$$\lambda\left(\Sigma_j, \Sigma_k\right) = \frac{l(j,k)}{\sqrt{z_j z_k}}.$$

(3.30)

∎

The relation (3.24) holds also among the $\{\lambda\left(\Sigma_j, \Sigma_k\right)\}$ and we can also compute the h-lengths associated with the ideal triangle $\sigma_{hyp}^2(k,h,j)$, decorated with the horospheres Σ_k, Σ_h, Σ_j, according to

$$H\left(\Sigma_k, \Sigma_h\right) = \frac{\lambda\left(\Sigma_k, \Sigma_h\right)}{\lambda\left(\Sigma_h, \Sigma_j\right)\lambda\left(\Sigma_j, \Sigma_k\right)},$$

(3.31)

$$H\left(\Sigma_h, \Sigma_j\right) = \frac{\lambda\left(\Sigma_h, \Sigma_j\right)}{\lambda\left(\Sigma_j, \Sigma_k\right)\lambda\left(\Sigma_k, \Sigma_h\right)},$$

(3.32)

$$H\left(\Sigma_j, \Sigma_k\right) = \frac{\lambda\left(\Sigma_j, \Sigma_k\right)}{\lambda\left(\Sigma_k, \Sigma_h\right)\lambda\left(\Sigma_h, \Sigma_j\right)}.$$

(3.33)

In particular, from (3.20) ÷ (3.22) we get

$$H\left(\Sigma_k, \Sigma_h\right) = \frac{l(k,h)}{l(h,j)l(j,k)}\, z_j,$$

(3.34)

$$H\left(\Sigma_h, \Sigma_j\right) = \frac{l(h,j)}{l(j,k)l(k,h)}\, z_k,$$

(3.35)

$$H\left(\Sigma_j, \Sigma_k\right) = \frac{l(j,k)}{l(k,h)l(h,j)}\, z_h.$$

(3.36)

3.5 Polyhedral Surfaces and Hyperbolic Surfaces with Boundaries

The geometrical analysis of the previous sections shows that to each of the $N_2(T)$ Euclidean triangles $\sigma^2(k,h,j)$ of a polyhedral surface (T_l, M) we can associate a corresponding ideal tetrahedron $\sigma_{hyp}^3(\infty,k,h,j) \in \mathbb{H}_{up}^{3,+}$, decorated with the horocyclic sectors induced by a choice of horospheres Σ_k, Σ_h, Σ_j. In turn, this decorated ideal tetrahedron generates an ideal triangle $\sigma_{hyp}^2(k,h,j)$ decorated by horocycle sectors, whose hyperbolic lengths can be recovered in terms of the Euclidean lengths $\{l(k,h)\}$ of $\sigma^2(k,h,j)$. Thus, we are naturally led to explore the possibility of glueing the ideal triangles $\{\sigma_{hyp}^2(k,h,j)\}$ in the same combinatorial pattern defined by (T_l, M). This must be done in such a way that the horosphere field on (T_l, M) provides a consistent horocyclical decoration of the vertices of the ideal triangulation defined by the hyperbolic triangles $\{\sigma_{hyp}^2(k,h,j)\}$. In performing such an operation one must take care of some basic facts:

(i) A conical star of Euclidean triangles $\{\sigma^2(k, h_\alpha, h_{\alpha+1})\}_{\alpha=1}^{q(k)}$ cannot be isometrically embedded in a single horosphere Σ_∞ unless the conical angle is trivial, i.e., $\Theta(k) = 2\pi$;

(ii) The ideal triangles $\{\sigma_{hyp}^2(k, h_\alpha, h_{\alpha+1})\}$ are rigid since any two of them are congruent;

(iii) The adjacent sides of any two ideal triangles in $\{\sigma_{hyp}^2(k, h_\alpha, h_{\alpha+1})\}$ can be identified up to the freedom of performing an arbitrary translation along the edges, (each edge $\sigma_{hyp}^1(k, h_\alpha)$ of an ideal triangle is isometric to the real line, its hyperbolic length being infinite, and two adjacent edges may freely slide one past another);

These degrees of freedom can be exploited in order to specify how the decoration provided by the horocyclic sectors in an ideal triangle is extended to the adjacent ideal triangle. If we denote by $K^0(T)$ the 0–skeleton of a triangulation (T, M) with a vertex set $V(T)$ and by $M/K^0(T)$ the corresponding open surface obtained by removing vertices from M, (see Remark 1.2), then we have

Theorem 3.1 *Let $\{\mathbb{L}_{(k)}^+\}$ denote N_0 copies of the future light cone $\mathbb{L}^+$ and let*

$$\left[(T_l, M); \{\vec{\xi}(k)\}\right] \in \mathcal{T}_{g,N_0}^{met}(M) \otimes_{k=1}^{N_0} \mathbb{L}_{(k)}^+ \tag{3.37}$$

be a polyhedral surface with a given set of future–directed null vectors $\{\vec{\xi}(k)\}$ assigned on its N_0 vertices. Then:

(a) *The decorated polyhedral surface $\left[(T_l, M); \{\vec{\xi}(k)\}\right]$ has a dual description as a ideal triangulation $\mathcal{H}\left((T_l, M); \{F_k(t)\}\right)$ of the open surface $M/K^0(T)$, endowed with a horocyclic foliation $\{t \mapsto F_k(t)\}$, $0 \leq t \leq \infty$, in a neighborhood of each ideal vertex $\{v^0(k)\}$. We denote by*

$$\mathcal{T}_{g,N_0}^{ideal}(M/K^0) \otimes_{k=1}^{N_0} \mathbb{L}_{(k)}^+, \tag{3.38}$$

the set of such horocyclically decorated ideal triangulations of M/K^0, (where now the future light cone $\mathbb{L}^+$ is thought of as representing the set of horospheres).

(b) *The mapping*

$$\mathcal{H} : \mathcal{T}_{g,N_0}^{met}(M) \longrightarrow \mathcal{T}_{g,N_0}^{ideal}(M/K^0) \otimes_{k=1}^{N_0} \mathbb{L}_{(k)}^+ \tag{3.39}$$

$$(T_l, M) \longmapsto (\mathcal{H}(T_l, M), \{F_k(t)\}),$$

characterizes an incomplete hyperbolic structure on $M/K^0(T)$ parametrized by a set of N_0 horocyclic Thurston invariants

$$d[F_k] = \ln \frac{\Theta(k)}{2\pi} \,, \quad \text{with} \quad \sum_{k=1}^{N_0} \left(e^{d[F_k]} - 1 \right) = 2g - 2. \qquad (3.40)$$

where $\{\Theta(k)\}$ are the conical angles of (T_l, M).

(c) *The hyperbolic structure on $\mathcal{H}\,(T_l, M)$ can be completed so as to generate a hyperbolic surface $\Omega(T_l, M)$ with N_0 geodesic boundaries $\partial\Omega_k$ of signed length $|\partial\Omega_k|_\pm = d[F_k] = \ln \frac{\Theta(k)}{2\pi}$, were the sign is chosen to be positive (with respect to the natural orientation of $\Omega(T_l, M)$) if $\Theta(k) > 2\pi$, negative if $\Theta(k) < 2\pi$, and $\Theta(k) = 2\pi$ is associated with a geodesic boundary $\partial\Omega_k$ of zero length, i.e., a cusp.*

(d) *Let $\mathfrak{T}^\partial_{g,N_0}$ denote the Teichmüller space of hyperbolic surfaces Ω of genus g with N_0 geodesic boundaries $\partial\Omega_k$, (see Sect. B.5), then*

$$\Omega : \mathcal{T}^{met}_{g,N_0}(M) \longrightarrow \mathfrak{T}^\partial_{g,N_0} \qquad (3.41)$$

$$(T_l, M) \longmapsto \Omega(T_l, M),$$

induces a natural bijection

$$\Omega_\Theta : POL_{g,N_0}(M, \{\Theta(k)\}, A(M)) \longrightarrow \mathfrak{M}_{g,N_0}(L) \times \mathbb{Z}_2^{N_0} \qquad (3.42)$$

between the space of polyhedral structures, $POL_{g,N_0}(M, \{\Theta(k)\}, A(M))$, with given area $A(M)$ and conical angles $\{\Theta(k)\}$, and the moduli space $\mathfrak{M}_{g,N_0}(L)$ of Riemann surfaces of genus g with N_0 boundary components of signed length vector $L \times \mathbb{Z}_2^{N_0} := \{L_k = |\ln \frac{\Theta(k)}{2\pi}|\}_{k=1}^{N_0} \times \{sign(\ln \frac{\Theta(k)}{2\pi})\}_{k=1}^{N_0}$.

Proof It is well-known[18] that by glueing ideal triangles along the pattern of a Euclidean triangulation, satisfying the 2π–flatness constraint around each vertex, we can generate a complete hyperbolic structures on the underlying surface M. Thus, in order to prove the theorem we need to control the effect that the presence of conical angles has on the completeness of the hyperbolic structure associated with (T_l, M). It is sufficient to carry such analysis on the generic vertex star $Star[\sigma^0(k)] \in (T_l, M)$.

Let us consider the star $Star[\sigma^0(k)]$ of a vertex $\sigma^0(k)$ over which $q(k)$ triangles $\sigma^2(k, h_\alpha, h_{\alpha+1})$ are incident. Let $\overrightarrow{\xi}(k)$ be the future–pointing null vector associated with $\sigma^0(k)$, and let $\Theta(k)$ be the conical angle of $Star[\sigma^0(k)]$. As usual, we denote by $\{\theta_\alpha(k)\}$ the corresponding sequence of $q(k)$ vertex angles $\theta_{h_\alpha, k, h_{\alpha+1}}$, with the index α defined modulo $q(k)$. If Σ_∞ is a horosphere centered at $v^0(0) := \lim_{t \nearrow \infty} t\,\overrightarrow{\xi}(k)$, then, as emphasized above, we cannot isometrically embed the whole $Star[\sigma^0(k)]$ in the chosen Σ_∞ unless $\Theta(k) = 2\pi$. Thus, rather than fixing our attention on a single horosphere, we consider the foliation (Fig. 3.7) of horospheres associated with the null direction $t\,\overrightarrow{\xi}(k)$, $0 \leq t \leq \infty$, and exploit the freedom in identifying the triangles $\{\sigma^2(k, h_\alpha, h_{\alpha+1})\}$, along their adjacent edges $\{\sigma^1(k, h_\alpha)\}$, modulo a translation along the (vertical) geodesic edges of the associated ideal hyperbolic triangles $\{\sigma^2_{hyp}(v^0(0) = \infty, k, h_\alpha)\}$. This translation generates in $\mathbb{H}^{3,+}_{up}$ a sequence of horospheres $\{\Sigma^\alpha_\infty\}$, $\alpha = 1, 2, \ldots$, centered at $v^0(0) = \infty$, and such that $\Sigma^{\alpha+1}_\infty \subset B^\alpha_\infty$, where $\{B^\alpha_\infty\}$ is the sequence of horoballs bounded by $\{\Sigma^\alpha_\infty\}$. We define inductively such a sequence according to the following prescription:

(i) Each horosphere Σ^α_∞, in the sequence $\{\Sigma^\alpha_\infty\}$, is represented by a corresponding horizontal plane $z = t_\alpha$ endowed with the Cartesian coordinates (O, x, y) induced by $\mathbb{H}^{3,+}_{up}$. We denote by $\pi_{\alpha+1,\alpha}$ the natural projection

$$\pi_{\alpha+1,\alpha} : \Sigma^{\alpha+1}_\infty \longrightarrow \Sigma^\alpha_\infty \tag{3.43}$$

$$(x, y, z = t_{\alpha+1}) \longmapsto (x, y, z = t_\alpha),$$

induced by the vertical geodesics in $\mathbb{H}^{3,+}_{up}$.

(ii) The α-th triangle $\sigma^2(k, h_\alpha, h_{\alpha+1})$ of $Star[\sigma^0(k)]$ is embedded in $\{\Sigma^\alpha_\infty\}$ in such a way that the (copy of the) vertex $\sigma^0(k)$ of $\sigma^2(k, h_\alpha, h_{\alpha+1})$ is located at the origin $(0, 0)$ of the plane $z = t_\alpha$.

(iii) Under the projection $\pi_{\alpha+1,\alpha} : \Sigma^{\alpha+1}_\infty \to \Sigma^\alpha_\infty$, the edge $\sigma^1(k, h_{\alpha+1})$ of the triangle $\sigma^2(k, h_{\alpha+1}, h_{\alpha+2}) \hookrightarrow \Sigma^{\alpha+1}_\infty$ goes into the edge $\sigma^1(k, h_{\alpha+1})$ of $\sigma^2(k, h_\alpha, h_{\alpha+1}) \hookrightarrow \Sigma^\alpha_\infty$, modulo a hyperbolic isometry, i.e. a similarity generated by the dilation

$$e^{d_{\mathbb{H}^3}(\Sigma^{\alpha+1}_\infty, \Sigma^\alpha_\infty)} \tag{3.44}$$

along the projecting geodesic, and by a rotation in the Σ^α_∞ plane.

(iv) Since the triangles $\{\sigma^2(k, h_\alpha, h_{\alpha+1})\}$ of $Star[\sigma^0(k)]$ are identified modulo $q(k)$ and generate a conical angle $\Theta(k) := \sum_{\alpha=1}^{q(k)} \theta_\alpha(k)$, we set

$$\frac{t_{\alpha+1}}{t_\alpha} = \begin{cases} \left(\frac{\Theta(k)}{2\pi}\right)^{\frac{1}{q(k)}} , & \text{if } \Theta(k) \geq 2\pi , \\[2em] \left(\frac{\Theta(k)}{2\pi}\right)^{-\frac{1}{q(k)}} , & \text{if } \Theta(k) \leq 2\pi , \end{cases} \tag{3.45}$$

in such a way that the horospheres $\Sigma^{\alpha+1}_\infty$ and Σ^α_∞ are at a constant hyperbolic distance

$$d_{\mathbb{H}^3}\left(\Sigma^{\alpha+1}_\infty, \Sigma^\alpha_\infty\right) = \ln\left(\frac{t_{\alpha+1}}{t_\alpha}\right) = \frac{1}{q(k)}\left|\ln\left(\frac{\Theta(k)}{2\pi}\right)\right|, \tag{3.46}$$

which reduces to zero when $\Theta(k) = 2\pi$. This spiral staircase construction (Fig. 3.7) allows to deal with the conical structure of the star of Euclidean triangles $\{\sigma^2(k, h_\alpha, h_{\alpha+1})\}_{\alpha=1}^{q(k)}$ and to glue, modulo hyperbolic isometries,

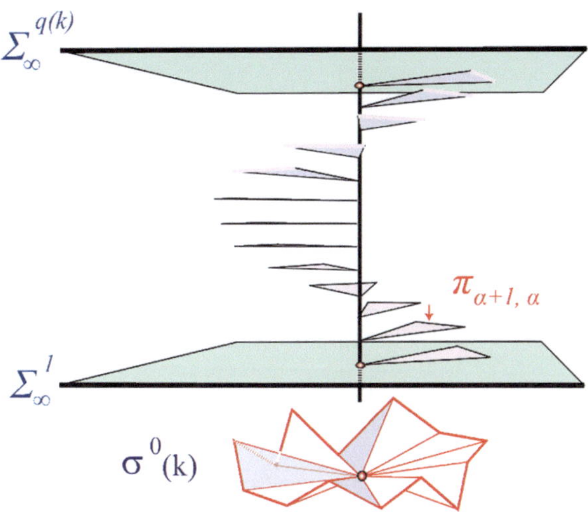

Fig. 3.7 The staircase construction with horospherical steps associated with a conical angle

the corresponding set of hyperbolic ideal triangles $\sigma^2_{hyp}(k, h_\alpha, h_{\alpha+1})$, $\alpha = 1, \ldots, q(k)$, ($h_\alpha = h_\beta$ if $\beta = \alpha \bmod q(k)$). Explicitly, the natural hyperbolic structure on

$$\bigcup_{\alpha=1}^{q(k)} \sigma^2_{hyp}(k, h_\alpha, h_{\alpha+1}) - \{v^0(k)\}, \tag{3.47}$$

($v^0(k)$ being the vertex associated with $\sigma^0(k)$), induces a similarity structure on the link associated with $v^0(k)$

$$link\,\left[v^0(k)\right] := \bigcup_{\alpha=1}^{q(k)} \sigma^1_{hyp}(h_\alpha, h_{\alpha+1}), \tag{3.48}$$

which characterizes, as k varies, the hyperbolic surface one gets by glueing the hyperbolic triangles $\sigma^2_{hyp}(k, h_\alpha, h_{\alpha+1})$. To determine such a similarity structure, let us consider the image of the above spiral staircase construction near the ideal vertex $v^0(k)$. We start with the hyperbolic triangle $\sigma^2_{hyp}(k, h_1, h_2)$. Let Σ^1_k denote a horosphere chosen sufficiently near $v^0(k)$ so that the oriented horocyclic segment F^1_k cut in $\sigma^2_{hyp}(k, h_1, h_2)$ by Σ^1_k has (Euclidean) length given by (Fig. 3.8)

$$|F^1_k|_{Euc} = \frac{\theta_1(k)}{2\pi}. \tag{3.49}$$

This horocyclic segment can be extended, in a counterclockwise order, to the other $q(k) - 1$ ideal triangles in the set $\{\sigma^2_{hyp}(k, h_\alpha, h_{\alpha+1})\}$ by requiring that

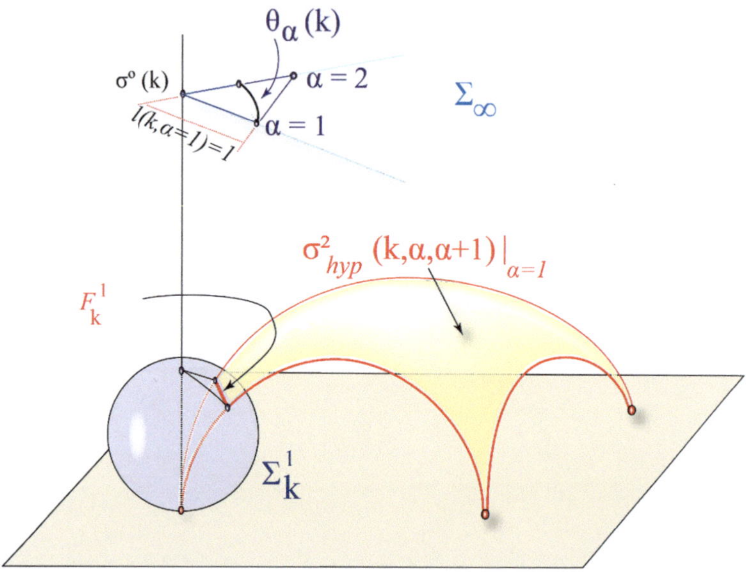

Fig. 3.8 The intersection of the ideal triangle $\sigma^2_{hyp}(k,\alpha,\alpha+1)$, (here for $\alpha = 1$), with the horosphere Σ^1_k defines a corresponding horocyclic segment F^1_k

the α-th segment F^α_k in such an extension has length $|F^\alpha_k|_{Euc} = \theta_\alpha(k)/2\pi$, and meets orthogonally the geodesic side $\sigma^1_{hyp}(k,h_{\alpha+1})$ of the ideal triangle $\sigma^2_{hyp}(k,h_{\alpha+1},h_{\alpha+2})$, along a corresponding horosphere Σ^α_k congruent to Σ^1_k. Such an extension procedure characterizes a sequence of congruent horospheres $\{\Sigma^\alpha_k\}^\infty_{\alpha=1}$. According to (3.46), we assume that

$$d_{\mathbb{H}^3}\left(\Sigma^{\alpha+1}_k,\Sigma^\alpha_k\right) = \frac{1}{q(k)}\left|\ln\left(\frac{\Theta(k)}{2\pi}\right)\right|. \tag{3.50}$$

Since the horospheres $\{\Sigma^\alpha_k\}$ are congruent and the identification between adjacent sides of the ideal triangles $\sigma^2_{hyp}(k,h_\alpha,h_{\alpha+1})$ is only defined up to a shift, (and a rotation around the vertical geodesic ending in $v^0(k)$), such an extension procedure generates a sequence of $q(k)$ horocyclic segments $\{F^\alpha_k\}$ which, for each $\alpha = 1 \sim \mod q(k)$, re-enters the triangle $\sigma^2_{hyp}(k,h_1,h_2)$ with a horocyclic segment $\widehat{F}^\alpha_k$ which will be parallel to F^1_k but not necessarily coincident with it. In particular, we can introduce the signed hyperbolic distance between the horocycle segments F^1_k and $\widehat{F}^{q(k)+1}_k$ according to

$$\mp d_{\mathbb{H}^3}(F^1_k,\widehat{F}^{q(k)+1}_k) = \mp \sum_{\alpha=1}^{q(k)} d_{\mathbb{H}^3}\left(\Sigma^\alpha_k,\Sigma^{\alpha+1}_k\right) = \ln\frac{\Theta(k)}{2\pi} \doteq d[F_k], \tag{3.51}$$

where the sign is chosen to be positive iff $\Theta(k) < 2\pi$, i.e. if the horodisk sector bounded by F_k^1 contains the sector bounded by $\widehat{F}_k^{q(k)+1}$. Thus, the horocycle curve $t \mapsto F_k(t)$, $0 \leq t \leq 1$, defined by the $q(k)$ horocyclic segments $\{F_k^\alpha\}$, closes up iff its Euclidean length $|F_k(t)|_{Euc} := \sum_{\alpha=1}^{q(k)} |F_k^\alpha|_{Euc}$ is 1. Note that, as soon as $d[F_k] \neq 0$, the horocyclic curve $t \mapsto F_k(t)$, can be extended to a curve $t \mapsto \widetilde{F}_k(t)$, $0 \leq t < \infty$, with bounded (Euclidean) length. However, its intersection with the hyperbolic edge $\sigma_{hyp}^1(k, h_1)$ generates a non–convergent Cauchy sequence of points along $\sigma_{hyp}^1(k, h_1)$, (the hyperbolic distance between such points being always $|d[v^0(k)]|$). Stated differently, the Euclidean similarity structure between the Euclidean triangle $\sigma^2(k, h_1, h_2)|_{\Sigma_k^1}$ cut by the horosphere Σ_k^1, and the Euclidean triangle $\sigma^2(k, h_1, h_2)|_{\Sigma_k^{q(k)+1}}$ cut by $\Sigma_k^{q(k)+1}$ does not generate a complete hyperbolic structure (Fig. 3.9). Such a similarity structure is completely characterized by the number $d[F_k]$, which does not depend from the initial choice of $F_{v^0(k)}^{h_1}$, and is an invariant only related to the conical defect $\Theta(k)$ supported at the vertex $\sigma^0(k)$ of (T_l, M). It can be identified with the holonomy invariant introduced by W. Thurston [18] in order to characterize the completeness of the hyperbolic structure of a surface obtained by gluing hyperbolic ideal triangles (the structure being complete iff the invariants $d[F_k]$ are all zero for each ideal vertex $v^0(k)$).

The topological constraint in (3.40) is an obvious rewriting of the computation (2.75) of the degree of the divisor associated with the conical singularities, (i.e., a restatement of the Gauss–Bonnet theorem–see Theorem 2.2). This completes the proof of part *(a)* and part *(b)* of the theorem. To prove part *(c)*, we exploit a classical result by W. Thurston, ([18], prop. 3.4.21), according to which the glueing of the $N_2(T)$ ideal triangles according to the procedure just described gives rise to an open hyperbolic surface Ω with geodesic boundaries (Fig. 3.10). Each boundary component $\partial\Omega_k$ is associated with a corresponding vertex $\sigma^0(k)$ of $|T_l| \to M$, and has a signed length provided by $\pm |\partial\Omega_k| = d[F_k]$, where the $\pm$ sign encodes the outward or inward, (with respect to the standard counterclockwise orientation induced by (T_l, M)), spiraling of the horocycle segments $t \mapsto F_k(t)$ around the given vertex. Thus

$$\{\pm |\partial\Omega_k|\} = |d[F_k]| = \{sign(\ln \frac{\Theta(k)}{2\pi})\}_{k=1}^{N_0} \times \{L_k = |\ln \frac{\Theta(k)}{2\pi}|\}_{k=1}^{N_0}.$$
$$(3.52)$$

The mapping between (T_l, M) and the corresponding hyperbolic surface with boundary $\Omega(T_l, M)$ characterizes (3.41). Recall that $POL_{g, N_0}(M, \{\Theta(k)\}, A(M))$, the space of polyhedral surfaces of given area $A(M)$ and conical angles $\{\Theta(k)\}_{k=1}^{N_0}$, is locally modeled by the polyhedral surfaces in $\mathcal{T}_{g, N_0}^{met}(M, \{\Theta(k)\})/Aut(T)$ with area $A(T_l, M) = A(M)$, (see (2.89)). For the generic hyperbolic surface $\Omega \in \mathcal{T}_{g, N_0}^\partial$, let $\mathfrak{Map}_{g, N_0}^\partial$ denote the mapping class group defined by the group of all the isotopy classes of orientation preserving homeomorphisms which leave each boundary component $\partial\Omega_j$ pointwise (and

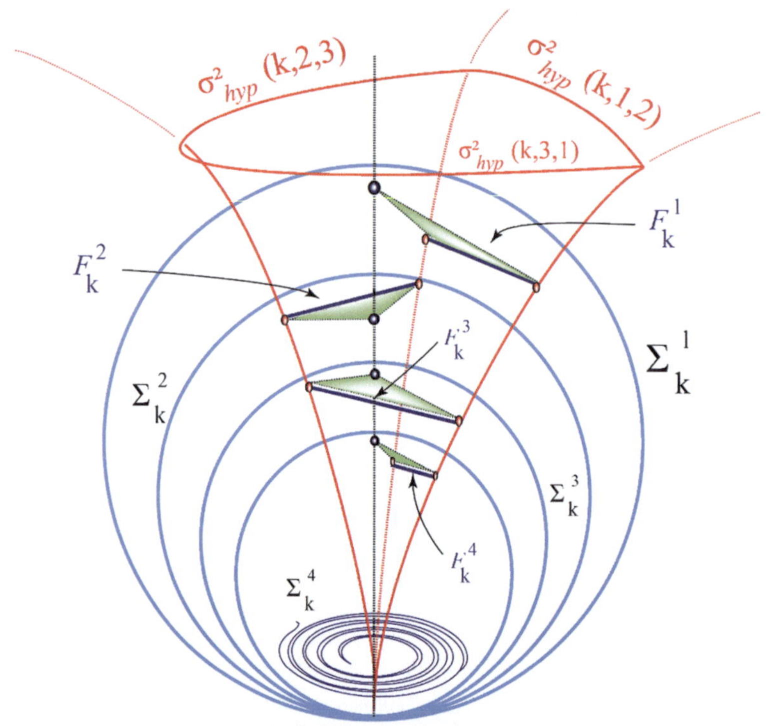

Fig. 3.9 The intersection of the ideal triangles $\{\sigma^2_{hyp}(k,\alpha,\alpha+1)\}$, (here for $\alpha = 1,2,3$, with $q(k)=3$), with a corresponding sequence of horospheres $\{\Sigma^\alpha_k\}$ generates the horocyclic segments F^α_k. The segment $F_k^{q(k)+1}$ re-enters $\sigma^2_{hyp}(k,1,2)$ at a signed distance $d[F_k]$ from F^1_k. The spiraling emphasizes that this mechanism goes on indefinitely

isotopy-wise) fixed. Since the hyperbolic surfaces $\Omega(T_l, M)$ are obtained by completion of ideal triangulations generated by (T_l, M), the automorphisms group $Aut(T)$ of (T_l, M) can be injected in $\mathfrak{Map}^\partial_{g,N_0}$. Thus, (3.41) naturally descends to a mapping between $POL_{g,N_0}(M, \{\Theta(k)\}, A(M))$ and the moduli space $\mathfrak{M}_{g,N_0}(L)$ of Riemann surfaces of genus g with N_0 boundary components of signed length vector $\mathbb{Z}_2^{N_0} \times L := \{\pm L_k = \ln \frac{\Theta(k)}{2\pi}\}_{k=1}^{N_0}$. Conversely, let Ω be a Riemann surface with N_0 geodesic boundaries $\{\partial\Omega_k\}$ of signed length vector $\{\pm L_k\}$, satisfying the topological constraint

$$\sum_{k=1}^{N_0} \left(e^{\pm L_k} - 1\right) = 2g - 2, \tag{3.53}$$

(see (3.40)), then $\{\frac{\Theta(k)}{2\pi} := e^{\pm L_k}\}$ is a coherent set of conical angles which, according to Troynov's theorem characterizes, up to homotety, a polyhedral

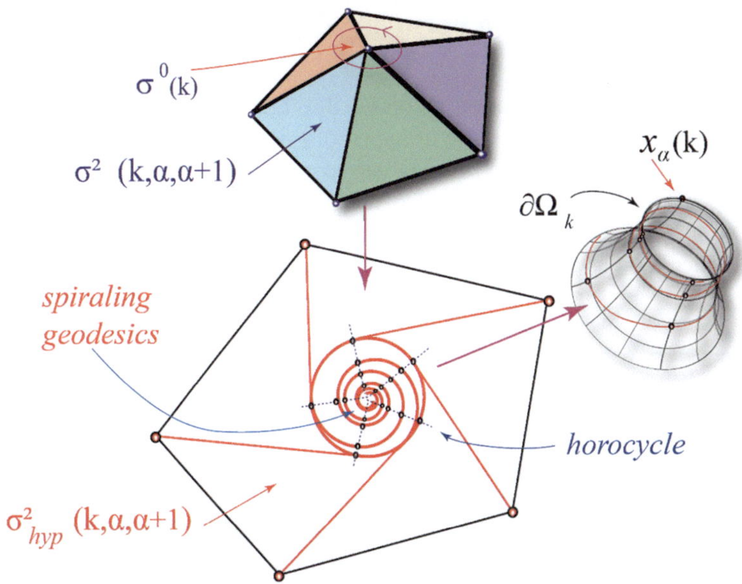

Fig. 3.10 A representation of how the conical singularity in the Euclidean star $star[\sigma^0(k)]$ generates a geodesic boundary $\partial\Omega_k$. Here the geodesic sides of the ideal triangles $\sigma^2_{hyp}(k, \alpha, \alpha + 1)$ are seen, from the intersecting horocicle curve, as spiraling down. Since adjacent sides of ideal triangles can glide with respect to each other, one can understand heuristically what goes on by drawing an analogy with a diaphragm shutter the blades of which are the ideal triangles

structure in $POL_{g, N_0}(M, \{\Theta(k)\}, A(M))$. This shows that the map Ω_Θ is a bijection, and completes the proof of Theorem 3.1. ∎

Remark 3.1 As discussed above, the completion defining the boundary component $\partial\Omega_k$ of $\Omega(T_l, M)$ is generated by adjoining, along the geodesics ending in $v^0(k)$, one limit point for each horocycle curve $t \mapsto F(t)$. Among these geodesics we have the $q(k)$ distinguished geodesics associated with the ideal edges $\{\sigma^1_{hyp}(k, h_\alpha)\}_{\alpha=1}^{q(k)}$. The corresponding limit points mark $q(k)$ distinguished points $\{x_\alpha(k)\}$ on the geodesic boundary $\partial\Omega_k$. Moreover, according to (3.46), these distinguished points are at a constant hyperbolic distance from each other,

$$d_{\mathbb{H}^3}(x_\alpha(k), x_{\alpha+1}(k)) = \frac{1}{q(k)} \left| \ln\left(\frac{\Theta(k)}{2\pi}\right) \right|, \tag{3.54}$$

where, as usual, the index α is defined modulo $q(k)$. $\square$

3.6   The Weil–Petersson Form on $\mathscr{T}^{met}_{g,N_0}(M, \{\Theta(k)\})$

According to the uniformization theorem, the Teichmüller space $\mathfrak{T}_{g,N_0}$ can be equivalently described as the space of hyperbolic metrics ds^2 of constant curvature -1 on $M/K^0(T)$, up to isometries isotopic to the identity relative to the marked points, (see Appendix B). Let us also recall that in such a hyperbolic setting, where points in $\mathfrak{T}_{g,N_0}$ are considered as hyperbolic surfaces with punctures, one can introduce a trivial bundle, Penner's decorated Teichmüller space,

$$\widetilde{\mathfrak{T}}_{g,N_0} \xrightarrow{\pi_{hor}} \mathfrak{T}_{g,N_0} \tag{3.55}$$

whose fiber over a punctured surface $\widetilde{\Omega}$ is the set of all N_0-tuples of horocycles in $\widetilde{\Omega}$, with one horocycle around each puncture. A section of this fibration is defined by choosing the total length of the horocycle assigned to each puncture in $\widetilde{\Omega}$.

In such a framework, an important role is played by ideal triangulations of the punctured surface $\widetilde{\Omega}$ and by the corresponding lambda-lengths associated with the decorated edges of the triangles σ^2_{hyp}. According to a seminal result by Penner ([16], Theorem 3.3.6), the pull-back $\pi^*_{hor}\omega_{WP}$ under the map $\pi_{hor} : \widetilde{\mathfrak{T}}_{g,N_0} \to \mathfrak{T}_{g,N_0}$ of the Weil-Petersson two-form ω_{WP} is explicitly given by

$$-2 \sum_{[\sigma^2_{hyp}]} d \ln \lambda_0 \wedge d \ln \lambda_1 + d \ln \lambda_1 \wedge d \ln \lambda_2 + d \ln \lambda_2 \wedge d \ln \lambda_0 , \tag{3.56}$$

where the sum runs over all ideal triangles σ^2_{hyp} whose ordered edges take the lambda-lengths $\lambda_0, \lambda_1, \lambda_2$. Note that (for dimensional reason) $\pi^*\omega_{WP}$ is a degenerate pre-symplectic form. As we have seen in the previous section, we can associate with a polyhedral surface $(T_l, M) \in \mathscr{T}^{met}_{g,N_0}(M)$ an incomplete hyperbolic structure defined by a ideal triangulation $\mathscr{H}(T_l, M)$ decorated by horocycles $\{F_k\}$. It easily follows that we can pull–back Penner's Kähler two-form $\pi^*_{hor}\omega_{WP}$ to $\mathscr{T}^{met}_{g,N_0}(M)$. We have

Theorem 3.2 *Let ω^{id}_{WP} denote the Penner representation of the Weil–Petersson form on the set of decorated ideal triangulations $\mathscr{T}^{ideal}_{g,N_0}(M/K^0) \otimes^{N_0}_{k=1} \mathbb{L}^+_{(k)}$, then to any polyhedral surface $(T_l, M) \in \mathscr{T}^{met}_{g,N_0}(M)$ we can associate the Weil–Petersson pre–symplectic form*

$$(T_l, M) \longmapsto \mathscr{H}^* \omega^{id}_{WP}(T_l) \tag{3.57}$$

$$= -2 \sum_{E(T)} \frac{d\,l(k,h) \wedge d\,l(h,j)}{l(k,h)l(h,j)} + \frac{d\,l(h,j) \wedge d\,l(j,k)}{l(h,j)l(j,k)} + \frac{d\,l(j,k) \wedge d\,l(k,h)}{l(j,k)l(k,h)} ,$$

where $\mathscr{H}$ is the mapping defined by Theorem 3.1, and $E(T)$ denotes the edge–set of the polyhedral surface (T_l, M).

Proof The Weil–Petersson form on the set of decorated ideal triangulations is explicitly given by

$$\omega^{id}_{WP} \tag{3.58}$$

$$= -2 \sum_{[\sigma^2_{hyp}]_{F(T)}} d \ln \lambda \left(\Sigma_k, \Sigma_h\right) \wedge d \ln \lambda \left(\Sigma_h, \Sigma_j\right)$$

$$+ d \ln \lambda \left(\Sigma_h, \Sigma_j\right) \wedge d \ln \lambda \left(\Sigma_j, \Sigma_k\right) + d \ln \lambda \left(\Sigma_j, \Sigma_k\right) \wedge d \ln \lambda \left(\Sigma_k, \Sigma_h\right).$$

The pull–back $\mathcal{H}^* \omega^{id}_{WP}(T_l)$ of ω^{id}_{WP} under the mapping $\mathcal{H}$, defined by Theorem 3.1, can be computed by exploiting the expressions (3.20), (3.21), (3.22) which provide the lambda-lengths in terms of the Euclidean edge-lengths $l(k, h)$ of the polyhedral surface $(T_l, M) \in \mathcal{T}^{met}_{g,N_0}(M)$. It is easily checked that under such a correspondence $\mathcal{H}^* \omega^{id}_{WP}(T_l)$ reduces to (3.57). ∎

According to Theorem 3.1, the hyperbolic structure that can be associated with a polyhedral surface generates, under completion, a hyperbolic surface $\Omega(T_l, M)$ with N_0 geodesic boundary components $\partial\Omega_k$ of length $|\partial\Omega_k| = |\ln(\Theta(k)/2\pi)|$. Thus, the delicate issue here is to extend (3.57) to the space $\mathcal{T}^{met}_{g,N_0}(M; \{\Theta(k)\})$ of polyhedral surfaces with a given set of conical angles $\{\Theta(k)\}$. Since the Weil–Petersson form (3.57) is degenerate, it is more appropriate to discuss this extension problem for the Poisson structure associated with (3.57), (see Sect. B.5). We have the following

Theorem 3.3 *To $\mathcal{T}^{met}_{g,N_0}(M)$ we can associate the Poisson structure generated by the Weil–Petersson bivector field whose local expression, at the generic polyhedral surface (T_l, M), is provided by*

$$\eta(T_l, M) = \frac{1}{8} \sum_k^{N_0} \sum_{1 \leq \alpha < \beta \leq q(k)} \frac{\sinh\left[\frac{1}{2}\left(1 - \frac{2(\beta-\alpha)}{q(k)}\right)\left|\ln\frac{\Theta(k)}{2\pi}\right|\right]}{\sinh\left[\frac{1}{2}\left|\ln\frac{\Theta(k)}{2\pi}\right|\right]} \times \tag{3.59}$$

$$\times \frac{\partial}{\partial \ln l(\alpha, \alpha + 1)} \wedge \frac{\partial}{\partial \ln l(\beta, \beta + 1)}.$$

Proof The proof of (3.59) rests on a remarkable result by G. Mondello [12] who has been able to extend to the case of hyperbolic surface with boundary the Poisson bivector associated with the Penner representation (3.56) of the Weil–Petersson form. Explicitly, we can adapt to our particular case the general Theorem 4.1 of [12], and consider the hyperbolic surface Ω, with N_0 geodesic boundaries $\partial\Omega_k$, generated by completing the ideal triangulation $(\mathcal{H}(T_l, M), \{F_k\})$ associated with (T_l, M). Recall that, according to Remark 3.1, each geodesic boundary component $\partial\Omega_k$ is decorated with the $q(k)$ points $\{x_\alpha(k)\}_{\alpha=1}^{q(k)}$ separated by a distance given by (3.54).

In such a case, (Theorem 4.1 of [12]), the Poisson bivector $\eta(\Omega)$ takes the explict form[2]

$$\eta(\Omega) = \frac{1}{8} \sum_{k}^{N_0} \sum_{1 \le \alpha < \beta \le q(k)} \frac{\sinh\left[\frac{1}{2}|\partial\Omega_k| - d_k(x_\alpha(k), x_\beta(k))\right]}{\sinh\left[\frac{1}{2}|\partial\Omega_k|\right]}$$

$$\times \frac{\partial}{\partial \ln \lambda(\alpha, \alpha + 1)} \wedge \frac{\partial}{\partial \ln \lambda(\beta, \beta + 1)},$$

(3.60)

where $d_k(x_\alpha(k), x_\beta(k))$ is the length of the geodesic arc running from $x_\alpha(k)$ to $x_\beta(k)$ along the oriented boundary $\partial\Omega_k$ in the positive direction, and where $\lambda(\alpha, \alpha + 1)$ denotes the λ–length associated with the ideal edge $\sigma^1_{hyp}(h_\alpha, h_{\alpha+1})$ dressed with the corresponding horocycles. Note that in [12], the above decorations are associated with the spine induced by a maximal system of disjoint simple geodesic arcs starting and ending perpendicularly at the boundaries. In our case, they are naturally generated by the horospherical correspondence $(T_l, M) \mapsto \mathscr{H}(T_l, M)$. Similarly to the computation leading to (3.57), we can exploit (3.20), (3.21), (3.22) providing the lambda-lengths in terms of the Euclidean edge-lengths $l(k, h)$ of the generating $(T_l, M) \in \mathscr{T}^{met}_{g,N_0}(M)$, so as to get

$$\frac{\partial}{\partial \ln \lambda(\alpha, \alpha + 1)} = \frac{\partial}{\partial \ln l(\alpha, \alpha + 1)}.$$

(3.61)

Moreover, since the points $\{x_\alpha(k)\}_{\alpha=1}^{q(k)}$ are separated by

$$d_{\mathbb{H}^3}(x_\alpha(k), x_{\alpha+1}(k)) = \frac{1}{q(k)} \left| \ln\left(\frac{\Theta(k)}{2\pi}\right) \right|,$$

(3.62)

(see (3.54)), we have

$$d_k(x_\alpha(k), x_\beta(k)) = \frac{(\beta - \alpha)}{q(k)} \left| \ln \frac{\Theta(k)}{2\pi} \right|,$$

(3.63)

for any $1 \le \alpha < \beta \le q(k)$. Introducing these expressions and the relation $|\partial\Omega_k| = \left| \ln \frac{\Theta(k)}{2\pi} \right|$ in (3.60) we easily get (3.59). ∎

The $\{\Theta(k)\}$–constant leaves of the Poisson structure defined by $\eta(T_l, M)$ are the spaces $\mathscr{T}^{met}_{g,N_0}(M; \{\Theta(k)\})$ of polyhedral surfaces with the given set of conical angles. We have the

---

[2]Since we use the λ–length and we ordered the summation over the indices α and β, (3.60) is off by a factor $1/4$ with respect to the expression for the Poisson bivector given in [12].

Corollary 3.1 *The restriction $\eta(T_l, M)|_{\{\Theta(k)\}}$ of the Poisson bivector to the $\{\Theta(k)\}$–constant leaves is non–degenerate and defines (by duality) a symplectic structure ω^{Θ}_{WP} on $\mathscr{T}^{met}_{g,N_0}(M; \{\Theta(k)\})$.*

Proof In analogy with the discussion of the conical symplectic form in Chap. 1, (see Sect. 1.9), let us consider the deformation induced in the edges $\{l(\alpha, \alpha + 1)\}$ by the variation, at constant $\Theta(k)$, of the vertex angles $\{\theta_\alpha(k)\}$, i.e.

$$(T_l, M) \;\longmapsto\; \left\{ l(k, \alpha + 1)\, \frac{\partial}{\partial \theta_\alpha(k)} \right\}^{q(k)-1}_{\alpha=1} \;,\quad k = 0, \ldots, N_0, \tag{3.64}$$

(see (1.118), with respect to which we have dropped the factor $\epsilon_\alpha(k)$, irrelevant for our purposes). Along such variations we have

$$\frac{\partial}{\partial \ln l(\alpha, \alpha + 1)} = \frac{l(\alpha, \alpha + 1)}{l(k, \alpha + 1)} \frac{\partial}{\partial \theta_\alpha(k)}, \tag{3.65}$$

and we can write (3.59) as

$$\eta(T_l, M) = \sum_{k}^{N_0} \sum_{1 \le \alpha < \beta \le q(k)} F_{(k)}(\alpha, \beta)\, \frac{\partial}{\partial \theta_\alpha(k)} \wedge \frac{\partial}{\partial \theta_\beta(k)}, \tag{3.66}$$

where

$$F_{(k)}(\alpha, \beta) := \frac{l(\alpha, \alpha + 1)\, l(\beta, \beta + 1)}{l(k, \alpha + 1)\, l(k, \beta + 1)} \times \frac{\sinh\left[\frac{1}{2}\left(1 - \frac{2(\beta-\alpha)}{q(k)}\right)\left|\ln \frac{\Theta(k)}{2\pi}\right|\right]}{\sinh\left[\frac{1}{2}\left|\ln \frac{\Theta(k)}{2\pi}\right|\right]}. \tag{3.67}$$

One can easily check that the following chain of identities holds

$$\sum_{1 \le \alpha < \beta \le q(k)} F_{(k)}(\alpha, \beta)\, \frac{\partial}{\partial \theta_\alpha(k)} \wedge \frac{\partial}{\partial \theta_\beta(k)} \tag{3.68}$$

$$= \sum_{v=1}^{q(k)-1} \sum_{\mu=v+1}^{q(k)} F_{(k)}(v, \mu)\, \frac{\partial}{\partial \theta_v(k)} \wedge \frac{\partial}{\partial \theta_\mu(k)}$$

$$= \sum_{v=1}^{q(k)-1} \sum_{\mu=v+1}^{q(k)-1} F_{(k)}(v, \mu)\, \frac{\partial}{\partial \theta_v(k)} \wedge \frac{\partial}{\partial \theta_\mu(k)}$$

$$+ \sum_{v=1}^{q(k)-1} F_{(k)}(v, q(k))\, \frac{\partial}{\partial \theta_v(k)} \wedge \frac{\partial}{\partial \theta_q(k)}$$

$$= \sum_{\nu=1}^{q(k)-1} \sum_{\mu=\nu+1}^{q(k)-1} F_{(k)}(\nu,\mu)\, \frac{\partial}{\partial\,\theta_\nu(k)} \wedge \frac{\partial}{\partial\,\theta_\mu(k)}$$

$$- \sum_{\eta=1}^{q(k)-1} \sum_{\nu=1}^{q(k)-1} F_{(k)}(\nu,q(k))\, \frac{\partial}{\partial\,\theta_\nu(k)} \wedge \frac{\partial}{\partial\,\theta_\eta(k)},$$

where, in the last line, we have exploited the fact that, since we work at constant $\Theta(k)$, we have

$$\frac{\partial}{\partial\theta_q(k)} = - \sum_{\eta=1}^{q(k)-1} \frac{\partial}{\partial\,\theta_\eta(k)}. \tag{3.69}$$

Moreover, we can write

$$\sum_{\eta=1}^{q(k)-1} \sum_{\nu=1}^{q(k)-1} F_{(k)}(\nu,q(k))\, \frac{\partial}{\partial\,\theta_\nu(k)} \wedge \frac{\partial}{\partial\,\theta_\eta(k)} \tag{3.70}$$

$$= \sum_{1\le\nu<\eta\le q(k)-1} \left[F_{(k)}(\nu,q(k)) - F_{(k)}(\eta,q(k)) \right] \frac{\partial}{\partial\,\theta_\nu(k)} \wedge \frac{\partial}{\partial\,\theta_\eta(k)},$$

so that we eventually get

$$\sum_{1\le\alpha<\beta\le q(k)} F_{(k)}(\alpha,\beta)\, \frac{\partial}{\partial\,\theta_\alpha(k)} \wedge \frac{\partial}{\partial\,\theta_\beta(k)} \tag{3.71}$$

$$= \sum_{1\le\alpha<\beta\le q(k)-1} \left[F_{(k)}(\alpha,\beta) - F_{(k)}(\alpha,q(k)) + F_{(k)}(\beta,q(k)) \right] \frac{\partial}{\partial\,\theta_\alpha(k)} \wedge \frac{\partial}{\partial\,\theta_\beta(k)}.$$

Introducing this result in (3.66) yields

$$\eta(T_l,M) = \sum_{k=1}^{N_0} \sum_{1\le\alpha<\beta\le q(k)-1} \left[F_{(k)}(\alpha,\beta) \right. \tag{3.72}$$

$$\left. - F_{(k)}(\alpha,q(k)) + F_{(k)}(\beta,q(k)) \right] \frac{\partial}{\partial\,\theta_\alpha(k)} \wedge \frac{\partial}{\partial\,\theta_\beta(k)}.$$

Thus, in order to prove the non–degeneracy of $\eta(T_l,M)$ over $\mathscr{T}^{met}_{g,N_0}(M;\{\Theta(k)\})$, we need to prove that, for $1 \le \alpha < \beta \le q(k)-1$, we always have

$$\left[F_{(k)}(\alpha,\beta) - F_{(k)}(\alpha,q(k)) + F_{(k)}(\beta,q(k)) \right] \ne 0, \tag{3.73}$$

as we move around the vertices of (T_l, M). An elegant way of providing such a check is to note that the function $F_{(k)}(\alpha, \beta)$, defined over $\mathscr{T}_{g,N_0}^{met}(M; \{\Theta(k)\})$, naturally extends to the Teichmüller space $\mathfrak{T}_{g,N_0}(M)$ of N_0–pointed closed Riemann surfaces of genus g, and then exploit the following deep property of equilateral triangulations

Theorem 3.4 (Belyĭ's theorem [2]) *The set of compact Riemann surfaces associated with equilateral polyhedral surfaces of genus g is a countable and dense subset of the Teichmüller space $\mathfrak{T}_{g,N_0}(M)$.*

Thus, according to *Belyĭ*'s theorem, we can limit our analysis to equilateral polyhedral surfaces. This considerably simplifies (3.73), and one easily checks that the vanishing of (3.73) necessarily implies that

$$e^{-d_k(x_\alpha(k), x_\beta(k))} + e^{-|\partial\Omega_k| + d_k(x_\alpha(k), x_q(k))} + e^{-d_k(x_\beta(k), x_q(k))} \tag{3.74}$$

$$= e^{-|\partial\Omega_k| + d_k(x_\alpha(k), x_\beta(k))} + e^{-d_k(x_\alpha(k), x_q(k))} + e^{-|\partial\Omega_k| + d_k(x_\beta(k), x_q(k))},$$

where $d_k(x_\alpha(k), x_\beta(k))$ is provided by (3.63), and $|\partial\Omega_k|$ is the geodesic boundary length given by (3.52). The condition (3.74) holds iff $\alpha = \beta$ and $|\partial\Omega_k| = 0$, thus under the stated hypotheses (3.73) never vanishes on the subset of equilateral triangulations $\subset \mathscr{T}_{g,N_0}^{met}(M; \{\Theta(k)\})$. By density, this extends to the whole $\mathscr{T}_{g,N_0}^{met}(M; \{\Theta(k)\})$. It follows that $\eta(T_l, M)|_{\{\Theta(k)\}}$ is non–degenerate and defines on $\mathscr{T}_{g,N_0}^{met}(M; \{\Theta(k)\})$ a symplectic structure. ∎

Remark 3.2 The density property of equilateral triangulated surfaces (Dynamical Triangulations) is strictly related to a result, due to V. A. Voevodskii and G. B. Shabat [19], which establishes a remarkable bijection between dynamical triangulations and curves over algebraic number fields. The proof in [19] exploits the characterization of the collection of algebraic curves defined over the algebraic closure $\overline{\mathbb{Q}}$ of the field of rational numbers, (i.e., over the set of complex numbers which are roots of non-zero polynomials with rational coefficients), provided by Belyĭ's theorem [2]. According to such a result, a nonsingular Riemann surface M has the structure of an algebraic curve defined over $\overline{\mathbb{Q}}$ if and only if there is a holomorphic map (a branched covering of M over the sphere)

$$f : M \to \mathbb{CP}^1 \tag{3.75}$$

that is ramified only at $0, 1$ and ∞, (such maps are known as Belyĭ maps). In other words, the Riemann surface associated with a dynamical triangulation (with edge-lengths normalized to $a = 1$) is, in a canonical way, a ramified covering of the Riemann sphere $\mathbb{CP}^1$ with ramification locus contained in $\{0, 1, \infty\}$. The triangulation (actually its barycentrically dual ribbon graph decorated with the corresponding quadratic differential) is the preimage of the interval $[0, 1]$, in particular the set of vertices appears as the preimage of 0 and the set of half-cylinders over the cells $\{\rho^2(k)\}$ as the preimage of ∞. The mid points of the edges $\{\rho^1(h)\}$, (identified with the barycenters of the edges of $|T_{a=1}| \to M$), correspond to the preimage of 1. Moreover, every branched covering of $\mathbb{CP}^1$ defining a Riemann surface over $\overline{\mathbb{Q}}$ arises in this fashion. It is worthwhile remarking that the inverse image of the line segment $[0, 1] \subset \mathbb{CP}^1$ under a Belyĭ map is a Grothendieck's *dessin d'enfant*, thus dynamically triangulated surfaces are eventually connected with the theory of the Galois group $Gal(\overline{\mathbb{Q}}/\mathbb{Q})$ action on the branched coverings $f : M \to \mathbb{CP}^1$. The correspondence between Belyĭ maps, *dessin d'enfant* and JS quadratic differentials has been analyzed in depth by M. Mulase and M. Penkava [14], an equally inspiring paper is [1] by M. Bauer and C. Itzykson.

3.7 The Symplectic Volume of the Space of Polyhedral Structures

As we have discussed in Sect. 1.11, the space of polyhedral structures $POL_{g,N_0}(M)$ is not a compact orbifold since triangulated surfaces can degenerate. In line with the relation with the stable compactification $\overline{\mathfrak{M}}_{g,N_0}$ of the Riemann moduli space, we have specifically considered those degenerations yielding for a suitable notion of stable polyhedral surface, introducing the space $\overline{POL}_{g,N_0}$, (see Theorem 1.7, and Definition 1.23). Thus, it is important to discuss how the symplectic form ω_{WP}^{Θ} extends to $\overline{POL}_{g,N_0}$. In full generality the detailed analysis of the behavior, under degenerations, of the combinatorial representative ω_{WP}^{Θ} of the Weil–Petersson form, as well as of the Euler classe ω_k associated with the circle bundle $(\mathcal{Q}_{(k)},\, \mathscr{T}_{g,N_0}^{met}(M, \{\Theta(k)\}))$, is extremely delicate, (for the Witten–Kontsevich theory see [8] and the detailed analysis in [10, 13]). For our purposes it is sufficient to discuss how the Weil–Petersson (W–P) bivector (3.59) $\eta(T_l, M)$ behaves in presence of a pinching node in a stable polyhedral surface. According to the characterization of stable polyhedral surface described in Theorem 1.7, and Definition 1.23, we have

Lemma 3.3 *Let* $(\widehat{T}_l, M) \in \overline{POL}_{g,N_0}$ *be the stable polyhedral surface we get from* (T_l, M) *by metrically collapsing, along a shrinking circle* S_k*, a polyhedral surface* (T_l, M)*. Then the W–P bivector* $\eta(T_l, M)$ *naturally extends to* $(\widehat{T}_l, M)$*.*

Proof Let $(\widehat{T}_l, M) \in \overline{POL}_{g,N_0}$ be the stable polyhedral surface we get from (T_l, M) by metrically collapsing the shrinking circle S_k, and let $\sigma_*^0(S_k)$ be the corresponding pinching point in $(\widehat{T}_l, M)$. Assume, for simplicity that the shrinking circle S_k is homotopically trivial, (i.e., we are not pinching a handle), so that at the nodal point $\sigma_*^0(S_k)$, we can separate $(\widehat{T}_l, M)$ into the two components $(\widehat{T}_l, M)_{(1)}$ and $(\widehat{T}_l, M)_{(2)}$ resulting from the pinching, with

$$(\widehat{T}_l, M) = (\widehat{T}_l, M)_{(1)} \sqcup_{\sigma_*^0(S_k)} (\widehat{T}_l, M)_{(2)}. \tag{3.76}$$

Let us denote by $\sigma_{*,1}^0(1)$ the conical vertex on $(\widehat{T}_l, M)_{(1)}$ associated with the pinching point $\sigma_*^0(S_k)$, similarly we denote by $\sigma_{*,2}^0(1)$ the corresponding vertex on $(\widehat{T}_l, M)_{(2)}$. To the component $(\widehat{T}_l, M)_{(1)}$ we can associate the corresponding Weil–

Petersson bivector, (see (3.66)),

$$
\eta(\widehat{T}_l, M)_{(1)} = F^1_{(1)}(\alpha, \beta) \, \frac{\partial}{\partial\,\theta^{(1)}_\alpha(1)} \wedge \frac{\partial}{\partial\,\theta^{(1)}_\beta(1)} \tag{3.77}
$$

$$
+ \sum_{k=2}^{N_0(1)} \sum_{1\le\alpha<\beta\le q(k)} F^1_{(k)}(\alpha, \beta) \, \frac{\partial}{\partial\,\theta^{(1)}_\alpha(k)} \wedge \frac{\partial}{\partial\,\theta^{(1)}_\beta(k)},
$$

where we have isolated the contribution coming from $\sigma^0_{*,1}(1)$. For the polyhedral surface $(\widehat{T}_l, M)_{(1)}$ we have an analogous expression

$$
\eta(\widehat{T}_l, M)_{(2)} = F^2_{(1)}(\alpha, \beta) \, \frac{\partial}{\partial\,\theta^{(2)}_\alpha(1)} \wedge \frac{\partial}{\partial\,\theta^{(2)}_\beta(1)} \tag{3.78}
$$

$$
+ \sum_{k=2}^{N_0(2)} \sum_{1\le\alpha<\beta\le q(k)} F^2_{(k)}(\alpha, \beta) \, \frac{\partial}{\partial\,\theta^{(2)}_\alpha(k)} \wedge \frac{\partial}{\partial\,\theta^{(2)}_\beta(k)}.
$$

The pinching prescription we have adopted in Theorem 1.7 and Definition 1.23 implies that, modulo the action of a finite number of flip moves, at the vertices $\sigma^0_{*,1}(1)$ and $\sigma^0_{*,2}(1)$ the corresponding vertex angles are pairwise equal, i.e. $\theta^{(2)}_\alpha(1) = \theta^{(2)}_\alpha(1)$, so that by identifying $\sigma^0_{*,1}(1)$ and $\sigma^0_{*,2}(1)$ via an orientation reversing homeomorphism we get that at the pinching node $\sigma^0_*(S_k)$ the contribution to the W–P bivector vanishes,

$$
F^1_{(1)}(\alpha, \beta) \, \frac{\partial}{\partial\,\theta^{(1)}_\alpha(1)} \wedge \frac{\partial}{\partial\,\theta^{(1)}_\beta(1)} - F^2_{(1)}(\alpha, \beta) \, \frac{\partial}{\partial\,\theta^{(2)}_\alpha(1)} \wedge \frac{\partial}{\partial\,\theta^{(2)}_\beta(1)} \equiv 0. \tag{3.79}
$$

Thus, the W–P bivector associated with the stable polyhedral surface $(\widehat{T}_l, M) \in \overline{POL}_{g,N_0}$ can be written as

$$
\eta(\widehat{T}_l, M) = \sum_{k=2}^{N_0(1)} \sum_{1\le\alpha<\beta\le q(k)} F^1_{(k)}(\alpha, \beta) \, \frac{\partial}{\partial\,\theta^{(1)}_\alpha(k)} \wedge \frac{\partial}{\partial\,\theta^{(1)}_\beta(k)} \tag{3.80}
$$

$$
+ \sum_{k=2}^{N_0(2)} \sum_{1\le\alpha<\beta\le q(k)} F^2_{(k)}(\alpha, \beta) \, \frac{\partial}{\partial\,\theta^{(2)}_\alpha(k)} \wedge \frac{\partial}{\partial\,\theta^{(2)}_\beta(k)},
$$

with $N_0(1)+N_0(2) = N_0$. The argument can be easily adapted to a handle–pinching. Thus, the Weil–Petersson bivector naturally extends to $\overline{POL}_{g,N_0}$. $\blacksquare$

The symplectic form ω^Θ_{WP} associated with the Weil–Petersson bivector $\eta(T_l, M)$ endows $\mathscr{T}^{met}_{g,N_0}(M; \{\Theta(k)\})$, and its stable closure $\overline{\mathscr{T}}^{\,met}_{g,N_0}(M; \{\Theta(k)\})$, with a well–defined notion of symplectic volume according to

$$Vol_{symp}\left[\mathscr{T}^{met}_{g,N_0}(M;\{\Theta(k)\})\right] := \int_{\overline{\mathscr{T}}^{met}_{g,N_0}(M;\{\Theta(k)\})} \exp\left(\omega^{\Theta}_{WP}\right), \qquad (3.81)$$

where $\exp\left(\omega^{\Theta}_{WP}\right)$ means that we are integrating over the maximal power of the Weil–Petersson form over $\overline{\mathscr{T}}^{met}_{g,N_0}(M;\{\Theta(k)\})$, i.e.

$$\exp\left(\omega^{\Theta}_{WP}\right) := \frac{\left(\omega^{\Theta}_{WP}\right)^m}{m!}, \qquad (3.82)$$

with $m = 2N_0 + 6g - 6 = \dim \mathscr{T}^{met}_{g,N_0}(M;\{\Theta(k)\})$, is the symplectic volume form.

Let us recall, (see (2.87)), that $\mathscr{T}^{met}_{g,N_0}(M\{\Theta(k)\})$ is a local model for the orbispace $POL_{g,N_0}(\Theta,A)$ of polyhedral structures of given area $A := A(M)$, and with a given sequence of conical angles $\Theta := \{\Theta\}^{N_0}_{k=1}$, (since notation wants to travel light we have dropped a few k's and M's with respect to (2.87)),

$$POL_{g,N_0}(\Theta,A) := \left.\bigsqcup_{[T]\in\mathscr{T}_{g,N_0}} \left.\frac{\mathbb{R}^{N_1}_{+}}{Aut(T)}\right|\right|_{(\Theta,A)}, \qquad (3.83)$$

where the union runs over the set $\mathscr{T}_{g,N_0}$ of equivalence classes[3] $[T]$ of *distinct* triangulations (T,M), (in the sense of Definition 1.7), with $N_0(T) = N_0$ labelled vertices satisfying the topological constraints $N_0 - N_1(T) + N_2(T) = 2 - 2g$, $2N_1(T) = 3N_2(T)$. Note that these equivalence classes $[T]$ in $\mathscr{T}_{g,N_0}$ label the distinct metrical orbicells $\mathbb{R}^{N_1}_{+}/Aut(T)$. It follows that we can extend (3.81) to an orbifold integration characterizing the symplectic volume of $POL_{g,N_0}(\Theta,A)$ according to

$$Vol_{symp}\left[POL_{g,N_0}(\Theta,A)\right] = \int_{\overline{POL}_{g,N_0}(\Theta,A)} \exp\left(\omega^{\Theta}_{WP}\right) \qquad (3.84)$$

$$:= \sum_{[T]\in\mathscr{T}_{g,N_0}} \frac{1}{|Aut(T)|} \int_{\overline{\mathscr{T}}^{met}_{g,N_0}(M;\Theta)\big|_{[T]}} \exp\left(\omega^{\Theta}_{WP}\right),$$

where the sum runs over the finite set of equivalence classes $[T]$ of distinct triangulations (T,M) in $\mathscr{T}_{g,N_0}$, and where

$$\overline{\mathscr{T}}^{met}_{g,N_0}(M;\Theta)\bigg|_{[T]} \qquad (3.85)$$

---

[3]Recall that any two triangulations $(T^{(1)},M)$ and $(T^{(2)},M)$ in $\mathscr{T}_{g,N_0}$ are considered equivalent iff they have the same incidence relations-see Definition 1.7.

denotes the set of stable polyhedral surfaces in $\overline{\mathscr{T}}_{g,N_0}^{\,met}(M;\Theta)$ whose incidence is in the equivalence class $[T]$ defined by (T,M). The above summation over $\mathscr{T}_{g,N_0}$ is a finite sum over a finite number of orbifold strata. Each stratum being labelled by the equivalence class $[T]$ *representative* of all the PL-equivalent triangulations with the same incidence relations of (T,M), (see also Theorem 1.3). Finally, in order to avoid overcounting the same stratum, we have divided by the order $|Aut(T)|$ of the automorphism group of the stratum $[T]$ corresponding to (T,M).

One may well argue that a more natural notion of volume to consider would have been that obtained by exploiting directly the Euclidean volume form

$$\bigwedge_{E(T)} dl(\alpha,\beta)\Bigg|_{(\Theta,A)}, \tag{3.86}$$

where the exterior product runs over the edge set of the triangulations $(T_l,M) \in \mathscr{T}_{g,N_0}^{met}(M;\Theta)$. In applications (typically Regge calculus) this is often the case. However, the Euclidean volume of $POL_{g,N_0}(\Theta,A)$ is very difficult to compute (even if we limit the analysis to a single orbifold strata as is, somewhat artificially, done in Regge calculus).

The relevance of the symplectic volume (3.84) over the Euclidean volume[4] of $POL_{g,N_0}(\Theta,A)$ stems from the relations (2.86) and (3.42), provided by the Theorems 2.4 and 3.1. These relations provide bijections between the spaces of polyhedral structures $POL_{g,N_0}(M)$, $POL_{g,N_0}(\Theta,A)$ and the (decorated) Riemann moduli spaces $\mathfrak{M}_{g,N_0}$ and $\mathfrak{M}_{g,N_0}(L)$, respectively. Both such bijections (2.86) and (3.42) extend to the respective stable compactifications, i.e.

$$\Upsilon \,:\, \overline{POL}_{g,N_0}(M) \longrightarrow \overline{\mathfrak{M}}_{g,N_0} \times \mathbb{R}_+ \times \mathbb{R}_{\geq 0}^{N_0-1}, \tag{3.87}$$

$$(T_l,M) \longmapsto \left[((M;N_0),\mathscr{C}),A,\{\Theta(k)\}^{N_0}\right],$$

$$\Omega_\Theta \,:\, \overline{POL}_{g,N_0}(\Theta(k),A) \longrightarrow \overline{\mathfrak{M}}_{g,N_0}(L) \times \mathbb{Z}_2^{N_0}, \tag{3.88}$$

and allow us to characterize a natural notion of volume, for the space of polyhedral structures POL_{g,N_0} and $POL_{g,N_0}(\Theta,A)$, in terms of the symplectic volumes of the Riemann moduli spaces $\mathfrak{M}_{g,N_0}$ and $\mathfrak{M}_{g,N_0}(L)$. We have

Theorem 3.5 *If $POL_{g,N_0}(A)$ and $POL_{g,N_0}(\Theta,A)$ respectively denote the set of polyhedral structures in $POL_{g,N_0}(M)$ with given polyhedral area A and with a given sequence of conical angles $\{\Theta(k)\}_{k=1}^{N_0}$, (always at fixed A), then the symplectic volumes of the Riemann moduli spaces $\mathfrak{M}_{g,N_0}$ and $\mathfrak{M}_{g,N_0}(L)$ induce natural volume forms on $POL_{g,N_0}(A)$ and $POL_{g,N_0}(\Theta,A)$ for which we compute*

---

[4]The interplay between the Euclidean volume and the symplectic volume, instrumental to the Witten–Kontsevich intersection theory over moduli space, is discussed in depth in [5].

$$Vol\left[POL_{g,N_0}(A)\right] = \frac{\sqrt{N_0}\left[2\pi\left(N_0 + 2g - 2\right)\right]^{(N_0-1)}}{(N_0 - 1)!\sqrt{2^{N_0-1}}}\, Vol_{WP}\left[\mathfrak{M}_{g,N_0}\right], \qquad (3.89)$$

and

$$Vol\left[POL_{g,N_0}(\Theta, A)\right] = Vol_{WP}\left[\mathfrak{M}_{g,N_0}(L)\right]. \qquad (3.90)$$

Proof We start proving (3.90). Since by construction $\omega_{WP}^{\Theta} = \Omega_{\Theta}^*[\omega_{WP}]$, under the action of the map Ω_{Θ} we can write

$$Vol_{WP}\left[\mathfrak{M}_{g,N_0}(L)\right] := \int_{\overline{\mathfrak{M}_{g,N_0}(L)}} \exp\left(\omega_{WP}\right) \qquad (3.91)$$

$$= \int_{\Omega_{\Theta}[\overline{POL_{g,N_0}(\Theta,A)}]} \exp\left(\omega_{WP}\right) = \int_{\overline{POL_{g,N_0}(\Theta,A)}} \exp\left(\omega_{WP}^{\Theta}\right)$$

$$:= Vol\left[POL_{g,N_0}(\Theta, A)\right],$$

as stated. Note that $Vol\left[POL_{g,N_0}(\Theta, A)\right]$ is the symplectic volume of $POL_{g,N_0}(\Theta, A)$ with respect to the natural symplectic form ω_{WP}^{Θ}. In order to prove (3.89), let us remark that the possible conical angles $\{\Theta(k)\}$ in $POL_{g,N_0}(A)$ are (only) subjected to the Gauss–Bonnet constraint (2.75), which we can rewrite as

$$\sum_{k}^{N_0} \Theta(k) = 2\pi\left(N_0 + 2g - 2\right). \qquad (3.92)$$

It follows that the set of possible $\{\Theta(k)\}$ in $POL_{g,N_0}(A)$ vary in the $(N_0 - 1)$-dimensional simplex in $\mathbb{R}^{N_0}$

$$\Delta_{\Theta} := \left\{ \Theta(k) \geq 0,\ \sum_{k=1}^{N_0-1} \Theta(k) = 2\pi\left(N_0 + 2g - 2\right)\right\}, \qquad (3.93)$$

(for simplicity, we allow also for degenerate angles $\Theta(k) = 0$). This can be considered as the $2\pi\left(N_0 + 2g - 2\right)$–dilated standard unit simplex

$$\Delta^{N_0-1} : \left\{(x^1, \ldots, x^{N_0}) \in \mathbb{R}_{\geq 0}^{N_0} \,\middle|\, \sum_{k=1}^{N_0} x^k = 1\right\} \qquad (3.94)$$

in $\mathbb{R}^{N_0}$. Note that Δ_{Θ} carries the natural volume form induced by the Euclidean measure $d\mu_E^{N_0-1}$ in the affine hyperplane $\sum_{k=1}^{N_0} x^k = 1$, $(x^1, \ldots, x^{N_0}) \in \mathbb{R}^{N_0}$. Since

$Vol_{WP}\left[\mathfrak{M}_{g,N_0}\right]$ does not depend on the actual distribution of the $\{\Theta(k)\}_{k=1}^{N_0}$, under the action of the map Υ, (see (2.86)), we can write

$$Vol_{WP}\left[\mathfrak{M}_{g,N_0}\right] \times Vol_E\left[\Delta_\Theta\right] := \int_{\Delta_\Theta} d\mu_E^{N_0-1} \int_{\overline{\mathfrak{M}}_{g,N_0}} \exp\left(\omega_{WP}\right) \tag{3.95}$$

$$= \int_{\Upsilon\left[\overline{POL}_{g,N_0}(A)\right]} d\mu_E^{N_0-1} \exp\left(\omega_{WP}\right) = \int_{\Delta_\Theta} d\mu_E^{N_0-1} \int_{\overline{POL}_{g,N_0}(\Theta,A)} \exp\left(\omega_{WP}^\Theta\right)$$

$$:= Vol\left[POL_{g,N_0}(A)\right],$$

where we have exploited the fact that the bijection Υ is the identity when considered as a map from the set of all possible conical angles $\{\Theta(k)\}_{k=1}^{N_0}$ to the simplex Δ_Θ, and that, at fixed $\{\Theta(k)\}_{k=1}^{N_0}$ we have

$$\Upsilon^*(\omega_{WP}) = \omega_{WP}^\Theta. \tag{3.96}$$

The Euclidean volume of the standard unit simplex Δ^{N_0-1} is given by

$$Vol_E\left[\Delta^{N_0-1}\right] = \frac{\sqrt{N_0}}{(N_0-1)!\sqrt{2^{N_0-1}}}. \tag{3.97}$$

Since Δ_Θ is the $2\pi(N_0+2g-2)$–dilated standard unit simplex, we immediately get

$$Vol_E\left[\Delta_\Theta\right] = \frac{\sqrt{N_0}\left[2\pi(N_0+2g-2)\right]^{(N_0-1)}}{(N_0-1)!\sqrt{2^{N_0-1}}}, \tag{3.98}$$

from which, once inserted in (3.95), the relation (3.89) follows. ∎

Remark 3.3 From the above analysis, in particular from the chain of passages (3.95) leading to the identification between $Vol_{WP}\left[\mathfrak{M}_{g,N_0}\right] \times Vol_E\left[\Delta_\Theta\right]$ and the volume of the space of polyhedral surfaces $POL_{g,N_0}(A)$, it follows that it is natural to introduce the following

Definition 3.1 If $F : POL_{g,N_0}(A) \rightarrow \mathbb{R}$ is a (measurable) function over $\overline{POL}_{g,N_0}(A)$, then we set

$$\int_{\overline{POL}_{g,N_0}(A)} d\mu_E^{N_0-1} \exp\left(\omega_{WP}\right) F[(T_l,M)] \tag{3.99}$$

$$:= \int_{\Delta_\Theta} d\mu_E^{N_0-1} \int_{\overline{POL}_{g,N_0}(\Theta,A)} \exp\left(\omega_{WP}^\Theta\right) F[(T_l,M)], \tag{3.100}$$

It follows, from this definition and Theorem 3.5, that the average value of F over $POL_{g,N_0}(A)$ defined by

$$\langle F \rangle_{POL_{g,N_0}(A)} := \frac{\int_{\overline{POL}_{g,N_0}(A)} d\mu_E^{N_0-1} \exp(\omega_{WP})\, F[(T_l, M)]}{Vol\left[POL_{g,N_0}(A)\right]} \tag{3.101}$$

is provided by

$$\langle F \rangle_{POL_{g,N_0}(A)} = \frac{(N_0-1)!\,\sqrt{2^{N_0-1}}}{\sqrt{N_0}\,[2\pi\,(N_0+2g-2)]^{(N_0-1)}\, Vol_{WP}\left[\mathfrak{M}_{g,N_0}\right]} \tag{3.102}$$

$$= \int_{\overline{POL}_{g,N_0}(A)} d\mu_E^{N_0-1} \exp(\omega_{WP})\, F[(T_l, M)].$$

Since the computation of the volumes of $POL_{g,N_0}(A)$ and $POL_{g,N_0}(\Theta, A)$ involve orbifold integrations, the above theorem and the relation (3.84) imply the following explicit representation of (3.89) and (3.90).

Theorem 3.6 *If $\overline{\mathcal{T}}_{g,N_0}^{met}(M;\Theta)$ denotes the set of stable polyhedral surfaces locally modeling $POL_{g,N_0}(\Theta, A)$, then*

$$Vol_{WP}\left[\mathfrak{M}_{g,N_0}(L)\right] = \sum_{[T]\in\mathcal{T}_{g,N_0}} \frac{1}{|Aut(T)|} \int_{\overline{\mathcal{T}}_{g,N_0}^{met}(M;\Theta)\Big|_{[T]}} \exp\left(\omega_{WP}^{\Theta}\right), \tag{3.103}$$

and

$$Vol_{WP}\left[\mathfrak{M}_{g,N_0}\right] = \frac{(N_0-1)!\,\sqrt{2^{N_0-1}}}{\sqrt{N_0}\,[2\pi\,(N_0+2g-2)]^{(N_0-1)}} \tag{3.104}$$

$$\times \sum_{[T]\in\mathcal{T}_{g,N_0}} \frac{1}{|Aut(T)|} \int_{\Delta_\Theta} d\mu_E^{N_0-1} \int_{\overline{\mathcal{T}}_{g,N_0}^{met}(M;\Theta)\Big|_{[T]}} \exp\left(\omega_{WP}^{\Theta}\right),$$

where the sum runs over the finite set of equivalence classes $[T]$ of distinct triangulations (T, M) in $\mathcal{T}_{g,N_0}$, and where $\overline{\mathcal{T}}_{g,N_0}^{met}(M;\Theta)\Big|_{[T]}$ denotes the set of stable polyhedral surfaces in $\overline{\mathcal{T}}_{g,N_0}^{met}(M;\Theta)$ whose incidence is in the equivalence class $[T]$ defined by (T, M).

Proof The representation (3.103) is an immediate consequence of the characterization (3.84) of orbifold integration over $\overline{POL}_{g,N_0}(\Theta, A)$ and of (3.90). The representation (3.104) follows similarly by exploiting (3.84) in the relation (3.95). ∎

Both these representations of the symplectic volumes of the Riemann moduli spaces $Vol_{WP}\left[\mathfrak{M}_{g,N_0}(L)\right]$ and $Vol_{WP}\left[\mathfrak{M}_{g,N_0}\right]$ in terms of the set of polyhedral surfaces

$\mathcal{T}_{g,N_0}^{met}(M;\Theta)$ will play a key role in the applications to quantum gravity and string theory that we describe in the next chapters.

References

1. Bauer, M., Itzykson, C.: Triangulations. In: Schneps, L. (ed.) The Grothendieck Theory of Dessins d'Enfants. London Mathematical Society Lecture Note Series, vol. 200, p. 179. Cambridge University Press, Cambridge (1994)
2. Belyi, G.V.: On Galois extensions of a maximal cyclotomic fields. Math. USSR Izv. **14**, 247–256 (1980)
3. Benedetti, R., Petronio, C.: Lectures on Hyperbolic Geometry. Universitext. Springer, New York (1992)
4. Carfora, M., Dappiaggi, C., Gili, V.L.: Triangulated surfaces in Twistor space: a kinematical set up for open/closed string duality. JHEP **0612**, 017 (2006). arXiv:hep-th/0607146
5. Chapman, M.K., Mulase, M., Safnuk, B.: The Kontsevich constants for the volume of the moduli of curves and topological recursion. arXiv:1009.2055 [math.AG]
6. Do, N.: The asymptotic Weil–Petersson form and intersection theory on $\overline{\mathfrak{M}}_{g,n}$. arXiv:1010.4126v1 [math.GT]
7. Gallego, E., Reventós, A., Solanes, G., Teufel, E.: Horospheres in hyperbolic geometry (Preprint 2008). http://mat.uab.cat/~egallego/pub.html
8. Looijenga, E.: Cellular decomposition of compactified moduli spaces of pointed curves. In: Dijkgraaf, R., Faber, C., van der Geer, G. (eds.) The Moduli Space of Curves, pp. 369–400. Birkhäuser, Boston (1995)
9. Manin, Y.I., Zograf, P.: Invertible cohomological filed theories and Weil-Petersson volumes. Annales de l' Institute Fourier **50**, 519 (2000). arXiv:math-ag/9902051
10. Mondello, G.: Combinatorial classes on the moduli space of Riemann surfaces are tautological. Int. Math. Res. Not. **44**, 2329–2390 (2004)
11. Mondello, G.: Riemann surfaces, arc systems and Weil-Petersson form. Bollettino U.M.I. **1**(9), 751–766 (2008)
12. Mondello, G.: Triangulated Riemann surfaces with boundary and the Weil-Petersson Poisson structure. J. Differ. Geom. **81**, 391–436 (2009)
13. Mondello, G.: Riemann surfaces, ribbon graphs and combinatorial classes. In: Papadopoulos, A. (ed.) Handbook of Teichmüller Theory, vol. 2, pp. 151–216. European Mathematical Society, Zurich (2009)
14. Mulase, M., Penkava, M.: Ribbon graphs, quadratic differentials on Riemann surfaces, and algebraic curves defined over $\overline{\mathbb{Q}}$. Asian J. Mater. Sci. **2**(4), 875–920 (1998). math-ph/9811024 v2
15. Penner, R.C.: The decorated Teichmüller space of punctured surfaces. Commun. Math. Phys. **113**, 299–339 (1987)
16. Penner, R.C.: Weil-Petersson volumes. J. Diff. Geom. **35**, 559 (1992)
17. Penner, R.C.: Lambda lengths. Lecture Notes from a Master's Class on Decorated Teichmuller Theory, Aarhus University (2006). Available on line at the www.ctqm.au.dk/research/MCS/lambdalengths.pdf
18. Thurston, W.P.: Three-Dimensional Geometry and Topology, vol. 1 (edited by S. Levy). Princeton University Press (1997). See also the full set of Lecture Notes (December 1991 Version), electronically available at the Math. Sci. Res. Inst. (Berkeley)
19. Voevodskii, V.A., Shabat, G.B.: Equilateral triangulations of Riemann surfaces, and curves over algebraic number fields. Sov. Math. Dokl. **39**, 38 (1989)
20. Zograf, P.: Weil-Petersson volumes of moduli spaces of curves and curves and the genus expansion in two dimensional. arXiv:math.AG/9811026

Chapter 4
The Quantum Geometry of Polyhedral Surfaces: Non–linear σ Model and Ricci Flow

In this chapter we discuss in great detail the connection between Non–Linear σ Model and Ricci Flow. In recent years, Ricci flow has been the point of departure and the motivating example for important developments in geometric analysis, most spectacularly for G. Perelman's proof of Thurston's geometrization program for three-manifolds and of the attendant Poincaré conjecture. For one of those strange circumstances not unusual in the history of Science, the Ricci flow, introduced in the early 1980s by Richard Hamilton, independently appeared on the scene also in Physics. Indeed, Daniel Friedan, studying the weak coupling limit of the renormalization group flow for non-linear sigma models, introduced what later on came to be known as the Hamilton–DeTurck version of the Ricci flow. This QFT avatar of the Ricci flow was largely ignored in geometry until G. Perelman acknowledged that in his groundbreaking analysis he was somewhat inspired by the role that the effective action plays in non–linear σ–model theory. This soon called attention to the fact that in QFT the Ricci flow is naturally embedded into a more general geometric flow, the renormalization group flow for non–linear σ models, which, even if mathematically ill-defined, provides an interpretation for the Ricci flow which is open to generalizations. Here we discuss this connection in great detail, pinpointing both the many mathematical as well as physical subtle points of the perturbative embedding of the Ricci flow in the renormalization group flow. The geometry of the dilaton fields is discussed in depth using Riemannian metric measure spaces and their role in connecting NLσM effective action to Perelman's $\mathscr{F}$ energy. We do not know of any other published account of these matter which is so detailed and informative.

© Springer International Publishing AG 2017

M. Carfora, A. Marzuoli, *Quantum Triangulations*, Lecture Notes in Physics 942,
https://doi.org/10.1007/978-3-319-67937-2_4

4.1 Introduction

Among the many significant ideas and developments that connect Mathematics with contemporary Physics one of the most intriguing is the role that Quantum Field Theory (QFT) plays in Geometry and Topology. We can argue back and forth on the relevance of such a role, but the perspective QFT offers is often surprising and far reaching. Examples abound, and a fine selection is provided by the revealing insights offered by Yang–Mills theory into the topology of 4–manifolds, by the relation between Knot Theory and topological QFT, and most recently by the interaction between Strings, Riemann moduli space, and enumerative geometry. Doubtless many of the most striking connections suggested by physicists failed to pass the censorship of the Department of Mathematics, and so do not appear in the above official list. As ill–defined these techniques may be, if we give them some degree of mathematical acceptance then the geometrical perspective they afford is always quite non–trivial and extremely rich. It is within such a framework that we shall examine in this and following chapters some aspects of the relation between an important class of QFTs and polyhedral surfaces. We start with a rather general introduction on geometrical aspects of QFT that will allow us to introduce naturally a notion of Quantum Geometry on polyhedral surfaces and provide a fertile ground for the interaction between Mathematics and Physics. As a significant case study illustrating such an interaction we discuss in detail the connection between non–linear sigma models associated to (triangulated) surfaces and *Ricci flow theory*. With the wisdom of hindsight prescient remarks on the relation of NLσMs with Ricci flow theory can be found in Polyakov's $O(n)$ sigma model (1975) [76], and in J. Honerkamp's Chiral multiloops paper (1972) [56]. Eventually the nature of this relation became fully manifest in Daniel Friedan's Ph.D. thesis[36] where the Ricci flow featured, along the lines described above, as the weak coupling limit of the renormalization group flow for NLσM. It must be noted that Friedan's approach [36–38] was very geometric and exploited a sophisticated control over the interplay between the renormalization group analysis and the diffeomorphism group, introducing notions, such as quasi–Einstein metrics and *Diff–modified* renormalization group equations for the metric coupling (V^n, g), that were to play a prominent role in Ricci flow theory (where they were to be independently introduced as Ricci solitons [51] and as the DeTurck version [28] of the Ricci flow, respectively). As hinted above, the rationale of the connection between NLσM and Ricci flow may be traced back to the common roots in the geometry of harmonic maps. However, a deeper understanding of this common structure has been very difficult to formalize since renormalization group techniques remain a challenge for mathematicians (and physicists as well). Hence, notwithstanding their common roots, it is not surprising that Ricci flow and renormalization group theory for non–linear σ model evolved independently until G. Perelman acknowledged the inspiring role played by the NLσM effective action in his groundbreaking paper [72]. This has drawn attention to the fact that renormalization group (*RG*) techniques may provide a framework for the Hamilton–Perelman theory which is open to generalizations [5–7, 18, 19, 25, 26, 29, 40, 42, 46, 71, 84].

4.2   Maps of Polyhedral Surfaces into Riemannian Manifolds

The analysis of the role of polyhedral surfaces in the development of quantum geometry is strictly related to the geometry of the non–linear space of maps

$$\phi : (T_l, M) \longrightarrow (V^n, g) \tag{4.1}$$

between a polyhedral surface (T_l, M) and a smooth compact n–dimensional Riemannian manifold (V^n, g) (Fig. 4.1). Any such map must be thought of as endowed with the minimal regularity allowing for the characterization of the corresponding (generalized) harmonic energy $E[\phi, g]_{(M,V^n)}$ (see below), since this is the building block of the typical action functionals considered in the physical applications of the theory. It must be stressed that this low regularity setting is not motivated by a desire of mathematical generality; rather, it is a necessary assumption when discussing the quantum geometry of surfaces from the viewpoint of geometric analysis. To draw a useful analogy one may think of the role in quantum mechanics and quantum field theory of non–smooth, nowhere differentiable paths and of the associated Wiener analysis. The low regularity of the maps $\phi : (T_l, M) \longrightarrow (V^n, g)$ we need to consider adds a layer of complications to our discussion. In particular it can make difficult to work in local charts over (T_l, M) and (V^n, g), even at a physical level of rigor. To circumvent this problem we make a number of geometrically natural hypothesis on the allowed class of maps between polyhedral surfaces and Riemannian manifolds.

Let (M, γ) denote the smooth orientable surface without boundary, endowed with the flat Riemannian metric γ with N_0 conical singularities associated to the

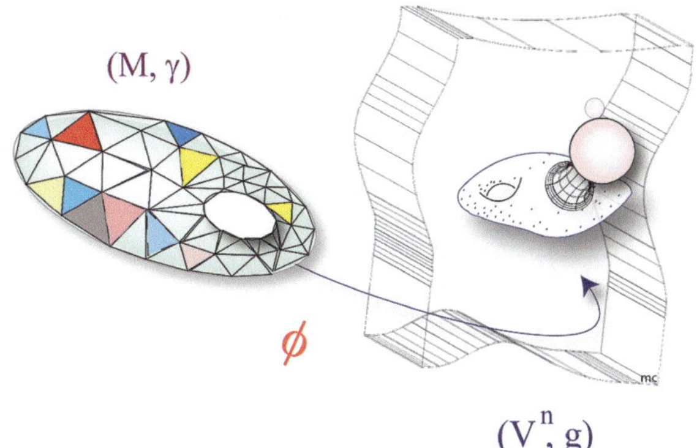

Fig. 4.1 The analysis of the role of polyhedral surfaces in the development of quantum geometry is strictly related to the geometry of the non–linear space of maps $\phi : (T_l, M) \longrightarrow (V^n, g)$

Fig. 4.2 There is a delicate issue in manipulating maps ϕ in the Sobolev space $\mathscr{H}^1(M, V^n)$ since they are not necessarily continuous. In particular, ϕ may not take values in any open subset $U_{(\alpha)} \subset V^n$, making problematic the use of local charts $(U_{(\alpha)}, \phi_{(\alpha)})$ to represent ϕ [55]

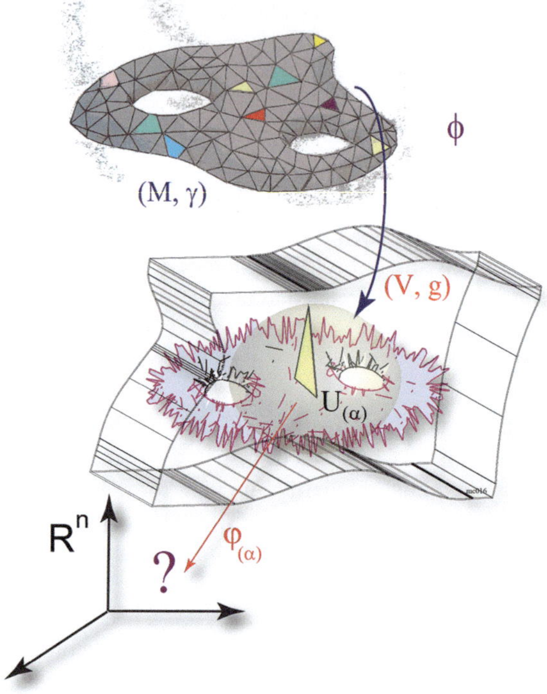

polyhedral surface (T_l, M). Recall (see Chap. 2) that to any such (M, γ) we can associate the Riemann surface $(M, [\gamma]) := ((M; N_0), \mathscr{C}_{sg})$, where $[\gamma]$ denotes the conformal class of the metric γ. The appropriate setting for maps $\phi : (M, \gamma) \longrightarrow (V^n, g)$ of low regularity which allow for the definition of a harmonic energy functional on (M, γ) is provided by the Sobolev space $\mathscr{H}^1(M, V^n)$ of maps which together with their first derivatives are square summable.

The characterization of $\mathscr{H}^1(M, V^n)$ itself is rather subtle since, to avoid problems with the lack of continuity of maps in $\mathscr{H}^1(M, V^n)$ (see below for details) one must use the Nash embedding theorem [49, 78], according to which any compact Riemannian manifold (V^n, g) can be isometrically embedded into some Euclidean space $\mathbb{E}^m := (\mathbb{R}^m, \delta)$ for m sufficiently large [49]. Explicitly, if $J : (V^n, g) \hookrightarrow \mathbb{E}^m$ is any such an embedding, we can define the Sobolev space of maps

$$\mathscr{H}^1_{(J)}(M, V^n) := \{\phi \in \mathscr{H}^1(M, \mathbb{R}^m) \mid \phi(M) \subset J(V^n)\}, \tag{4.2}$$

where $\mathscr{H}^1(M, \mathbb{R}^m)$ is the Hilbert space of square summable $\varphi : M \to \mathbb{R}^m$, with (first) distributional derivatives $\in L^2(M, \mathbb{R}^m)$, endowed with the norm

$$\| \phi \|_{\mathscr{H}^1} := \int_M \left(\phi^a(x)\, \phi^b(x)\, \delta_{ab} + \gamma^{\mu\nu}(x)\, \frac{\partial \phi^a(x)}{\partial x^\mu} \frac{\partial \phi^b(x)}{\partial x^\nu}\, \delta_{ab} \right) d\mu_\gamma. \tag{4.3}$$

Here $a, b = 1, \ldots, m$ label coordinates in $(\mathbb{R}^m, \delta)$, and $d\mu_\gamma$ denotes the Riemannian measure on (M, γ). As long as V^n is compact the spaces of maps $\mathscr{H}^1_{(J_1)}(M, V^n)$ and $\mathscr{H}^1_{(J_2)}(M, V^n)$, associated with two isometric embeddings J_1 and J_2, are homeomorphic [55] and one can simply write $\mathscr{H}^1(M, V^n)$.

As mentioned above, in dimension $\dim M = 2$ there is a well–known delicate issue in manipulating $\phi \in \mathscr{H}^1(M, V^n)$ (Fig. 4.2). This is related to the observation that even if the space $C^\infty(M, V^n)$ is dense [77] in the Sobolev space $\mathscr{H}^1(M, V^n)$, maps of class $\mathscr{H}^1(M, V^n)$ are not necessarily continuous[1] [68]. To carry out explicit local computations in such a case we introduce the open cover $\{U_{(i)}\}_{i=1}^{N_0}$ of the polyhedral surface (M, γ) defined by the open stars $\{U_{(i)}\}_{i=1}^{N_0}$ of the N_0 vertexes $\{\sigma^0(i)\}$ of (M, γ) (see Chap. 1). If $\phi \in \mathscr{H}^1(M, V^n)$ is interpreted as a map from $(M \simeq \{U_{(i)}\}_{i=1}^{N_0}, \gamma)$ into (V^n, g) we require that any such a map is localizable, (cf. [58], Sect. 8.4), and of bounded geometry. Explicitly, we assume that for every vertex $\sigma^0(i)$, $i = 1, \ldots, N_0$, (and hence for any $x_0 \in U_{(i)} \subset M$), there exists a corresponding metric disks $D(\sigma^0(i), \delta) := \{x \in U_{(i)} | d_\gamma(\sigma^0(i), x) \leq \delta\} \subset M$, of radius $\delta > 0$ and a metric ball $B(p(i), r) := \{z \in V^n | d_g(p(i), z) \leq r\} \subset (V^n, g)$ centered at $p(i) := \phi(\sigma^0(i)) \in V^n$, of radius $r > 0$ such that $\phi(D(\sigma^0(i), \delta)) \subset B(p(i), r)$, with

$$r < r_0 := \min\left\{ \frac{1}{3}\, inj\,(V^n),\ \frac{\pi}{6\sqrt{\kappa}} \right\}, \tag{4.4}$$

where $inj\,(V^n)$ and κ respectively denote the injectivity radius of (V^n, g), and the upper bound to the sectional curvature of (V^n, g), (we are adopting the standard convention of defining $\pi/2\sqrt{\kappa} \doteq \infty$ when $\kappa \leq 0$). Under such assumptions, one can use local coordinates also for maps in $\mathscr{H}^1(M, V^n)$ (Fig. 4.3). In particular for any $\phi \in \mathscr{H}^1(D(\sigma^0(i), \delta), B(p(i), r))$ we can introduce local coordinates x^α, for points in $(D(\sigma^0(i), \delta), U_{(i)})$, and $y^k = \phi^k(x)$, $k = 1, \ldots, n$, for the corresponding image points in $\phi(D(\sigma^0(i), \delta)) \subset B(p(i), r)$. These assumptions allow us to work locally and coordinate-wise in the "smooth" framework provided by the space of maps

$$Map\,(M, V^n) := \left\{ \phi \in \mathscr{H}^1(M \simeq \{U_{(i)}\}_{i=1}^{N_0}, V^n) \,\middle|\, \phi \text{ localizable} \right\}. \tag{4.5}$$

---

[1] Recall that maps $\phi : (M, \gamma) \longrightarrow (V^n, g)$ which are, together with their first derivative, in $L^p(M, \mathbb{R}^m)$ are continuous for $p > \dim M$. Hence maps which are close in the $\mathscr{H}^1(M, V^n)$ topology can be thought of as small pointwise deformation of each other. This picture fails when $p \leq \dim M$, in particular for $\phi \in \mathscr{H}^1(M, V^n)$ when, as in our case, $\dim M = 2$.

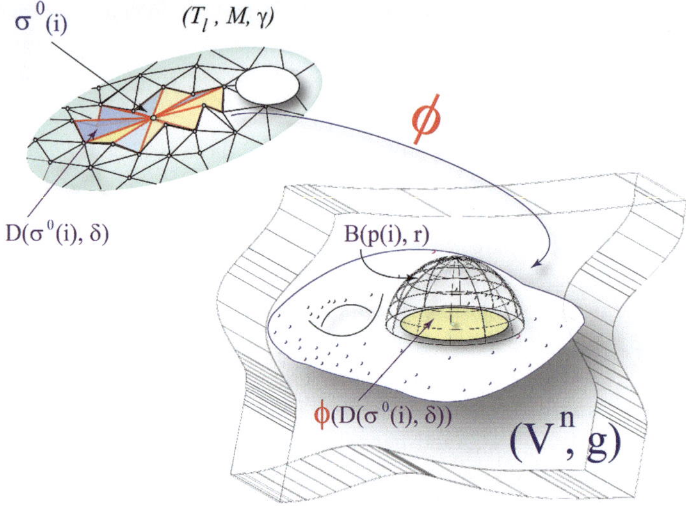

Fig. 4.3 We assume that the map $\phi \in \mathscr{H}^1(M, V^n)$ is localizable and of bounded geometry. In such a case one can use local coordinates to describe ϕ

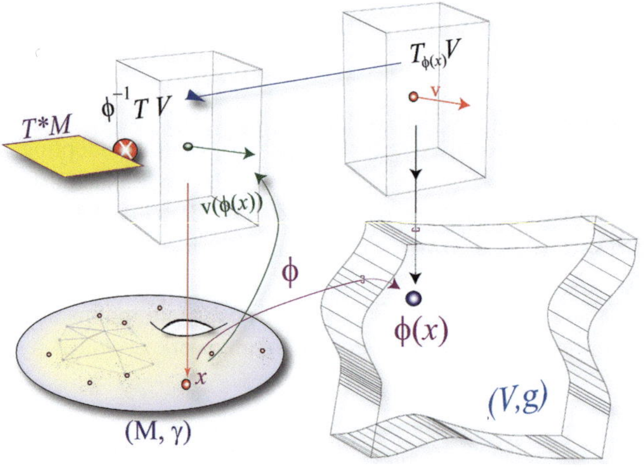

Fig. 4.4 The bundle $T^*M \otimes \phi^{-1}TV^n$ over the surface M associated with the localizable map $\phi \in \mathscr{H}^1(M, V^n)$

4.3   The Harmonic Map Energy Functional

As already stressed, the quantum geometry of (polyhedral) surfaces is strictly related to harmonic map theory. To set notation let $\phi \in Map(M, V^n)$ be a map which is localizable in the sense specified above (see (4.5)). Under such regularity hypotheses, we can introduce the pull–back bundle $\phi^{-1}TV^n$ (Fig. 4.4) whose sections $\mathsf{v} \equiv \phi^{-1}V := V \circ \phi$, $V \in C^\infty(V^n, TV^n)$, are the vector fields over M covering the map ϕ. In local coordinates we can write any such section as[2]

$$M \ni x \longmapsto \mathsf{v}(x) = \mathsf{v}^k(x) \frac{\partial}{\partial \phi^k(x)} \in \phi^{-1}TV^n\big|_x . \tag{4.6}$$

In particular, if T^*M denotes the cotangent bundle to (M, γ), we can locally introduce the differential

$$d\phi = \frac{\partial \phi^i}{\partial x^\alpha} dx^\alpha \otimes \frac{\partial}{\partial \phi^i} , \tag{4.7}$$

and interpret it as a section of the product bundle $T^*M \otimes \phi^{-1}TV^n$. The Riemannian metric in the fiber $\left(\phi^{-1}TV^n\right)_x$ over $x \in M$ is provided by $g(\phi(x))$, hence the tensor bundle $T^*M \otimes \phi^{-1}TV^n$ over the surface M is endowed with the pointwise inner product

$$\langle \cdot, \cdot \rangle_{T^*M \otimes \phi^{-1}TV^n} := \gamma^{-1}(x) \otimes g(\phi(x))(\cdot, \cdot) , \tag{4.8}$$

where $\gamma^{-1}(x) := \gamma^{\mu\nu}(x) \, \partial_\mu \otimes \partial_\nu$ is the metric tensor in T^*M_x. The corresponding Levi-Civita connection will be denoted by $\nabla^{(\cdot)}$. Explicitly, if $W = W^i_\alpha \, dx^\alpha \otimes \frac{\partial}{\partial \phi^i}$ is a section of $T^*M \otimes \phi^{-1}TV^n$, the covariant derivative of W in the direction $\frac{\partial}{\partial x^\beta}$ is provided by

$$\nabla^{(\cdot)}_\beta W = \nabla^{(\cdot)}_\beta \left(W^i_\alpha \, dx^\alpha \otimes \frac{\partial}{\partial \phi^i} \right) \tag{4.9}$$

$$= \frac{\partial}{\partial x^\beta} W^i_\alpha \, dx^\alpha \otimes \frac{\partial}{\partial \phi^i} + W^i_\alpha \left(\nabla^M_\beta \, dx^\alpha \right) \otimes \frac{\partial}{\partial \phi^i}$$

$$+ W^i_\alpha \, dx^\alpha \otimes \nabla^{\phi^{-1}TV^n}_\beta \left(\frac{\partial}{\partial \phi^i} \right) ,$$

where ∇^M denotes the Levi–Civita connection on (M, γ), and $\nabla^{\phi^{-1}TV^n}$ is the pull back on $\phi^{-1}TV^n$ of the Levi–Civita connection of (V^n, g). If $\Gamma^\alpha_{\beta\nu}(\gamma)$ and $\Gamma^k_{ij}(g)$ respectively denote the Christoffel symbols of (M, γ) and (M, g), then $\nabla^M_\beta \, dx^\alpha = -\Gamma^\alpha_{\beta\nu}(\gamma) \, dx^\nu$ and $\nabla^{\phi^{-1}TV^n}_\beta \left(\frac{\partial}{\partial \phi^i} \right) = \frac{\partial \phi^j}{\partial x^\beta} \Gamma^k_{ij}(g) \frac{\partial}{\partial \phi^k}$, and one computes

$$\nabla^{(\cdot)}_\beta W = \left(\frac{\partial}{\partial x^\beta} W^i_\alpha - W^i_\nu \Gamma^\nu_{\beta\alpha}(\gamma) + W^k_\alpha \frac{\partial \phi^j}{\partial x^\beta} \Gamma^i_{kj}(g) \right) dx^\alpha \otimes \frac{\partial}{\partial \phi^i} . \tag{4.10}$$

---

[2] In what follows we freely refer to the excellent [58] for a detailed analysis of the geometry of the computations involved in harmonic map theory.

The Hilbert–Schmidt norm of $d\phi$, in the bundle metric is provided by (see e.g. [58]),

$$\langle d\phi, d\phi \rangle_{T^*M \otimes \phi^{-1}TV^n} = \gamma^{\mu\nu}(x) \frac{\partial \phi^i(x)}{\partial x^\mu} \frac{\partial \phi^j(x)}{\partial x^\nu} g_{ij}(\phi(x)) = tr_{\gamma(x)} (\phi^* g) .$$

(4.11)

The corresponding density

$$e(\gamma, \phi; g) \, d\mu_\gamma := \frac{1}{2} \langle d\phi, d\phi \rangle_{T^*M \otimes \phi^{-1}TV^n} \, d\mu_\gamma = \frac{1}{2} tr_{\gamma(x)} (\phi^* g) \, d\mu_\gamma ,$$

(4.12)

where $d\mu_\gamma$ is the volume element of the Riemannian surface (M, γ), is conformally invariant under two-dimensional conformal transformations

$$(M, \gamma_{\mu\nu}) \mapsto (M, e^{-\psi} \gamma_{\mu\nu}) , \qquad \psi \in C^\infty(M, \mathbb{R}) ,$$

(4.13)

and defines the harmonic map energy density associated with the localizable $\phi \in Map(M, V^n)$. The corresponding harmonic map energy functional is defined by

$$E[\gamma, \phi; g]_{(M,V^n)} := \int_M e(\gamma, \phi; g) \, d\mu_\gamma .$$

(4.14)

Its critical points characterize the harmonic maps of the Riemann surface $(M, [\gamma])$ into (V^n, g), where $[\gamma]$ denotes the conformal class of the metric γ. More generally, for $\phi \in \mathscr{H}^1(M, V^n)$, the critical points of (4.14) define weakly harmonic maps $(M, \gamma) \to (V^n, g)$. Note that for surfaces weakly harmonic maps are harmonic [54]. In terms of the possible geometrical characterization of (M, γ) and (V^n, g), important examples of harmonic maps include harmonic functions, geodesics, isometric minimal immersions, holomorphic (and anti–holomorphic) maps of Kähler manifolds. It is worthwhile to observe that in such a rich panorama, also the seemingly trivial case of constant maps plays a basic role for the quantum geometry of (polyhedral) surfaces.

It is also worthwhile to note that under an overall rescaling of the metric of (V^n, g), $g_{ik} \longmapsto \lambda \, g_{ik}$, $\lambda \in \mathbb{R}_{>0}$ (see Appendix A) the harmonic energy functional scales according to

$$E[\gamma, \phi; \lambda \, g]_{(M,V^n)} = \lambda \, E[\gamma, \phi; g]_{(M,V^n)}$$

(4.15)

as is immediate to verify from the expression of the energy density (4.12).

4.4 Harmonic Energy as a Center of Mass Functional

To better grasp the geometrical and physical meaning of $E[\gamma, \phi; g]_{(M,V^n)}$ we interpret it as a center of mass functional [31, 58, 63] in terms of the metric geometry of (V^n, d_g), where d_g denotes the Riemannian distance on (V^n, g). It must be stressed that this interpretation plays, somewhat in disguise (see e.g. [39]), a basic role when considering the perturbative quantization of actions associated with $E[\gamma, \phi; g]_{(M,V^n)}$.

We begin by specializing the scale over which we consider the map ϕ is localizable in the sense described in the previous paragraphs.

Definition 4.1 Let $\{U_{(i)}\}_{i=1}^{N_0}$ be the open cover of the polyhedral surface (M, γ) defined by the open stars of the N_0 vertexes $\{\sigma^0(i)\}$ of (M, γ). A map

$$\phi : (M \simeq \{U_{(i)}\}_{i=1}^{N_0}, \gamma) \longrightarrow (V^n, g) \tag{4.16}$$

is said to be localizable at the (area) scale a if for every vertex $\sigma^0(i)$, $i = 1, \ldots, N_0$, (and hence for any $x_0 \in U_{(i)} \subset M$), there exists a corresponding metric disks $D(\sigma^0(i), \delta) := \{x \in U_{(i)} | d_\gamma(\sigma^0(i), x) \le \delta\} \subset M$, of radius $\delta > 0$ and a metric ball $B(p(i), \sqrt{a}) := \{z \in V^n | d_g(p(i), z) \le \sqrt{a}\} \subset (V^n, g)$ centered at $p(i) := \phi(\sigma^0(i)) \in V^n$, of radius $r = \sqrt{a}$ such that $\phi(D(\sigma^0(i), \delta)) \subset B(p(i), a^{1/2})$, with

$$a < r_0^2, \tag{4.17}$$

where r_0 is the geometric bound defined by (4.4).

For a given $x \in U_{(i)} \subset M$, and for $0 < \epsilon < 1$ small enough, let $\mathfrak{B}_x^{(i)}(\epsilon)$ be the disk of radius ϵ in the tangent space $T_x M$, and denote by

$$\exp_x^{(\gamma)} : T_x M \longrightarrow (M \simeq \{U_{(i)}\}_{i=1}^{N_0}, \gamma) \tag{4.18}$$

the exponential mapping based at x. Without loss in generality we can assume that $\exp_x^{(\gamma)}(\mathfrak{B}_x^{(i)}(\epsilon))$ is a convex set contained in the metric disks $D_{(i)}(x, \delta) := \{y \in U_{(i)} | d_\gamma(x, y) \le \delta\}$. Let us consider the map describing the fluctuations of the field ϕ at the point $x \in M$, induced by the vector $\upsilon \in \mathfrak{B}_x^{(i)}(\epsilon) \subset T_x M$, since ϕ is assumed to be localizable at the scale a, we can write (Fig. 4.5)

$$\widetilde{\phi}_x : T_x M \supset \mathfrak{B}_x^{(i)}(\epsilon) \longrightarrow B(\phi(x), a^{1/2}) \subset V^n \tag{4.19}$$

$$\upsilon \longmapsto \widetilde{\phi}_x(\upsilon) := \phi\left(\exp_x^{(\gamma)}(\upsilon)\right).$$

In particular, we can characterize the relative strength of the fluctuation of the field ϕ, with respect to the localization scale a, according to

$$(x, \upsilon) \longmapsto a^{-1} d_g^2\left(\phi(x), \phi\left(\exp_x^{(\gamma)}(\upsilon)\right)\right), \tag{4.20}$$

where d_g is the distance function in (V^n, g).

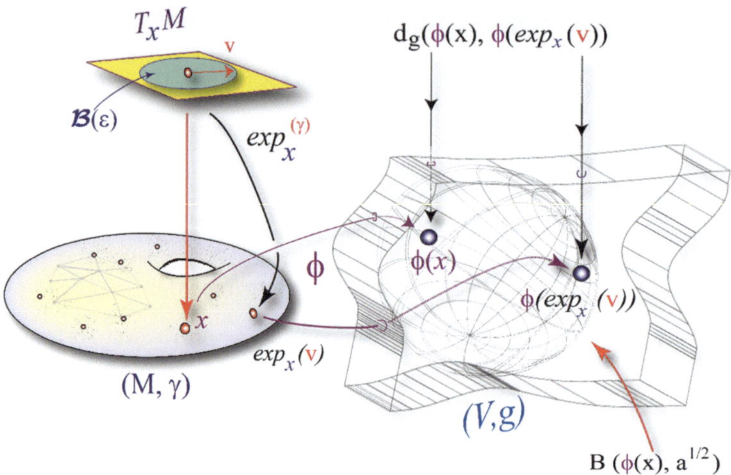

Fig. 4.5 The geometry of the map $\widetilde{\phi}_x \; : \; T_x M \supset \mathcal{B}_x^{(i)}(\epsilon) \longrightarrow B(\phi(x), a^{1/2}) \subset V^n$

If we endow $\mathcal{B}_x^{(i)}(\epsilon)$ with the pull–back measure $d\widehat{\mu}_\gamma := \left(\exp_x^{(\gamma)} \right)^* d\mu_\gamma$, then we can compute, for almost all $x \in M$, the average of (4.20) over the disk $\mathcal{B}_x^{(i)}(\epsilon) \subset T_x M$ according to

$$x \longmapsto G\left[\phi(x), \, \mathcal{B}_x^{(i)}(\epsilon) \right]$$

$$(4.21)$$

$$= \frac{1}{\left| \mathcal{B}_x^{(i)}(\epsilon) \right|} \int_{\mathcal{B}_x^{(i)}(\epsilon)} a^{-1} d_g^2 \left(\phi(x), \, \phi \left(\exp_x^{(\gamma)}(v) \right) \right) \, d\widehat{\mu}_\gamma(v) \, ,$$

where $\left| \mathcal{B}_x^{(i)}(\epsilon) \right| = 4\pi\epsilon^2$ is the Euclidean area of the disk $\mathcal{B}_x^{(i)}(\epsilon)$. Notice that this average can be interpreted as a center of mass function (evaluated at x) associated to the distribution of the maps $\widetilde{\phi}(v)$ with respect to the push–forward measure $d\widetilde{\phi}_\sharp(\widehat{\mu}_\gamma)$. This latter is defined by

$$\int_{B(\phi(x),r)\subset V^n} f(y) \, d\widetilde{\phi}_\sharp(\widehat{\mu}_\gamma)(y) := \int_{T_x M} f\left(\widetilde{\phi}(v) \right) \, d\widehat{\mu}_\gamma(v) \, , \qquad (4.22)$$

for every bounded Borel function $f \; : \; V^n \longrightarrow \mathbb{R}$. In particular

$$\widetilde{\phi}_\sharp \, \widehat{\mu}_\gamma \left[\widetilde{\phi} \left(\mathcal{B}_x^{(i)}(\epsilon) \right) \right] = \widehat{\mu}_\gamma \left(\mathcal{B}_x^{(i)}(\epsilon) \right) = \left| \mathcal{B}_x^{(i)}(\epsilon) \right| \, , \qquad (4.23)$$

and, according to (4.22), we can write

$$G\left[\phi(x),\,\mathfrak{B}_x^{(i)}(\epsilon)\right] \tag{4.24}$$

$$= \frac{1}{\widetilde{\phi}_\sharp\,\widehat{\mu}_\gamma\left[\widetilde{\phi}\left(\mathfrak{B}_x^{(i)}(\epsilon)\right)\right]}\int_{\widetilde{\phi}\left(\mathfrak{B}_x^{(i)}(\epsilon)\right)} a^{-1}\,d_g^2\left(\phi(x),\,y\right)\,d\widetilde{\phi}_\sharp(\widehat{\mu}_\gamma)\,,$$

which is indeed the center of mass function of the points $y \,=\, \phi\left(\exp_x^{(\gamma)}(\upsilon)\right) \in \widetilde{\phi}\left(\mathfrak{B}_x^{(i)}(\epsilon)\right)$ with respect to $d\widetilde{\phi}_\sharp(\widehat{\mu}_\gamma)$. In particular, we shall say that the map $\phi \in \mathscr{H}^1\left(M,\,V^n\right)$ is the ϵ–center of mass of the subset $\phi \circ \exp_x^{(\gamma)}\left(\mathfrak{B}_x^{(i)}(\epsilon)\right)$, with respect to the push–forward measure $d\widetilde{\phi}_\sharp(\widehat{\mu}_\gamma)$, if the point $\phi(x) \in V^n$ minimizes the function

$$z \longmapsto G\left[z,\,\mathfrak{B}_x^{(i)}(\epsilon)\right] \,:=\, \frac{1}{\left|\mathfrak{B}_x^{(i)}(\epsilon)\right|}\int_{\mathfrak{B}_x^{(i)}(\epsilon)} a^{-1}\,d_g^2\left(z,\,\widetilde{\phi}(\upsilon)\right)\,d\widehat{\mu}_\gamma(\upsilon)\,, \tag{4.25}$$

for almost every $z \in B(\phi(x),r)$. From these remarks it follows that the center of mass function $G\left[\phi(x),\,\mathfrak{B}_x^{(i)}(\epsilon)\right]$ describes the energetic cost of fluctuations $\widetilde{\phi}_x$, averaged over length scales $\leq \sqrt{a}$, of the *background field* ϕ. By letting the point x vary over the surface M we can characterize an approximate harmonic energy functional according to the following definition.

Definition 4.2 Let $\phi : (M,\gamma) \to (V^n,\,d_g)$ be a square summable map, and let

$$E_\epsilon\left[\gamma,\phi;\,a^{-\frac{1}{2}}\,d_g\right] \,:=\, \frac{1}{2}\int_{M\simeq\{U_{(i)}\}_{i=1}^{N_0}} \frac{G\left[\phi(x),\,\mathfrak{B}_x^{(i)}(\epsilon)\right]}{\epsilon^2}\,d\mu_\gamma(x)\,, \tag{4.26}$$

denote the ϵ–approximate energy associated with the center of mass function $x \longmapsto G\left[\phi(x),\,\mathfrak{B}_x^{(i)}(\epsilon)\right]$, then the L^2–energy of the map ϕ is defined by

$$E\left[\gamma,\phi;\,a^{-\frac{1}{2}}\,d_g\right] \,:=\, \lim_{\epsilon\to 0} E_\epsilon\left[\gamma,\phi;\,a^{-\frac{1}{2}}\,d_g\right] \in \mathbb{R} \cup \{+\infty\}\,, \tag{4.27}$$

where the limit is in the sense of weak convergence of measures.

It is elementary to prove (as in [58], Lemma 8.4.1) that the ϵ–approximate energy functional $E_\epsilon\left[\gamma,\phi;\,a^{-\frac{1}{2}}\,d_g\right]$ is minimized iff $\phi(x)$ is a center of mass for almost all $x \in M$. The basic observation in this general setting is that the center of mass functional $\mathscr{E}\left[\gamma,\phi;\,a^{-\frac{1}{2}}\,d_g\right]$ reduces to the standard harmonic map energy $E\left[\gamma,\phi;\,a^{-1}\,g\right]_{(\Sigma,V^n)}$ whenever the map ϕ is regular enough. In particular we have the basic result.

Proposition 4.1 *If ϕ is a localizable map $\in \mathcal{H}^1 (M, V^n)$, then*

$$\lim_{\epsilon \to 0} E_\epsilon [\gamma, \phi; a^{-\frac{1}{2}} d_g] = E [\gamma, \phi; a^{-1} g]_{(\Sigma, V^n)} = \frac{1}{a} E [\gamma, \phi; g]_{(\Sigma, V^n)} . \tag{4.28}$$

Proof Since

$$a^{-1} d_g^2 \left(\phi(x), \phi \left(\exp_x^{(\gamma)} (\upsilon) \right) \right) = d_{a^{-1} g}^2 \left(\phi(x), \phi \left(\exp_x^{(\gamma)} (\upsilon) \right) \right) , \tag{4.29}$$

(see (A.12) in Appendix A), we can write $E_\epsilon [\gamma, \phi; a^{-\frac{1}{2}} d_g] = E_\epsilon [\gamma, \phi; d_{a^{-1} g}]$ where we have rescaled the metric tensor g of (V^n, g) according to

$$a^{-1} g = a^{-1} g_{ik} dx^i \otimes dx^k . \tag{4.30}$$

The equality $\lim_{\epsilon \to 0} E_\epsilon [\gamma, \phi; a^{-\frac{1}{2}} d_g] = E [\gamma, \phi; a^{-1} g]_{(\Sigma, V^n)}$ follows by observing that $\phi \in L^2(M, V^n)$ is localizable and the functional $E [\gamma, \phi; a^{-1} g]_{(M, V^n)}$ can be defined in the local chart associated to the localization adopted (see e.g. Theor. 8.4.1 of [58]). Finally, the last identity in (4.28) immediately follows from the scaling relation (see (4.15))

$$E[\gamma, \phi; a^{-1} g]_{(M, V^n)} = a^{-1} E[\gamma, \phi; g]_{(M, V^n)} . \tag{4.31}$$

4.5   The Non Linear σ Model Action

The above center of mass analysis provides the rationale for interpreting the map

$$\widetilde{\phi} : TM \supset \mathfrak{B}_x^{(i)} (\epsilon) \longrightarrow V^n \tag{4.32}$$

$$(x, \upsilon) \longmapsto \quad \widetilde{\phi}(x, \upsilon) := \widetilde{\phi}_x(\upsilon) ,$$

(see (4.19)) as a *fluctuating field*, averaged over length scales $\leq \sqrt{a}$, pointwise interacting with the *background* field $\phi(x)$ through the point dependent coupling $a^{-1} g(\phi(x))$. This interpretation of $E_\epsilon [\gamma, \phi; a^{-\frac{1}{2}} d_g]$ characterizes the Non–Linear σ Model (NMσM) action, a basic player in discussing the quantum geometry of polyhedral surfaces.

Definition 4.3 (The NLσM action) Let (M, γ) be the orientable surface without boundary, endowed with the flat Riemannian metric γ with N_0 conical singularities describing the polyhedral surface (T_l, M), and let $(V^n, a^{-1} g)$ be the smooth compact n–dimensional ($n \geq 3$) Riemannian manifold associated with the rescaled metric $a^{-1} g$. The ϵ–regularized non–linear σ model action describing the dynamics, on the polyhedral surface (M, γ), of the field $\phi : (M, \gamma) \to (V^n, d_g)$, averaged over length scales $\leq \sqrt{a}$, is defined by[3]

$$S_\epsilon[\gamma, \phi; \, a^{-1} d_g] := 2 E_\epsilon [\gamma, \phi; \, a^{-\frac{1}{2}} d_g]$$

(4.33)

$$= \int_{M \simeq \{U_{(i)}\}_{i=1}^{N_0}} \frac{G\left[\phi(x), \, \mathfrak{B}_x^{(i)}(\epsilon)\right]}{\epsilon^2} \, d\mu_\gamma(x) \, .$$

If the field ϕ is described by a localizable map $\phi \in \mathcal{H}^1(M, V^n)$ then the corresponding smooth non–linear σ model action is defined, in the sense of weak convergence of measures, by the limit $\lim_{\epsilon \to 0} S_\epsilon[\gamma, \phi; \, a^{-1} d_g]$, and is provided by

$$S[\gamma, \phi; \, a^{-1} g] := 2 E[\gamma, \phi; \, a^{-1} g]_{(M, V^n)}$$

(4.34)

$$= \int_M \gamma^{\mu\nu}(x) \frac{\partial \phi^i(x)}{\partial x^\mu} \frac{\partial \phi^j(x)}{\partial x^\nu} \left(a^{-1} g\right)_{ij} (\phi(x)) \, d\mu_\gamma \, .$$

Remark 4.1 It must be stressed that the geometric equivalence between (4.29) and (4.30) implies that we cannot consider a as a separate parameter from g: if we rescale the metric g according to $g \mapsto \lambda \, g$, $\lambda \in \mathbb{R}_{>0}$, then a rescales according to $a \mapsto \lambda^{-1} a$ so as to leave $a^{-1} g$ unchanged (see also [38]). Nonetheless, it is customary to write the NLσM action as

$$S[\gamma, \phi; \, a^{-1} g] = \frac{1}{a} \int_M \gamma^{\mu\nu}(x) \frac{\partial \phi^i(x)}{\partial x^\mu} \frac{\partial \phi^j(x)}{\partial x^\nu} g_{ij}(\phi(x)) \, d\mu_\gamma \, , \tag{4.35}$$

a writing that emphasizes the role of the parameter a, rather than of $a^{-1} g$, as the (classical) coupling constant of the theory (a notable exception is Friedan's original paper [38]). According to the scaling (4.31), both (4.35) are trivially equivalent expressions at the level of the classical theory. However when discussing the effect of the fluctuations of the field ϕ around a fiducial configuration, (say a given harmonic map), the role of true coupling pertains to $a^{-1} g$. In other words, the perturbative parameter characterizing the strength of the fluctuations of ϕ at various wavelengths, fluctuations which generate a scale–dependent effective action out of $S[\gamma, \phi; \, a^{-1} g]$, is to be extracted from the geometry of $a^{-1} g$. This is well known in the literature on NLσM where it gives rise to the rather sophisticated phenomenology of the model. Nonetheless the idiosyncratic attitude emphasizing the role of the parameter a as a coupling constant still persists and often obscures

---

[3]The numerical factor 2 appearing in the relation (4.33) connecting $S_\epsilon[\gamma, \phi; \, a^{-1} d_g]$ to $E_\epsilon [\gamma, \phi; \, a^{-\frac{1}{2}} d_g]$ is inserted for later convenience when applying the theory to Ricci flow.

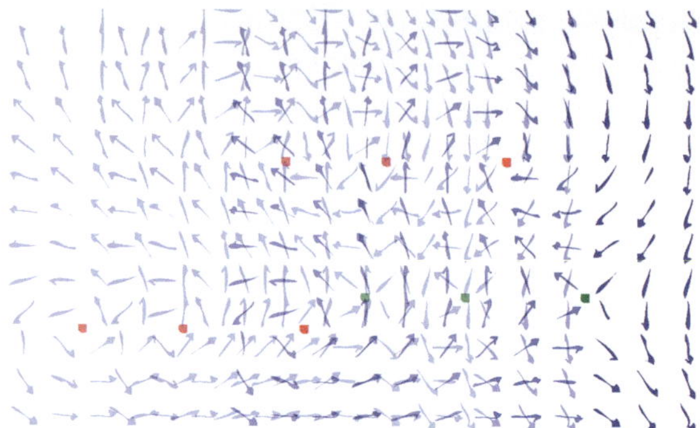

Fig. 4.6 The NLσM action (4.34) provides an ubiquitous toy model in theoretical physics finding applications ranging from condensed matter to string theory. Typical examples are to be found in magnetism: here depicted as a polyhedral surface (a lattice) whose vertices are decorated with *spins* represented by unit vectors interacting with their nearest-neighbors

the geometrical interpretation of the results obtained if the reader is not fully familiar with the subject.

The NLσM action (4.34) provides an ubiquitous toy model in theoretical physics finding applications ranging from condensed matter to string theory. By selecting the appropriate target manifold $(V^n, a^{-1}g)$ (often a homogeneous space or a Lie group) it characterizes effective field theories which describe the low energy limit (probed by the long wavelength fluctuations of the appropriate field ϕ) of many distinct microscopic models in condensed matter physics and string theory. Typical examples are to be found in magnetism, e.g. the Heisenberg magnet theory as described by a NLσM associated with a lattice whose vertices are decorated with *spins* represented by unit vectors interacting with their nearest-neighbors (Fig. 4.6). In string theory NLσM does provide a remarkable way to perturbatively generate Riemannian geometry out of the long wavelength limit of the string *matter fields* ϕ. The rationale involved in this ubiquitous role of NLσM is largely connected to the feedback between the geometry of the target manifold $(V^n, a^{-1}g)$ and the deformation of the map ϕ, a feedback that we start exploring geometrically at the classical level before moving to the renormalization group techniques called upon by the quantum theory.

4.6   The Geometry of $S[\gamma, \phi; \, a^{-1}g]$ Deformations

To explore the classical aspects of NLσM which are relevant to our analysis we consider the one–parameter (t) variations $\phi_t : [0,1] \times (M, \gamma) \to (V^n, a^{-1}g)$ induced, via the exponential mapping, by a vector field $\xi \in \phi^{-1}TV^n$ covering a fiducial map $\phi : (M, \gamma) \to (V^n, a^{-1}g)$. Explicitly, let

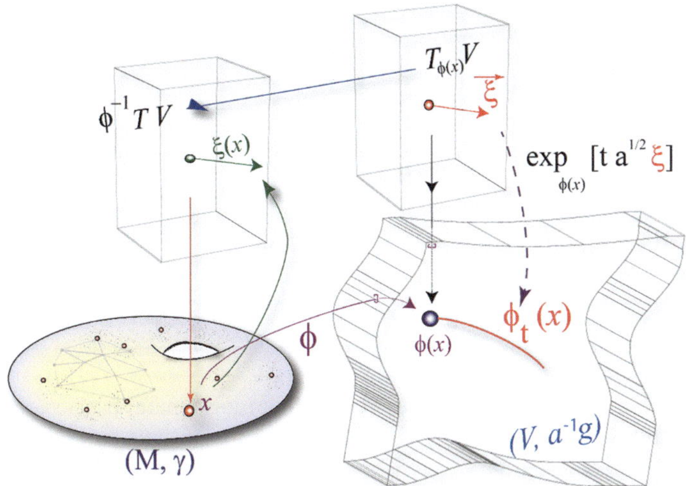

Fig. 4.7 The one–parameter (t) variations $\phi_t :\ [0,1] \times (M, \gamma) \to (V^n, a^{-1}g)$ induced, via the exponential mapping, by a vector field $\xi \in \phi^{-1}TV^n$ covering a fiducial map $\phi :\ (M, \gamma) \to (V^n, a^{-1}g)$

$$\Upsilon_{\xi}^{(a^{-1}g)} :\ [0,1] \longrightarrow (V^n, a^{-1}g) \tag{4.36}$$

be the *unit speed* geodesic in $(V^n, a^{-1}g)$ issued from $\phi(x)$ with initial velocity

$$\dot{\Upsilon}_{\xi}^{(a^{-1}g)}\big|_{t=0} = \xi = \xi^i(x)\,\frac{\partial}{\partial \phi^i(x)} \in \phi^{-1}TV^n\ . \tag{4.37}$$

Since $a^{-1}g_{ik}\xi^i\xi^k = 1$, the geodesics in (V^n, g) are traveled with speed $g_{ik}\xi^i\xi^k = a$, and under the scaling $g \longmapsto a^{-1}g$ we have (Fig. 4.7)

$$\exp_{\phi(x)}^{(a^{-1}g)} := \Upsilon_{\xi}^{(a^{-1}g)}(1) = \Upsilon_{a^{\frac{1}{2}}\xi}^{(g)}(1) =: \exp_{\phi(x)}^{(g)} [a^{\frac{1}{2}}\xi(x)]\ , \tag{4.38}$$

where $\exp_{\phi(x)}^{(a^{-1}g)}$ and $\exp_{\phi(x)}^{(g)}$ respectively are the exponential map in $(V^n, a^{-1}g)$ and (V^n, g) based at the point $\phi(x)$ (see (A.31) in Appendix A). The one–parameter (t) variations $\phi_t : [0,1] \times (M, \gamma) \to (V^n, g)$ induced by the vector field ξ, are given by

$$[0,1] \times M \longrightarrow (V^n, g) \tag{4.39}$$

$$(t, x) \longmapsto \phi_t(x) := \exp_{\phi(x)}^{(g)} [t\,a^{\frac{1}{2}}\xi(x)]\ .$$

We compute

$$e(\gamma, \phi_t; g) := \frac{1}{2} \langle d\phi_t, d\phi_t \rangle_{T^*M \otimes \phi^{-1}TV^n} \tag{4.40}$$

$$= \frac{1}{2} \left\langle d \, \exp^{(g)}_{\phi(x)} [t \, a^{\frac{1}{2}} \xi(x)], d \, \exp^{(g)}_{\phi(x)} [t \, a^{\frac{1}{2}} \xi(x)] \right\rangle_{T^*M \otimes \phi^{-1}TV^n},$$

as the small $t \in [0, 1]$ expansion

$$e(\gamma, \phi_t; g) = e(\gamma, \phi; g) + \frac{d}{dt} e(g, \phi_t) \Big|_{t=0} t + \frac{1}{2} \frac{d^2}{dt^2} e(g, \phi_t) \Big|_{t=0} t^2 + \dots . \tag{4.41}$$

To this end we consider $\frac{\partial}{\partial x^\alpha}$ and $\frac{\partial}{\partial t}$ as commuting (coordinate) basis vectors on the product manifold $M_t := M \times [0, 1]$ endowed with the product metric $\gamma + dt \otimes dt$. The Levi–Civita connection $\nabla^{(\cdot,\cdot)}$ on $(T^*M, \gamma^{-1}) \otimes (\phi^{-1}TV^n, g(\phi))$ can be trivially extended to $T^*M_t \otimes \phi^{-1}TV^n$ by setting $\nabla^{(\cdot,\cdot)}_{\partial/\partial t} dx^\alpha = 0$. By a slight abuse of notation, we still denote by $\nabla^{(\cdot,\cdot)}$ this extended connection, and compute

$$\frac{d}{dt} d\phi_t = \nabla^{(\cdot,\cdot)}_{\partial/\partial t} \left(\frac{\partial \phi_t^i}{\partial x^\alpha} \frac{\partial}{\partial \phi^i} \otimes dx^\alpha \right) \tag{4.42}$$

$$= \nabla^{(\cdot,\cdot)}_{\partial/\partial t} \left(\frac{\partial \phi_t^i}{\partial x^\alpha} \frac{\partial}{\partial \phi^i} \right) \otimes dx^\alpha + \left(\frac{\partial \phi_t^i}{\partial x^\alpha} \frac{\partial}{\partial \phi^i} \right) \otimes \nabla^{(\cdot,\cdot)}_{\partial/\partial t} dx^\alpha$$

$$= \nabla^{\phi^{-1}TV^n}_{\partial/\partial t} \left(\frac{\partial \phi_t^i}{\partial x^\alpha} \frac{\partial}{\partial \phi^i} \right) \otimes dx^\alpha ,$$

where we have exploited (4.9). Since $\frac{\partial}{\partial x^\alpha}$ and $\frac{\partial}{\partial t}$ commute we can exchange $\nabla^{\phi^{-1}TV^n}_{\partial/\partial t} \left(\frac{\partial \phi_t^i}{\partial x^\alpha} \frac{\partial}{\partial \phi^i} \right)$ with $\nabla^{\phi^{-1}TV^n}_{\partial/\partial x^\alpha} \left(\frac{\partial \phi_t^i}{\partial t} \frac{\partial}{\partial \phi^i} \right)$ and write (4.42) as

$$\frac{d}{dt} d\phi_t(x) = \nabla^{\phi^{-1}TV^n}_{\partial/\partial x^\alpha} \left(\frac{\partial \phi_t^i(x)}{\partial t} \frac{\partial}{\partial \phi^i(x)} \right) \otimes dx^\alpha , \tag{4.43}$$

By evaluating this expression for $t = 0$ and taking into account (4.39), we eventually get

$$\frac{d}{dt} d\phi_t(x) \Big|_{t=0} = \nabla^{\phi^{-1}TV^n}_{\partial/\partial x^\alpha} \left(\frac{\partial \phi_t^i(x)}{\partial t} \Big|_{t=0} \frac{\partial}{\partial \phi^i(x)} \right) \otimes dx^\alpha \tag{4.44}$$

$$= a^{\frac{1}{2}} \nabla^{\phi^{-1}TV^n}_{\partial/\partial x^\alpha} \left(\xi^i(x) \frac{\partial}{\partial \phi^i(x)} \right) \otimes dx^\alpha ,$$

where we used the fact that the derivative of $\phi_t(x) = \exp^{(g)}_{\phi(x)}[t \, a^{\frac{1}{2}} \xi]$ for $t = 0$, (i.e. at the base point $\phi(x)$), is the identity map on $\phi^{-1}TV^n$. Hence,

$$\frac{d}{dt}\, d\phi_t\bigg|_{t=0} = a^{\frac{1}{2}}\, d\xi\,, \tag{4.45}$$

which in components reads

$$\frac{d}{dt}\, d\phi_t\bigg|_{t=0} = a^{\frac{1}{2}}\, \nabla^{\phi^{-1}TV^n}_{\partial/\partial x^\alpha}\left(\xi^i\,\frac{\partial}{\partial\phi^i}\right)\otimes dx^\alpha \tag{4.46}$$

$$= a^{\frac{1}{2}}\left(\frac{\partial\xi^k}{\partial x^\alpha} + \xi^i\,\Gamma^k_{ij}(g)\,\frac{\partial\phi^j}{\partial x^\alpha}\right)\frac{\partial}{\partial\phi^k}\otimes dx^\alpha\,.$$

We can similarly compute $\frac{d^2}{dt^2}\, d\phi_t$ according to

$$\frac{d^2}{dt^2}\, d\phi_t = \nabla^{\phi^{-1}TV^n}_{\partial/\partial t}\nabla^{\phi^{-1}TV^n}_{\partial/\partial x^\alpha}\left(\frac{\partial\phi_t}{\partial t}\right)\otimes dx^\alpha \tag{4.47}$$

$$= \left[\nabla^{\phi^{-1}TV^n}_{\partial/\partial x^\alpha}\nabla^{\phi^{-1}TV^n}_{\partial/\partial t}\left(\frac{\partial\phi_t}{\partial t}\right) + \mathrm{Rm}(g)\left(\frac{\partial\phi_t}{\partial t},\frac{\partial\phi_t}{\partial x^\alpha}\right)\frac{\partial\phi_t}{\partial t}\right]\otimes dx^\alpha\,,$$

where $\mathrm{Rm}(g)$ is the Riemann tensor of (V^n, g), and where for notational ease we have set $\frac{\partial\phi_t}{\partial t} = \frac{\partial\phi^k_t}{\partial t}\frac{\partial}{\partial\phi^k}$, $\frac{\partial\phi_t}{\partial x^\alpha} = \frac{\partial\phi^i_t}{\partial x^\alpha}\frac{\partial}{\partial\phi^i}$. Since the variation (4.39) is geodesic for every x, we have $\nabla^{\phi^{-1}TV^n}_{\partial/\partial t}\left(\frac{\partial\phi_t}{\partial t}\right) \equiv 0$ and for $t = 0$ we can write

$$\frac{d^2}{dt^2}\, d\phi_t\bigg|_{t=0} = a\,\mathrm{Rm}(g)\,(\xi, d\phi)\,\xi\,, \tag{4.48}$$

where we have exploited (4.39). Hence, from (4.41), (4.45), and (4.48) we get

$$e\,(\gamma, \phi_t; g) = e(\gamma, \phi; g) + a^{\frac{1}{2}}\,\langle d\xi, d\phi\rangle^\gamma_g\, t \tag{4.49}$$

$$+ \frac{1}{2}\,a\,\langle d\xi, d\xi\rangle^\gamma_g\, t^2 + \frac{1}{2}\,a\,\langle\mathrm{Rm}(g)\,(\xi, d\phi)\,\xi, d\phi\rangle^\gamma_g\, t^2 + \mathscr{O}(a^{\frac{3}{2}}t^3)\,,$$

where to simplify notation we used the shorthand $\langle\cdot,\cdot\rangle^\gamma_g$ for $\langle\cdot,\cdot\rangle_{T^*M\otimes\phi^{-1}TV^n}$. Further higher order terms in the expansion (4.41) can be (painstakingly!) obtained along a similar pattern by using (4.44) and by iterating the Riemann commutation relation (4.47). Explicitly we get up to 4th order[4]

$$e\,(\gamma, \phi_t; g) = e(\gamma, \phi; g) + a^{\frac{1}{2}}\,\langle d\xi, d\phi\rangle^\gamma_g\, t \tag{4.50}$$

---

[4]Explicit computations of this type can often be found in the vast literature on non–linear σ model theory. A useful reference to start with is provided by [1]; A brief and geometrically clear presentation by Jacques Distler is available on line at https://golem.ph.utexas.edu/~distler/blog/archives/002532.html

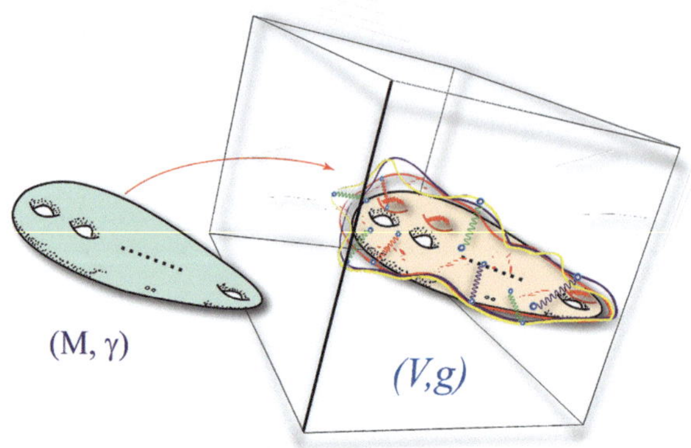

Fig. 4.8 In harmonic map theory it is usually sufficient to consider the deformation of $e\,(\gamma,\phi_t;g)$ up to terms of second order in the deformation parameter t

$$+ \;\frac{1}{2}\,a\,\langle d\xi,\,d\xi\rangle_g^\gamma\,t^2 \;+\; \frac{1}{2}\,a\,\langle \mathrm{Rm}(g)\,(\xi,d\phi)\,\xi,\,d\phi\rangle_g^\gamma\,t^2$$

$$+ \;\frac{2}{3}\,a^{\frac{3}{2}}\,\langle \mathrm{Rm}(g)\,(\xi,d\phi)\,\xi,\,d\xi\rangle_g^\gamma\,t^3 \;+\; \frac{1}{6}\,a^{\frac{3}{2}}\,\langle (\nabla\mathrm{Rm}(g))\,(\xi,\xi,d\phi)\,\xi,\,d\phi\rangle_g^\gamma\,t^3$$

$$+ \;\frac{1}{4}\,a^2\,\langle (\nabla\mathrm{Rm}(g))\,(\xi,\xi,d\phi)\,\xi,\,d\xi\rangle_g^\gamma\,t^4 \;+\; \frac{1}{6}\,a^2\,\langle \mathrm{Rm}(g)\,(\xi,d\xi)\,\xi,\,d\xi\rangle_g^\gamma\,t^4$$

$$+ \;\frac{1}{6}\,a^2\,\langle \mathrm{Rm}(g)\,(\xi,d\phi)\,\xi,\,\mathrm{Rm}(g)\,(\xi,d\phi)\,\xi\rangle_g^\gamma\,t^4$$

$$+ \;\frac{1}{24}\,a^2\,\langle (\nabla\nabla\mathrm{Rm}(g))\,(\xi,\xi,\xi,d\phi)\,\xi,\,d\phi\rangle_g^\gamma\,t^4 \;+\; \mathcal{O}(a^{\frac{5}{2}}t^5)\,,$$

The corresponding expression for the harmonic energy functional $E[\gamma,\phi_t;g]_{(M,V^n)}$ is readily obtained by integrating over the surface (M,γ). In addressing harmonic map theory it is usually sufficient to consider the deformation of $e\,(\gamma,\phi_t;g)$ up to terms of second order in the deformation parameter t (Fig. 4.8). In particular, by integrating (4.49), we get

$$E[\gamma,\phi_t;g]_{(M,V^n)} \;=\; E[\gamma,\phi;g]_{(M,V^n)} \;+\; a^{\frac{1}{2}}t\int_M \langle d\xi,\,d\phi\rangle_g^\gamma\,d\mu_\gamma$$

$$\tag{4.51}$$

$$+ \;\frac{1}{2}\,at^2\left[\int_M \langle d\xi,\,d\xi\rangle_g^\gamma\,d\mu_\gamma \;+\; \int_M \langle \mathrm{Rm}(g)\,(\xi,d\phi)\,\xi,\,d\phi\rangle_g^\gamma\,d\mu_\gamma\right] \;+\; \mathcal{O}(a^{\frac{3}{2}}t^3)\,.$$

Integration by parts of the term $\int_M \langle d\xi,\,d\phi\rangle_g^\gamma\,d\mu_\gamma$ provides

$$\int_M \langle d\xi,\,d\phi\rangle_g^\gamma\,d\mu_\gamma \;=\; -\int_M g_{ki}(\phi(x))\,\xi^k\,(\mathrm{tr}_\gamma\nabla d\phi)^i\,d\mu_\gamma \tag{4.52}$$

$$= -\int_M \langle \xi, \tau(\phi)\rangle_g^\gamma \, d\mu_\gamma \, ,$$

where

$$\tau^i(\phi) := \left(\mathrm{tr}_\gamma \nabla d\phi\right)^i = \triangle_{(\gamma)}\phi^i + \gamma^{\mu\nu}\, \Gamma^i_{jk}\frac{\partial\phi^j}{\partial x^\mu}\frac{\partial\phi^k}{\partial x^\nu} \, , \tag{4.53}$$

is the *tension field* of the map ϕ and

$$\triangle_{(\gamma)} := \frac{1}{\sqrt{\det \gamma}}\,\frac{\partial}{\partial x^\mu}\left(\sqrt{\det \gamma}\,\gamma^{\mu\nu}\,\frac{\partial}{\partial x^\nu}\right) \tag{4.54}$$

is the scalar Laplace–Beltrami operator on (M, γ). Hence,

$$E[\gamma, \phi_t; g]_{(M,V^n)} = E[\gamma, \phi; g]_{(M,V^n)} - a^{\frac{1}{2}}t \int_M \langle \xi, \tau(\phi)\rangle_g^\gamma \, d\mu_\gamma$$

$$\tag{4.55}$$

$$+ \frac{1}{2}\,at^2\left[\int_M \langle d\xi, d\xi\rangle_g^\gamma \, d\mu_\gamma + \int_M \langle \mathrm{Rm}(g)\,(\xi, d\phi)\,\xi, d\phi\rangle_g^\gamma \, d\mu_\gamma\right] + \mathscr{O}(a^{\frac{3}{2}}t^3) \, .$$

If the fiducial map $\phi : (M, \gamma) \longrightarrow (V^n, g)$ is harmonic its tension field vanishes, $\tau(\phi) = 0$, and (4.51) reduces to

$$E[\gamma, \phi_t; g]_{(M,V^n)} = E[\gamma, \phi; g]_{(M,V^n)}\big|_{\text{harmonic}}$$

$$\tag{4.56}$$

$$+ \frac{1}{2}\,at^2\left[\int_M \langle d\xi, d\xi\rangle_g^\gamma \, d\mu_\gamma + \int_M \langle \mathrm{Rm}(g)\,(\xi, d\phi)\,\xi, d\phi\rangle_g^\gamma \, d\mu_\gamma\right] + \mathscr{O}(a^{\frac{3}{2}}t^3) \, .$$

Note that since $E[\gamma, \phi;\ a^{-1}g] = a^{-1}E[\gamma, \phi;\ g]$ (see (4.15)), $\langle \circ, \circ\rangle_g^\gamma = a\,\langle \circ, \circ\rangle_{a^{-1}g}^\gamma$, and $\mathrm{Rm}(a^{-1}g) = \mathrm{Rm}(g)$ (see Appendix A) we can equivalently rewrite (4.56) as

$$E[\gamma, \phi_t;\ a^{-1}g]_{(M,V^n)} = E[\gamma, \phi;\ a^{-1}g]_{(M,V^n)}\big|_{\text{harmonic}}$$

$$\tag{4.57}$$

$$+ \frac{1}{2}\,at^2\left[\int_M \langle d\xi, d\xi\rangle_{a^{-1}g}^\gamma \, d\mu_\gamma + \int_M \langle \mathrm{Rm}(a^{-1}g)\,(\xi, d\phi)\,\xi, d\phi\rangle_{a^{-1}g}^\gamma \, d\mu_\gamma\right]$$

$$+ \mathscr{O}(a^{\frac{3}{2}}t^3) \, .$$

The second order deformation of the functional $E[\gamma, \phi_t; g]_{(M,V^n)}$ around a (harmonic) fiducial map ϕ is often sufficient to analyze in depth the interplay between the curvature of the target manifold (V^n, g) and harmonic map theory. However it does not provide the appropriate framework for discussing the (perturbative)

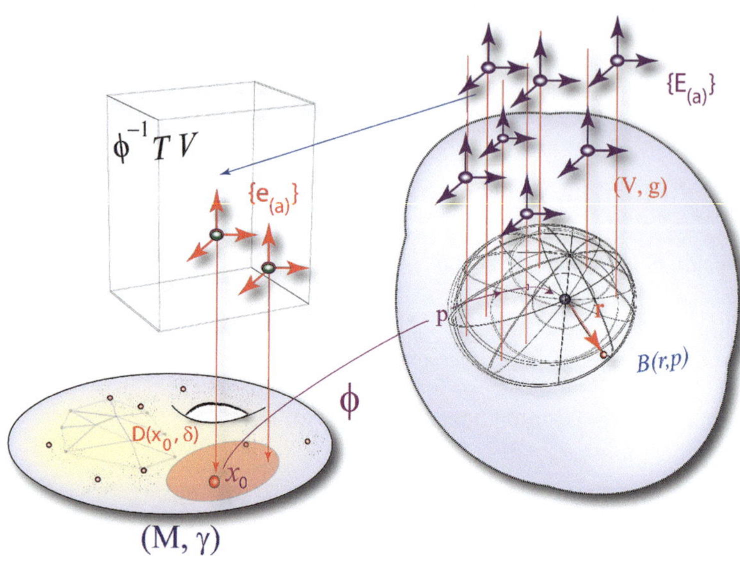

Fig. 4.9 The quadratic term $\int_M \langle d\xi, d\xi \rangle^{\gamma}_{a^{-1}g}\, d\mu_\gamma$ in the expression for the deformed harmonic energy $E[\gamma, \phi_t; g]_{(M,V^n)}$ depends explicitly from the metric $g(\phi)$. This quadratic term characterizes the formal functional measure and the associated propagators when discussing the quantum theory. The explicit dependence of these propagators from $g(\phi)$ makes their structure non standard and the fields ξ cannot be easily promoted to quantum fields. This problem is overcome by the introduction of local orthonormal frames over metric balls $B(p, r) \subset (V^n, g)$

quantum fluctuations of the map ϕ. Indeed, as the parameter t varies, the dependence of $E[\gamma, \phi_t; g]_{(M,V^n)}$ from the target manifold metric g manifests itself not just in the $\mathcal{O}(at^2)$ term $\mathrm{Rm}(g)$ appearing in (4.56), but rather in a full collection of $\mathcal{O}(a^{\frac{k}{2}}t^k)$ terms, $k \geq 3$, involving powers of $\mathrm{Rm}(g)$ and of its derivatives up to order $(k-2)$. For instance, the relevant expression for $E[\gamma, \phi_t; g]_{(M,V^n)}$ (needed for computing the two-loops contribution in the renormalization group flow for NLσM) up to the $\mathcal{O}(a^2t^4)$ terms can be readily obtained by integrating (4.50).

To put the physical role of the various terms in (4.56) to the fore explicitly it is convenient to introduce [1] local orthonormal frames over (V^n, g) (Fig. 4.9). Globally this may be not possible if the V^n is not parallelizable, however, since we are interested in localizable maps ϕ, we may limit ourselves to introduce orthonormal frames over metric balls in (V^n, g). As discussed in Sect. 4.2, for every $x_0 \in M$ let $D(x_0, \delta) := \{x \in M \,|\, d_\gamma(x_0, x) \leq \delta\} \subset M$ denote a metric disks, of radius $\delta > 0$ and let $B(p, r) := \{z \in V^n \,|\, d_g(p, z) \leq r\} \subset (V^n, g)$ be a corresponding metric ball centered at $p = \phi(x_0) \in V^n$, of radius $r > 0$ such that $\phi(D(x_0, \delta)) \subset B(p, r)$, with $r < \min\left\{\frac{1}{3} inj\,(V^n), \frac{\pi}{6\sqrt{\kappa}}\right\}$ (see (4.4)). Let $\{E_{(a)}\}_{a=1,\ldots n}$ be a smooth orthonormal tangent frame field over $(B(p, r), g)$, so that for $y \in B(p, r)$, $\left(E_{(1)}(y), \ldots, E_{(n)}(y)\right)$ is an orthonormal basis of $T_y V^n$. If $y = \phi(x)$, we denote by $\{e_{(a)}(x)\}_{a=1,\ldots n}$ the orthonormal basis of $\phi^{-1}TV^n\big|_x$ at $x \in D(x_0, \delta)$ obtained by pulling back $\{E_{(a)}\}_{a=1,\ldots n}$ over $D(x_0, \delta) \subset M$. We denote by $\{\theta^{(a)}\}_{a=1,\dots n}$, with $\theta^{(a)}\left(e_{(b)}\right) = \delta^a_b$ the (orthonormal) basis dual to $\{e_{(a)}(x)\}_{a=1,\dots n}$. In terms of the $\{\frac{\partial}{\partial \phi^k(x)}\}_{k=1,\dots n}$ we can write

$$e_{(a)}(x) = \iota^k_a(x)\,\frac{\partial}{\partial \phi^k(x)}\,, \qquad \frac{\partial}{\partial \phi^k(x)} = \iota^b_k(x)\,e_{(a)}(x)\,, \tag{4.58}$$

with $\iota^k_a(x)\,\iota^b_k(x) = \delta^b_a, \iota^k_a(x)\,\iota^a_h(x) = \delta^k_h$. If $\mathrm{Aut}\left(\phi^{-1}TV^n\right)$ denotes the bundle with fiber over $x \in M$ the group $SO(n)$ of orthogonal transformations of the fiber $\phi^{-1}TV^n\big|_x$, then the frame $\{e_{(a)}(x)\}_{a=1,\dots n}$ is defined up to the *gauge transformation* provided by the restriction to $D(x_0, \delta) \subset M$ of a section of $\mathrm{Aut}\left(\phi^{-1}TV^n\right)$, i.e.

$$e'_{(a)}(x) = \lambda^b_a(x)\,e_{(b)}(x)\,, \tag{4.59}$$

with

$$\lambda^b_a : D(x_0, \delta) \longrightarrow \mathrm{Aut}\left(\phi^{-1}TV^n\right)\big|_{D(x_0,\delta)} \tag{4.60}$$

$$x \longmapsto \lambda^b_a(x) \in SO(n)\,.$$

Let

$$A^c_a := (A_\nu\,dx^\nu)^c_a = \left(\iota^h_a \iota^c_k\,\Gamma^k_{hj}(g)\,\frac{\partial \phi^j}{\partial x^\nu} + \iota^c_k\,\frac{\partial \iota^k_a}{\partial x^\nu}\right)dx^\nu\,, \tag{4.61}$$

be the components, in the orthonormal basis $\{e_c \otimes \theta^a\}$, of the pull back over $D(x_0, \delta)$ of the Levi–Civita connection of $(B(p, r),\, g)$. Under the gauge transformation (4.60) the $\mathfrak{so}(n)$–valued one–form $(A_\nu\,dx^\nu)^c_a$ so defined transforms as an $SO(n)$ gauge connection on $\left(\phi^{-1}TV^n\right)\big|_{D(x_0,\delta)}$. In particular, if $\widehat{\xi}^a = \theta^a(\xi)$ are the components of the vector field ξ in the orthonormal basis $\{e_a(x)\}$, then in terms of the *gauge field* $A = (A_\nu\,dx^\nu)^a_b\,e_a \otimes \theta^b$, we can write (see (4.45) and (4.48))

$$d\xi = \left(d\widehat{\xi}^a + A^a_b\,\widehat{\xi}^b\right)e_a\,, \tag{4.62}$$

where, for each fixed index a, $d\widehat{\xi}^a := \frac{\partial \theta^a(\xi)}{\partial x^\nu}\,dx^\nu$ is the differential of the scalar function $\theta^a(\xi) = \widehat{\xi}^a$. Hence, $d\xi$ transforms as a vector under local $SO(n)$ rotations of the orthonormal basis $\{e_a(x)\}$ and behaves as a scalar under coordinate transformations [1]. By introducing the explicit expression (4.62) in the term $\int_M \langle d\xi, d\xi\rangle^\gamma_g\,d\mu_\gamma$ appearing in (4.56) we get

$$\int_M \langle d\xi, d\xi\rangle^\gamma_g\,d\mu_\gamma = \int_M \delta_{ac}\,\gamma^{\nu\sigma}\left(\frac{\partial \widehat{\xi}^a}{\partial x^\nu} + (A_\nu)^a_b\,\widehat{\xi}^b\right)\left(\frac{\partial \widehat{\xi}^c}{\partial x^\sigma} + (A_\sigma)^c_b\,\widehat{\xi}^b\right)d\mu_\gamma$$

$$\tag{4.63}$$

$$= \int_M \delta_{ac}\, \gamma^{v\sigma}\, \frac{\partial \widehat{\xi}^a}{\partial x^v}\, \frac{\partial \widehat{\xi}^c}{\partial x^\sigma}\, d\mu_\gamma + 2 \int_M \delta_{ac}\, \gamma^{v\sigma}\, (A_v)^a_b\, \widehat{\xi}^b\, \frac{\partial \widehat{\xi}^c}{\partial x^\sigma}\, d\mu_\gamma$$

$$+ \int_M \delta_{ac}\, \gamma^{v\sigma}\, (A_v)^a_b\, \widehat{\xi}^b\, (A_\sigma)^c_d\, \widehat{\xi}^d\, d\mu_\gamma$$

$$= - \int_M \delta_{ac}\, \widehat{\xi}^a\, \triangle_{(\gamma)} \widehat{\xi}^c\, d\mu_\gamma + 2 \int_M \delta_{ac}\, \gamma^{v\sigma}\, (A_v)^a_b\, \widehat{\xi}^b\, \frac{\partial \widehat{\xi}^c}{\partial x^\sigma}\, d\mu_\gamma$$

$$+ \int_M \delta_{ac}\, \gamma^{v\sigma}\, (A_v)^a_b\, \widehat{\xi}^b\, (A_\sigma)^c_d\, \widehat{\xi}^d\, d\mu_\gamma\ ,$$

where, by taking into account the local characterization (4.54) of the Laplace-Beltrami operator, we have integrated by parts the squared gradient term according to

$$\int_M \delta_{ac}\, \gamma^{v\sigma}\, \frac{\partial \widehat{\xi}^a}{\partial x^v}\, \frac{\partial \widehat{\xi}^c}{\partial x^\sigma}\, d\mu_\gamma = \int_M \frac{1}{\sqrt{\det \gamma}}\, \frac{\partial}{\partial x^v} \left[\delta_{ac}\, \sqrt{\det \gamma}\, \gamma^{v\sigma}\, \widehat{\xi}^a\, \frac{\partial \widehat{\xi}^c}{\partial x^\sigma} \right] d\mu_\gamma$$

$$\tag{4.64}$$

$$- \int_M \delta_{ac}\, \widehat{\xi}^a\, \frac{1}{\sqrt{\det \gamma}}\, \frac{\partial}{\partial x^v} \left[\sqrt{\det \gamma}\, \gamma^{v\sigma}\, \frac{\partial \widehat{\xi}^c}{\partial x^\sigma} \right] d\mu_\gamma = - \int_M \delta_{ac}\, \widehat{\xi}^a\, \triangle_{(\gamma)}\, \widehat{\xi}^c\, d\mu_\gamma\ .$$

If we denote by $\langle \circ, \circ \rangle^\gamma_\delta$ the inner product on $(T * M, \gamma^{-1}) \otimes (\phi^{-1} T V^n, \{e_{(a)}\})$ associated to the chosen g–orthonormal basis $\{e_{(a)}\}^n_{a=1}$ of $\phi^{-1} T V^n$, then we can write

$$\delta_{ac}\, \widehat{\xi}^a\, \triangle_{(\gamma)}\, \widehat{\xi}^c = \left\langle \widehat{\xi},\, \triangle_{(\gamma)}\, \widehat{\xi} \right\rangle^\gamma_\delta\ , \tag{4.65}$$

$$\delta_{ac}\, \gamma^{v\sigma}\, (A_v)^a_b\, \widehat{\xi}^b\, \frac{\partial \widehat{\xi}^c}{\partial x^\sigma} = \left\langle A\widehat{\xi},\, d\widehat{\xi} \right\rangle^\gamma_\delta\ , \tag{4.66}$$

$$\delta_{ac}\, \gamma^{v\sigma}\, (A_v)^a_b\, \widehat{\xi}^b\, (A_\sigma)^c_d\, \widehat{\xi}^d = \left\langle A\widehat{\xi},\, A\widehat{\xi} \right\rangle^\gamma_\delta\ , \tag{4.67}$$

$$\gamma^{v\sigma}\, \widehat{R}^a_{bcd}(g)\, \frac{\partial \phi^j}{\partial x^v}\, \frac{\partial \phi^l}{\partial x^\sigma}\, \iota^b_j \iota^d_l\, \widehat{\xi}_a \widehat{\xi}^c = \left\langle \widehat{\mathrm{Rm}}(g) \left(\widehat{\xi},\, d\phi \right) \widehat{\xi},\, d\phi \right\rangle^\gamma_\delta\ , \tag{4.68}$$

where $\widehat{R}_{abcd}$ denotes the component of the Riemann tensor in the orthonormal basis $\{e_a\}$ (to avoid the explicit presence of the ϕ–dependent factors $\iota^b_j \iota^d_l$ in the above expression one may use the mixed notation $\widehat{R}_{ajbl} := R_{ijkl} \iota^i_a \iota^k_b$). Hence, in terms of the g–orthonormal frames $\{e_{(a)}\}^n_{a=1}$ we have

$$E[\gamma, \phi_t; g]_{(M, V^n)} = E[\gamma, \phi; g]_{(M, V^n)} \big|_{\mathrm{harmonic}}$$

$$\tag{4.69}$$

$$+ \frac{1}{2} a t^2 \left[- \int_M \left\langle \widehat{\xi}, \ \triangle_{(\gamma)} \widehat{\xi} \right\rangle_\delta^\gamma \, d\mu_\gamma \ + \ 2 \int_M \left\langle A \widehat{\xi}, \ d\widehat{\xi} \right\rangle_\delta^\gamma \right.$$

$$\left. + \int_M \left\langle A \widehat{\xi}, \ A \widehat{\xi} \right\rangle_\delta^\gamma \, d\mu_\gamma \ + \ \int_M \left\langle \widehat{Rm}(g) \left(\widehat{\xi}, d\phi \right) \widehat{\xi}, d\phi \right\rangle_\delta^\gamma \, d\mu_\gamma \right] \ + \ \mathcal{O}(a^{\frac{3}{2}} t^3) \ .$$

By taking into account (4.34) and(4.56), we can rewrite the NLσM action associated with the fluctuating field $\widehat{\xi} \in \phi_{har}^{-1} \, TV^n$ as

$$S[\gamma, \phi_t; \ a^{-1}g] \ = \ S[\gamma, \ \exp_\psi^{(a^{-1}g)} (t \xi_{(j)}); \ a^{-1} \, g] \ = \ S[\gamma, \phi; \ a^{-1}g] \big|_{\mathrm{har}}$$

$$\tag{4.70}$$

$$+ t^2 \left[- \int_M \left\langle \widehat{\xi}, \ \triangle_{(\gamma)} \widehat{\xi} \right\rangle_\delta^\gamma \, d\mu_\gamma \ + \ 2 \int_M \left\langle A \widehat{\xi}, \ d\widehat{\xi} \right\rangle_\delta^\gamma \right.$$

$$\left. + \int_M \left\langle A \widehat{\xi}, \ A \widehat{\xi} \right\rangle_\delta^\gamma \, d\mu_\gamma \ + \ \int_M \left\langle \widehat{Rm}(a^{-1}g) \left(\widehat{\xi}, d\phi \right) \widehat{\xi}, d\phi \right\rangle_\delta^\gamma \, d\mu_\gamma \right] \ + \ \mathcal{O}(a^{\frac{3}{2}} t^3) \ .$$

It follows from these remarks that the leading $\mathcal{O}(t^2)$ corrections to $S[\gamma, \phi_t; \ a^{-1}g]$, around a given harmonic map ϕ_{har}, are governed by the kinetic term

$$\int_M \left\langle \widehat{\xi}, \ \triangle_{(\gamma)} \widehat{\xi} \right\rangle_\delta^\gamma \, d\mu_\gamma \ , \tag{4.71}$$

and by the quadratic interaction terms

$$2 \int_M \left\langle A \widehat{\xi}, \ d\widehat{\xi} \right\rangle_\delta^\gamma \ + \ \int_M \left\langle A \widehat{\xi}, \ A \widehat{\xi} \right\rangle_\delta^\gamma \, d\mu_\gamma$$

$$\tag{4.72}$$

$$+ \int_M \left\langle \widehat{Rm}(a^{-1}g) \left(\widehat{\xi}, d\phi \right) \widehat{\xi}, d\phi \right\rangle_\delta^\gamma \, d\mu_\gamma \ ,$$

describing the coupling of $\widehat{\xi}$ with the $SO(n)$ gauge field A and with the curvature term $\widehat{Rm}$. It must be stressed that in this field theoretic picture the gauge field $A(a^{-1}g)$, associated with the pulled back *spin* connection (4.61), and the curvature $\widehat{Rm}(a^{-1}g)$ feature as point-dependent coupling parameters describing the relative strength of these interaction terms.[5]

---

[5]This interpretation extends also to the higher order curvature dependent terms $\mathcal{O}(a^{\frac{k}{2}} t^k), k \geq 3$, that can be obtained from the normal coordinate expansion (4.50).

4.7   The Dilaton and Riemannian Metric Measure Spaces

There is another basic player in the quantum geometry of polyhedral surfaces: the *dilaton* field. To set the stage for the study of its geometrical properties let us start by observing that we can naturally extend the harmonic map setting described above by considering maps between the polyhedral surface (M, γ) and n–dimensional compact *Riemannian metric measure spaces*[6] $(V^n, g, d\omega)$ [44, 45], i.e. a smooth orientable manifold, without boundary, endowed with a Riemannian metric g and a positive Borel measure $d\omega$, absolutely continuous with respect to the Riemannian volume element $d\mu_g$, i.e. $d\omega << d\mu_g$, (see Fig. 4.10).

The set $\mathrm{Meas}(V^n)$ of all smooth Riemannian metric measure spaces on V^n can be characterized as

$$\mathrm{Meas}(V^n) := \{(V^n, g; d\omega) \mid g \in \mathcal{M}\mathrm{et}(V^n),\ d\omega \in \mathcal{B}(V^n, g)\}\ , \tag{4.73}$$

where $\mathcal{B}(V^n, g)$ is the set of positive Borel measure on (V^n, g) with $d\omega << d\mu_g$. Since in the compact–open C^∞ topology $\mathcal{M}\mathrm{et}(V^n)$ is contractible, the space $\mathrm{Meas}(V^n)$ fibers trivially over $\mathcal{M}\mathrm{et}(V^n)$. In particular, the fiber $\pi^{-1}(V^n, g)$, where $\pi : (V^n, g, d\omega) \longmapsto (V^n, g)$ is the natural projection from $\mathrm{Meas}(V^n)$ to $\mathcal{M}\mathrm{et}(V^n)$, can be identified with the set of all (orientation preserving) measures $d\omega << d\mu_g$

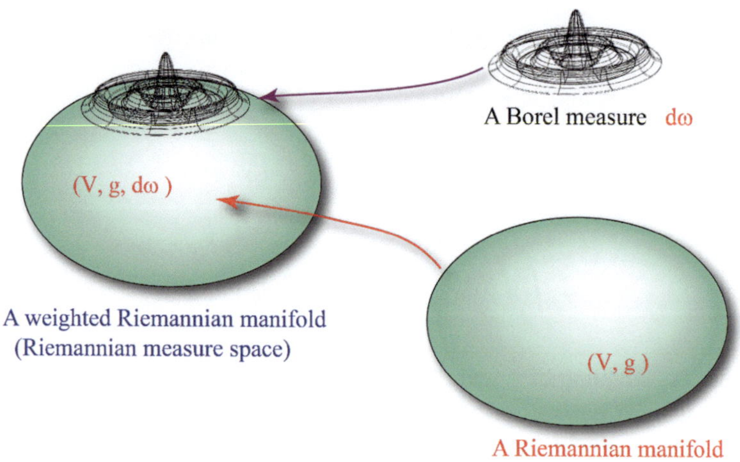

Fig. 4.10 To discuss the geometry of the dilaton we can extend the harmonic map setting by considering n–dimensional compact *weighted Riemannian manifold* $(V^n, g, d\omega)$

---

[6]Strictly speaking, $(V^n, g, d\omega)$ characterizes a *weighted Riemannian manifold* (or *Riemannian manifold with density*) the corresponding metric measure space being actually defined by $(V^n, d_g(\cdot, \cdot), d\omega)$, where $d_g(x, y)$ denotes the Riemannian distance on (V^n, g). By a slight abuse of notation, we shall use $(V^n, g, d\omega)$ and $(V^n, d_g(\cdot, \cdot), d\omega)$ interchangeably.

over the given (V^n, g),

$$\mathrm{Meas}(V^n, g) := \left\{ d\omega \in \mathrm{Meas}(V^n) \, : \, d\omega << d\mu_g \right\} , \tag{4.74}$$

endowed with the topology of weak convergence. Let us recall that the natural action of the group of diffeomorphisms $\mathscr{D}iff(V^n)$ on the space $\mathrm{Meas}(V^n)$ is defined by

$$\mathscr{D}iff(V^n) \times \mathrm{Meas}(V^n) \; \longrightarrow \; \mathrm{Meas}(V^n) \tag{4.75}$$

$$(\varphi; \, g, \, d\omega) \qquad \longmapsto \quad (\varphi^* g, \, \varphi^* d\omega) ,$$

where $(\varphi^* g, \, \varphi^* d\omega)$ is the pull–back under $\varphi \in \mathscr{D}iff(V^n)$. The Radon–Nikodym derivative $\wp(g, \, d\omega) := \frac{d\omega}{d\mu_g}$ is a local *Riemannian measure space invariant* [21] under this action, i.e.

$$\wp(\varphi^* g, \, \varphi^*(d\omega)) = \varphi^* \, \wp(g, \, d\omega) = \frac{d\omega}{d\mu_g} \circ \varphi , \qquad \forall \varphi \in \mathscr{D}iff(V^n) . \tag{4.76}$$

We can consider the measure $d\omega$ dimensionless and proportional to the normalized Riemannian volume element $\frac{d\mu_g}{\int_{V^n} d\mu_g}$ so as to introduce the dilaton field according to the following characterization.

Definition 4.4 (The dilaton field) The geometrical dilaton field associated with the Riemannian metric measure space $(V^n, g, \, d\omega)$ is defined by the map

$$f \, : \, \mathrm{Meas}(V^n) \; \longrightarrow \; C^\infty(V^n, \mathbb{R}) \tag{4.77}$$

$$(V^n, g, \, d\omega) \; \longmapsto \; f(V^n, g, \, d\omega) := - \ln \left(V_g(V^n) \, \frac{d\omega}{d\mu_g} \right) ,$$

where $V_g(V^n) := \int_{V^n} d\mu_g$.

Remark 4.2   According to (4.76) we can parametrize $\mathrm{Meas}(V^n, g)$ according to

$$\mathrm{Meas}(V^n, g) = \left\{ d\omega = e^{-f} \, \frac{d\mu_g}{V_g(V^n)} : f \in C^\infty(V^n, \mathbb{R}) \right\} , \tag{4.78}$$

however sometimes it is necessary to restrict the action of $\mathscr{D}iff(V^n)$ to the metric g alone and leave the measure $d\omega$ fixed (cf. section 5 in [21]). Absolute continuity of $d\omega$ with respect to $d\mu_g$ implies that in such a case f is not a scalar and under the action of a diffeomorphism $\varphi \in \mathscr{D}iff(V^n)$ it transforms as a density according to

$$e^{-f(\varphi^* g, \, d\omega)} = \left(Jac_{d\mu_g} \varphi^{-1} \right) e^{-f} , \tag{4.79}$$

where $Jac_{d\mu_g} \varphi^{-1}$ denotes the Jacobian of φ^{-1} with respect to $d\mu_g$. Yet, a subtler interpretation is possible [21] which still preserves the invariant nature of the dilaton and the characterization (4.78). To wit, according to (4.79) and Moser theorem [69], one can consider $\varphi \in \mathscr{D}iff(V^n)$ as generating an active transformation on

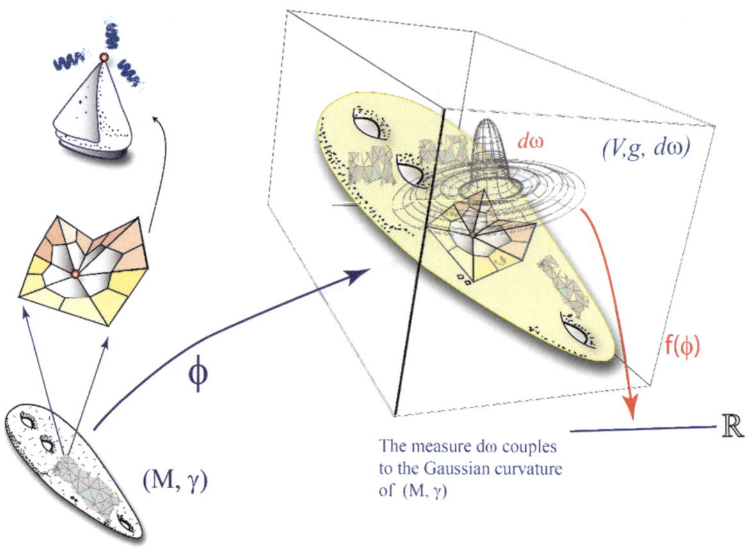

Fig. 4.11 The Borel measure $d\omega = e^{-f}\,d\mu_g/V_g(V^n)$ couples, via the dilaton f with the Gaussian curvature $\mathcal{K}_\gamma$ of the surface (M, γ). For a polyhedral surface this coupling is realized in terms of the deficit angles $\{\epsilon(k)\}_{k=1}^{N_0}$ associated with the conical vertexes $\{p_k\}$ of the surface

Meas(V^n, g) mapping the measure $d\omega$ into another measure $d\omega' := (\varphi^{-1})^*(d\omega)$ (with the same total mass $\int_{V^n} d\omega$). By composing this mapping on Meas(V^n, g) with the induced pull–back action on the resulting measure we get from (4.76) that the Radon–Nikodym derivative $\wp(g, d\omega)$ transforms according to

$$\wp(g, d\omega) \longmapsto \wp(\varphi^*g, \varphi^*(d\omega')) = \varphi^* \wp(g, d\omega') \tag{4.80}$$

$$= \varphi^* \wp(g, (\varphi^{-1})^*(d\omega)) = \left(Jac_{d\mu_g}\varphi^{-1}\right)\frac{d\omega}{d\mu_g}\,,$$

which is consistent both with (4.79) and the local Riemannian measure space invariant nature of the dilaton. If not otherwise stated we shall consider the natural action on $(V^n, g, d\omega)$ given by (4.75).

With these remarks along the way, we can introduce the dilatonic Non–Linear σ Model (NLσM) action associated with a polyhedral surface (M, γ) according to the following geometrical characterization (Fig. 4.11).

Theorem 4.1 (The dilatonic NLσM action for polyhedral surfaces) *Let (M, γ) be a smooth orientable surface without boundary, endowed with the flat Riemannian metric γ with N_0 conical singularities associated to a polyhedral surface (T_l, M), and let $(V^n, g, d\omega)$ be the Riemannian measure space associated with the dilatonic* *measure* $d\omega = e^{-f}\frac{d\mu_g}{V_g(V^n)}$. *Denote by* $\{p_k\}_{k=1}^{N_0} \in (M; N_0)$ *the marked points associated with the vertices* $\{\sigma^0(k)\} \in (T_l, M)$ *and let* $\{\varepsilon(k)\}_{k=1}^{N_0} \in (M; N_0)$ *be the corresponding set of deficit angles. Then, the dilatonic non–linear* σ *model action associated with a localizable map* $\phi : (M, \gamma) \longrightarrow (V^n, g, d\omega)$ *is provided by*

$$S[\gamma, \phi; \, a^{-1} g, f] := \frac{2}{a} E[\gamma, \phi; \, g]_{(M,V^n)} + 2\pi \sum_{k=1}^{N_0} \left(\frac{\varepsilon(k)}{2\pi}\right) f\left(\phi(p_k)\right), \qquad (4.81)$$

where the dilaton $f\left(\phi(p_k)\right)$, *characterizing the strength of the interaction with the localized curvature defects of* (M, γ), *plays together with* $a^{-1} g$ *the role of a point dependent coupling of the theory.*

Proof According to the results of Sect. 2.4, the Riemann surface $((M; N_0), \mathscr{C}_{sg})$ associated with the polyhedral surface (M, γ) carries the distinguished *uniformization* (as a singular Euclidean surface) which associates to any marked point p_k, (corresponding to the vertex $\sigma^0(k) \in (T_l, M)$) the complex coordinate $t(k) \in \mathbb{C}$, defined in the open disk

$$U_{\rho^2(k)} \doteq \{t(k) \in \mathbb{C} \mid |t(h)| < 1, \qquad t(k)[p_k] = 0\}, \qquad (4.82)$$

where $\rho^2(k)$ denotes the polytopal 2–cell in (P_T, M) baricentrically dual to the vertex $\sigma^0(k) \in (T_l, M)$. The metric γ locally characterizing the conical Euclidean structure of angle $\Theta(k) := 2\pi - \varepsilon(k)$ around p_k can be written as (see (2.52) and (2.53))

$$\gamma_{(k)} = e^{2V(k)} \, dt(k) \otimes d\bar{t}(k), \qquad (4.83)$$

where the conformal factor $V(k)$ is provided by

$$V(k) := -\frac{\varepsilon(k)}{2\pi} \ln |t(k)|. \qquad (4.84)$$

Hence, the metric γ associated to the polyhedral manifold (T_l, M) is the collection of locally conical metrics

$$\gamma := \left\{\left(U_{\rho^2(k)}, \, \gamma_{(k)} = \exp\left[-\frac{\varepsilon(k)}{2\pi} \ln |t(k)|^2\right] |dt(k)|^2\right)\right\}_{k=1}^{N_0} \qquad (4.85)$$

(compare with the explicit construction described in Sect. 2.4). The dilatonic action associated with the localizable map $\phi : (M, \gamma^{(0)}) \longrightarrow (V^n, g, d\omega)$ is provided by

$$S[\gamma, \phi; \, a^{-1} g, f] := a^{-1} \int_M tr_{\gamma(x)}(\phi^* g) \, d\mu_\gamma + \int_M \mathscr{K}_\gamma f(\phi) \, d\mu_\gamma, \qquad (4.86)$$

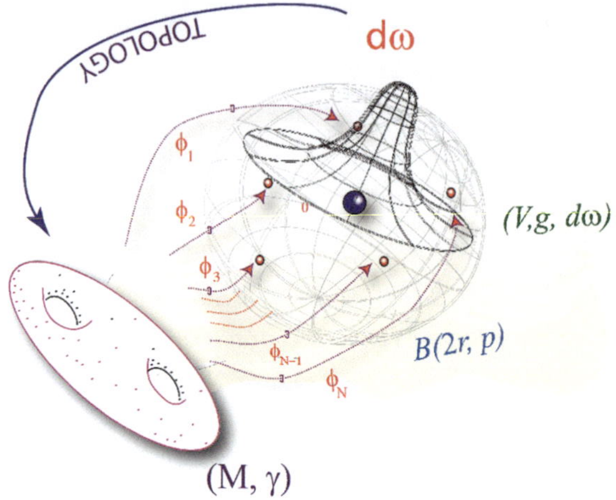

Fig. 4.12 When we consider constant maps, the dilatonic action $S[\gamma, \phi;\ a, g, f]$ calls into play the topology of the surface M

where the Gaussian curvature $\mathcal{K}_\gamma$, smoothly defined only on $M - \{p_k\}_{k=1}^{N_0}$, can be extended to the whole open surface M by locally interpreting it as the $(1, 1)$ current (see (2.84))

$$\mathcal{K}_\gamma\, d\mu_\gamma\big|_{U_{\rho^2(k)}} \;=\; \frac{\varepsilon(k)}{2\pi}\, d * d\,(\ln |t(k)|) \tag{4.87}$$

$$= 2\pi\left(\frac{\varepsilon(k)}{2\pi}\right)\delta_{p_k}\frac{\sqrt{-1}}{2}\,dt(k) \wedge d\bar{t}(k)\ ,$$

where δ_{p_k} is the Dirac distribution supported on the marked point p_k. Inserting this expression in the dilatonic term $\int_M \mathcal{K}_\gamma f(\phi)\, d\mu_\gamma$ associated with (4.86) we get

$$\int_M \mathcal{K}_\gamma f(\phi)\, d\mu_\gamma \;=\; 2\pi \sum_{k=1}^{N_0} \left(\frac{\varepsilon(k)}{2\pi}\right) f\left(\phi(p_k)\right)\ , \tag{4.88}$$

from which (4.81) immediately follows.

It must be remarked that while the dilaton field *couples* with the discrete *charges* $\{\varepsilon(k)\}_{k=1}^{N_0}$ describing the localized curvatures of the surface (M, γ), the harmonic energy functional $E[\gamma, \phi;\ g]_{(M,V^n)}$, because of conformal invariance, is insensitive to the presence of such curvature charges. As first stressed by E. Fradkin and A. Tseytlin [34], the role of the dilatonic term in (4.81) is to restore the conformal invariance of $E[\phi, g]$ which is broken upon quantization. Another important consequence of the structure of (4.81) comes about when we consider constant maps (Fig. 4.12)

$$\phi\ :\ M \longrightarrow V^n \tag{4.89}$$

$$x \longmapsto \phi(x) = y_0\ , \qquad \forall\, x \in M\ .$$

In such a case the harmonic energy $E[\gamma, \phi; g]_{(M,V^n)} \equiv 0$ and the dilatonic action $S[\gamma, \phi; a, g, f]$ calls explicitly into play the topology of the surface M according to

$$S[\gamma, \phi; a^{-1} g, f] = 2\pi f(y_0) \sum_{k=1}^{N_0} \left(\frac{\varepsilon(k)}{2\pi} \right) \tag{4.90}$$

$$= 2\pi f(y_0) \chi(M) ,$$

where we exploited the relation (2.72) connecting the deficit angles $\{\varepsilon(k)\}$ to the Euler characteristic of the surface M.

It is also instructive to consider the one–parameter variation, induced by the field fluctuations (4.39), of the dilatonic term $2\pi \sum_{k=1}^{N_0} \left(\frac{\varepsilon(k)}{2\pi} \right) f(\phi(p_k))$. If we Taylor expand the ϕ_t pull–back $f \circ \phi_t$ of the dilaton field

$$f(\phi_t) = f(\phi) + \frac{d}{dt} f(\phi_t) \Big|_{t=0} t + \frac{1}{2} \frac{d^2}{dt^2} f(\phi_t) \Big|_{t=0} t^2 + \mathcal{O}(t^3) , \tag{4.91}$$

and take into account that $\phi_t(x)$ is a geodesic issued from $\phi(x)$ with tangent vector $a^{\frac{1}{2}} \xi$, we easily get

$$f(\phi_t) = f(\phi) + a^{\frac{1}{2}} t \frac{\partial f}{\partial \phi^i} \xi^i + \frac{1}{2} a t^2 \operatorname{Hess}_{ik} f \, \xi^i \xi^k + \mathcal{O}(a^{\frac{3}{2}} t^3) . \tag{4.92}$$

Hence we can write

$$2\pi \sum_{k=1}^{N_0} \left(\frac{\varepsilon(k)}{2\pi} \right) f(\phi_t(p_k)) = 2\pi \sum_{k=1}^{N_0} \left(\frac{\varepsilon(k)}{2\pi} \right) \left[f(\phi(p_k)) + a^{\frac{1}{2}} t \frac{\partial f(\phi(p_k))}{\partial \phi^i} \xi^i(p_k) \right.$$

$$\left. + \frac{1}{2} a t^2 \operatorname{Hess}_{ij} f(\phi(p_k)) \, \xi^i(p_k) \xi^j(p_k) + \mathcal{O}(a^{\frac{3}{2}} t^3) \right] . \tag{4.93}$$

4.8 Couplings and the Space of Actions

From the above analysis it follows that the action $S[\gamma, \phi; a^{-1} g, f]$ characterizes on the space of maps $Map\ (M, V^n)$ a *family* of two–dimensional field theories parametrized by the Riemannian metrics $\frac{1}{a} g$ on M and by the dilaton field[7] f,

---

[7]Note that further position dependent coupling terms could have been added to the above action (in particular $a^{-1} \int_M U(\phi) \, d\mu_\gamma$ and $a^{-1} \int_M \phi^* \omega$ where $U \in C^\infty(V^n, \mathbb{R})$ and $\omega \in C^\infty(V^n, \wedge^2 TV^*)$ is a 2-form on V^n) but we limit our analysis mostly to (4.86) which exhibits a very subtle geometric structure.

or equivalently by the corresponding metric measure spaces $(V^n, \frac{1}{a} g, d\omega) \in$ Meas(V^n, g). In such a sense target metric measure spaces provide a natural geometrical setting for discussing, from a unified viewpoint, the *point dependent coupling parameters* (Fig. 4.13)

$$\widetilde{\alpha} = a^{-1} \, (g, af) \, , \qquad (4.94)$$

controlling the strength of the fluctuations of the action $S[\gamma, \phi; a, g, f]$ around a given map $\phi \in Map \, (M, V^n)$. Note that the set $\mathscr{C}$ of coupling fields $\alpha = a^{-1} \, (g, af)$ can be identified with the infinite dimensional space

$$\mathscr{C} := \mathbb{R}_{>0} \times \text{Meas}(V^n) \, , \qquad (4.95)$$

where the factor $\mathbb{R}_{>0}$ accounts for the running parameter a and Meas(V^n), defined by (4.73), parametrizes the couplings g and f associated with $(V^n, g, d\omega)$. This geometrical origin of $\mathscr{C}$ is what makes the study of the quantum geometry of (polyhedral) surfaces such a fertile ground for interaction between Mathematics and Physics.

Let us introduce the following notation

$$S[\gamma, \phi; \alpha] := S[\gamma, \phi; a^{-1} g, f = 0],$$

$$\qquad (4.96)$$

$$S[\gamma, \phi; \widetilde{\alpha}] := S[\gamma, \phi; a^{-1} g, f] \, ,$$

respectively associated with the NLσM action and its dilatonic extension. To illustrate the general strategy in handling the space of couplings $\mathscr{C}$, we consider explicitly the deformation $S[\gamma, \phi_t; \alpha]$ of $S[\gamma, \phi; \alpha]$. According to Sect. 4.6, $S[\gamma, \phi_t; \alpha]$ describes the field fluctuations $\phi_t(x) := \exp_{\phi_{\text{har}}(x)} [t \, a^{\frac{1}{2}} \xi(x)]$ around a given fiducial

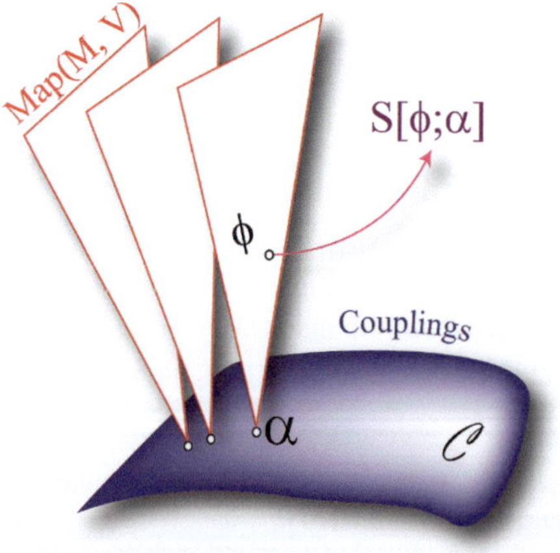

Fig. 4.13 Spaces of maps $Map(M, V^n)$ parametrized by the space of couplings $\mathscr{C}$

harmonic map $\phi_{\text{har}} \in Map(M, V^n)$. It can be represented as a deformation of a *fiducial action* $\widehat{S}[\gamma, \phi_{\text{har}}, \widehat{\xi}; \alpha_0]$ according to

$$S[\gamma, \phi_t; \alpha] = \widehat{S}[\gamma, \phi_{\text{har}}; \alpha_0] + \sum_{i \geq 1} \int_M \mathscr{O}_i(\phi, \alpha_i) \,, \tag{4.97}$$

where we have denoted by $\mathscr{O}_i(\phi, \alpha_i)$ the distinct terms associated with the fluctuations–induced deformations α_i of the couplings α. Explicitly (see (4.70)) these terms are provided by

$$\widehat{S}[\gamma, \phi_{\text{har}}; \alpha_0] := \left. S[\gamma, \phi; \, a^{-1}g] \right|_{\text{har}} - t^2 \int_M \left\langle \widehat{\xi}, \, \Delta_{(\gamma)} \widehat{\xi} \right\rangle^{\gamma}_{\delta} \, d\mu_{\gamma} \,, \tag{4.98}$$

and by

$$\sum_{i \geq 1} \int_M \mathscr{O}_i(\phi, \alpha_i) := t^2 \left[2 \int_M \left\langle A\widehat{\xi}, \, d\widehat{\xi} \right\rangle^{\gamma}_{\delta} + \int_M \left\langle A\widehat{\xi}, \, A\widehat{\xi} \right\rangle^{\gamma}_{\delta} \, d\mu_{\gamma} \right.$$

$$\tag{4.99}$$

$$\left. + \int_M \left\langle \widehat{\text{Rm}}(a^{-1}g) \left(\widehat{\xi}, d\phi \right) \widehat{\xi}, d\phi \right\rangle^{\gamma}_{\delta} \, d\mu_{\gamma} \right] \,,$$

It is somewhat natural to interpret the coupling fields so introduced

$$\alpha_1 := t^2 \, \delta_{ac} \, \gamma^{\nu\sigma} \, (A_\nu)^a_b \,, \tag{4.100}$$

$$\alpha_2 := t^2 \, \delta_{ac} \, \gamma^{\nu\sigma} \, (A_\nu)^a_b \, (A_\sigma)^c_d \,, $$

$$\alpha_3 := t^2 \, \gamma^{\nu\sigma} \, \widehat{R}^a_{bcd}(g(\phi)) \frac{\partial \phi^j}{\partial x^\nu} \frac{\partial \phi^l}{\partial x^\sigma} \, \iota^b_j \iota^d_l \,, \tag{4.101}$$

as a sort of *coordinates* for $S[\gamma, \phi; \alpha]$ in a neighborhood of the fiducial $\widehat{S}[\gamma, \phi_{\text{har}}; \alpha_0]$ (Fig. 4.14). A similar parametrization holds also for the dilatonic NLσM action $S[\gamma, \phi; \, a^{-1} g, f]$. In a highly speculative way we may think that this coordinatization provides a differentiable structure of sort on the formal space $\mathscr{ACT}(M, V^n; \mathscr{C})$ of actions associated with $Map(M, V^n) \times \mathscr{C}$, i.e.,

$$\mathscr{ACT}(M, V^n; \mathscr{C})|_{S_0 - Patch} \,\, "{:=}" \tag{4.102}$$

$$\left\{ (\ldots, \mathscr{O}_i(\phi, \alpha_i) \ldots) \mid S[\gamma, \phi; \alpha] = \widehat{S}[\gamma, \phi_{\text{har}}; \alpha_0] + \sum_{i \geq 1} \int_M \mathscr{O}_i(\phi, \alpha_i) \right\} \,.$$

This space, as formal as it may appear, relates naturally to (Euclidean) Quantum Field Theory where one is forced, by the very nature of the quantization procedure, to introduce a running energy scale parameter $a^{1/2} t$ and a collection of special

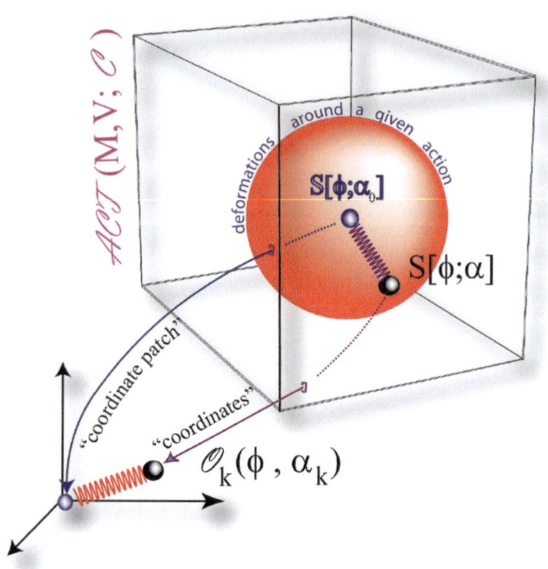

Fig. 4.14 A *coordinatization* of the deformations of a given fiducial action

tangent vectors $\frac{\partial}{\partial t}\alpha_i$, describing the variations of the coupling fields with the energy scale $a^{1/2}t$, which formally provide a distinguished semi–flow in $\mathscr{C}$.

4.9 The Quantum Theory: General Remarks

In classical physics one is typically interested in deformations of the field $\phi \in Map(M, V^n)$ providing information on the geometry of the *graph* $(\phi, S[\gamma, \phi; \alpha])$ around the critical points of the action $S[\gamma, \phi; \alpha]$. In quantum theory the situation changes drastically. Quantum fluctuations of ϕ call into play an inherent element of randomness in the relevant deformations of the field ϕ. This randomness takes the form of quantum interference (QFT proper) or stochastic diffusion (Euclidean QFT). In (a very few) favorable cases, the two mechanisms are related (axiomatic QFT) resulting in a formalism which is rather appealing in its mathematical structure. A good example which is worthwhile to keep in mind, and which to some extent fits nicely in our geometrical set up is the case when M is the circle and (V^n, g, x_0) is a pointed Riemannian manifold, i.e., the loop space $Map(\mathbb{S}^1, V^n)$. This space is naturally endowed with a probability measure, the pinned Wiener measure $\mathscr{W}_{x_0}(V^n)$ on continuous paths in V^n starting and ending at some fixed point $x_0 \in V^n$, and it is the framework appropriate for blending quantum mechanics with the Riemannian geometry of (V^n, g) (Fig. 4.15). If the target manifold V^n is a compact Lie group, (endowed with a bi–invariant Riemannian metric), then the measure space $\{Map(\mathbb{S}^1, V^n),\ \mathscr{W}_{x_0}(V^n)\}$ is invariant under the flow induced by vector fields on $Map(\mathbb{S}^1, V^n)$ [30]. This is basically an extension of the Cameron–Martin theorem

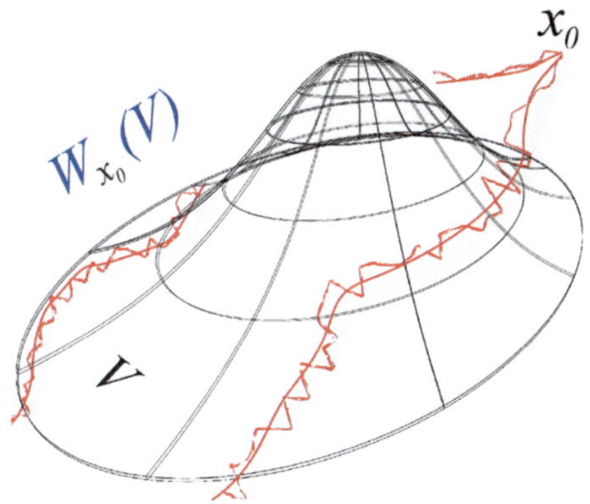

Fig. 4.15 Wiener measure on path space over a Riemannian manifold

which characterizes the path space $\mathscr{P} := \{\eta \in C([0,1], \mathbb{R}^n) \mid \eta(0) = 0\}$, endowed with the standard Wiener measure $\mathscr{W}_0$, and according to which the mapping $\mathscr{P} \to \mathscr{P}$, defined by $\eta \mapsto \eta + f$ with $f \in \mathscr{P}, f(0) = 0$, preserves (up to a density) the measure space $\{\mathscr{P}, \mathscr{W}_0\}$ iff $\int_0^1 |\frac{d}{ds} f|^2 \, ds < \infty$. Explicitly, one defines $\mathscr{H} := \{f \in \mathscr{P} \mid \frac{d}{ds} f \text{ exists a.e. and satisfies } \int_0^1 |\frac{d}{ds} f|^2 \, ds < \infty\}$. This Hilbert space is densely embedded in $\mathscr{P}$, (however $\mathscr{W}_0(\mathscr{H}) = 0$), and can be identified with the tangent space $T_\eta \mathscr{P}$ to $\{\mathscr{P}, \mathscr{W}_0\}$.

In our case, the Quantum Field Theory (QFT) associated with the action $S[\gamma, \phi; \alpha]$ is characterized by the set of correlations, induced by the quantum fluctuations, among the values $\{\phi(x_i)\} \in (V^n)^k$ that the field attain at distinct marked points $x_1, \ldots, x_k \in M$. These correlations are formally defined by

$$Z[\gamma, \phi(x_i); \alpha] \ " \doteq " \ \frac{1}{Z_0} \int_{\{Map(M,V^n)\}} D_\alpha[\phi] \, (\phi(x_1) \ldots \phi(x_i) \ldots) \, e^{-S[\gamma, \phi; \alpha]} \,,$$

$$(4.103)$$

where $D_\alpha[\phi]$ is a functional measure on $Map(M, V^n)$, possibly depending on the couplings $\alpha \in \mathscr{C}$, and Z_0 is a *normalization constant* (Fig. 4.16). Although bona–fide functional measures on the map spaces $Map(M, V^n)$ can be introduced (cf. [64, 81, 86]), the above expression is actually well–defined only as a formal power series in the coupling parameters α, i.e. as a perturbative quantum field theory. In particular, if, according to (4.97), we consider $S[\gamma, \phi; \alpha]$ as a deformation of a fiducial $\widehat{S}[\gamma, \phi; \alpha_0]$, i.e., $S[\gamma, \phi; \alpha] = \widehat{S}[\gamma, \phi; \alpha_0] + \sum_{a \geq 1} \int_M \mathscr{O}_a(\phi, \alpha_a)$, then we can write

$$\int_{\{Map(M,V^n)\}} D_\alpha[\phi] \, e^{-S[\gamma, \phi; \alpha]} = \int_{\{Map(M,V^n)\}} D_\alpha[\phi] e^{-\widehat{S}[\gamma, \phi; \alpha_0]} \prod_a e^{-\int_M \mathscr{O}_a(\phi, \alpha_a)} \,.$$

Fig. 4.16 Correlations over fluctuating surfaces

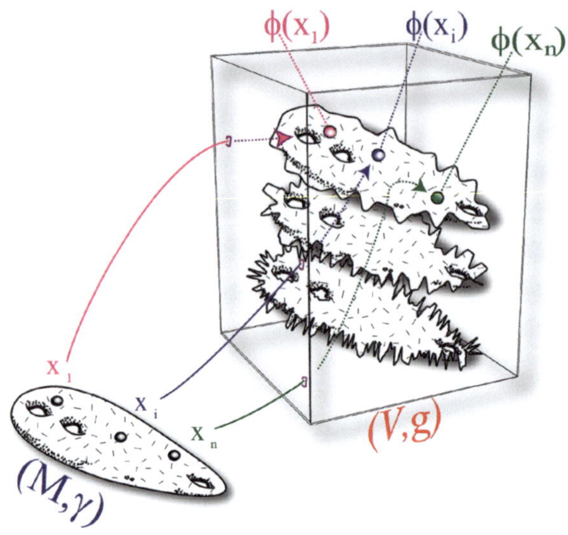

In such a setting one typically assumes the *partition function*

$$Z_0 := \int_{\{Map(M,V^n)\}} D_\alpha[\phi]e^{-\widehat{S}[\gamma,\phi;\alpha_0]} \tag{4.104}$$

as the normalization factor introduced in (4.103). Notice also that the probability measure so defined formally induces on the coupling space $\mathscr{C}$ a covariance

$$\mathscr{G}(\alpha_i,\alpha_j) := \frac{1}{Z_0} \int_{\{Map(M,V^n)\}} D_\alpha[\phi]e^{-\widehat{S}[\gamma,\phi;\alpha_0]} \int_M \mathscr{O}_i(\phi,\alpha_i) \int_M \mathscr{O}_j(\phi,\alpha_j)\,, \tag{4.105}$$

which, if positive, turns $\mathscr{C}$ into a measure space, $\{\mathscr{C};\, D[\mathscr{G}]\}$, (this covariance is often called the Zamolodchikov metric in 2D QFT).

It must be stressed that the somewhat fanciful expressions defined above hardly make sense by themselves, even at a physical level of rigor, if we do not devise a way of controlling the spectrum of fluctuations of the fields $\phi \in Map(M,V^n)$ (Fig. 4.17). In particular, one fundamental problem concerning (4.103) is to introduce in the QFT $\{Map(M,V^n)\,,\,e^{-S[\gamma,\phi;\alpha]}\,D_\alpha[\phi]\}$ a filtration parametrized by a length scale, (the only scale of measurement significant in a relativistic quantum theory), which allows to control the way (4.103) behaves under scale–dependent transformations of the fields $\phi \in Map(M,V^n)$ and of the couplings $\alpha \in \mathscr{C}$. This basic need calls into play the circle of ideas that rather generically go under the name of *Wilsonian Renormalization Group*: a conceptual framework which connects the dynamical behavior of a physical theory to the analysis of flows in the space of its coupling parameters, and which must be adapted case by case, often with very different and sophisticated mathematical techniques.

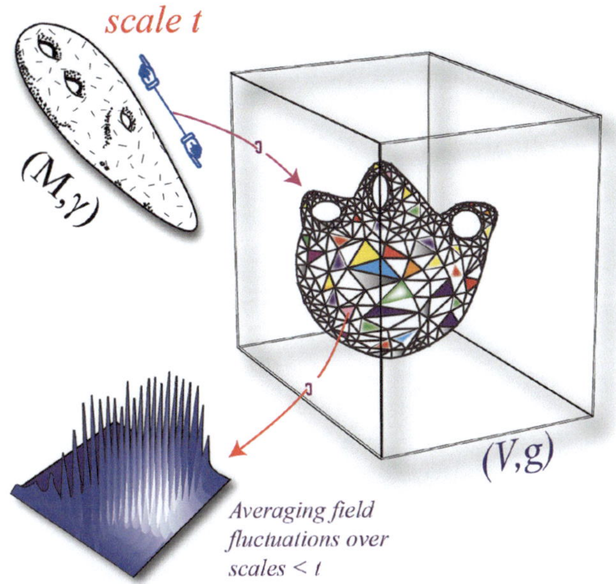

Fig. 4.17 Averaging fluctuations over length scales $< t$

4.10   An Informal Geometrical View to Renormalization

According to the previous remarks, a basic ingredient in discussing how the QFT formalism takes shape for the NLσM action $S[\gamma, \phi; \alpha]$ is the search for a flow, (*renormalization group flow*),

$$\mathscr{RG}_t : [Map(M, V^n) \times \mathscr{C}] \longrightarrow [Map(M, V^n) \times \mathscr{C}] \qquad (4.106)$$

$$(\phi, \alpha) \qquad \longmapsto \mathscr{RG}_t(\phi, \alpha) = (\phi_t; \alpha(t)) ,$$

which, as we vary the scale of distances t at which we probe the Riemannian surface M, allows to tame the energetics of the fluctuations of the fields $\phi : M \to V^n$ in terms of the (running) couplings $\alpha \mapsto \alpha(t)$ (Fig. 4.17). In order to describe this procedure in physical terms, select two scales of distances, say Λ^{-1} and Λ'^{-1}, (one can equivalently interpret Λ and Λ' as the respective scales of momentum in the spectra of field fluctuations), with $\Lambda'^{-1} > \Lambda^{-1}$. The general idea, central in K.G. Wilson's analysis of the renormalization group flow, is to assume that if the action $S[\gamma, \phi_\Lambda; \alpha(\Lambda)] \in \mathscr{ACT}(M, V^n; \mathscr{C})$ describes the theory at a cut–off scale Λ^{-1}, then there is a map (Fig. 4.18)

$$\widetilde{\mathscr{RG}}_{\Lambda\Lambda'} : \mathscr{ACT}(M, V^n; \mathscr{C}) \longrightarrow \mathscr{ACT}(M, V^n; \mathscr{C}) , \qquad (4.107)$$

$$S[\gamma, \phi_\Lambda; \alpha(\Lambda)] \longmapsto S[\gamma, \phi_{\Lambda'}; \alpha(\Lambda')] \doteq \widetilde{\mathscr{RG}}_{\Lambda\Lambda'} S[\gamma, \phi_\Lambda; \alpha(\Lambda)]$$

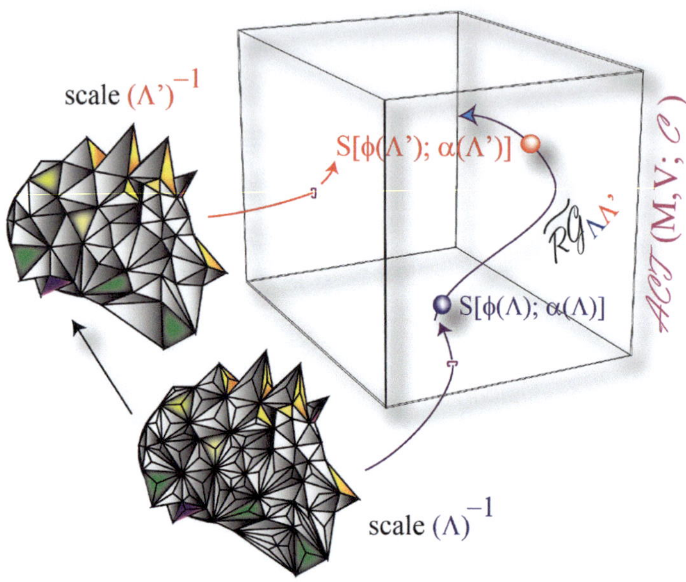

Fig. 4.18 Renormalization as a map in the space of actions

such that the action $S[\gamma, \phi_{\Lambda'}; \alpha(\Lambda')] \doteq \widetilde{\mathcal{RG}}_{\Lambda\Lambda'} S[\gamma, \phi_{\Lambda}; \alpha(\Lambda)]$ provides the effective theory at scale Λ'^{-1}, obtained upon suitably averaging field–fluctuations in moving from the distance scale Λ^{-1} to the scale Λ'^{-1}. Such a map is required to satisfy the semigroup property $\widetilde{\mathcal{RG}}_{\Lambda\Lambda''} = \widetilde{\mathcal{RG}}_{\Lambda'\Lambda''} \circ \widetilde{\mathcal{RG}}_{\Lambda\Lambda'}$ for all $\Lambda'' < \Lambda'$. This formal (semi)-flow, if exists, induces a corresponding flow on $Map(M, V^n) \times \mathcal{C}$

$$\mathcal{RG}_{\Lambda\Lambda'} : [Map(M, V^n) \times \mathcal{C}] \longrightarrow [Map(M, V^n) \times \mathcal{C}] \tag{4.108}$$

$$(\phi_{\Lambda}, \alpha(\Lambda)) \longmapsto \mathcal{RG}_{\Lambda\Lambda'}(\phi_{\Lambda}, \alpha(\Lambda)) = (\phi_{\Lambda'}; \alpha(\Lambda')) ,$$

by requiring that the natural commutativity relation

$$\widetilde{\mathcal{RG}}_{\Lambda\Lambda'} S[\gamma, \phi_{\Lambda}; \alpha(\Lambda)] = S[\gamma, \mathcal{RG}_{\Lambda\Lambda'}(\phi_{\Lambda}, \alpha(\Lambda))] ,$$

holds. According to these heuristic properties, the local map $\widetilde{\mathcal{RG}}_{\Lambda\Lambda'}$ defined by (4.107) should partition the space $Map(M, V^n) \times \mathcal{C}$ in scale–invariant QFTs associated with the fixed points of the renormalization group flow and in the complementary set of cut–off dependent QFTs. Note that formally, under the action of $\mathcal{RG}_{\Lambda\Lambda'}$ we can either pull–back or push–forward the measure $e^{-S[\gamma,\phi;\alpha]} D_{\alpha}[\phi]$. Given the physical meaning of the renormalization group flow, the push–forward should be perhaps more appropriate in a measure–theoretic sense, however given the informal level of this paragraph, we shall use for simplicity the pull–back measure $\mathcal{RG}^*_{\Lambda\Lambda'}(D_{\alpha}[\phi])\ e^{-\widetilde{\mathcal{RG}}_{\Lambda\Lambda'} S[\gamma,\phi;\alpha]}$. Indeed, in order to characterize the flow $\mathscr{RG}_{\Lambda\Lambda'}$ we *assume* that it should leave the measure space $\{Map(M, V^n) \times \mathscr{C}, e^{-S[\gamma,\phi;\alpha]} D_\alpha[\phi]\}$ quasi–invariant in a suitable sense. The idea is roughly the following: suppose that, at least for $(\Lambda' \setminus \Lambda)$ small enough, we can put the pulled–back functional measure $\mathscr{RG}^*_{\Lambda\Lambda'}(D_\alpha[\phi]) e^{-\widetilde{\mathscr{RG}}_{\Lambda\Lambda'} S[\gamma,\phi;\alpha]}$ in the same form as the original functional measure $D_\alpha[\phi] e^{-S[\gamma,\phi;\alpha]}$, except for a small modification of the couplings α. Explicitly, let $\Lambda' = e^{-\epsilon} \Lambda$, with $0 < \epsilon < 1$, and assume that for every such ϵ there exists a corresponding coupling $\alpha + \delta\alpha$ such that the following identity holds

$$\mathscr{RG}^*_\epsilon (D_\alpha[\phi]) \, e^{-\widetilde{\mathscr{RG}}_\epsilon S[\gamma,\phi;\alpha]} = D_{\alpha+\delta\alpha}[\phi] \, e^{-S[\gamma,\phi;\alpha+\delta\alpha]} \, , \qquad (4.109)$$

where we have denoted $\widetilde{\mathscr{RG}}_\epsilon$ the action of the map $\widetilde{\mathscr{RG}}_{\Lambda\Lambda'}$ for $\Lambda' = e^{-\epsilon} \Lambda$. In other words, we assume that an infinitesimal change in the cutoff can be completely *absorbed* in an infinitesimal change of the couplings. If this equation is valid at least to some order in ϵ, we can iteratively use it to see how α is affected by a finite change of the cutoff. If we set $t \doteq -\ln(\Lambda' \setminus \Lambda)$, then the map $\mathscr{RG}_t$ so induced by $\widetilde{\mathscr{RG}}_{\Lambda\Lambda'}$ on $[Map(M, V^n) \times \mathscr{C}]$, as t varies, is the renormalization group flow $\mathscr{RG}_t$ introduced above, (see (4.108)). Since $[Map(M, V^n) \times \mathscr{C}]$ is non–linear, the infinitesimal quasi–invariance described by (4.109) is what we may reasonably expect to replace the quasi–invariance characterizing linear measure spaces such as $\{\mathscr{P}, \mathscr{W}_0\}$. This infinitesimal quasi–invariance yields for what is basically an integration by parts formula characterizing a set of distinguished tangent vector fields to the measure space $\{\mathscr{C}; D[\mathscr{G}]\}$. To show how this comes about, let us consider a scale interval $-\epsilon \leq t \leq \epsilon$, for $\epsilon > 0$, and assume that the associated functional measure $D_\alpha[\phi] \, e^{-S[\gamma,\phi;\alpha]}$ has natural transformation properties under $\mathscr{RG}_t$, i.e.,

$$\int_{\mathscr{RG}_t\{Map(M,V^n)\}} D_{\alpha(t)}[\phi_t] \, e^{-S[\gamma,\phi_t;\alpha(t)]} \qquad (4.110)$$

$$= \int_{\{Map(M,V^n)\}} \mathscr{RG}^*_t (D_\alpha[\phi]) \, e^{-\widetilde{\mathscr{RG}}_t S[\gamma,\phi;\alpha]} \, .$$

The strategy is to exploit (4.109) by evaluating, along the $\mathscr{RG}_t$ map, the flow derivative $\frac{d}{dt} Z[\alpha(t)]$ at the generic scale t, where

$$Z[\alpha(t)] \doteq \int_{\mathscr{RG}_t\{Map(M,V^n)\}} D_{\alpha(t)}[\phi_t] \, e^{-S[\gamma,\phi_t;\alpha(t)]} \, .$$

Denoting, from notational ease, $(M, V^n)_t := \mathscr{RG}_t\{Map(M, V^n)\}$, we compute at a very (in)formal level

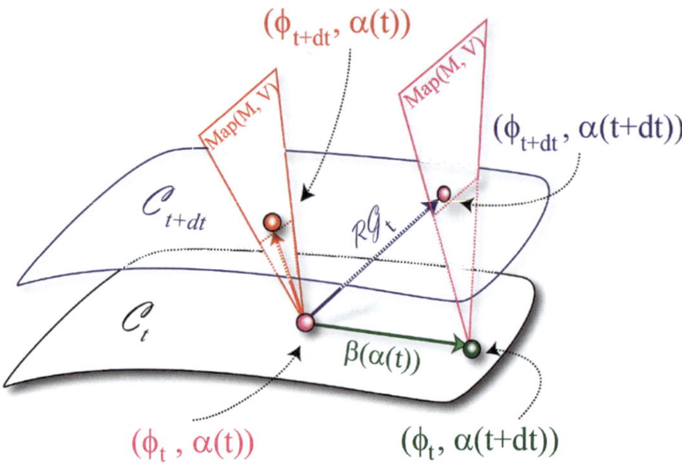

Fig. 4.19 The geometry of the beta function

$$\frac{d}{dt} Z[\alpha(t)] = \lim_{\epsilon \to 0} \frac{1}{\epsilon} \left[\int_{(M,V^n)_{t+\epsilon}} D_{\alpha(t)}[\phi_t] \, e^{-S[\gamma,\phi_t;\alpha(t)]} \right. \tag{4.111}$$

$$\left. - \int_{(M,V^n)_t} D_{\alpha(t)}[\phi_t] \, e^{-S[\gamma,\phi_t;\alpha(t)]} \right]$$

$$= \lim_{\epsilon \to 0} \frac{1}{\epsilon} \left[\int_{\mathcal{RG}_\epsilon((M,V^n)_t)} D_{\alpha(t)}[\phi_t] \, e^{-S[\gamma,\phi_t;\alpha(t)]} \right.$$

$$\left. - \int_{(M,V^n)_t} D_{\alpha(t)}[\phi_t] \, e^{-S[\gamma,\phi_t;\alpha(t)]} \right]$$

$$= \int_{(M,V^n)_t} \lim_{\epsilon \to 0} \frac{1}{\epsilon} \left[\mathcal{RG}_\epsilon^*[D_{\alpha(t)}[\phi_t]] \, e^{-\widetilde{\mathcal{RG}}_\epsilon S[\gamma,\phi_t;\alpha(t)]} \right.$$

$$\left. - D_{\alpha(t)}[\phi_t] \, e^{-S[\gamma,\phi_t;\alpha(t)]} \right]$$

$$= - \int_{(M,V^n)_t} \beta(\alpha(t)) \frac{\partial}{\partial \alpha(t)} \left(D_{\alpha(t)}[\phi_t] \, e^{-S[\gamma,\phi_t;\alpha(t)]} \right) ,$$

where we have introduced the β–flow *vector field* on the space of couplings $\mathcal{C}$ (Fig. 4.19)

$$\beta(\alpha(t)) \doteq - \frac{\partial}{\partial t} \alpha(t) , \tag{4.112}$$

and where we have exploited the semigroup property of the flow and the scaling hypothesis (4.109). Since the integration is over $\mathscr{RG}_t\{Map(M, V^n)\}$, we can formally extract the operator $\beta(\alpha(t))\frac{\partial}{\partial\alpha(t)}$ from the functional integral, and rewrite the relation (4.111) as

$$\left\{\frac{d}{dt} + \beta(\alpha(t))\frac{\partial}{\partial\alpha(t)}\right\} Z_t[\alpha] = 0 \ . \tag{4.113}$$

Roughly speaking (4.113) says is that if we rescale distances in M by a factor e^t and at the same time we flow in the space of couplings along β for a *time t*, the theory we obtain looks the same as before. If the theory is, along the lines sketched above, renormalizable by a renormalization of the couplings, many of its properties can be desumed by the analysis of (4.112). In the framework so described, the renormalization group formally appears as a natural geometrical flow on the measure space $\{Map(M, V^n) \times \mathscr{C}; D_{\alpha(t)}[\phi_t]\}$ and the β–flow *vector fields* (4.112) play a role analogous to the role played by the Cameron–Martin vector fields for Wiener measure space. In particular, since $\{\mathscr{C}; D[\mathscr{G}]\}$ is a measure space when endowed with the $D_{\alpha(t)}[\phi_t]$–induced covariance defined by the *Zamolodchikov metric* $\mathscr{G}_{ij}$, (see (4.105)), it is natural to argue whether the β– vector fields are gradient of a suitable functional $\Phi(\alpha)$ with respect to $\mathscr{G}$: i.e., if $\beta(\alpha) = \mathscr{G}(\mathscr{D}\Phi(\alpha) \circ v)$, (here $\mathscr{D}\Phi(\alpha) \circ v$ denotes the linearization of $\Phi(\alpha)$ in the direction of the coupling perturbation v). This is a fascinating open issue, (see e.g. [16], and the recent [60]), deeply connected with a geometrical understanding of the renormalization group in its role of averaging out fluctuations: being a gradient flow would avoid recurrent behaviors like limit cycles and strange attractors. It would also imply that any such a functional $\Phi(\alpha)$ is monotonically non–increasing along the flow and that the renormalization group flow is, as intuitively expected, irreversible. A somewhat weaker form of such irreversibility is associated with the celebrated Zamolodchikov's *c–theorem* [88], which, under a unitarity condition, states that for 2D QFT there exists a function $c(\alpha)$ which is monotonically non–increasing along the renormalization group flow. The associated fixed points of the RG flow are conformal field theories (CFT), corresponding to which $c(\alpha)$ reduces to the corresponding central charge c. This picture is considered quite well-understood [43] for 2–dimensional QFTs, like the NLσM discussed in this chapter. Conversely, the situation remains rather conjectural for higher dimensional theories. Even if there have been recent results [62] to a proof of a weak form of the c–theorem in dimension 4, it is fair to say that there are still very delicate points in the argument. In this connection, it is worthwhile to stress that there are very interesting indications [60] that the holographic correspondence may play a relevant role in disclosing the nature of the RG flow for higher dimensional QFTs.

4.11   The QFT Analysis of the NLσM

A major technical difficulty in implementing the renormalization group procedure described above is that the space of maps $Map(M, V^n)$ is a non–linear functional space, and the actual geometrical characterization of the renormalization group analysis of $S[\gamma, \phi; \alpha]$ is quite subtle. A basic observation here is that to the (point dependent) coupling parameter of the theory $a^{-1} g$ we can associate the (adimensional) length scale defined by

$$\left| Rm \left(a^{-1} g \right) \right| , \tag{4.114}$$

where $\left| Rm(a^{-1} g) \right| := [R^{iklm}(a^{-1} g) \, R_{iklm}(a^{-1} g)]^{1/2}$ is the pointwise norm of the Riemann tensor of the rescaled metric $a^{-1} g$. Note that an elementary computation exploiting the scaling property of the Riemann tensor (see Appendix A) provides $\left| Rm \left(a^{-1} g \right) \right| = |Rm(g)| \, a$. From the expression of the curvature *interaction term* (see (4.72))

$$\int_M \left\langle \widehat{Rm}(a^{-1}g) \left(\widehat{\xi}, d\phi \right) \widehat{\xi}, d\phi \right\rangle_\delta^\gamma \, d\mu_\gamma , \tag{4.115}$$

in the NLσM action(4.70) associated with the fluctuating field $\widehat{\xi} \in \phi_{har}^{-1} TV^n$, we get (assuming, just for the sake of the argument, that $\int_M d\mu_\gamma = 1$)

$$\left| \int_M \left\langle \widehat{Rm}(a^{-1}g) \left(\widehat{\xi}, d\phi \right) \widehat{\xi}, d\phi \right\rangle_\delta^\gamma \, d\mu_\gamma \right| \tag{4.116}$$

$$\leq \left| \int_M \left| \widehat{Rm}(a^{-1} g(\phi)) \right|^2 d\mu_\gamma \right|^{\frac{1}{2}} \left| \int_M \left| \gamma^{v\sigma} \partial_v \phi^j \partial_\sigma \phi^l \, \iota_j^b \iota_l^d \, \widehat{\xi}^a \widehat{\xi}^c \right|^2 d\mu_\gamma \right|^{\frac{1}{2}}$$

$$\leq \sup_{x \in M} \left| \widehat{Rm}(a^{-1} g(\phi(x))) \right| \left| \int_M \left| \gamma^{v\sigma} \partial_v \phi^j \partial_\sigma \phi^l \, \iota_j^b \iota_l^d \, \widehat{\xi}^a \widehat{\xi}^c \right|^2 d\mu_\gamma \right|^{\frac{1}{2}} .$$

For the gauge interaction terms present in (4.70) we get similar bounds, expressed in terms of $\sup_{x \in M} \left| A(a^{-1}g) \right|$, where $A(a^{-1}g)$ denotes the connection (4.61) on $\phi_{har}^{-1} TV^n$. Roughly speaking, these gauge terms are not relevant in a point–like limit (the gauge field being locally eliminable), and from the point of view of a perturbative renormalization group analysis, it is the term $|Rm(a^{-1} g)|$ which plays the role of the geometric coupling governing the fluctuations of the fields ξ (Fig. 4.20). This remark implies that when the curvature of target Riemannian manifold $(M, a^{-1}g)$ is small with respect to the length scale $a^{1/2}$ over which the field ϕ is localizable (see Definition 4.1), the measure $e^{-S[\gamma, \phi; \alpha]} D_\alpha[\phi]$ is concentrated around the minima of the action $S[\gamma, \phi; \alpha]$, i.e. the constant maps. In such a case we can control the spectrum of fluctuations of the fields (Fig. 4.10) by large deviation techniques. With a slight abuse of language, it is usual to describe this point–like

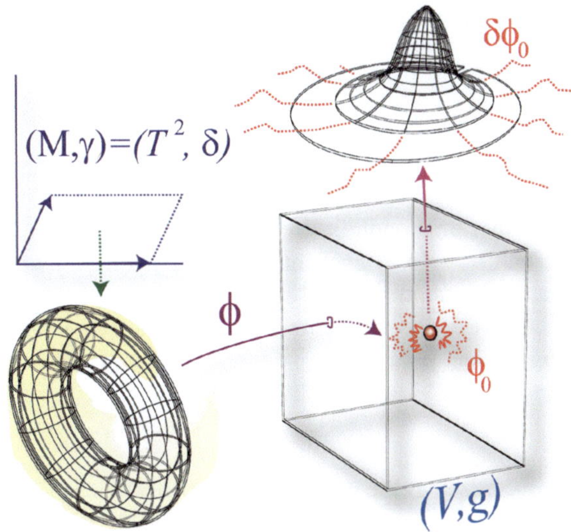

Fig. 4.20 The point–like limit: nearly Gaussian fluctuations near constant maps

limit regime as characterizing the perturbative renormalization group analysis of $S[\gamma, \phi; \alpha]$ for small a (actually meaning for small $|Rm(a^{-1} g)|$).

In discussing the point–like limit we can exploit the center of mass interpretation of the NLσM action (see Sects. 4.4 and 4.5). The idea (due to Gawedzki [39] in its field theoretic implementation) is to extract the behavior of the (perturbative) quantum fluctuations of the field ϕ by describing it as a random field resulting from the induced distribution of the center of mass of a large number of independent and identically distributed random constant maps $\{\phi_{(k)}\}_{k=1\ldots,q}$: $(M, \gamma) \longrightarrow (V^n, a^{-1} g)$. The resulting formalism is quite natural when (M, γ) is a polyhedral surface. Let $\{\sigma^0(k)\}$, $k = 1,\ldots, q \leq N_0$, denote a subset of the N_0 vertexes of (M, γ), and let us consider a corresponding collection $\{\phi_{(k)}\}_{k=1\ldots,q} \in Map\,(M, V^n)$ of constant maps, (hence harmonic), which associates to the whole M the value $\phi_{(k)}$ that the field ϕ takes at the vertex $\sigma^0(k) \in M$. Since in the point–like limit the map ϕ : $(M, \gamma) \longrightarrow (V^n, a^{-1} g)$ collapses the surface M to a point, one may heuristically think of $\phi_{(k)}$ as the *spin* that one sees at the collapsed point. This spin value is extracted out of the possible collection of distinct *spins* $\{\phi_{(k)}\}_{k=1}^{q}$ associated with the chosen set of vertexes $\{\sigma^0(k)\}$, $k = 1,\ldots, q$ in M. Explicitly we assume that each $\phi_{(k)}$ takes value in the interior of a metric ball $B_{a^{-1}g}(p, r) :=$ $\{z \in V^n | d_{a^{-1}g}(p, z) \leq r\}$, centered at a generic point $p \in (V^n, a^{-1} g)$, i.e.

$$\phi_{(k)} : M \longrightarrow \quad B_{a^{-1}g}(p, r) \setminus \partial B_{a^{-1}g}(p, r) \subset (V^n, a^{-1} g) \qquad (4.117)$$

$$x \longmapsto \phi_{(k)}(x) = y_{(k)}, \qquad \forall x \in M, \qquad k = 1, \ldots, q .$$

The idea is that quantum fluctuations of the constant maps averaged over a controlled length scale, (say that set by the ball $B_{a^{-1}g}(p, r)$), generate an effective QFT for ϕ on the given scale. To implement this strategy we need first to localize the fluctuations of the constant maps $\phi_{(k)}$ and to this end we assume (see (4.4))

$$r \; < \; \frac{\pi}{6\,\sqrt{\kappa(a^{-1}g)}} \; , \tag{4.118}$$

$$\mathrm{inj}_{a^{-1}g}(y) \; > \; 3r, \qquad \forall\, y \in B_{a^{-1}g}(p, r). \tag{4.119}$$

Note that the condition (4.118) is scale invariant: under the map $g \longmapsto a^{-1}g$, the ball $B_{a^{-1}g}(p, r)$ goes into the ball $B_g(p, a^{1/2}r)$ whereas the (upper bound to the) sectional curvature $\kappa(a^{-1}g)$ rescales as $a\kappa(g)$. We also assume that $B_{a^{-1}g}(p, 2r)$ is convex and consider in $(V^n, a^{-1}g)$ the *center of mass* [12] of the maps $\{\phi_{(k)}\}$ (Fig. 4.21),

$$\phi_{cm} \; := \; cm\,\{\phi_{(1)}, \ldots, \phi_{(q)}\} \; , \tag{4.120}$$

characterized as the minimizer of the function

$$F(y;\, q) \; \doteq \; \frac{1}{2} \sum_{k=1}^{q} d^2_{a^{-1}g}(y, y_{(k)}) \; , \tag{4.121}$$

where $d^2_{a^{-1}g}(\cdot\,,\cdot)$ denotes the distance in $(V^n, a^{-1}g)$.

Lemma 4.1 *Under the stated hypotheses the minimizer*

$$\psi \; := \; \phi_{cm}(x) \; = \; \inf_{y \in B_{a^{-1}g}(p, 2r)} \left[\frac{1}{2} \sum_{j=1}^{q} d^2_{a^{-1}g}(y,\, \phi_{(j)}(x)) \right] \tag{4.122}$$

exists, is unique and $\psi \in B_{a^{-1}g}(p, 2r)$.

Proof This is a standard property of the center of mass [12]. See for instance [24] (Chap.4, p.175). One may think of the center of mass ψ as an average of the spins $\phi_{(k)}$ over the given collection of vertexes $\{\sigma^0(k)\}$.

Let us denote by $\exp_\psi : T_\psi V^n \to B_{a^{-1}g}(p, 2r)$ the exponential map, with respect to the rescaled metric $a^{-1}g$, based at ψ. Since $B_{a^{-1}g}(p, 2r)$ is assumed convex, we can introduce a collection of q vectors $X_{(j)} \in T_\psi V^n$ and parametrize the constant maps $\{\phi_{(j)}\}_{j=1}^{q}$ according to

$$\phi_{(j)}(x) \; = \; \exp_\psi(X_{(j)} \circ \psi(x)) \; = \; \exp_\psi(\xi_{(j)}(x)) \; , \tag{4.123}$$

where, corresponding to the vectors $X_{(j)} \in T_\psi V^n$, we have introduced the sections

$$\xi_{(j)} \, : \, M \; \longrightarrow \; \psi^{-1}\, TV^n, \tag{4.124}$$

$$x \; \longmapsto \; \xi_{(j)}(x) \; = \; X^k_{(j)}(\psi(x))\frac{\partial}{\partial \psi^k(x)} \; ,$$

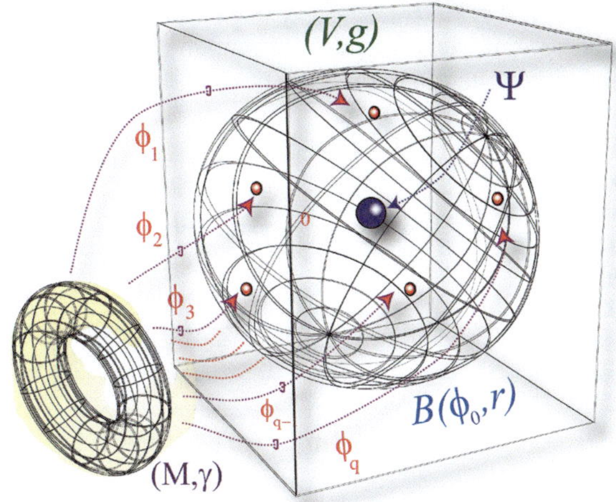

Fig. 4.21 In the point–like limit we can define the center of mass ψ of N independent copies of the mapping ϕ

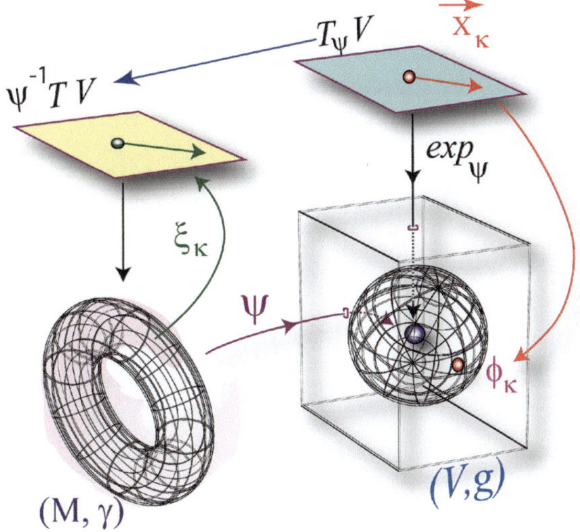

Fig. 4.22 The geometrical set up for discussing fluctuations around the background field ψ in terms of the sections $\xi_{(k)} := X_{(k)} \circ \psi$

of the pull–back bundle $\psi^{-1} TV^n$ (Fig. 4.22). Note that the $\{\xi_j\}$ satisfy the center of mass constraint $\sum_{j=1}^{q} \xi_{(j)}(x) = 0$. More generally, for $\xi \in \psi^{-1} TV^n$ we shall write $\phi_{(\xi)}(x) = \exp_\psi(\xi(x))$. The strategy in the (perturbative) QFT analysis of non–linear σ model is to let the sections $\xi_{(j)}$ fluctuate around a *background* defined by their center of mass map. The fluctuations are generated by assuming that the fields $\xi_{(j)}$, subject to the center–of–mass constraint $\sum_{j=1}^{q} \xi_{(j)}(x) = 0$, are vector valued random variables distributed on $Map(M, \psi^{-1} TV^n)$ according to the measure

$$e^{-S[\gamma,\, \exp_\psi(\xi_{(j)});\, a^{-1} g]} D\left[\exp_\psi(\xi_{(j)})\right], \qquad (4.125)$$

where $S[\gamma,\ \exp_\psi(\xi_{(j)});\ a^{-1}g]$ is the NLσM action associated with the fluctuating fields $\xi_{(j)}$. If we take into account the geometric decomposition (4.70) we can explicitly write (up to order $\mathcal{O}(|\xi_{(j)}|^3)$)

$$e^{-S[\gamma,\ \exp_\psi(\xi_{(j)});\ a^{-1}g]}\,D\left[\exp_\psi(\xi_{(j)})\right] = e^{-S[\gamma,\ \psi;\ a^{-1}g]} \tag{4.126}$$

$$\times D\left[\exp_\psi(\xi_{(j)})\right]\exp\left[\int_M \left\langle \widehat{\xi},\, \triangle_{(\gamma)}\,\widehat{\xi}\right\rangle_\delta^\gamma\, d\mu_\gamma\right]$$

$$\times \exp\left[-2\int_M \left\langle A\widehat{\xi},\, d\widehat{\xi}\right\rangle_\delta^\gamma - \int_M \left\langle A\widehat{\xi},\, A\widehat{\xi}\right\rangle_\delta^\gamma\, d\mu_\gamma\right.$$

$$\left. -\int_M \left\langle \widehat{\mathrm{Rm}}(a^{-1}g)\left(\widehat{\xi},d\psi\right)\widehat{\xi},d\psi\right\rangle_\delta^\gamma\, d\mu_\gamma\right].$$

where we have absorbed the parameter t in the fluctuating fields $\xi_{(j)}$. Hence the QFT for each quantum field $\xi_{(j)}$, $j=1,\dots,q$, is described, around the classical background provided by the center of mass (associated to the term $e^{-S[\gamma,\ \psi;\ a^{-1}g]}$ in (4.126)), by the reference Gaussian measure

$$D\left[\exp_\psi(\xi_{(j)})\right]\exp\left[\int_M \left\langle \widehat{\xi},\, \triangle_{(\gamma)}\,\widehat{\xi}\right\rangle_\delta^\gamma\, d\mu_\gamma\right], \tag{4.127}$$

with respect to which the interaction terms

$$\exp\left[-2\int_M \left\langle A\widehat{\xi},\, d\widehat{\xi}\right\rangle_\delta^\gamma - \int_M \left\langle A\widehat{\xi},\, A\widehat{\xi}\right\rangle_\delta^\gamma\, d\mu_\gamma\right.$$

$$\left. -\int_M \left\langle \widehat{\mathrm{Rm}}(a^{-1}g)\left(\widehat{\xi},d\psi\right)\widehat{\xi},d\psi\right\rangle_\delta^\gamma\, d\mu_\gamma\right]. \tag{4.128}$$

are weighted. The quantum fluctuations of the fields $\{\xi_{(j)}\}$ generate also and induced spectrum of fluctuations around the classical center of mass background field ψ. Explicitly, since $\{\phi_{(j)} = \exp_\psi(\xi_{(j)})\}_{j=1}^q$ are, up to the center of mass constraint $\sum_{j=1}^q \xi_{(j)} = 0$, independent and identically distributed random fields, their center of mass (4.120) is distributed according to the law (Fig. 4.23)

$$P_q(\psi)\,D[\psi] := \int_{Map(M,\psi^{-1}TV^n)} \delta\left[\sum_{j=1}^q \xi_{(j)}\right]\prod_{k=1}^q e^{-S[\gamma,\ \exp_\psi(\xi_{(j)});\ a^{-1}g]}\,D\left[\exp_\psi(\xi_{(j)})\right], \tag{4.129}$$

where $\delta[\sum_{j=1}^q \xi_{(j)}]$ is the Dirac measure, based at the zero map, on the linear space $Map\,(M,\psi^{-1}TV^n)$. The functional measure $P_q(\psi)\,D[\psi]$ so defined characterize the

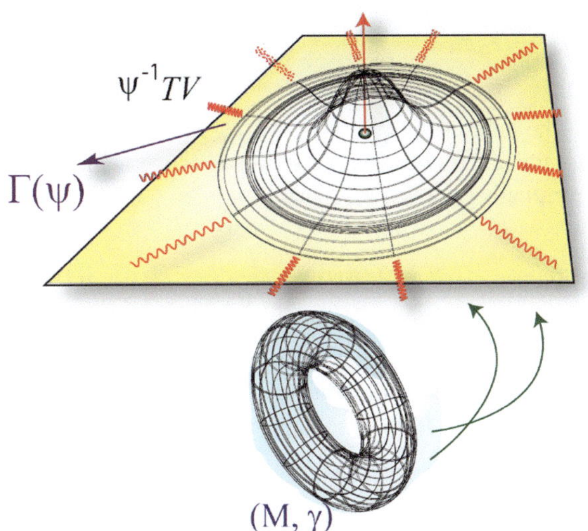

Fig. 4.23 Nearly Gaussian distribution of the fields η in the linear space of maps $Map(\Sigma, \psi^{-1}TM)$. This distribution generates the effective action for the background field ψ

(perturbative) QFT associated with the NLσM action on the given scale (in the approximation considered, this perturbative scale is set by $|Rm(a^{-1}\,g)| \ll 1$).

To discuss the properties of $P_q(\psi)\,D[\psi]$ we start by exploiting the formal Fourier representation of the functional Dirac–$\delta[\circ]$ so as to write

$$\delta[\sum_{j=1}^{q}\xi_{(j)}] = \int_{Map^*(M,\psi^{-1}TV^n)} [DJ]\,\exp i \ll J\cdot \sum_{j=1}^{q}\xi_{(j)} \gg, \qquad (4.130)$$

where the pairing between $Map\left(M,\psi^{-1}TV^n\right)$ and its dual $Map^*\left(M,\psi^{-1}TV^n\right)$ is defined by the L^2 inner product

$$\ll J\cdot \sum_{j=1}^{q}\xi_{(j)} \gg := \int_{\psi^{-1}TV^n} (\psi^*g)_{\mu\nu}\,J^\mu \sum_{j=1}^{q}\xi_j^\nu \left(\psi^*d\mu_g\right). \qquad (4.131)$$

Hence, we can express (4.129) as

$$P_q(\psi)\,D[\psi] := \int_{Map^*(M,\psi^{-1}TV^n)} [DJ] \int_{Map(M,\psi^{-1}TV^n)} D\left[\exp_\psi(\xi_{(j)})\right]$$

$$\qquad (4.132)$$

$$\times\, e^{i\ll J\cdot\sum_{j=1}^{q}\xi_{(j)}\gg} \prod_{k=1}^{q} e^{-S[\gamma,\,\exp_\psi(\xi_{(j)});\,a^{-1}g]}.$$

The formal integration over $D\left[\exp_\psi(\xi_{(j)})\right]$ in the above expression provides the characteristic functional

$$e^{W_\psi(J)} \doteq \int_{Map(M,\psi^{-1}\,TV^n)} e^{i \ll J\cdot\eta \gg} e^{-S[\gamma,\,\exp_\psi(\eta);\,a^{-1}g]} D\left[\exp_\psi(\eta)\right], \qquad (4.133)$$

of the measure $e^{-S[\gamma,\,\exp_\psi(\xi_{(j)});\,a^{-1}g]} D\left[\exp_\psi(\xi_{(j)})\right]$, where we have denoted by $\eta :=$ $\xi_{(j)}$ the integrated field (each one of which is identically distributed). Hence, we can eventually express (4.129) as

$$P_q(\psi)\,D[\psi] = \int_{Map^*(M,\psi^{-1}\,TV^n)} [DJ]\, e^{q\,W_\psi(J)}. \qquad (4.134)$$

If we take into account the explicit characterization of the functional measure (4.126) defining the QFT associated with fields $\eta := \xi_{(j)}$, we can write (up to $\mathscr{O}(|\eta|^3)$ terms)

$$e^{W_\psi(J)} = e^{-S[\gamma,\,\psi;\,a^{-1}g]} \qquad (4.135)$$

$$\times \int_{Map(M,\psi^{-1}\,TV^n)} \exp\left[i \ll J\cdot\eta \gg -2\int_M \langle A\widehat{\eta},\, d\widehat{\eta}\rangle^\gamma_\delta\, d\mu_\gamma - \int_M \langle A\widehat{\eta},\, A\widehat{\eta}\rangle^\gamma_\delta\, d\mu_\gamma \right.$$

$$\left. -\int_M \left\langle \widehat{Rm}(a^{-1}g)\,(\widehat{\eta},\,d\psi)\,\widehat{\eta},\,d\psi \right\rangle^\gamma_\delta\, d\mu_\gamma \right] D\left[\exp_\psi(\eta)\right] \exp\left[\int_M \langle \widehat{\eta},\, \triangle_{(\gamma)}\,\widehat{\eta}\rangle^\gamma_\delta\, d\mu_\gamma\right].$$

According to (4.134) the large q asymptotics of the distribution $P_q(\psi)\,D[\psi]$ of the background field ψ is provided by

$$P_q(\psi)\,D[\psi] = e^{q\,\inf_J\,W_\psi(J)\,+\,o(q)},$$

where the inf is over all $J \in Map^*(\Sigma,\psi^{-1}\,TM)$. As emphasized in [39],

$$\Gamma_0(\psi) := \sup_J\,[\ll \sum_{j=1}^{q} \xi_j,\,J \gg -W_\psi(J)] \qquad (4.136)$$

is the large deviation functional governing the $\mathscr{O}(q)$–fluctuations in the distribution of $\{\xi_k\}_{k=1,...,q}$, around $\sum_{j=1}^{q} \xi_j$, as compared to the $\mathscr{O}(\sqrt{q})$ Gaussian fluctuations sampled by the central limit theorem, (Fig. 4.13). Since $\sup_J\,[\ll \eta,\,J \gg -W_\psi(J)]$ is the Legendre transform of $W_\psi(J)$, it follows, from standard QFT, that $\Gamma_0(\psi)$ can be identified with the *effective action* associated with $W_\psi(J)$, i.e. with the action functional whose corresponding partition function gives, at tree (classical) level, the full characteristic functional $W_\psi(J)$. According to the center of mass constraint $\{\xi_k\}_{k=1,...,q} = 0$, and the *background field effective action* is provided by

$$\Gamma_0(\psi) = \sup_J\,[-W_\psi(J)]. \qquad (4.137)$$

Hence, we have

$$\Gamma_0(\psi) = S[\gamma, \psi; a^{-1}g] - \ln \left\{ \int_{Map(M,\psi^{-1}TV^n)} \exp\left[-2\int_M \langle A\widehat{\eta}, d\widehat{\eta}\rangle_\delta^\gamma \, d\mu_\gamma \right.\right.$$

$$(4.138)$$

$$-\int_M \langle A\widehat{\eta}, A\widehat{\eta}\rangle_\delta^\gamma \, d\mu_\gamma - \int_M \left\langle \widehat{Rm}(a^{-1}g)(\widehat{\eta}, d\psi)\widehat{\eta}, d\psi\right\rangle_\delta^\gamma \, d\mu_\gamma \right]$$

$$e^{\int_M \langle \widehat{\eta}, \Delta_{(\gamma)}\widehat{\eta}\rangle_\delta^\gamma \, d\mu_\gamma} \, D\left[\exp_\psi(\eta)\right]\right\} \, .$$

From a formal QFT point of view we can associate to the Gaussian measure,

$$e^{\int_M \langle \widehat{\eta}, \Delta_{(\gamma)}\widehat{\eta}\rangle_\delta^\gamma \, d\mu_\gamma} \, D\left[\exp_\psi(\eta)\right] , \qquad (4.139)$$

a massless field propagator $\langle \eta^a(x)\,\eta^b(y)\rangle$, and treat the remaining η–dependent terms in (4.138) as interactions. The propagator $\langle \eta^a(x)\,\eta^b(y)\rangle$ is ill–defined since, in dimension two, the Green function of $\Delta_{(\gamma)}$ is both infrared (logarithmically) and ultraviolet divergent and needs to be regularized in order to be used as the covariance function of a Gaussian measure. Dimensional regularization is often the choice (see e.g. [39]), but the assumed localizability for the field fluctuations η suggests to handle these divergences by introducing reference length scales Λ'^{-1} and Λ^{-1} respectively controlling the long range decay law and the short range behavior of $\langle \eta^a(x)\,\eta^b(y)\rangle$. This has a geometrical origin in the fact that we wish to average the η $(= \{\xi_{(j)}\})$ fluctuating fields over a controlled set of amplitudes (because of the requirement of bounded geometry) and over length scales which confine the corresponding center of mass ψ in the geodesic ball $B\left(p, \ln\left(\frac{\Lambda'}{\Lambda}\right) a^{1/2}\right)$. The usual strategy is to consider the appropriate long-distance and short-distance limits, $\left(\frac{\Lambda'}{\Lambda}\right) \longrightarrow (\infty, 0)$ after regularization. Hence, for $x \neq y$, we set

$$\langle \eta^a(x)\eta^b(y)\rangle_{\Lambda'/\Lambda} := \frac{\delta^{ab}}{\pi} \int d^2k \, \frac{e^{ik\cdot(x-y)}}{k^2 + \Lambda'^2}$$

$$(4.140)$$

$$= 4\pi\,\delta^{ab} \left(\frac{2\,\Lambda'}{\pi\,|x-y|}\right)^{\frac{1}{2}} e^{-\frac{|x-y|}{\Lambda'}} \left(1 + \mathcal{O}\left(\frac{\Lambda'}{|x-y|}\right)\right) ,$$

describing an exponential decay law associated with a characteristic length $= \Lambda'^{-1}$, whereas in the coincidence limit $x \longrightarrow y$ we define

$$\lim_{x\to y} \langle \eta^a(x)\eta^b(y)\rangle_{\Lambda'/\Lambda} := \frac{\delta^{ab}}{\pi} \int \frac{d^2k}{k^2 + \Lambda'^2} = 2\,\delta^{ab} \ln\left(\frac{\Lambda}{\Lambda'}\right) . \qquad (4.141)$$

Expanding (4.138) in Feynman graphs, using the propagator (4.140) on internal lines, we find three potentially divergent graphs Υ at 1 loop (i.e., to first order in a) (Fig. 4.24). They correspond to the 2-legs vertices: *(i)* $A^\mu \partial_\mu$, *(ii)* $A^\mu A_\mu$ and *(iii)* $Rm\ \partial^\mu \psi \partial_\mu \psi$. The first two, $\Upsilon_{(i)}$ and $\Upsilon_{(ii)}$ have amplitudes which are divergent, however their sum is finite. Thus, we need to discuss only the amplitude associated to the graph, $\Upsilon_{(iii)}$, which is divergent in the coincidence limit $x \to y$. According the prescription (4.141) for the propagator we regularize this amplitude by setting

$$
F(\Upsilon_{(iii)}) := \int_\Sigma d\mu_\gamma(x)\widehat{R}_{ajbl}(\psi(x))\partial^\mu \psi^j(x)\partial_\mu \psi^l(x) \lim_{x\to y}\langle \eta^a(x)\eta^b(y)\rangle_{\Lambda'/\Lambda}
$$

(4.142)

$$
= -2\ln\left(\frac{\Lambda}{\Lambda'}\right)\int_\Sigma R_{ij}(g(\psi))\partial^\mu \psi^i \partial_\mu \psi^j d\mu_\gamma
$$

$$
= 2\ln\left(\frac{\Lambda'}{\Lambda}\right)\int_\Sigma R_{ij}(g(\psi))\partial^\mu \psi^i \partial_\mu \psi^j d\mu_\gamma \ ,
$$

(the minus sign above comes from our sign convention on the Ricci tensor according to which $R_{jl} = -\widehat{R}_{ajbl}\,\delta^{ab}$). Note that this regularized amplitude is still divergent in the long-distance and short-distance limits, $\left(\frac{\Lambda'}{\Lambda}\right) \longrightarrow (\infty, 0)$. To take care of these divergences let us insert (4.142) in the effective action (4.138). To leading order we get

$$
\Gamma_{(0)}(\psi) = a^{-1}\int_M g_{ij}(\psi)\partial^\mu \psi^i \partial_\mu \psi^j d\mu_\gamma
$$

(4.143)

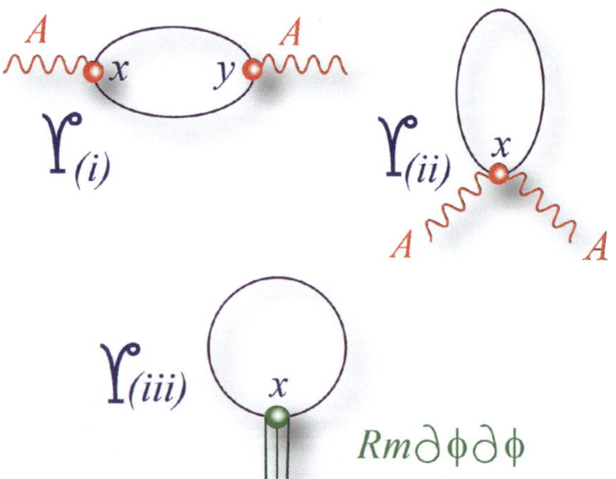

Fig. 4.24 The three graphs contributing to the effective action at 1–loop

$$+ a^{-1} \int_M 2a \, \ln\left(\frac{\Lambda'}{\Lambda}\right) R_{ij}(g(\psi)) \partial^\mu \psi^i \partial_\mu \psi^j d\mu_\gamma + \text{finite } \text{ part } + \mathcal{O}(a^2) \,,$$

where *finite part* indicates terms that are not singular in the limit $\left(\frac{\Lambda'}{\Lambda}\right) \longrightarrow (\infty, 0)$. The standard procedure to take care of the $\ln\left(\frac{\Lambda'}{\Lambda}\right)$ divergent term in $\Gamma_{(0)}(\psi)$ is *minimal subtraction*, which consists in regarding the coupling metric $a^{-1} g$ in the first term of (4.143) as formally infinite and extracting from it the divergent part so to cancel the 1-loop singularity:

$$a^{-1} g_{ij}(\psi) = a^{-1} g_{ij}(\Lambda'/\Lambda) - 2 \, \ln(\Lambda'/\Lambda) R_{ij}\left(a^{-1} g(\psi)\right) + \mathcal{O}(a|\mathrm{Rm}|). \qquad (4.144)$$

The metric $a^{-1} g(\psi)$ in the left hand side is the *bare metric coupling* and $a^{-1} g(\Lambda'/\Lambda)$ is the *renormalized metric coupling*. $R_{ij}(a^{-1} g(\psi)) = R_{ij}(g(\psi))$ (by the scale invariance of the Ricci curvature) are the components of the Ricci tensor of the bare metric, but *we can as well substitute it with that of the renormalized metric*

$$R_{ij}(g(\psi)) \Leftrightarrow R_{ij}[g(\Lambda'/\Lambda)] \,, \qquad (4.145)$$

since the two metrics are equal to order 0 in $a|\mathrm{Rm}|$. Substituting (4.144) into (4.143) we finally get

$$\Gamma_{(0)}(\psi) = \int_\Sigma g_{ij}(\Lambda'/\Lambda) \, \partial^\mu \psi^i \partial_\mu \psi^j d\mu_\gamma + a \, (\text{finite } \text{ part}) + \mathcal{O}(a^2) \qquad (4.146)$$

Notice that this procedure does not depend explicitly on the point p in the geodesic neighborhood of which we are working, i.e. the splitting (4.144) of the bare metric can be extended smoothly to all V^n. Thus, one can extend the above result to the full nonlinear sigma model (that is to background fields ψ taking values in a geodesic neighborhood $B(p, 2r)$ of any point $p \in V^n$).

The long and short distance properties (i.e. its renormalizability) of the theory depends on the behavior of $a^{-1} g(\Lambda'/\Lambda)$ as the ratio Λ'/Λ varies; this behavior is described by the beta function (4.112), that we can easily compute from (4.144). Indeed, by defining

$$t := a \, \ln(\Lambda/\Lambda') \qquad (4.147)$$

so that $\frac{\partial}{\partial t}$ has the same dimension of the Ricci tensor, we immediately get

$$0 = \frac{\partial}{\partial t} g_{ij}(\psi) = \frac{\partial}{\partial t} g_{ij}(t) + 2 R_{ij}(g(t)) + \mathcal{O}(a) \,. \qquad (4.148)$$

Hence, one can conclude that the RG flow of the nonlinear sigma model at one loop is [36]

$$\frac{\partial}{\partial t} g(t) = -2Ric(g(t)) + \mathcal{O}(a) .$$
(4.149)

At this point, it is important to recall that a more detailed analysis at two–loop [36] would have produced[8]

$$\frac{\partial}{\partial t} g_{ik}(t) = -2 R_{ik}(t) - \frac{a}{2} (R_{ilmn}R_k^{lmn}) + \mathcal{O}(a^2) .$$
(4.150)

Both these RG flow expressions, in the weak coupling limit $a\,|\mathrm{Rm}| \to 0$, yield the Ricci flow (R. Hamilton, '82) [50]

$$\frac{\partial}{\partial t} g_{ab}(t) = -2R_{ab}(t) , \quad g_{ab}(t=0) = g_{ab} .$$
(4.151)

We can also take into account the dilatonic coupling f along the same lines. In the point–like limit f fluctuates around the constant value $f_0 = f(\psi)$ it takes corresponding to the center of mass ψ of the constant maps $\{\phi_{(j)}\}$. The associated renormalization group analysis at leading order provides [38, 82, 83]

$$\frac{\partial}{\partial t} g_{ik}(t) = -2 (R_{ik}(t) + \nabla_i \nabla_k f(t)) ,$$
(4.152)

$$a \frac{\partial}{\partial t} f(t) = -c_0 + \frac{a}{4} (\Delta f(t) - |\nabla f|^2) ,$$
(4.153)

where for notational convenience we have used the unscaled variable $a^{-1}t = \ln\left(\frac{\Lambda}{\Lambda'}\right)$ in for the dilaton evolution so as to emphasize the role of the parameter c_0, the central charge of the conformal field theory described by the NLσM when the metric beta function vanishes (conformal invariance), i.e. when $R_{ik}(t) + \nabla_i \nabla_k f(t) = 0$. In (4.152) one recognizes Hamilton's Ricci flow deformed [28] by the action of the t–dependent diffeomorphism generated by the vector field $W^i(t) := g^{ik}\nabla_k f(t)$.

---

[8]In the standard NLσM literature, the two–loop expression or the beta function is provided by $-T R_{ik}(t') - \frac{T^2}{2} (R_{ilmn}R_k^{lmn})$ where T denotes the coupling parameter. This corresponds to a NLσM action [36] defined according to $\frac{1}{2T} \int_M \gamma^{\mu\nu} \frac{\partial\phi^i}{\partial x^\mu} \frac{\partial\phi^j}{\partial x^\nu} g_{ij}(\phi) d\mu_\gamma$ and to a different normalization t' of the running coupling parameter (4.147). It must be noted that the situation of the correct normalization of the geometric flow (4.150), (the RG-2 flow), is somewhat confusing. If one considers the parameter a as a constant then (4.150) is not scale invariant (whereas the Ricci flow is scale invariant under constant rescaling of the metric). On the other hand, if one more correctly consider $a^{-1}g_{ik}$ as the coupling of the theory, (as originally advocated by D. Friedan), then (4.150) is not the appropriate expression for the RG-2 flow.

4.12   Conformal Invariance and Quasi-Einstein Metrics

As often stressed, the dilatonic NLσM (4.35) is not conformally invariant for a generic Riemannian metric measure space $(V^n, g, d\omega)$, however the structure of (4.152) and (4.153) shows that the pertubative QFT defined on $Map(M, V^n)$ by a dilatonic NLσM is conformally invariant as long as we choose the metric g and the dilaton f couplings so that the corresponding β–functions vanish at the given perturbative order. From (4.152) and (4.153) we get

$$R_{ik}(t) + \nabla_i \nabla_k f(t) = 0 \,, \tag{4.154}$$

and

$$\frac{a}{4} \left(\Delta f(t) - |\nabla f|^2 \right) - c_0 = 0 \,. \tag{4.155}$$

More generally, the condition (4.154) (and (4.155)) defining the (one–loop perturbative) fixed points of the metric beta function can be extended by considering the combined action of scaling (A.11) and diffeomorphisms on the metric coupling $a^{-1} g$. To leading order (one loop), this characterizes [36–38] the generalized fixed points for the renormalization group of NLσM. Geometrically these fixed points, as we shall explicitly show below, characterize the notion of *quasi–Einstein metrics*, namely those metrics whose Ricci tensor can be written as

$$\mathrm{Ric}(g) = \rho_{(g)}\, g - \frac{1}{2} \mathscr{L}_{V_{(g)}}\, g \,, \tag{4.156}$$

for some constant $\rho_{(g)}$ and some complete vector field $V_{(g)} \in C^\infty(V^n, TV^n)$. It is worthwhile to note that this class of metrics, of basic importance in Ricci flow theory and in such a framework introduced by R. Hamilton [51], was indeed presciently introduced by D. Friedan in his landmark analysis of NLσM [36–38].

In order to show explicitly that the metrics (4.156) provide generalized fixed points of the (one–loop perturbative) metric beta function, let us start by recalling that a Riemannian metric g is Einstein if its Ricci tensor $\mathrm{Ric}(g) = \rho_{(g)}\, g$ for some constant $\rho_{(g)}$. The notation $\rho_{(g)}$ emphasizes the fact that the Einstein constant $\rho_{(g)}$ scales non–trivially under a rescaling $g \mapsto \lambda g$ of the metric (see (A.11)): we have $\rho_{(\lambda g)} \longmapsto \lambda^{-1} \rho_{(g)}$. More generally, the scale invariance of $\mathrm{Ric}(g)$ implies that under a metric rescaling $g \mapsto \lambda g$ the constant $\rho_{(g)}$ and the vector field $V_{(g)}$, for which (4.156) holds, scale according to

$$\rho_{(\lambda g)} \longmapsto \lambda^{-1} \rho_{(g)} \tag{4.157}$$

$$V_{(\lambda g)} \longmapsto \lambda^{-1} V_{(g)} \,.$$

The interaction of this scaling with the diffeomorphisms group is quite subtle. To put it to the fore[9] let us consider for each point $p \in V^n$ the one–parameter family of diffeomorphisms $\phi_\beta : V^n \longrightarrow V^n$ arising as the solution of the non–autonomous ordinary differential equation

$$\frac{\partial}{\partial \beta} \phi_\beta(p) \;=\; \frac{1}{1 - 2\rho_{(g)}\,\beta}\, V_{(g)}(\phi_\beta(p)) \tag{4.158}$$

$$\phi_{\beta=0} \;=\; \mathrm{id}_{V^n}\,, \quad 0 \le \beta < \epsilon < \frac{1}{2\rho_{(g)}}\,.$$

Since $V_{(g)}/(1 - 2\rho_{(g)}\,\beta)$ is, for $0 \le \beta < \epsilon$, a smooth family of β–dependent vector fields on the compact manifold V^n, the one–parameter family of diffeomorphisms ϕ_β solving of (4.158) exists and is unique for $0 \le \beta < \epsilon$, (for a proof of this result, often used in a Ricci flow setting, see e.g. Lemma 3.15 in [22]). Note that $V_{(g)}(\phi_\beta(p))/(1 - 2\rho_{(g)}\,\beta)$ can be thought of as the vector field obtained from $V_{(g)}(\phi_\beta(p))$ by rescaling the pulled-back metric $\phi_\beta^* g$ according to

$$\phi_\beta^* g \;\longmapsto\; \lambda(\beta)\,\phi_\beta^* g\,, \tag{4.159}$$

where

$$\lambda(\beta) \;:=\; (1 - 2\rho_{(g)}\,\beta)\,. \tag{4.160}$$

According to this remark it is natural to consider the one–parameter family of metrics defined by

$$g(\beta) \;:=\; \lambda(\beta)\,\phi_\beta^* g\,, \quad 0 \le \beta < \epsilon\,, \tag{4.161}$$

with $g|_{\beta=0} = g$. Let us also introduce the corresponding constant $\rho_{(g(\beta))}$ and vector field $V_{(g(\beta))}$, (the latter thought of as a section of the pulled–back bundle $\phi^* TV^n$), viz.

$$\rho_{(g(\beta))} \;:=\; \lambda(\beta)^{-1}\,\rho_{(g)} \tag{4.162}$$

$$V_{(g(\beta))}(p) \;:=\; \lambda(\beta)^{-1}\left(\phi_\beta^* V_{(g)}\right)(p) \;:=\; \lambda(\beta)^{-1}\left(\phi_\beta^{-1}\right)_*\left(V_{(g)}(\phi_\beta(p))\right)\,.$$

By scale invariance and $\mathrm{Diff}(V^n)$–equivariance, we easily compute

$$\mathrm{Ric}(g(\beta)) \;=\; \mathrm{Ric}\left(\lambda(\beta)\,\phi_\beta^* g\right) \tag{4.163}$$

---

[9]For further details see [15, 27, 32, 75].

$$= \mathrm{Ric}\left(\phi_\beta^* g\right) = \phi_\beta^* \left(\mathrm{Ric}(g)\right) = \rho_{(g)}\,\phi_\beta^* g$$

$$-\frac{1}{2}\,\phi_\beta^*\left(\mathscr{L}_{V_{(g)}}\,g\right) = \lambda(\beta)^{-1}\,\rho_{(g)}\,\lambda(\beta)\,\phi_\beta^* g$$

$$-\frac{1}{2}\,\lambda(\beta)^{-1}\,\phi_\beta^*\left(\mathscr{L}_{V_{(g)}}\,\lambda(\beta)\,g\right)$$

$$= \rho_{(g(\beta))}\,g(\beta) - \frac{1}{2}\,\mathscr{L}_{V(g(\beta))}\,g(\beta)\,,$$

where we have exploited the $\mathrm{Diff}(V^n)$–equivariance of the Lie derivative, $\phi_\beta^*\left(\mathscr{L}_{V_{(g)}}\,g\right) = \mathscr{L}_{\phi_\beta^*\,V_{(g)}}\,\phi_\beta^* g$, and (4.162). In a similar way we can compute

$$\frac{\partial}{\partial\beta}\,g(\beta) = \left(\frac{d}{d\beta}\lambda(\beta)\right)\phi_\beta^* g + \left(\frac{d}{d\beta}\phi_\beta^* g\right)\lambda(\beta) \tag{4.164}$$

$$= -2\,\rho_{(g)}\,\phi_\beta^* g + \lambda(\beta)\,\phi_\beta^*\left(\mathscr{L}_{\lambda_\beta^{-1}V_{(g)}}\,g\right)$$

$$= -2\,\lambda(\beta)^{-1}\,\rho_{(g)}\,\lambda(\beta)\,\phi_\beta^* g + \mathscr{L}_{\lambda_\beta^{-1}\phi_\beta^* V_{(g)}}\left(\lambda(\beta)\,\phi_\beta^* g\right)$$

$$= -2\,\rho_{(g(\beta))}\,g(\beta) + \mathscr{L}_{V(g(\beta))}\,g(\beta)$$

$$= -2\,\mathrm{Ric}(g(\beta))\,.$$

Hence, under the combined action of the family of diffeomorphisms ϕ_β and of the scaling (4.159), the quasi–Einstein metric g generates indeed a self–similar solution $g(\beta) := \lambda(\beta)\phi_\beta^* g$, $0 \leq \beta < \epsilon$, (cf. (4.161)), of (4.149) (hence of the *Ricci flow* (4.151)), the *Ricci solitons* [51]

$$\frac{\partial}{\partial\beta}g_{ab}(\beta) = -2\,\mathrm{R}_{ab}(\beta),$$

$$g_{ab}|_{\beta=0} = g_{ab}\,, \quad 0 \leq \beta < \frac{1}{2\rho_{(g)}}\,. \tag{4.165}$$

As generalized fixed points both for the perturbative renormalization group for NLσM as well as for Ricci flow, Quasi–Einstein metrics play a fundamental role and in this connection it is worthwhile to recall some of their properties. First we introduce some basic terminology:

- If, in the expression (4.156) for the Ricci tensor, $\rho_{(g)} = 0$, then g is a *steady* quasi–Einstein metric, (since it generates a steady or translating Ricci soliton);
- If $\rho_{(g)} > 0$ then g is a *shrinking* quasi–Einstein metric, (since it generates a shrinking Ricci soliton);
- If $\rho_{(g)} < 0$ then g is an *expanding* quasi–Einstein metric, (since it generates an expanding Ricci soliton);
- If in the expression (4.156) for the Ricci tensor, $V_{(g)}$ is a gradient vector field, i.e. $V_{(g)} = g^{-1}(df)$ for some $f \in C^\infty(V^n, \mathbb{R})$, then g is a *gradient* quasi–Einstein

metric (or a gradient Ricci soliton) with potential f, and the condition (4.156) can be equivalently written as

$$\mathrm{Ric}(g) \,+\, \mathrm{Hess}_g f \,=\, \rho_{(g)}\, g \,, \tag{4.166}$$

where $\mathrm{Hess}_g f := \nabla df$ denotes the Hessian of f.

Being a quasi–Einstein metric is a condition generating a rather complex landscape. In particular we have the following facts

- Quasi–Einstein metrics on a compact manifold V^n are always gradient [72];
- Steady or expanding quasi–Einstein metrics on a compact manifold V^n, (hence gradient), are necessarily Einstein [52];
- In dimension $n = 3$, any compact shrinking quasi–Einstein metric must have positive sectional curvature [57];
- In dimension $n \geq 4$, there exist nontrivial (i.e. non–Einstein) compact (gradient) shrinking quasi–Einstein metrics [10, 14, 61], (often these are Kähler, whereas non–Kähler and non–Einstein metrics are easily generated by taking products);
- There are non–compact complete quasi–Einstein metrics which are not Einstein [11, 14, 33]

This shows that the quasi–Einstein condition is a non–trivial extension of Einstein manifolds. There is a vast and active mathematical literature on these metrics, and the reader who wants more background might profitably consult the survey articles [15, 27, 32, 75] and the references therein.

The conditions (4.154) and (4.155) defining the (one–loop perturbative) fixed points of the metric beta function formally have the structure of *2nd order partial differential equations of motion* for the coupling fields (g, f). This latter interpretation takes shape in the fact that (4.154) and (4.155) can be also obtained as extremals of the *effective action* functional on $(V^n, g, d\omega)$ given by

$$\mathscr{F}_{c_0}[g(t), f(t)] := \int_{V^n} \left[\frac{a}{4}\left(\mathrm{R}(g) + |\nabla f|^2\right) + c_0 \right] e^{-f}\, d\mu_g \,. \tag{4.167}$$

This observation, which plays a fundamental role in the string theory applications of NLσM (see [34, 35], and [59] for a fast introduction to this vast subject), has unexpected implications also in Ricci flow theory, ultimately suggesting the introduction of Perelman's $\mathscr{F}[g, f]$–energy functional and of its generalizations [72–74]. Conversely, mathematical developments in Ricci flow, in particular in its interpretation as a dynamical system on the space of Riemannian metrics $\mathscr{M}et(V^n)$ and on the set $\mathrm{Meas}(V^n)$ of Riemannian metric measure spaces, generating non–trivial deformations of the metric geometry (i.e. of the metric coupling $a^{-1}\,g$) has shed light on the geometrical nature of NLσM renormalization group flow.

4.13   Ricci Flow as a Dynamical System on $\mathcal{M}et(V^n)$

The Ricci tensor $\mathrm{Ric}(g)$, being a symmetric bilinear form, naturally appears as a *distinguished tangent vector* to $\mathcal{M}et(V^n)$ at the given *point* g, i.e. $\mathrm{Ric}(g) \in T_{(V^n,g)}\mathcal{M}et(V^n)$. Roughly speaking, $\mathrm{Ric}(g)$ can be seen as a *natural infinitesimal variation* of the metric, and the example of quasi–Einstein metrics discussed above shows that self–similar $\mathrm{Diff}(V^n)$–equivariant flows can be associated to $\mathrm{Ric}(g)$. However these would appear as fixed points in the orbit space $\frac{\mathcal{M}et(V^n)}{\mathbb{R}^* \times \mathrm{Diff}(V^n)}$, (cf. (A.35)), rather than a genuine geometric deformation, and establishing the existence of a non–trivial flow associated with $\mathrm{Ric}(g)$ is a rather delicate problem. In his Ph.D. thesis [8] (see also [3] and[9]) J.-P. Bourguignon was able to prove that the Ricci tensor $\mathrm{Ric}(g)$ does generate families of directions which are generically transversal to the $\mathrm{Diff}(V^n)$–orbits $\mathcal{O}_g$, and as such it may give rise to non–trivial (infinitesimal) deformations of g. However, even restricting to the Banach manifold of Riemannian metrics $\mathcal{M}et^s(V^n)$ of Sobolev class $s > \frac{n}{2}$, where we have available the local existence and uniqueness theorem for the flow of vector fields $\in \mathcal{T}\mathcal{M}et^s(V^n)$, we are faced with the fact that the Ricci curvature, seen as a system of partial differential equations acting on the components of the metric tensor, is a highly degenerate non–linear operator. Degeneracy comes from the $\mathrm{Diff}(V^n)$–equivariance of $\mathrm{Ric}(g)$, *viz.* $\mathrm{Ric}(\phi^*g) = \phi^*\mathrm{Ric}(g)$, for any $\phi \in \mathrm{Diff}(V^n)$. In particular, the linearization of the Ricci tensor is elliptic only when restricted to the directions transversal to the $\mathrm{Diff}(V^n)$–orbits $\mathcal{O}_g$. A direct test of this remark comes from the expression (A.55) of the components of the Ricci tensor in harmonic coordinates

$$\mathrm{R}_{ik}\underset{(harmonic)}{=} -\frac{1}{2}\Delta_g\,(g_{ik}) + Q_{ik}\left(g^{-1},\,\partial g\right)\,, \qquad (4.168)$$

which shows that from the point of view of geometric analysis we can associate a flow to Ricci curvature only in the direction $-\mathrm{Ric}(g) \in \mathcal{T}_g\mathcal{M}et(V^n)$, yet another motivation, independent from the renormalization group analysis discussed above, for considering the weakly-parabolic diffusion–reaction PDE [50]

$$\tfrac{\partial}{\partial\beta}g_{ab}(\beta) = -2\,\mathrm{R}_{ab}(\beta),$$

$$(4.169)$$

$$g_{ab}|_{\beta=0} = g_{ab}\,,\ \ 0 \le \beta < T_0\,,$$

where $\mathrm{R}_{ab}(\beta)$ is the Ricci tensor of the evolving metric $g_{ik}(\beta)$. Because of the $\mathrm{Diff}(V^n)$–equivariance of $\mathrm{Ric}(g)$, this flow is not manifestly parabolic, (this is the origin of the terminology *weakly*–parabolic), and its analysis requires a delicate application of Nash–Moser iteration method [50]. (By considering $\mathrm{Diff}(V^n)$–equivariance as a solution rather than a problem D. DeTurck [28] was able to connect (4.169) to a standard local existence problem for systems of nonlinear parabolic partial differential equations.) In the hands of R. Hamilton this approach

Fig. 4.25 A heuristic rendering of a non–singular and non–collapsing Ricci flow evolution in the space of Riemannian metric $\mathcal{M}et(\Sigma)$

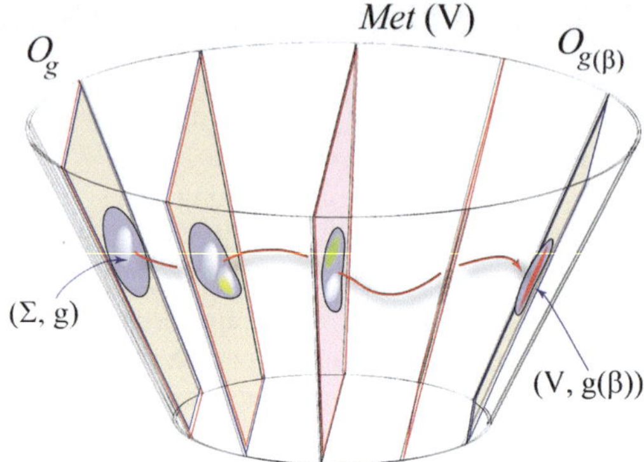

worked out beautifully showing that the degeneracy of (4.169) is caused only by the Diff(V^n)–equivariance of Ric(g), and that for the Ricci flow one always has short–time existence and uniqueness according to [50] (Fig. 4.25).

Theorem 4.2 *If (V^n, g) is a closed Riemannian manifold, there exists a unique solution $(V^n, g) \mapsto (V^n, g(\beta))$ to the Ricci flow (4.169) on a maximal interval $0 \leq \beta < T_0$, for some $T_0 \leq \infty$. Moreover, ([22], Corollary 7.7), if the initial manifold (V^n, g) is such that*

$$|\mathrm{Rm}(g)|_g := \left(\mathrm{Rm}(g)_{abcd}\, \mathrm{Rm}(g)^{abcd}\right)^{1/2} \leq K_0 , \tag{4.170}$$

then the solution $(V^n, g) \mapsto (V^n, g(\beta))$ exists at least for $\beta \in [0, c_n/K_0]$, where $c_n > 0$ is a constant depending only on the dimension n of (V^n, g).

The most direct geometric intuition (Fig. 4.26) on the nature of the Ricci flow comes by looking at the evolution of the scalar curvature $\mathrm{R}(g(\beta))$ along $(V^n, g) \mapsto (V^n, g(\beta))$, (for a leisurely introduction to these matters see [80]). Recall that for a germ of a curve of metrics $\eta \longmapsto g_{ab}(\eta) := g_{ab} + \eta\, h_{ab}, \eta \in [0, \epsilon)$, defined by a metric variation $h \in \mathscr{T}_g\mathcal{M}et(V^n)$, one has (see e.g. [22])

$$\frac{d}{d\eta}\, (\mathrm{R}(g(\eta)))\bigg|_{\eta=0} = -\,\triangle_g\, (\mathrm{tr}_g\, h) - h^{ab}\, \mathrm{R}_{ab}(g) + \nabla^a\nabla^b\, h_{ab} . \tag{4.171}$$

In particular, for $h = -2\,\mathrm{Ric}(g(\beta))$ we compute

$$-\,\triangle_g\, (\mathrm{tr}_g\, h) - h^{ab}\, \mathrm{R}_{ab}(g) + \nabla^a\nabla^b\, h_{ab} = \triangle_g\, \mathrm{R}(g) + 2\, |\mathrm{Ric}(g)|_g^2 , \tag{4.172}$$

where we have exploited the relation $2\,\nabla^a\nabla^b\, \mathrm{R}_{ab} = \triangle_g\, \mathrm{R}$, a consequence of the contracted Bianchi identity $\nabla^b\left(\mathrm{R}_{ab} - \frac{1}{2}\,g_{ab}\,\mathrm{R}\right) \equiv 0$. Hence, along the Ricci flow (4.169), the scalar curvature of the metric $g(\beta)$ evolves according to the non-linear reaction–diffusion PDE

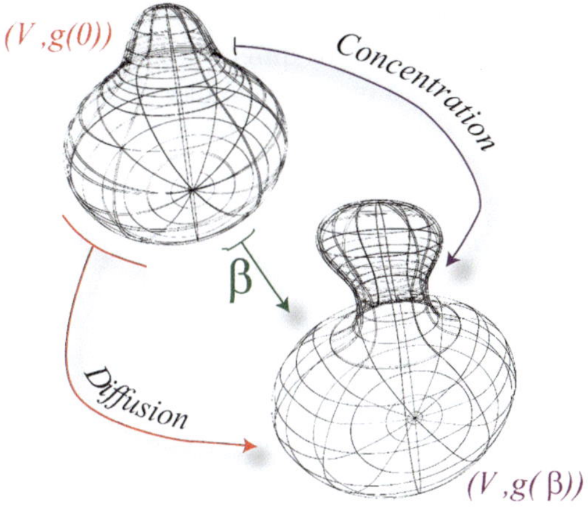

Fig. 4.26 The Ricci flow evolution is governed by a diffusion–reaction mechanism which is manifest in the evolution of the scalar curvature $R(g(\beta))$: the term $\triangle_{g(\beta)} R(g(\beta))$ tends to move curvature fluctuations to the average curvature, whereas the quadratic term $2\,|Ric(g(\beta))|^2_{g(\beta)}$ tends to concentrate curvature

$$\frac{\partial}{\partial\beta} R(g(\beta)) = \triangle_{g(\beta)} R(g(\beta)) + 2\,|Ric(g(\beta))|^2_{g(\beta)}$$

(4.173)

$$= \triangle_{g(\beta)} R(g(\beta)) + \frac{2}{n} [R(g(\beta))]^2 + 2\left|\widehat{Ric}(g(\beta))\right|^2_{g(\beta)} ,$$

where $\widehat{Ric}(g(\beta))$ is the trace–free part of the Ricci curvature. Over time the diffusive term $\triangle_{g(\beta)} R(g(\beta))$ tends to move curvature fluctuations to the average curvature, whereas the (non–negative) quadratic term $2\,|Ric(g(\beta))|^2_{g(\beta)}$ tends to concentrate curvature (and favor positive *vs.* negative curvature). In particular, from the parabolic maximum principle which holds for (4.173), we immediately get

$$R_{min}(g(\beta)) \geq \frac{R_{min}(g(0))}{1 - \frac{2\beta}{n} R_{min}(g(0))} ,$$

(4.174)

where $R_{min}(g(\beta))$ and $R_{min}(g(0))$ denote the minimum value, at time β and $\beta = 0$ respectively, of the scalar curvature over V^n. This bound clearly shows that if initially $R_{min}(g(0)) \geq c > 0$ then concentration wins, to the effect that the Ricci flow cannot be extended beyond the critical time $T_{crit} = \frac{n}{2c}$. More generally, if the interval of existence T_0 in Theorem 4.2 is finite, then [50, 52]

$$\lim_{\beta\nearrow T_0} \left[\sup_{x\in V^n} |Rm(x,\beta)|\right] = \infty ,$$

(4.175)

where $Rm(\beta)$ is the Riemann tensor of $(V^n, g(\beta))$. Conversely, if $T_0 \nearrow \infty$, then the flow $(V^n, g) \mapsto (V^n, g(\beta))$, evolves towards generalized fixed points [52] associated with the action of $\mathrm{Diff}(V^n) \times \mathbb{R}_+$ on $\mathscr{M}et(V^n)$, where $\mathbb{R}_+$ acts by rescaling. These are the Ricci solitons $-2R_{ab}(\beta) = \mathscr{L}_{\mathbf{v}(\beta)} g_{ab} + \varepsilon\, g_{ab}$. Here $\mathscr{L}_{\mathbf{v}(\beta)}$ denotes the

Lie derivative along the β–dependent (complete) vector field $\mathbf{v}(\beta)$ generating a β–dependent curve of diffeomorphisms $\beta \mapsto \varphi(\beta) \in \mathrm{Diff}(V^n)$, and where, up to rescaling, we may assume that $\varepsilon = -1,\ 0,\ 1$, (respectively yielding for the shrinking, steady, and expanding solitons).

The rescaling action in (4.169) can be easily removed by volume normalizing the flow. To this end, let us recall that along a germ of curve of Riemannian metrics $[0,\ \epsilon) \ni \eta \longmapsto g(\eta)$, the corresponding curve of volume forms $d\mu_{g(\eta)}$ has a tangent provided by

$$\frac{\partial}{\partial \eta}\, d\mu_{g(\eta)} \;=\; \frac{1}{2}\, \mathrm{tr}_{g(\eta)}\left(\frac{\partial}{\partial \eta} g(\eta)\right) d\mu_{g(\eta)} \; . \tag{4.176}$$

Hence, we have that the Riemannian measure $d\mu_{g(\beta)}$ and the associated volume of $(V^n, g(\beta))$ evolve along the Ricci flow (4.169) according to

$$\frac{\partial}{\partial \beta}\, d\mu_{g(\beta)} \;=\; -\,\mathrm{R}(g(\beta))\, d\mu_{g(\beta)} \; ,$$

$$\tag{4.177}$$

$$\frac{d}{d\beta}\, \log \mathrm{Vol}_{g(\beta)}(V^n) \;=\; -\,\langle \mathrm{R}(g(\beta)) \rangle_{g(\beta)} \; , \qquad \beta \in [0, T_0)\,,$$

where

$$\langle \mathrm{R}(g(\beta)) \rangle_{g(\beta)} \;=\; \left[\mathrm{Vol}_{g(\beta)}(V^n)\right]^{-1} \int_{V^n} \mathrm{R}(g(\beta))\, d\mu_{g(\beta)} \tag{4.178}$$

is the average scalar curvature of $(V^n, g(\beta))$. It easily follows that if we rescale the solution of (4.169) according to

$$g(\beta) \longmapsto \widetilde{g}\left(\widetilde{\beta}\right) := c(\beta)\, g(\beta)\,, \tag{4.179}$$

where $\widetilde{\beta} := \int_0^\beta c(\tau)\, d\tau$, and

$$c(\beta) \;=\; \exp\left(\frac{2}{n} \int_0^\beta \langle \mathrm{R}(g(\tau)) \rangle_{g(\tau)}\, d\tau\right)\,, \tag{4.180}$$

then $(\widetilde{\beta},\, g) \longmapsto \widetilde{g}(\widetilde{\beta})$ is a solution of the *volume normalized* Ricci flow

$$\frac{\partial}{\partial \widetilde{\beta}}\, g_{ab}(\widetilde{\beta}) = -2\,\mathrm{R}_{ab}\left(\widetilde{g}(\widetilde{\beta})\right) + \frac{2}{n}\left\langle \mathrm{R}\left(\widetilde{g}(\widetilde{\beta})\right)\right\rangle_{\widetilde{g}(\widetilde{\beta})} \widetilde{g}_{ab}(\widetilde{\beta})$$

$$\tag{4.181}$$

$$\widetilde{g}\,|_{\widetilde{\beta}=0} = g\,, \quad 0 \le \widetilde{\beta} < \widetilde{T}_0\,,$$

along which we have

$$\frac{d}{d\widetilde{\beta}}\,\mathrm{Vol}_{\widetilde{g}(\widetilde{\beta})}(V^n) = 0\,. \tag{4.182}$$

A deep interplay between the Ricci flow and its volume normalized version occurs in the celebrated Hamilton's *rounding theorem* [50] according to which if the initial metric on a compact Riemannian 3–manifold (V^n, g) has strictly positive Ricci curvature, (i.e. if $\mathrm{Ric}_g(v, v) \geq C\,g(v, v)$, with $C \in \mathbb{R}_{>0}$ for all $v \in C^\infty(V^n, TV^n)$), then along the Ricci flow $(\beta, g) \longmapsto g(\beta)$ the metric develops a singularity in finite time while becoming rounder and rounder, (the ratio between the largest and the smallest eigenvalues of the Ricci curvature tends to 1 as the singularity is approached). In particular if $(\beta, g) \longmapsto g(\beta)$ is rescaled so as to evolve along the volume normalized Ricci flow (4.181), then as $\widetilde{\beta} \nearrow \infty$ the flow approaches a constant positive sectional curvature 3–geometry (i.e. $\lim_{\widetilde{\beta}\nearrow\infty} (V^n, \widetilde{g}(\widetilde{\beta}))$ is either the round 3–sphere or a spherical space form). Volume normalization (and its variants) provides a basic example of the parabolic rescaling and of the blow up techniques necessary for dealing with Ricci flow singularities.

The expression (4.168) seems to suggest, as a convenient way of avoiding the necessity of using a Nash–Moser iteration method, a rewriting of (4.169) in $(g(\beta)$–dependent) harmonic coordinates. If that were possible, we would formally get

$$\frac{\partial}{\partial\beta}g_{ab}(\beta) = \Delta_\beta\,(g_{ab}(\beta)) - 2\,Q_{ab}\left(g^{-1}(\beta),\,\partial g(\beta)\right)\,, \tag{4.183}$$

a semi–linear system of parabolic equations in the functions $g_{ab}(\beta)$ for which we have a standard local existence theorem. Alas, this strategy badly fails since harmonic coordinates do not propagate naturally along (4.169). A way out, that captures the key aspect of this strategy, shifts the emphasis from harmonic coordinates to the interplay of the diffeomorphism group $\mathrm{Diff}(V^n)$ with (4.169). It is rather instructive to see this latter remark at work since it features prominently in the analysis of the renormalization group flow of non-linear σ models as discussed in these lecture notes.

We start by recovering a β–dependent $\mathrm{Diff}(V^n)$–equivariance of (4.169), which is not manifest. To this end, let us describe the kinematics of the flow (4.169) in the *parabolic spacetime* $M^{n+1}_{Par} \doteq I \times V^n$, $I \doteq [0, T_0) \subset \mathbb{R}$ where the diffeomorphism of $I \times V^n$ onto M^{n+1}_{Par} is the identity map (Fig. 4.27)

$$F : I \times V^n \longrightarrow M^{n+1}_{Par};\quad (\beta, x) \mapsto F(\beta, x) := (\beta,\, id_{V^n}(x))\,, \tag{4.184}$$

and where M^{n+1}_{Par} carries the product metric $^{(n+1)}g_{par}$ defined (in the $I \times V^n$ coordinates) by

$$(F^*\,^{(n+1)}g_{par}) = g_{ab}(\beta)dx^a \otimes dx^b + d\beta \otimes d\beta\,. \tag{4.185}$$

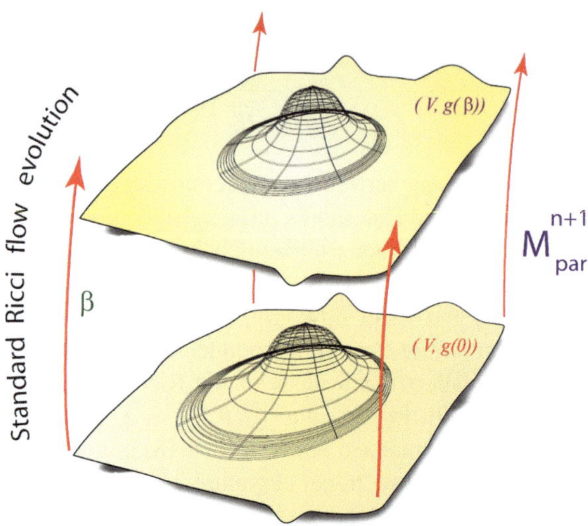

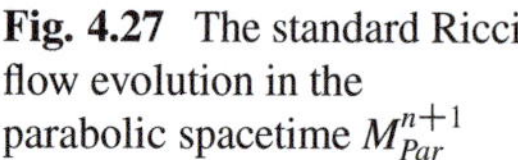

Fig. 4.27 The standard Ricci flow evolution in the parabolic spacetime M^{n+1}_{Par}

In such a framework, $\frac{\partial}{\partial\beta} : V^n \rightarrow TM^{n+1}_{Par}$, can be interpreted as a vector field, transversal (actually, $^{(n+1)}g_{par}$–normal) to the leaves $V^n_\beta = F(\beta, V^n)$, describing the Ricci flow evolution as seen by observers at rest on V^n_β. The evolution of the metric $g(\beta)$ can be equivalently described by observers in *motion* on V^n_β. To this end, consider a curve of diffeomorphisms $I \ni \beta \mapsto \varphi(\beta) \in \text{Diff}(V^n)$, (with the initial condition $\varphi|_{\beta=0} = id_{V^n}$), and define the vector field $X_\varphi : V^n_\beta \rightarrow TV^n_\beta$, $X_\varphi = \frac{\partial}{\partial\beta}\varphi(\beta)$, generating $\beta \mapsto \varphi(\beta)$. Such a β–dependent X_φ provides the velocity field of these non–static observers. Thus,

$$\frac{d}{d\beta} F_\varphi = \frac{\partial}{\partial\beta} + X_\varphi : V^n_\beta \longrightarrow T M^{n+1}_{Par}, \tag{4.186}$$

is the space–time vector field covering the diffeomorphism defined by the curve of embeddings as

$$F_\varphi : I \times V^n \longrightarrow M^{n+1}_{Par}; \quad (\beta, x) \mapsto F_\varphi(\beta, x) := (\beta, \varphi^i(\beta, x)), \tag{4.187}$$

and defining space–time coordinates $(\beta, y^i = \varphi^i(\beta, x))$. In terms of the coordinates (β, y^i) we can write

$$(F_\varphi^{*\ (n+1)}g_{par}) = \breve{g}_{ab}(\beta)(dy^a + X^a_\varphi d\beta) \otimes (dy^b + X^b_\varphi d\beta) + d\beta \otimes d\beta, \tag{4.188}$$

where the metric $\breve{g}_{ab}(\beta)$ is β–propagated according to the Hamilton–DeTurck flow

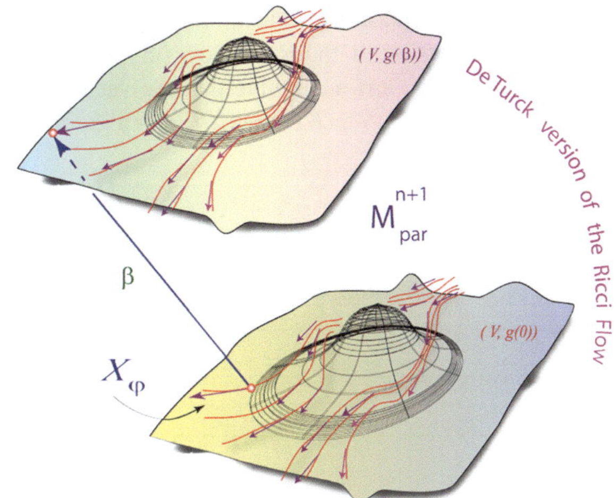

Fig. 4.28 The DeTurck version of the Ricci flow evolution in the parabolic spacetime M^{n+1}_{Par}

$$\frac{\partial}{\partial \beta} \check{g}_{ab}(\beta) = -2\check{R}_{ab}(\beta) - \mathscr{L}_{X_\varphi} \check{g}_{ab}(\beta),$$

(4.189)

$$\check{g}_{ab}|_{\beta=0} = g_{ab}, \quad 0 \le \beta < T_0.$$

The connection between (4.189) and (4.169) is most easily established if we proceed as in the mechanics of continuous media, when shifting from the *body* (Lagrangian) to the *space* (Eulerian) point of view (Fig. 4.28). To this end, let us introduce the *substantial* derivative $\frac{D}{D\beta} \doteq \frac{\partial}{\partial \beta} + \mathscr{L}_{X_\varphi}$ associated with the convective action defined by X_φ. Since $\frac{D}{D\beta} \check{g}_{ab}(\beta) = (\varphi^*)^{-1} \frac{\partial}{\partial \beta} [\varphi^* \check{g}]_{ab}$ and $\varphi^* \operatorname{Ric}(\check{g}) = \operatorname{Ric}(\varphi^* \check{g})$, it follows from (4.189) that the pull–back $\beta \mapsto (\varphi^* \check{g})$ of the flow $\beta \mapsto \check{g}_{ik} dy^i \otimes dy^k$, under the action of the β–dependent diffeomorphism $\varphi(\beta)$, solves (4.169). The non–trivial action of the diffeomorphism group described above is the rationale underlying DeTurck's technique for fixing a gauge F_φ which makes the evolution $\beta \mapsto (TV^n, g(\beta))$ of the metric in the tangent bundle manifestly parabolic [28]. In particular, one can couple this DeTurck's technique with the harmonic map flow (see e.g. [22]) to provide a rather direct and elegant proof of the existence and uniqueness of the local flow solution of (4.169), and which geometrically captures the harmonic coordinate philosophy expressed by (4.183).

4.14   Ricci Flow on Riemannian Metric Measure Spaces

As we have seen in Sects. 4.7 and 4.11 when dealing with the dilatonic NLσM we need to consider n–dimensional compact *Riemannian metric measure spaces* as the natural geometric setting of the theory. This gives rise to a far reaching interplay among Ricci flow, scaling, and $\operatorname{Diff}(V^n)$ equivariance epitomized in (4.152) and in the notion of quasi–Einstein metric. We start by looking more carefully into the DeTurck version of the Ricci flow (cf. (4.189)) which feature prominently in the dilaton setting. Let us assume that the vector X_φ, generating the β–dependent diffeomorphism $\beta \longmapsto \varphi(\beta) \in \mathrm{Diff}(V^n)$, is a gradient vector field

$$X_\varphi = \nabla_{g(\beta)} f(\beta) \,, \tag{4.190}$$

for some smooth family of functions $f(\beta) \in C^\infty(V^n, \mathbb{R})$. Then, we can formally rewrite (4.189), (omitting the $\check{}$ upperscript), as

$$\frac{\partial}{\partial \beta} g(\beta) = -2\,\mathrm{Ric}(\beta) - 2\,\mathrm{Hess}_{g(\beta)} f(\beta),$$

$$\tag{4.191}$$

$$g|_{\beta=0} = g\,, \quad 0 \leq \beta < T_0\,,$$

where the β–dependent function $f \in C^\infty(V^n \times \mathbb{R}, \mathbb{R})$ is, at this formal stage of the discussion, thought of as having being chosen in advance. The presence of the Bakry–Emery Ricci tensor $\mathrm{Ric}(\beta) + \mathrm{Hess}_{g(\beta)} f(\beta)$, suggests that (4.191) can be naturally interpreted as geometric flow in the set of Riemannian metric measure spaces

$$\mathrm{Meas}(V^n) := \{(V^n, g; d\omega) \mid (V^n, g) \in \mathscr{M}et(V^n),\ d\omega \in \mathrm{Prob}_{ac}(V^n, g)\}\,.$$

$$\tag{4.192}$$

To describe explicitly the evolution of the measure $d\omega := e^{-f}\, d\mu_g$, induced by the given dilaton field $f \in C^\infty(V^n \times \mathbb{R}, \mathbb{R})$, we impose the somewhat unphysical normalization

$$\int_{V^n} d\omega = \int_{V^n} e^{-f}\, d\mu_g = 1\,, \tag{4.193}$$

i.e., we assume that $d\omega$ is a probability measure on (V^n, g). This assumption as well as the preservation of the measure $d\omega := e^{-f} d\mu_g$ along the relevant Ricci flow that we will consider below strongly constraint the dilatonic field, however they are instrumental to Perelman's analysis and also allow for a deeper insight into the geometry of the dilatonic field. Let us note that along the flow (4.191) we have

$$\frac{\partial}{\partial \beta} d\mu_{g(\beta)} = \frac{1}{2} g^{ab}(\beta) \frac{\partial}{\partial \beta} g_{ab}(\beta)\, d\mu_{g(\beta)}$$

$$= -\left(\mathrm{R}(g(\beta)) + \Delta_{g(\beta)} f(\beta)\right) d\mu_{g(\beta)}\,, \tag{4.194}$$

(cf. (4.177)). Hence one computes

$$\frac{\partial}{\partial \beta} d\omega(\beta) = \frac{\partial}{\partial \beta} \left(e^{-f(\beta)} d\mu_{g(\beta)}\right)$$

$$= \left(-\frac{\partial f(\beta)}{\partial \beta} + \frac{1}{2} g^{ab}(\beta) \frac{\partial}{\partial \beta} g_{ab}(\beta) \right) d\omega(\beta)$$

$$= - \left(\frac{\partial f(\beta)}{\partial \beta} + R(g(\beta)) + \triangle_{g(\beta)} f(\beta) \right) d\omega(\beta) , \quad (4.195)$$

where we have set $d\omega(\beta) := e^{-f(\beta)} d\mu_{g(\beta)}$. It follows that to (4.191) we can associate the coupled flows, parametrized by the given choice of $f \in C^\infty(V^n \times \mathbb{R}, \mathbb{R})$, and defined by

$$\frac{\partial}{\partial \beta} g(\beta) = -2 \operatorname{Ric}(\beta) - 2 \operatorname{Hess}_{g(\beta)} f(\beta) ,$$

$$\frac{\partial}{\partial \beta} d\omega(\beta) = - \left(\frac{\partial f(\beta)}{\partial \beta} + R(g(\beta)) + \triangle_{g(\beta)} f(\beta) \right) d\omega . \quad (4.196)$$

Among all such flows we can select in particular those preserving the normalized measure $d\omega$, *viz.* those satisfying

$$\frac{\partial}{\partial \beta} d\omega = 0 \implies \frac{\partial f(\beta)}{\partial \beta} = - \triangle_{g(\beta)} f(\beta) - R(g(\beta)) . \quad (4.197)$$

Hence, for a given initial choice of $d\omega$, the associated measure preserving (DeTurck) Ricci flow is described by the coupling between (4.191) and a backward parabolic evolution for $f(\beta)$, i.e.

$$\frac{\partial}{\partial \beta} g(\beta) = -2 \operatorname{Ric}(\beta) - 2 \operatorname{Hess}_{g(\beta)} f(\beta) ,$$

$$\frac{\partial f(\beta)}{\partial \beta} = - \triangle_{g(\beta)} f(\beta) - R(g(\beta)) . \quad (4.198)$$

This interplay between a forward and a backward parabolic dynamics in $\operatorname{Meas}(V^n)$ is related to the fact that by fixing the measure $d\omega$, we lose the full $\operatorname{Diff}(V^n)$ equivariance of (4.191), (the action of $\operatorname{Diff}(V^n)$ is reduced to $\operatorname{Diff}_\omega(V^n)$, the group of $d\omega$–preserving diffeomorphisms). Thus, we cannot interpret (4.198) at face value, as a flow evolving the initial (g, f). We can only solve (4.198) *instantaneously* at any given length scale β. An elegant way out to this computationally delicate situation is provided by recovering the full $\operatorname{Diff}(V^n)$ equivariance from (4.198) by introducing a β–dependent $\operatorname{Diff}(V^n)$ action on $(V^n, g(\beta), d\omega(\beta))$. This action will generate out of the original fixed $d\omega$ a flow of distinct (probability) measures $\beta \longmapsto d\omega(\beta)$, much in the spirit of the remarks accompanying (4.76). Explicitly, let

$$\beta \longmapsto \left(\bar{g}(\beta), \bar{f}(\beta) \right) \quad (4.199)$$

be such that (4.198) holds for $0 \leq \beta < T_0$. Correspondingly we define a curve of diffeomorphisms

$$[0, T_0) \longrightarrow \mathrm{Diff}(V^n) \tag{4.200}$$

$$\beta \longmapsto \Psi(\beta) \, ,$$

solving the non–autonomous ODE

$$\frac{\partial}{\partial \beta} \, \Psi(\beta) \; = \; \nabla_{\overline{g}(\beta)} \overline{f}(\beta) \, ,$$

$$\Psi|_{\beta=0} \; = \; id_V^n \, , \qquad 0 \leq \beta < T_0 \, , \tag{4.201}$$

(cf. (4.158) and [22]). If

$$g(\beta) \; := \; \psi^*(\beta) \, \overline{g}(\beta)$$

$$f(\beta) \; := \; \overline{f}(\beta) \circ \Psi(\beta) \, , \tag{4.202}$$

then, proceeding as in Sect. A.4, we get (cf. (4.164) and Section 2.2 of [23]), with $\overline{V}_\beta := \mathrm{grad}_{\overline{g}(\beta)} \overline{f}(\beta)$,

$$\frac{\partial}{\partial \beta} \, g(\beta) \; = \; \Psi^*(\beta) \left(\frac{\partial}{\partial \beta} \, \overline{g}(\beta) \right) + \Psi^*(\beta) \left(\mathscr{L}_{\overline{V}_\beta} \, \overline{g}(\beta) \right) \tag{4.203}$$

$$= \Psi^*(\beta) \left(-2 \, \mathrm{Ric}(\overline{g}(\beta)) - 2 \, \mathrm{Hess}_{\overline{g}(\beta)} \overline{f}(\beta) + 2 \, \mathrm{Hess}_{\overline{g}(\beta)} \overline{f}(\beta) \right)$$

$$= -2 \, \Psi^*(\beta) \mathrm{Ric}(\overline{g}(\beta)) \; = \; -2 \, \mathrm{Ric}(g(\beta)) \, ,$$

and

$$\frac{\partial}{\partial \beta} f(\beta) \; = \; \frac{\partial}{\partial \beta} \left(\overline{f}(\beta) \circ \Psi(\beta) \right)$$

$$= \left(- \, \triangle_{\overline{g}} \overline{f} - \mathrm{R}(\overline{g}(\beta)) \right) \circ \Psi(\beta) + \left| (\nabla_{\overline{g}} \overline{f}) \circ \Psi(\beta) \right|^2_{\overline{g}}$$

$$= - \, \triangle_g f - \mathrm{R}(g(\beta)) + \left| \nabla_g f \right|^2_g \, .$$

Hence the pair $\left(g(\beta) := \psi^*(\beta) \, \overline{g}(\beta), \, f(\beta) := \overline{f}(\beta) \circ \Psi(\beta) \right)$ satisfies the system

$$\frac{\partial}{\partial \beta} \, g(\beta) \; = \; -2 \, \mathrm{Ric}(g(\beta)) \, ,$$

$$\frac{\partial}{\partial \beta} f(\beta) \; = \; - \, \triangle_g f - \mathrm{R}(g(\beta)) + \left| \nabla_g f \right|^2_g \, . \tag{4.204}$$

Note that $f(\beta)$ no longer appears in the metric evolution and we can interpret (4.204) first by solving for the Ricci flow forward in time, say for $\beta \in [0, T_1]$, and then solve for the flow of the function f along the time–reversed Ricci evolution $\eta \longmapsto g(\eta)$, $0 \le \eta := T_1 - \beta \le T_1$,

$$\frac{\partial}{\partial \eta} f(\eta) \;=\; \triangle_g f + \mathrm{R}(g(\eta)) - \left|\nabla_g f\right|_g^2 \,,$$

$$f\big|_{\eta=0} = f \,. \tag{4.205}$$

This computational point of view will be adopted throughout this set of notes, (for further details on this parabolic see-saw game, see Section 2.2 of [23]). Note that if we set

$$d\omega(\eta) \;:=\; e^{-f(\eta)}\, d\mu_{g(\eta)} \,, \tag{4.206}$$

then we can rewrite (4.205) as

$$\frac{\partial}{\partial \eta}\, d\omega(\eta) \;=\; -e^{-f(\eta)} \frac{\partial f(\eta)}{\partial \eta}\, d\mu_{g(\eta)} + e^{-f(\eta)} \frac{\partial}{\partial \eta}\, d\mu_{g(\eta)}$$

$$= -\left(\triangle_{g(\eta)} f + \mathrm{R}(g(\eta)) - \left|\nabla_g f\right|_{g(\eta)}^2\right) e^{-f(\eta)}\, d\mu_{g(\eta)}$$

$$+ e^{-f(\eta)}\, \mathrm{R}(g(\eta))\, d\mu_{g(\eta)} \,, \tag{4.207}$$

where we exploited (4.205) and the relation

$$\frac{\partial}{\partial \eta}\, d\mu_{g(\eta)} \;=\; -\frac{1}{2} g^{ab}(\beta) \frac{\partial}{\partial \beta}\, g_{ab}(\beta)\, d\mu_{g(\beta)} \;=\; \mathrm{R}(g(\eta))\, d\mu_{g(\eta)} \,, \tag{4.208}$$

a consequence of the Ricci flow equation in (4.204). For any function $h \in C^2(V^n, \mathbb{R})$ we have the identity

$$\triangle_g \left(e^{-h}\right) \;=\; e^{-h} \left|\nabla_g h\right|_g^2 - e^{-h}\, \triangle_g\, h \,, \tag{4.209}$$

so that we get from (4.207)

$$\frac{\partial}{\partial \eta}\, d\omega(\eta) \;=\; \left(\triangle_{g(\eta)} e^{-f(\eta)}\right) d\mu_{g(\eta)} \,. \tag{4.210}$$

Since the Riemannian measure $d\mu_{g(\eta)}$ is covariantly constant with respect to the Levi-Civita connection $\nabla_{g(\eta)}$, we can equivalently write (4.210) as

$$\frac{\partial}{\partial \eta}\, d\omega(\eta) \;=\; \triangle_{g(\eta)}\, d\omega(\eta) \,. \tag{4.211}$$

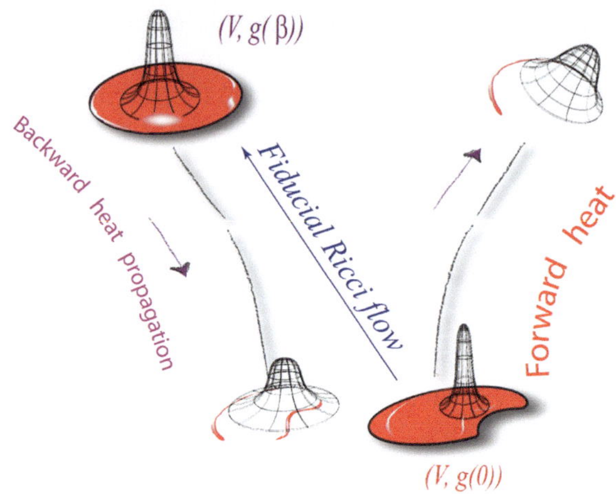

Fig. 4.29 The Ricci flow on a Riemannian metric measure spaces $(V^n, g, d\omega)$ is naturally coupled with a backward heat evolution $\eta \longmapsto d\omega(\eta)$ for the density

In other words, whereas the measure $d\overline{\omega} = e^{-\overline{f}(\beta)} d\mu_{\overline{g}(\beta)}$ is constant along the DeTurck Ricci flow (4.191), the pulled-back (probability) measure

$$d\omega(\beta) := \Psi^*(\beta)\, d\overline{\omega} = e^{-\overline{f}\circ\Psi(\beta)}\, \Psi^*\left(d\mu_{\overline{g}(\beta)}\right) = e^{-f(\beta)}\, d\mu_{g(\beta)} \qquad (4.212)$$

is no longer constant but evolves according to a backward heat equation coupled with the standard Ricci flow, (Fig. 4.29)

$$\frac{\partial}{\partial\beta}\, g(\beta) = -2\,\mathrm{Ric}(g(\beta))\,, \quad g|_{\beta=0} = g\,,$$

$$(4.213)$$

$$\frac{\partial}{\partial\eta}\, d\omega(\eta) = \triangle_{g(\eta)}\, d\omega(\eta)\,, \quad d\omega|_{\eta=0} = e^{-f(T_1)} d\mu_{g(T_1)}\,.$$

Notice that in terms of the variable $\eta := T_1 - \beta \in [0, T_1]$ the Radon–Nikodym derivative $\frac{d\omega}{d\mu_g} = e^{-f}$ evolves according to

$$\bigcirc^*_{g(\eta)}\, e^{-f(\eta)} := \left(\frac{\partial}{\partial\eta} - \triangle_{g(\eta)} + \mathrm{R}(g(\eta))\right)\left(e^{-f(\eta)}\right) = 0\,. \qquad (4.214)$$

in which one recognizes the backward heat equation conjugated, with respect to the $L^2(d\mu_{g(\eta)} d\eta)$ inner product on $V^n \times [0, T_1]$, to the forward heat operator defined by

$$\bigcirc_{g(\beta)} := \left(\frac{\partial}{\partial\beta} - \triangle_{g(\beta)}\right)\,. \qquad (4.215)$$

Hence we may equivalently write (4.213) as

$$\frac{\partial}{\partial \beta} g(\beta) = -2\operatorname{Ric}(g(\beta)), \quad g|_{\beta=0} = g,$$

$$(4.216)$$

$$\bigcirc^{*}_{g(\eta)} e^{-f(\eta)} = 0, \quad e^{-f}\big|_{\eta=0} = e^{-f(T_1)}.$$

It must be stressed that the scalar heat operator $\bigcirc_g$ it is not the only natural heat operator admitting a $L^2(d\mu_{g(\eta)}d\eta)$ conjugate $\bigcirc^{*}_g$ along the Ricci flow. $\bigcirc_g$ and $\bigcirc^{*}_g$ come in a multiplet together with two other pairs of heat operators (and their conjugates) associated with the vector Laplacian $\triangle_{vec}$ and the tensor Lichnerowicz Laplacian $\triangle_L$. This multiplet of operators and their parabolic conjugates plays an important role in the interplay between Ricci flow theory and renormalization group. Their properties will be discussed in detail in Sect. 4.16.

Even if (4.216) appears much easier to handle than (4.198), a salient feature of the latter, (not so easily established in a direct way for (4.216)), is that it can be interpreted as a gradient flow of a suitable functional with respect to the $L^2(d\overline{\omega})$ inner product on $\mathscr{TM}et(V^n)$. The argument, introduced by Perelman in his classic paper [72], is well-known but rather subtle, (again Diff(V^n) and scaling at work!), and it is important to describe it in some detail since, as emphasized in Sect. 4.12, it motivates our detailed geometrical analysis of the Ricci flow, and connects quite directly with the role that the effective action (4.167) plays in NLσM.

4.15   Perelman's Entropy Generating Functional $\mathscr{F}$

Let us explicitly write the coupled flow (4.198) in terms of the pair $(\overline{g}(\beta), \overline{f}(\beta))$ defined by (4.199), i.e.

$$\frac{\partial}{\partial \beta} \overline{g}(\beta) = -2\operatorname{Ric}(\overline{g}(\beta)) - 2\operatorname{Hess}_{\overline{g}(\beta)} \overline{f}(\beta),$$

$$(4.217)$$

$$\frac{\partial \overline{f}(\beta)}{\partial \beta} = -\triangle_{\overline{g}(\beta)} \overline{f}(\beta) - \operatorname{R}(\overline{g}(\beta)),$$

and let us consider a symmetric bilinear form $\overline{h}_{ab}(\beta)$ covering $\overline{g}_{ab}(\beta)$, for all $\beta \in [0, T_1]$, i.e. $\overline{h}(\beta) \in \mathscr{T}_{(V^n, \overline{g}(\beta))} \mathscr{M}et(V^n)$. We can think of any such $\overline{h}(\beta)$ as representing an infinitesimal variation

$$\overline{g}^{(t)}_{ab}(\beta) = \overline{g}_{ab}(\beta) + t\,\overline{h}_{ab}(\beta), \quad t \in (-\varepsilon, \varepsilon), \qquad (4.218)$$

of the flow $\beta \longmapsto \overline{g}_{ab}(\beta)$ defined by (4.217). It will be useful, also for later use, to recall the expression of the linearization of the Ricci curvature in the direction $\overline{h}(\beta)$. A lengthy but otherwise standard computation, (see e.g., [20, 22, 24]), provides

$$\mathrm{DRic}(\overline{g}^{(t)}) \circ \overline{h}_{ab}(\beta) \tag{4.219}$$

$$:= \left. \frac{d}{dt}\left(\mathrm{R}_{ab}(\overline{g}^{(t)})\right)\right|_{t=0} = -\frac{1}{2}\overline{\Delta}_L\overline{h}_{ab} - \left[\delta^*_{\overline{g}}\,\delta_{\overline{g}}\,G(\overline{g},\overline{h})\right]_{ab}$$

$$= -\frac{1}{2}\overline{\Delta}_L\overline{h}_{ab} + \frac{1}{2}\left(\overline{\nabla}_a\overline{\nabla}^c\overline{h}_{cb} + \overline{\nabla}_b\overline{\nabla}^c\overline{h}_{ca}\right) - \frac{1}{2}\overline{\nabla}_a\overline{\nabla}_b\,tr_{\overline{g}}\,(\overline{h})\;,$$

where for notational ease we have dropped the explicit β-dependence, $\overline{\nabla}$ denotes the Levi–Civita connection on $(V^n,\ \overline{g})$, and we have introduced the Einstein–conjugate $G(\overline{g},\overline{h}) \doteq \overline{h} - \frac{1}{2}\left(tr_{\overline{g}}\,\overline{h}\right)\overline{g}$ of $\overline{h} \in C^\infty(V^n,\otimes^2 T^*\,V^n)$, $(G(\overline{h})$ for short, if it is clear, from the context, with respect to which metric we are conjugating). The operators $\delta^*_{\overline{g}}$ and $\delta_{\overline{g}}$, respectively half of the Lie derivative of $\overline{g}$ and the negative of the $\overline{g}$–divergence, are defined in the Appendix A (cf. (A.43) and (A.46)). Finally, $\overline{\Delta}_L\ :\ C^\infty(V^n,\otimes^2 T^*\,V^n) \to C^\infty(V^n,\otimes^2 T^*\,V^n)$ is the Lichnerowicz-DeRham Laplacian on symmetric bilinear forms on $(V^n,\ \overline{g})$ defined by

$$\overline{\Delta}_L\overline{h}_{ab} \doteq \Delta_{\overline{g}}\,\overline{h}_{ab} - R_{as}(\overline{g})\,\overline{h}^s_b - R_{bs}(\overline{g})\,\overline{h}^s_a + 2R_{asbt}(\overline{g})\,\overline{h}^{st}, \tag{4.220}$$

where $\Delta_{\overline{g}} \doteq \overline{g}^{ab}(\beta)\,\overline{\nabla}_a\overline{\nabla}_b$ is the rough Laplacian associated with $\overline{g}(\beta)$. From (4.219) one immediately computes also the linearization of the scalar curvature $\mathrm{R}(\overline{g}(\beta))$ according to

$$\mathrm{DR}(\overline{g}^{(t)}) \circ \overline{h} \tag{4.221}$$

$$:= \left.\frac{d}{dt}\left(\mathrm{R}(\overline{g}^{(t)})\right)\right|_{t=0} = \left[\frac{d}{dt}\left(\overline{g}^{ab}_{(t)}\right)\mathrm{R}_{ab}(\overline{g}) + \overline{g}^{ab}\frac{d}{dt}\left(\mathrm{R}_{ab}(\overline{g}^{(t)})\right)\right]_{t=0}$$

$$= -\overline{h}^{ab}\mathrm{R}_{ab}(\overline{g}) + \overline{g}^{ab}\left(-\frac{1}{2}\overline{\Delta}_L\overline{h}_{ab} - \left[\delta^*_{\overline{g}}\,\delta_{\overline{g}}\,G(\overline{g},\overline{h})\right]_{ab}\right)$$

$$= -\Delta_{\overline{g}}\left(tr_{\overline{g}}\,\overline{h}\right) - \overline{h}^{ab}\mathrm{R}_{ab}(\overline{g}) + \overline{\nabla}^a\overline{\nabla}^b\overline{h}_{ab}\;.$$

With these preliminary remarks along the way, let us consider, for the *given choice* of the measure $d\overline{\omega}$, the $\mathrm{Diff}_{\overline{\omega}}(V^n)$–invariant functional $\mathscr{F}^{\overline{\omega}}\ :\ \mathscr{M}et(V^n) \longrightarrow \mathbb{R}$ defined by

$$\mathscr{F}^{\overline{\omega}}[\overline{g}] := \int_{V^n}\left(\mathrm{R}(\overline{g}) + \left|\nabla\log\left(\frac{d\mu_{\overline{g}}}{d\overline{\omega}}\right)\right|^2_{\overline{g}}\right)d\overline{\omega}\;. \tag{4.222}$$

If for the given $d\overline{\omega}$, we consider the DeTurck modified Ricci flow

$$\frac{\partial}{\partial\beta}\,\overline{g}(\beta) = -2\,\mathrm{Ric}(\overline{g}(\beta)) - 2\,\mathrm{Hess}\,_{\overline{g}(\beta)}\,\log\left(\frac{d\mu_{\overline{g}(\beta)}}{d\overline{\omega}}\right), \tag{4.223}$$

and introduce the relative entropy functional of the β–evolving Riemannian measure $d\mu_{\overline{g}(\beta)}$ with respect to the fixed measure $d\overline{\omega}$,

$$\mathscr{S}[d\overline{\omega}\,|\,d\mu_{\overline{g}(\beta)}] := -\int_{V^n}\log\left(\frac{d\mu_{\overline{g}(\beta)}}{d\overline{\omega}}\right)d\overline{\omega}, \tag{4.224}$$

then one easily checks [24, 70] that,

$$\frac{d}{d\beta}\,\mathscr{S}[d\overline{\omega}\,|\,d\mu_{\overline{g}(\beta)}] = \mathscr{F}^{\overline{\omega}}[\overline{g}]. \tag{4.225}$$

In other words, $\mathscr{F}^{\overline{\omega}}[\overline{g}]$ is the (relative) entropy generating functional, (also known as the relative Fisher information of $d\mu_{\overline{g}(\beta)}$ with respect to $d\overline{\omega}$), along (4.223). It is important to stress that $\mathscr{F}^{\overline{\omega}}$ is invariant only under the group of $d\overline{\omega}$–preserving diffeomorphisms $\mathrm{Diff}_{\overline{\omega}}(V^n)$. This is the reason why in the expression for $\mathscr{F}^{\overline{\omega}}$ we have used the (logarithm of the) Radon–Nikodym derivative rather than the function $e^{-\overline{f}}$ expressing the absolute continuity condition $d\overline{\omega} \ll d\mu_{\overline{g}}$. Indeed, the parametrization $d\overline{\omega} = e^{-\overline{f}}\,d\mu_{\overline{g}}$ must keep track of the fact that, at fixed $d\overline{\omega}$, the function $e^{-\overline{f}}$ is not a scalar but rather a density, (cf. (4.79)), and we would be forced to vary $\overline{f}$ along with (4.218), imposing the preservation of $d\overline{\omega}$ as a constraint. All this can be easily done (cf. Chap. 5 of [23]), but it is conceptually simpler to avoid it by using $\frac{d\mu_{\overline{g}}}{d\overline{\omega}}$ directly. The parametrization $d\overline{\omega} = e^{-\overline{f}}\,d\mu_{\overline{g}}$ can be exploited a posteriori when interpreting the results obtained from the analysis of $\mathscr{F}^{\overline{\omega}}$.

To compute the linearization of the functional $\mathscr{F}^{\overline{\omega}}$ under the metric variation (4.218), at *fixed reference measure $d\overline{\omega}$*, we start by computing the linearization of the weighted Einstein–Hilbert action

$$\int_{V^n} \mathrm{R}(\overline{g})\,e^{-\log\left(\frac{d\mu_{\overline{g}}}{d\overline{\omega}}\right)}\,d\mu_{\overline{g}} = \int_{V^n} \mathrm{R}(\overline{g})\,d\overline{\omega}. \tag{4.226}$$

According to (4.221) we get

$$\frac{d}{dt}\int_{V^n}\mathrm{R}(\overline{g}_{(t)})\,d\overline{\omega}\bigg|_{t=0} \tag{4.227}$$

$$= \int_{V^n}\left(-\,\triangle_{\overline{g}}\,(tr_{\overline{g}}\,\overline{h}) - \overline{h}^{ab}\,\mathrm{R}_{ab}(\overline{g}) + \overline{\nabla}^a\overline{\nabla}^b\,\overline{h}_{ab}\right)d\overline{\omega}.$$

Note that for $d\omega \in \mathrm{Prob}_{ac}(V^n)$ we can write, after a few obvious integrations by parts,

$$
\int_{V^n} \triangle_{\overline{g}} \left(tr_{\overline{g}}\,\overline{h}\right) d\overline{\omega} = \int_{V^n} \triangle_{\overline{g}} \left(tr_{\overline{g}}\,\overline{h}\right) e^{-\log\left(\frac{d\mu_{\overline{g}}}{d\overline{\omega}}\right)} d\mu_{\overline{g}}
$$

$$
= \int_{V^n} tr_{\overline{g}}\,\overline{h} \left(\triangle_{\overline{g}}\, e^{-\log\left(\frac{d\mu_{\overline{g}}}{d\overline{\omega}}\right)}\right) d\mu_{\overline{g}}
$$

$$
= -\int_{V^n} tr_{\overline{g}}\,\overline{h} \left[\triangle_{\overline{g}} \log\left(\frac{d\mu_{\overline{g}}}{d\overline{\omega}}\right) - \left|\nabla \log\left(\frac{d\mu_{\overline{g}}}{d\overline{\omega}}\right)\right|^2_{\overline{g}}\right] d\overline{\omega},
$$

$$(4.228)$$

and

$$
\int_{V^n} \overline{\nabla}^a \overline{\nabla}^b\, \overline{h}_{ab}\, d\overline{\omega} = \int_{V^n} \overline{\nabla}^a \overline{\nabla}^b\, \overline{h}_{ab}\, e^{-\log\left(\frac{d\mu_{\overline{g}}}{d\overline{\omega}}\right)} d\mu_{\overline{g}}
$$

$$
= -\int_{V^n} \overline{\nabla}^b\, \overline{h}_{ab}\, \overline{\nabla}^a \left(e^{-\log\left(\frac{d\mu_{\overline{g}}}{d\overline{\omega}}\right)}\right) d\mu_{\overline{g}}
$$

$$
= \int_{V^n} \overline{h}_{ab}\, \overline{\nabla}^b \overline{\nabla}^a \left(e^{-\log\left(\frac{d\mu_{\overline{g}}}{d\overline{\omega}}\right)}\right) d\mu_{\overline{g}}
$$

$$
= \int_{V^n} \left[\overline{\nabla}^a \left(\log \frac{d\mu_{\overline{g}}}{d\overline{\omega}}\right) \overline{\nabla}^b \left(\log \frac{d\mu_{\overline{g}}}{d\overline{\omega}}\right) - \overline{\nabla}^a \overline{\nabla}^b \left(\log \frac{d\mu_{\overline{g}}}{d\overline{\omega}}\right)\right] \overline{h}_{ab}\, d\overline{\omega}.
$$

$$(4.229)$$

By inserting these two identities in (4.227) we get

$$
\frac{d}{dt} \int_{V^n} \mathrm{R}(\overline{g}_{(t)})\, d\overline{\omega}\bigg|_{t=0}
$$

$$
:= \int_{V^n} \left[\overline{g}^{ab} \left(\triangle_{\overline{g}} \log\left(\frac{d\mu_{\overline{g}}}{d\overline{\omega}}\right) - \left|\nabla \log\left(\frac{d\mu_{\overline{g}}}{d\overline{\omega}}\right)\right|^2_{\overline{g}}\right) - R^{ab}(\overline{g})\right.
$$

$$(4.230)$$

$$
\left.+ \overline{\nabla}^a \left(\log \frac{d\mu_{\overline{g}}}{d\overline{\omega}}\right) \overline{\nabla}^b \left(\log \frac{d\mu_{\overline{g}}}{d\overline{\omega}}\right) - \overline{\nabla}^a \overline{\nabla}^b \left(\log \frac{d\mu_{\overline{g}}}{d\overline{\omega}}\right)\right] \overline{h}_{ab}\, d\overline{\omega}.
$$

By proceeding similarly we can also compute the linearization of the remaining term $\int_{V^n} |\nabla \log (\frac{d\mu_{\bar{g}}}{d\bar{\omega}})|^2 \, d\bar{\omega}$, present in the expression (4.222) for $\mathscr{F}^{\bar{\omega}}$. We get

$$\frac{d}{dt} \int_{V^n} \left| \nabla \log \left(\frac{d\mu_{\bar{g}_{(t)}}}{d\bar{\omega}} \right) \right|^2_{\bar{g}_{(t)}} d\bar{\omega} \Bigg|_{t=0}$$

$$(4.231)$$

$$= \int_{V^n} \frac{d}{dt} \left| \nabla \log \left(\frac{d\mu_{\bar{g}_{(t)}}}{d\bar{\omega}} \right) \right|^2_{\bar{g}_{(t)}} \Bigg|_{t=0} e^{-\log \left(\frac{d\mu_{\bar{g}}}{d\bar{\omega}} \right)} d\mu_{\bar{g}}$$

$$= \int_{V^n} \left[-\bar{h}^{ab} \overline{\nabla}_a \left(\log \frac{d\mu_{\bar{g}}}{d\bar{\omega}} \right) \overline{\nabla}_b \left(\log \frac{d\mu_{\bar{g}}}{d\bar{\omega}} \right) \right.$$

$$\left. + 2 g^{ab} \overline{\nabla}_a \left(\log \frac{d\mu_{\bar{g}}}{d\bar{\omega}} \right) \overline{\nabla}_b \frac{d}{dt} \Bigg|_{t=0} \left(\log \frac{d\mu_{\bar{g}_{(t)}}}{d\bar{\omega}} \right) \right] e^{-\log \left(\frac{d\mu_{\bar{g}}}{d\bar{\omega}} \right)} d\mu_{\bar{g}}$$

$$= \int_{V^n} \left[-\bar{h}^{ab} \overline{\nabla}_a \left(\log \frac{d\mu_{\bar{g}}}{d\bar{\omega}} \right) \overline{\nabla}_b \left(\log \frac{d\mu_{\bar{g}}}{d\bar{\omega}} \right) \right.$$

$$\left. + g^{ab} \overline{\nabla}_a \left(\log \frac{d\mu_{\bar{g}}}{d\bar{\omega}} \right) \overline{\nabla}_b \left(tr_{\bar{g}} \bar{h} \right) \right] e^{-\log \left(\frac{d\mu_{\bar{g}}}{d\bar{\omega}} \right)} d\mu_{\bar{g}}$$

$$= \int_{V^n} \left[-\bar{h}^{ab} e^{-\log \left(\frac{d\mu_{\bar{g}}}{d\bar{\omega}} \right)} \overline{\nabla}_a \left(\log \frac{d\mu_{\bar{g}}}{d\bar{\omega}} \right) \overline{\nabla}_b \left(\log \frac{d\mu_{\bar{g}}}{d\bar{\omega}} \right) \right.$$

$$\left. + \left(tr_{\bar{g}} \bar{h} \right) \Delta_{\bar{g}} \left(e^{-\log \left(\frac{d\mu_{\bar{g}}}{d\bar{\omega}} \right)} \right) \right] d\mu_{\bar{g}}$$

$$= - \int_{V^n} \left[\overline{\nabla}_a \left(\log \frac{d\mu_{\bar{g}}}{d\bar{\omega}} \right) \overline{\nabla}_b \left(\log \frac{d\mu_{\bar{g}}}{d\bar{\omega}} \right) \right.$$

$$\left. + \bar{g}_{ab} \left(\Delta_{\bar{g}} \log \left(\frac{d\mu_{\bar{g}}}{d\bar{\omega}} \right) - \left| \nabla \log \left(\frac{d\mu_{\bar{g}}}{d\bar{\omega}} \right) \right|^2_{\bar{g}} \right) \right] \bar{h}^{ab} \, d\bar{\omega} .$$

It is immediate to check that the Laplacian and gradient terms in (4.231) cancel with the corresponding terms present in (4.230) and we eventually obtain

$$\frac{d}{dt} \mathscr{F}^{\bar{\omega}}[\bar{g}_{(t)}] \Bigg|_{t=0} = - \int_{V^n} \left(\mathrm{R}^{ab}(\bar{g}) + \overline{\nabla}^a \overline{\nabla}^b \log \left(\frac{d\mu_{\bar{g}}}{d\bar{\omega}} \right) \right) \bar{h}_{ab} \, d\bar{\omega} . \qquad (4.232)$$

From here it follows that if there exists a solution $\beta \longmapsto \overline{g}(\beta)$ of the DeTurck modified Ricci flow (4.223) for the given choice of fixed measure $d\overline{\omega}$, then such a solution is an $L^2(d\overline{\omega})$ gradient flow of the functional $\mathscr{F}^{\overline{\omega}}$, and we have the celebrated Perelman's monotonicity formula

$$\frac{d}{d\beta}\,\mathscr{F}^{\overline{\omega}}[\overline{g}(\beta)] = 2 \int_{V^n} \left| \mathrm{Ric}(\overline{g}(\beta)) + Hess_{\overline{g}(\beta)} \log\left(\frac{d\mu_{\overline{g}(\beta)}}{d\overline{\omega}}\right) \right|^2_{\overline{g}(\beta)} d\overline{\omega}\,.$$

(4.233)

Since a solution of (4.223) is also solution of (4.217) for $\overline{f}(\beta) = \log\left(\frac{d\mu_{\overline{g}(\beta)}}{d\overline{\omega}}\right)$, (and vice versa), the gradient flow nature of (4.223) extends to the coupled equations (4.217) with respect to the functional

$$\mathscr{F}[\overline{g}(\beta);\overline{f}(\beta)] := \int_{V^n} \left(\mathrm{R}(\overline{g}) + \left|\nabla\overline{f}(\beta)\right|^2_{\overline{g}}\right) e^{-\overline{f}(\beta)} d\mu_{\overline{g}(\beta)}\,,$$

(4.234)

and along a solution $\beta \longmapsto (\overline{g}(\beta), \overline{f}(\beta))$ of (4.217) one has

$$\frac{d}{d\beta}\,\mathscr{F}[\overline{g}(\beta);\overline{f}(\beta)] = 2 \int_{V^n} \left| \mathrm{Ric}(\overline{g}(\beta)) + Hess_{\overline{g}(\beta)}\overline{f}(\beta) \right|^2_{\overline{g}(\beta)} e^{-\overline{f}(\beta)} d\mu_{\overline{g}(\beta)}\,.$$

(4.235)

Note that $\mathscr{F}[\overline{g}(\beta);\overline{f}(\beta)]$ is associated to the pair $(\overline{g}(\beta);\overline{f}(\beta))$ where $\overline{f}(\beta)$ is a $C^\infty(V^n \times [0,T_1], \mathbb{R})$ function. Hence, unlike $\mathscr{F}^{\overline{\omega}}[\overline{g}(\beta)]$, the functional $\mathscr{F}[\overline{g}(\beta);\overline{f}(\beta)]$ is $\mathrm{Diff}(V^n)$–equivariant. In particular, under the β–dependent diffeomorphims $\Psi(\beta)$, solving (4.201), we have $\mathscr{F}[\overline{g}(\beta);\overline{f}(\beta)] = \mathscr{F}[g(\beta);f(\beta)]$, where (f, g) are defined by (4.202), and we eventually get

$$\frac{d}{d\beta}\,\mathscr{F}[g(\beta);f(\beta)] = 2 \int_{V^n} \left| \mathrm{Ric}(g(\beta)) + Hess_{g(\beta)}f(\beta) \right|^2_{g(\beta)} e^{-f(\beta)} d\mu_{g(\beta)}\,,$$

(4.236)

along the Ricci flow coupled with the conjugate backward heat Eq. (4.216). In Ricci flow theory one further proceeds in eliminating also the dependence from the choice of the function f so as to obtain from $\mathscr{F}[g(\beta);f(\beta)]$ a monotonic quantity depending only on the Riemannian geometry of (V^n, g). This is done by minimizing $\mathscr{F}[g(\beta);f(\beta)]$ over all probability measures $d\omega = e^{-f} d\mu_g \in \mathrm{Prob}_{ac}(V^n)$. The interested reader can consult Chap. 5 of [23] for details.

4.16 Factorization and Ricci Flow Conjugation

As we have recalled, when the initial data for the Ricci flow is a round n–dimensional sphere $(V^n, g), = (\mathbb{S}^n, g_{can})$ the flow ends in a finite time collapsing $(\mathbb{S}^n, g_{can})$ to a point, hence developing a curvature singularity. This behavior strongly contrasts with what we expect from the action of the renormalization group for NLσM since for $(\mathbb{S}^n, g_{can})$ the flow should be eternal. This basic remark suggests that the embedding of the Ricci flow in the renormalization group flow should be more amenable to a systematic study if we restrict our attention to a particular subclass of Ricci flow metrics. In particular, if we assume that the curvature does not grow unboundedly large during the Ricci evolution, then we can use the Ricci flow to interpolate between two given Riemannian manifolds (V^n, g) and $(V^n, \bar{g})$, and study in detail the dissipative/non–dissipative properties in a (functional) neighborhood of this fiducial flow. Note that this is not a stability analysis around a (generalized) fixed point of the flow but rather a form of entropic stability pinpointing in which direction the flow has *dissipative or non–dissipative* properties and can mimic a renormalization group flow. These remarks motivate the following.

Definition 4.5 (Interpolating fiducial Ricci flow of bounded geometry) A fiducial Ricci flow of bounded geometry interpolating between the two Riemannian manifolds (V^n, g) and $(V^n, \bar{g})$ is a non–collapsing solution of the weakly–parabolic initial value problem

$$\frac{\partial}{\partial \beta}\, g_{ab}(\beta) = -2\, \mathrm{R}_{ab}(\beta),$$

$$g_{ab}|_{\beta=0} = g_{ab}, \quad 0 \le \beta \le \beta^*,$$

(4.237)

such that $g_{ab}(\beta^*) = \bar{g}_{ab}$, and such that there exists constants $C_k > 0$ for which $\left|\nabla^k Rm(\beta)\right| \le C_k,\ k \in \mathbb{N} \cup \{0\}$, for $0 \le \beta \le \beta^*$. That is the Riemann curvature and its covariant derivatives of each order have uniform bounds for all metrics along the evolution $\beta \longmapsto g(\beta)$. We assume that any such a flow is contained in a corresponding maximal solution, $\beta \to g_{ab}(\beta), 0 \le \beta \le \beta^* < T_0 \le \infty$, with the same initial metric $g_{ab}(\beta = 0) = g_{ab}$. $\square$

Let us recall (see e.g. [23]) that given a constant $\kappa > 0$, a length scale r_0, a "spacetime" point (β_0, x_0), and a metric ball $B_{g(\beta_0)}(x_0, r_0)$ of radius r_0 centered at $x_0 \in (V^n, g(\beta_0))$, a Ricci flow solution of (4.237) with $[\beta_0 - r_0^2, \beta_0] \subset [0, \beta^*]$, is said to be κ–non–collapsed at scale r_0 at (β_0, x_0) if the norm of the Riemannian curvature tensor, $|\mathrm{R}m(\beta, x)|_{g(\beta)}$, is $\le r_0^{-2}$ for all $(\beta, x) \in [\beta_0 - r_0^2, \beta_0] \times B_{g(\beta_0)}(x_0, r_0)$ and the ball $B_{g(\beta_0)}(x_0, r_0)$ has volume $\mathrm{Vol}_{g(\beta_0)} \ge \kappa r_0^n$. This volume non–collapsing allows to control the evolution of the injectivity radius of $(V^n, g(\beta))$. Note that the assumption of non–collapsing is necessary since torus bundles over the circle admit smooth Ricci flows with bounded geometry which exist for all $\beta \in [0, \infty)$, and collapse as $\beta \to \infty$ [53].

Let $\beta \longmapsto g_{ab}(\beta)$ be a fiducial Ricci flow of bounded geometry on $V^n \times [0, \beta^*]$ in the sense of Definition 4.5. The hypothesis of bounded geometry implies that we can apply the Berger–Ebin splitting (A.49) to generate an induced affine slice parametrization in a neighborhood of the given Ricci flow (Fig. 4.30).

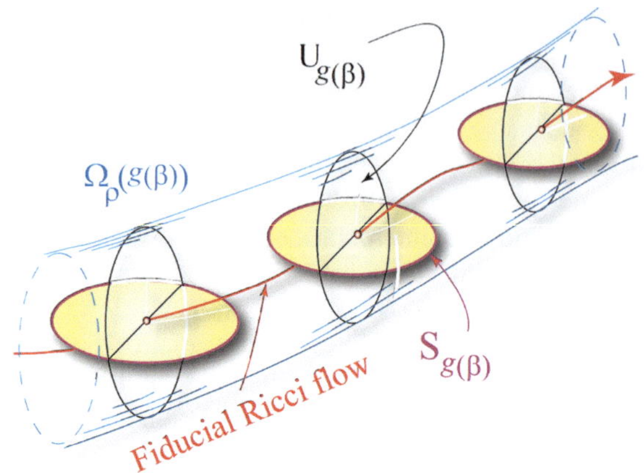

Fig. 4.30 The affine Berger–Ebin slice theorem along a non–collapsing fiducial Ricci flow of bounded geometry

Lemma 4.2 *Let $\beta \mapsto g(\beta)$ be a fiducial Ricci flow $\beta \mapsto g(\beta)$ of bounded geometry on $V^n \times [0, \beta^*]$, then there exists an affine slice parametrization of a tubular neighborhood, $\Omega_\rho(g(\beta))$, of $\beta \longmapsto g_{ab}(\beta)$ such that*

$$\Omega_\rho(g(\beta)) \doteq \left(\mathcal{U}_{g(\beta)} \times \mathcal{S}_{g(\beta)} \right) \times [0, \beta^*] \,, \tag{4.238}$$

where, for each given $\beta \in [0, \beta^*]$, $\mathcal{U}_{g(\beta)} \subset \mathcal{O}_{g(\beta)}$ is an open neighborhood of the Ricci flow metric $g(\beta)$ in the $\mathrm{Diff}(V^n)$*–orbit $\mathcal{O}_{g(\beta)}$, and*

$$\mathcal{S}_{g(\beta)} \doteq \left\{ g(\beta) + h^T \mid h^T \in \ker \, \delta_{g(\beta)} \,, \, \|h^T\|_{L^2(V^n, d\mu_{g(\beta)})} < \rho \right\} \,, \tag{4.239}$$

is, for $\rho > 0$ independent from β and small enough, the associated affine slice through $g(\beta)$.

Proof Along $\beta \longmapsto g(\beta), 0 \leq \beta \leq \beta^*$, the Berger–Ebin decomposition

$$T_{g(\beta)}\mathcal{M}et(V^n) \cong \ker \, \delta_{g(\beta)} \, \oplus \, \mathrm{Im} \, \delta^*_{g(\beta)} \tag{4.240}$$

is well–defined since the fiducial Ricci flow is of bounded geometry. At each given β, the corresponding affine slice is provided by,

$$\widetilde{\mathcal{S}}_{g(\beta)} = \{ g(\beta) + \widetilde{\mathcal{B}}_{\rho(\beta)}(g(\beta))\} \,, \tag{4.241}$$

where the open ball $\widetilde{\mathcal{B}}_{\rho(\beta)}(g(\beta)) \subset \mathcal{T}_{(V^n, g(\beta))}\mathcal{M}et(V^n)$ of divergence–free tensors is defined according to

$$\widetilde{\mathcal{B}}_{\rho(\beta)}(g(\beta)) \doteq \left\{ h^T(\beta) \in \ker \, \delta_{g(\beta)} \mid \|h^T(\beta)\|_{L^2(d\mu_{g(\beta)})} < \rho(\beta) \right\} \,. \tag{4.242}$$

The hypothesis of bounded geometry implies that the set of $\{\rho(\beta) : 0 \leq \beta \leq \beta^*\}$ is uniformly bounded away from zero by some positive constant $\rho := \inf_{0 \leq \beta \leq \beta^*}\{\rho(\beta)\}$. We correspondingly define

$$\mathscr{B}_\rho(g(\beta)) \doteq \left\{ h^T(\beta) \in \ker \delta_{g(\beta)} \mid \|h^T(\beta)\|_{L^2(d\mu_{g(\beta)})} < \rho \right\} . \qquad (4.243)$$

and set

$$\mathscr{S}_{g(\beta)} \doteq \left\{ g(\beta) + \mathscr{B}_\rho(g(\beta)) \right\} . \qquad (4.244)$$

Since $\mathscr{S}_{g(\beta)} \subseteq \widetilde{\mathscr{S}}_{g(\beta)}$, the slice map associated to $\widetilde{\mathscr{S}}_{g(\beta)}$ restricts naturally to $\mathscr{S}_{g(\beta)}$, to the effect that for each $\beta \in [0, \beta^*]$ there is a neighborhood $\mathscr{U}_{g(\beta)}$ of $g(\beta) \in \mathscr{O}_{g(\beta)}(V^n)$ and a section $\chi_\beta : \mathscr{U}_{g(\beta)} \to \mathrm{Diff}(V^n)$, $g' \mapsto \chi_\beta(g') \in \mathrm{Diff}(V^n)$, such that the map $\Upsilon_\beta : \mathscr{U}_{g(\beta)} \times \mathscr{S}_{g(\beta)} \longrightarrow \mathscr{M}et(V^n)$

$$\Upsilon_\beta(g', g(\beta) + h^T(\beta)) \doteq \chi_\beta(g')^* (g(\beta) + h^T(\beta)) , \qquad (4.245)$$

is a local homeomorphism onto a neighborhood of $g(\beta)$ in $\mathscr{M}et(V^n)$.

The Ricci flow interacts with the slice map Υ_β in a rather sophisticated way: a perturbation of the Ricci flow which propagates an $h \in \mathscr{T}_{(V^n, g(0))}\mathscr{M}et(V^n) \cap \mathscr{U}_{g(0)}$, with $g(0) \equiv g(\beta)|_{\beta=0}$, will give rise to a perturbed Ricci flow evolution in $\mathscr{U}_{g(\beta)}$, whereas an $h \in \mathscr{T}_{(V^n, g(0))}\mathscr{M}et(V^n) \cap \mathscr{S}_{g(\beta)}$ in general fails to evolve in $\mathscr{S}_{g(\beta)}$. Naively, this can be attributed to the dissipative (weakly–parabolic) nature of the Ricci flow, however the underlying rationale is quite subtler and holds a few surprises. To discuss this point, let $\epsilon \mapsto g_{ab}^{(\epsilon)}(\beta)$, $0 \leq \epsilon \leq 1$, be a smooth one–parameter family of Ricci flows in the tubular neighborhood $\Omega_\rho(g(\beta))$ defined above. For $\epsilon \searrow 0$, this set $\{g_{ab}^{(\epsilon)}(\beta)\}$ is locally characterized by the tangent vector $h_{ab}(\beta)$ in $T_{g(\beta)}\mathscr{M}et(V^n)$, covering the fiducial curve $\beta \mapsto g_{ab}(\beta)$, $0 \leq \beta \leq \beta^*$, and defined by the first jet $h_{ab}(\beta) \doteq \frac{d}{d\epsilon}g_{ab}^{(\epsilon)}(\beta)|_{\epsilon=0}$ of $g_{ab}^{(\epsilon)}(\beta)$. Any such $h_{ab}(\beta)$ satisfies the linearized Ricci flow equation

$$\frac{\partial}{\partial \beta} h_{ab}(\beta) = -2 \frac{d}{d\epsilon} \mathrm{R}_{ab}^{(\epsilon)}(\beta)\Big|_{\epsilon=0} ,$$
$$h_{ab}|_{\beta=0} = h_{ab} , \quad 0 \leq \beta \leq \beta^* . \qquad (4.246)$$

According to (4.219), the linearization (4.246) characterizes the flow $\beta \mapsto h_{ab}(\beta)$ as a solution of the initial value problem

$$\frac{\partial}{\partial \beta} h_{ab}(\beta) = \Delta_L h_{ab}(\beta) + 2 \left[\delta_{g(\beta)}^* \, \delta_{g(\beta)} \, G(h(\beta)) \right]_{ab} ,$$
$$h_{ab}|_{\beta=0} = h_{ab} , \quad 0 \leq \beta < T_0 , \qquad (4.247)$$

where Δ_L is the Lichnerowicz-DeRham Laplacian on $(V^n, g(\beta))$ defined by (4.220). As already stressed, this linearization is not parabolic due to the equivariance of the Ricci flow under diffeomorphisms, and we need to factorize (4.247) into a strictly parabolic flow and a $Diff(V^n)$ generating term. There are various distinct ways of implementing such a decomposition, all eventually related to the DeTurck trick [28]. For the convenience of the reader, here we describe a well–known factorization [2] in a form particularly suited to our purposes, (to the best of our knowledge, such a factorization appeared first explicitly in [65]). Further details can be found in Chap. 2 of [23].

Let us consider a given symmetric bilinear form $\widetilde{h}_{ab}^{(0)} \in T_g \mathcal{M}et(V^n)$. Along the Ricci flow of metrics $\beta \longmapsto g_{ab}(\beta)$, $g_{ab}(\beta = 0) = g_{ab}$, $0 \leq \beta < T_0$, look for solutions $\beta \mapsto h_{ab}(\beta)$ of the associated linearized flow in the form

$$h_{ab}(\beta) = \widetilde{h}_{ab}(\beta) + \nabla_a w_b(\beta) + \nabla_b w_a(\beta), \tag{4.248}$$

with $\widetilde{h}_{ab}|_{\beta=0} = \widetilde{h}_{ab}$, and where the β-dependent vector field $w^a(\beta)$ is associated with β-dependent infinitesimal $Diff(V^n)$ reparametrizations of the Riemannian structure associated with $g_{ab}(\beta)$. Since $\widetilde{h}_{ab}(\beta) + \mathcal{L}_w g_{ab}$ must satisfy the linearized Ricci flow, we get

$$\frac{\partial}{\partial \beta} \widetilde{h}_{ab} + \frac{\partial}{\partial \beta} \mathcal{L}_w g_{ab} \tag{4.249}$$

$$= \Delta_L \widetilde{h}_{ab} + \Delta_L \mathcal{L}_w g_{ab} + 2 \left[\delta_g^* \delta_g \, G(\widetilde{h}) \right]_{ab} + 2 \left[\delta_g^* \delta_g \, G(\mathcal{L}_w g) \right]_{ab} .$$

From the relations

$$\frac{\partial}{\partial \beta} \mathcal{L}_{w(\beta)} g_{ab}(\beta) = \mathcal{L}_{w(\beta)} \frac{\partial}{\partial \beta} g_{ab}(\beta) + \mathcal{L}_{\frac{\partial}{\partial \beta} w(\beta)} g_{ab}(\beta), \tag{4.250}$$

$$\mathcal{L}_{w(\beta)} \frac{\partial}{\partial \beta} g_{ab}(\beta) = -2 \mathcal{L}_{w(\beta)} R_{ab}(\beta) \tag{4.251}$$

(in the latter we have exploited the fact that $\beta \longmapsto g_{ab}(\beta)$ evolves along the Ricci flow), and

$$\mathcal{L}_{w(\beta)} R_{ab}(\beta) = -\frac{1}{2} \Delta_L \mathcal{L}_{w(\beta)} g_{ab}(\beta) - \left[\delta_{g(\beta)}^* \delta_{g(\beta)} \, G(\mathcal{L}_{w(\beta)} g(\beta)) \right]_{ab} , \tag{4.252}$$

(consequence of the $Diff(V^n)$-equivariance of the Ricci tensor) we obtain (see (4.219))

$$\frac{\partial}{\partial \beta} \mathcal{L}_w g_{ab} = \mathcal{L}_{\frac{\partial}{\partial \beta} w} g_{ab} + \Delta_L \mathcal{L}_w g_{ab} + 2 \left[\delta_g^* \delta_g \, G(\mathcal{L}_w g) \right]_{ab} , \tag{4.253}$$

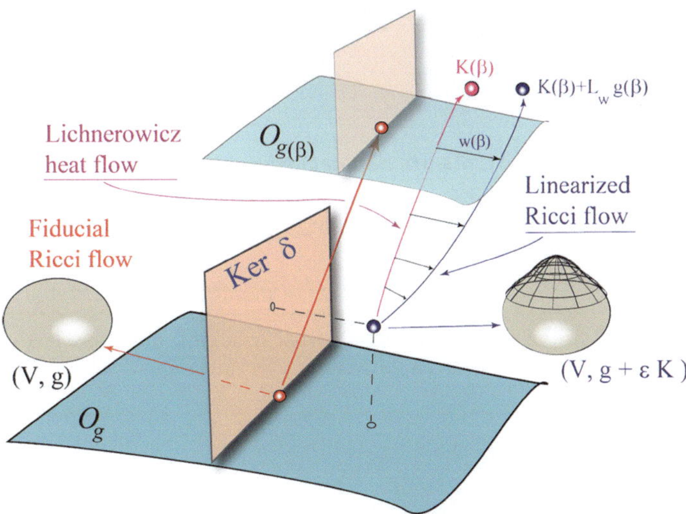

Fig. 4.31 The geometry of the linearized Ricci flow is related to the observation that if $K(\beta)$ is a solution of the Lichnerowicz heat equation, then $\widetilde{K}(\beta) := K(\beta) + \mathscr{L}_{w(\beta)} g(\beta)$ is a solution of the linearized Ricci flow (with the same initial datum K) as long as the covector field is such that $\frac{\partial}{\partial \beta} w(\beta) = \delta_{g(\beta)} G(K(\beta))$, with $w(\beta = 0) = 0$, and where $G(K(\beta))$ denotes the Einstein conjugate of $K(\beta)$ with respect to the metric $g(\beta)$

where for notational ease we have dropped the explicit β–dependence. Inserting this latter in (4.249) we have

$$\frac{\partial}{\partial \beta} \widetilde{h}_{ab} + \mathscr{L}_{(\frac{\partial}{\partial \beta} w_k - \delta_g G(\widetilde{h}))} \, g_{ab} = \Delta_L \widetilde{h}_{ab} \, . \tag{4.254}$$

As an immediate consequence of the structure of this relation it follows that, under the stated hypotheses, we can naturally factorize the linearized Ricci flow according to the following result (see e.g. [23]) (Fig. 4.31).

Lemma 4.3 (The Reduced Linearized Ricci Flow) *Let* $\beta \longmapsto \widetilde{h}_{ab}(\beta)$, $\beta \in [0, T_0)$, *denote the flow solution of the parabolic initial value problem*

$$\frac{\partial}{\partial \beta} \widetilde{h}_{ab} = \Delta_L \widetilde{h}_{ab}$$

$$\widetilde{h}_{ab}|_{\beta=0} = h_{ab}, \tag{4.255}$$

and let $\beta \longmapsto w_a(\beta)$, $\beta \in [0, T_0)$, *be the β-dependent (co)vector field solution of the initial value problem*

$$\frac{\partial}{\partial \beta}\, w_a(\beta) = -\nabla^b \left(\widetilde{h}_{ab} - \tfrac{1}{2}\widetilde{h}\, g_{ab}\right),$$

$$(4.256)$$

$$w_a|_{\beta=0} = 0 \;.$$

Then the flow $\beta \longmapsto h_{ab}(\beta)$, $\beta \in (0, \beta_0)$, defined by

$$h_{ab}(\beta) \doteq \widetilde{h}_{ab}(\beta) + \mathscr{L}_{w(\beta)} g_{ab}(\beta), \qquad (4.257)$$

solves the linearized Ricci flow (4.247) with initial datum $h_{ab}|_{\beta=0} = h_{ab}$.

Proof The proof of the lemma amounts to backtracking the steps leading to the identity (4.254). Explicitly, from (4.249) and (4.253), we get

$$\frac{\partial}{\partial \beta}\, \widetilde{h}_{ab} + \frac{\partial}{\partial \beta}\, \mathscr{L}_w\, g_{ab} \qquad (4.258)$$

$$= \Delta_L \widetilde{h}_{ab} + \mathscr{L}_{\frac{\partial}{\partial\beta} w}\, g_{ab} + \Delta_L \mathscr{L}_w\, g_{ab} + 2\left[\delta_g^* \delta_g\, G(\mathscr{L}_w g)\right]_{ab} \;.$$

Moreover, from (4.256), we have

$$\mathscr{L}_{\frac{\partial}{\partial\beta} w}\, g_{ab} + 2\left[\delta_g^* \delta_g\, G(\mathscr{L}_w g)\right]_{ab} = 2\left[\delta_g^* \delta_g\, G(\widetilde{h} + \mathscr{L}_w g)\right]_{ab} \;. \qquad (4.259)$$

By inserting (4.259) in (4.258), and gathering terms, we get that $h_{ab}(\beta) \doteq \widetilde{h}_{ab}(\beta) + \mathscr{L}_{w(\beta)} g_{ab}(\beta)$, solves the linearized Ricci flow (4.247) with initial datum $h_{ab}|_{\beta=0} = h_{ab}$.

4.16.1   The Hodge–DeRham–Lichnerowicz Heat Operator

According to Lemma 4.3, the linearized Ricci flow (4.246) is equivalent to the dynamical system $\beta \mapsto h_{ab}(\beta) \in T_{g(\beta)}\mathscr{M}et(V^n)$ defined, along the fiducial Ricci flow $\beta \mapsto g(\beta)$, by the Lichnerowicz heat equation

$$\bigcirc_L h_{ab}(\beta) \doteq \left(\frac{\partial}{\partial \beta} - \Delta_L\right) h_{ab}(\beta) = 0 \;,$$

$$(4.260)$$

$$h_{ab}|_{\beta=0} = h_{ab}\,, \;\; 0 \le \beta \le \beta^* \;.$$

Henceforth, when discussing the linearized Ricci flow (4.246) we will explicitly refer to the reduced version (4.260) where we have fixed the β–dependent Diff(V^n)– gauge freedom according to (4.256). In this connection we have the following basic observation relating all natural elliptic operators, (and their attendant parabolic counterparts), defined along a fiducial Ricci flow (Fig. 4.32).

Lemma 4.4 (The Hodge–DeRham–Lichnerowicz heat operator) *If*

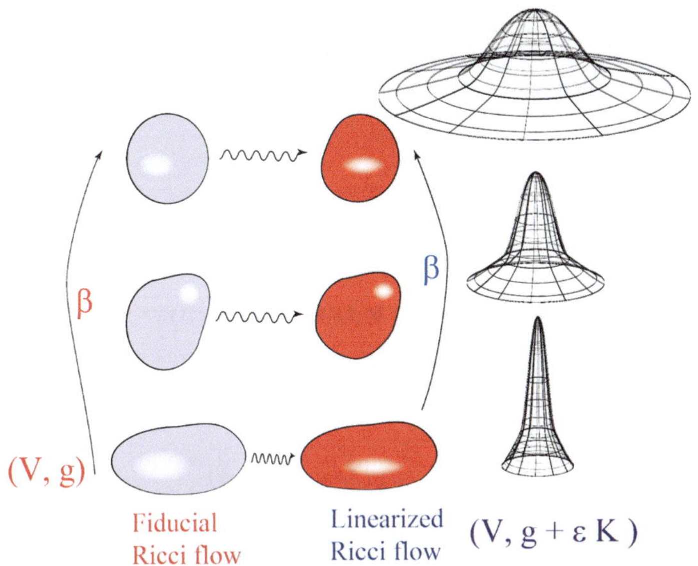

Fig. 4.32 The Hodge–DeRham–Lichnerowicz representation of the linearized Ricci flow is a geometrical heat equation which unifies the scalar, vector, and tensor heat equations which we often use to analyze specific properties of the Ricci flow

$$\triangle_{d(\beta)} \doteq -\left(d\,\delta_{g(\beta)} + \delta_{g(\beta)}\,d\right) \qquad (4.261)$$

denotes the $g(\beta)$–Hodge–DeRham Laplacian along a Ricci flow of bounded geometry $\beta \mapsto g_{ab}(\beta)$, $0 \le \beta \le \beta^*$, then the scalar heat flow $\left(\frac{\partial}{\partial \beta} - \Delta_{g(\beta)}\right)\omega(\beta) = 0$, the covector heat flow $\left(\frac{\partial}{\partial \beta} - \Delta_{vec(\beta)}\right)v_a(\beta) = 0$, and the linearized Ricci flow $\left(\frac{\partial}{\partial \beta} - \Delta_{L(\beta)}\right)h_{ab}(\beta) = 0$, along $\beta \mapsto g_{ab}(\beta)$, $0 \le \beta \le \beta^*$, can be compactly represented by the kernel of the Hodge–DeRham–Lichnerowicz (HDRL) heat operator

$$\bigcirc_{d(\beta)} \doteq \frac{\partial}{\partial \beta} - \Delta_{d(\beta)} , \qquad (4.262)$$

thought of as acting on the appropriate parabolic space of β–dependent sections: $C^\infty(V^n \times \mathbb{R}, \mathbb{R})$ for the scalar heat operator, $C^\infty(V^n \times \mathbb{R}, T^*V^n)$ for the covector heat operator, and finally $C^\infty(V^n \times \mathbb{R}, \otimes_S^2 T^*V^n)$ for the Lichnerowicz heat equation.

Proof The elliptic operators defined, along the fiducial Ricci flow $\beta \longmapsto g_{ab}(\beta)$, by: (i) The standard Laplace–Beltrami operator acting on scalar functions

$$\Delta_{g(\beta)} := g^{ab}(\beta)\,\nabla_a^{(g(\beta))}\,\nabla_b^{(g(\beta))} ; \qquad (4.263)$$

(ii) The vector Laplacian acting on (co)vector fields,

$$\triangle_{vec(\beta)} \doteq \triangle_{g(\beta)} - \mathrm{Ric}(g(\beta)) \; ; \tag{4.264}$$

and (iii) The Lichnerowicz–DeRham Laplacian acting on symmetric bilinear forms (cf. (4.220))

$$\triangle_{L(\beta)} = \triangle_{g(\beta)} + \mathscr{E}(\beta) \; , \tag{4.265}$$

where $\mathscr{E}(\beta)$ the local section of $End\left(\otimes^2 T^* V^n\right)$, $h_{ik} \mapsto \mathscr{E}_{ab}^{ik}(\beta)\, h_{ik}$, defined by

$$\mathscr{E}_{ab}^{ik}(\beta) := -\, \mathrm{R}_a^i(\beta)\delta_b^k - \mathrm{R}_b^k(\beta)\delta_a^i + 2\mathrm{R}_{ab}^{ik}(\beta) \; , \tag{4.266}$$

can all be identified with the $g(\beta)$–Hodge–DeRham Laplacian acting on differential p–forms

$$\triangle_{d(\beta)} \doteq -\,(d\,\delta_{g(\beta)} + \delta_{g(\beta)}\,d) \; . \tag{4.267}$$

For the Laplace–Beltrami operator $\triangle_{g(\beta)}$ and for the vector Laplacian $\triangle_{vec(\beta)}$ the statement is obvious. For the Lichnerowicz–DeRham Laplacian $\triangle_{L(\beta)}$, it directly follows by observing, (see e.g. [23]), that the Hodge–DeRham Laplacian $\triangle_d$ acts on 2–forms in the same way that $\triangle_L$ acts on symmetric 2–tensors. Thus, along a Ricci flow of bounded geometry $\beta \mapsto g_{ab}(\beta)$, $0 \leq \beta \leq \beta^*$, both the scalar, the vector, and the (symmetric) 2–tensor heat equation have the representation (4.262). Hence, for any $\omega \in C^\infty(V^n \times \mathbb{R}, \mathbb{R})$, $v(\beta) \in C^\infty(V^n \times \mathbb{R}, T^* V^n)$, and $h(\beta) \in C^\infty(V^n \times \mathbb{R}, \otimes_S^2 T^* V^n)$, we can use for (4.262) the compact notation

$$\bigcirc_d \begin{pmatrix} \omega(\beta) \\ v_i(\beta) \\ h_{ab}(\beta) \end{pmatrix} := \begin{pmatrix} (\frac{\partial}{\partial\beta} - \triangle_g)\,\omega(\beta) \\ (\frac{\partial}{\partial\beta} - \triangle_{vec})\,v_i(\beta) \\ (\frac{\partial}{\partial\beta} - \triangle_L)\,h_{ab}(\beta) \end{pmatrix} , \tag{4.268}$$

where, for notational ease, we dropped the explicit β–dependence in the operators, (a convention we shall follow henceforth if there is no danger of confusion), and denoted $\bigcirc_{d(\beta)}$, $\triangle_{g(\beta)}$, $\triangle_{vec(\beta)}$, and $\triangle_{L(\beta)}$ simply by $\bigcirc_d$, $\triangle_g$, $\triangle_{vec}$, and $\triangle_L$, respectively.

The above remarks extend naturally also to the parabolic conjugation $\bigcirc_d^*$ of the operator $\bigcirc_d$ (Fig. 4.33). To see this explicitly and set the appropriate notation, let $\beta \mapsto (V^n, g_{ab}(\beta))$, $0 \leq \beta \leq \beta^*$, $\beta^* \in [0, T_0)$ be a given Ricci flow metric of bounded geometry interpolating between $(V^n, g(\beta = 0) := g)$ and $(V^n, \overline{g} := g(\beta^*))$. Let $(M_{Par}^{n+1} \simeq V^n \times [0, \beta^*], {}^{(n+1)}g_{par})$ denote the corresponding parabolic spacetime. Through the diffeomorphism $F_\beta^{-1} : M_{par}^{n+1} \to I \times V^n$, (see (4.184)), any $(\beta, x) \mapsto B(\beta, x)$, with $B(\beta, x) \in C^\infty(V^n \times \mathbb{R}, \otimes_S^2 T^* V^n)$, can be seen as a bilinear form in $C^\infty(M_{Par}^{n+1}, \otimes_S^2 T^* V^n)$. Since the volume form on M_{Par}^{n+1} is given by the product measure $d\mu_g(\beta)\, d\beta$, we can consider the natural pairing

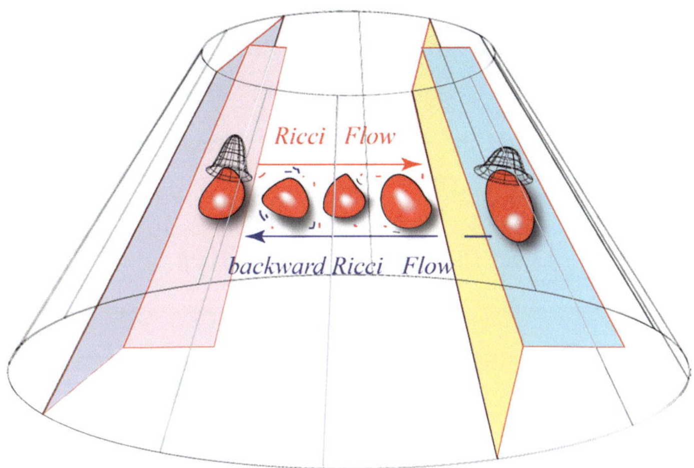

Fig. 4.33 Together with the given fiducial Ricci flow $\beta \longmapsto g(\beta)$, $\beta \in [0, \beta^*]$, we need also to consider the associated time–reversed flow $\eta \longmapsto g(\eta)$, where $\eta := \beta^* - \beta$

$$\int_0^{\beta^*} \int_{V^n} H^{ab}(\beta)\, B_{ab}(\beta)\, d\mu_g(\beta)\, d\beta \;, \tag{4.269}$$

between $H^{ab}(\beta) \in C^\infty(M_{Par}^{n+1}, \otimes_S^2 TV^n)$ and $B_{ab}(\beta) \in C^\infty(M_{Par}^{n+1}, \otimes_S^2 T^*V^n)$. Let us consider the operator

$$\bigcirc_L \doteq \frac{\partial}{\partial \beta} - \triangle_L \;, \tag{4.270}$$

and let us compute its $L^2(M_{Par}^{n+1}, d\mu_g(\beta)\, d\beta)$ conjugate $\bigcirc_L^*$, thought of as acting on the space of symmetric two–tensors with compact support. From the relation

$$\int_0^{\beta^*} \int_{V^n} H^{ab}(\beta)\frac{\partial}{\partial \beta} B_{ab}(\beta) d\mu_g(\beta) d\beta = \int_0^{\beta^*} \frac{d}{d\beta} \int_{V^n} H^{ab} B_{ab} d\mu_g(\beta) d\beta \tag{4.271}$$

$$+ \int_0^{\beta^*} \int_{V^n} B_{ab}\left(-\frac{\partial}{\partial \beta} H^{ab} + \mathrm{R}\, H^{ab}\right) d\mu_{g(\beta)}\, d\beta$$

$$= \int_0^{\beta^*} \int_{V^n} B_{ab}\left(-\frac{\partial}{\partial \beta} H^{ab} + \mathrm{R}\, H^{ab}\right) d\mu_{g(\beta)}\, d\beta \;,$$

(where we have exploited the Ricci flow evolution for $d\mu_g$), and the relation

$$\int_0^{\beta^*} \int_{V^n} \left[H^{ab}(\beta) \, \triangle_L \left(B_{ab}(\beta) \right) - B_{ab}(\beta) \, \triangle_L \left(H^{ab}(\beta) \right) \right] d\mu_g(\beta) \, d\beta \; = \; 0 \; ,$$

$$(4.272)$$

(where we have exploited the fact that $\triangle_L$ is formally self–adjoint on each $(V^n, g(\beta))$), we compute

$$\int_0^{\beta^*} \int_{V^n} H^{ab}(\beta) \; \bigcirc_L \left(B_{ab}(\beta) \right) \, d\mu_g(\beta) \, d\beta \tag{4.273}$$

$$= \int_0^{\beta^*} \int_{V^n} H^{ab} \left(\frac{\partial}{\partial \beta} - \triangle_L \right) \left(B_{ab}(\beta) \right) \, d\mu_g \, d\beta$$

$$= \int_0^{\beta^*} \int_{V^n} B_{ab} \left(-\frac{\partial}{\partial \beta} - \triangle_L + \mathrm{R} \right) \left(H^{ab}(\beta) \right) \, d\mu_g \, d\beta$$

$$= \int_0^{\beta^*} \int_{V^n} B_{ab}(\beta) \; \bigcirc_L^* \; H^{ab}(\beta) \, d\mu_g(\beta) \, d\beta \; .$$

Thus,

$$\bigcirc_L^* \; \doteq \; -\frac{\partial}{\partial \beta} - \triangle_L + \mathrm{R} \; . \tag{4.274}$$

One can proceed similarly for characterizing the conjugate scalar and vector heat operators, to the effect that if we consider the $L^2(V^n \times \mathbb{R}, d\mu_{g(\beta)} d\beta)$ parabolic pairing between the spaces $C^\infty(V^n \times \mathbb{R}, \otimes_S^p T^* V^n)$ and $C^\infty(V^n \times \mathbb{R}, \otimes_S^p T V^n)$, $(p = 0, 1, 2)$, we can introduce [17] the backward L^2–conjugated flow associated with (4.260), generated by the operator (Fig. 4.34)

$$\bigcirc_d^* \; \doteq \; -\frac{\partial}{\partial \beta} - \triangle_d + \mathrm{R} \; , \tag{4.275}$$

acting on the appropriate space of β-dependent sections $C^\infty(V^n \times \mathbb{R}, \otimes_S^p T V^n)$, $(p = 0, 1, 2)$. In a more compact form, paralleling the notation (4.268), we can write

$$\bigcirc_d^* \begin{pmatrix} \varpi(\beta) \\ W^i(\beta) \\ H^{ab}(\beta) \end{pmatrix} := \begin{pmatrix} \left(-\frac{\partial}{\partial \beta} - \triangle_g + \mathrm{R} \right) \varpi(\beta) \\ \left(-\frac{\partial}{\partial \beta} - \triangle_{vec} + \mathrm{R} \right) W^i(\beta) \\ \left(-\frac{\partial}{\partial \beta} - \triangle_L + \mathrm{R} \right) H^{ab}(\beta) \end{pmatrix} \; , \tag{4.276}$$

for any $\varpi \in C^\infty(V^n \times \mathbb{R}, \mathbb{R})$, $W \in C^\infty(V^n \times \mathbb{R}, T V^n)$, and $H \in C^\infty(V^n \times \mathbb{R}, \otimes_S^2 T V^n)$.

The role of $\bigcirc_d$ and $\bigcirc_d^*$ in Ricci flow conjugation is connected with their interaction with the Berger–Ebin splitting of $\mathscr{T}_{(V^n, g(\beta))} \mathscr{M} et(V^n)$. To make this interaction explicit, we organize in a unique pattern a number of commutation rules among $\bigcirc_d$, $\bigcirc_d^*$, and the action of $\delta_{g(\beta)}$, and $\delta_{g(\beta)}^*$. Some of these relations are rather familiar, a few others extend, in a non trivial way, known properties of the conjugate scalar heat flow.

Lemma 4.5 (Commutation rules) *If $\bigcirc_d$ and $\bigcirc_d^*$ respectively denote the Hodge–DeRham–Lichnerowicz heat operator and its conjugate along a fiducial Ricci flow $\beta \longmapsto g_{ab}(\beta)$ on $V^n \times [0, \beta^*]$, then for any $v(\beta) \in C^\infty(V^n \times \mathbb{R}, T^*V^n)$, $h(\beta) \in C^\infty(V^n \times \mathbb{R}, \otimes_S^2 T^*V^n)$, and $H(\beta) \in C^\infty(V^n \times \mathbb{R}, \otimes_S^2 TV^n)$, we have*

$$\frac{d}{d\beta} \, \langle h(\beta), H(\beta)\rangle_{L^2(d\mu_{g(\beta)})} := \frac{d}{d\beta} \int_{V^n} h_{ab}(\beta)\, H^{ab}(\beta)\, d\mu_{g(\beta)}$$

$$(4.277)$$

$$= \int_{V^n} \left[H^{ab}(\beta)\, \bigcirc_d\, h_{ab}(\beta) - h_{ab}(\beta)\, \bigcirc_d^*\, H^{ab}(\beta)\right] d\mu_{g(\beta)} \,.$$

Moreover the following set of commutation rules hold:

$$tr_{g(\beta)}\, (\bigcirc_d h(\beta)) = \bigcirc_d \left(tr_{g(\beta)}\, h(\beta)\right) - 2\, \mathrm{R}^{ik}(\beta)\, h_{ik}(\beta) \,, \tag{4.278}$$

$$tr_{g(\beta)}\, (\bigcirc_d^* H(\beta)) = \bigcirc_d^* \left(tr_{g(\beta)}\, H(\beta)\right) - 2\, \mathrm{R}_{ik}(\beta)\, H^{ik}(\beta) \,, \tag{4.279}$$

$$\bigcirc_d \left(\delta_{g(\beta)}^*\, v^\sharp(\beta)\right) = \delta_{g(\beta)}^* \left(\bigcirc_d v(\beta)\right)^\sharp \,, \tag{4.280}$$

$$\bigcirc_d^* \left(\delta_{g(\beta)}\, H(\beta)\right) = \delta_{g(\beta)} \left(\bigcirc_d^* H(\beta)\right) \,, \tag{4.281}$$

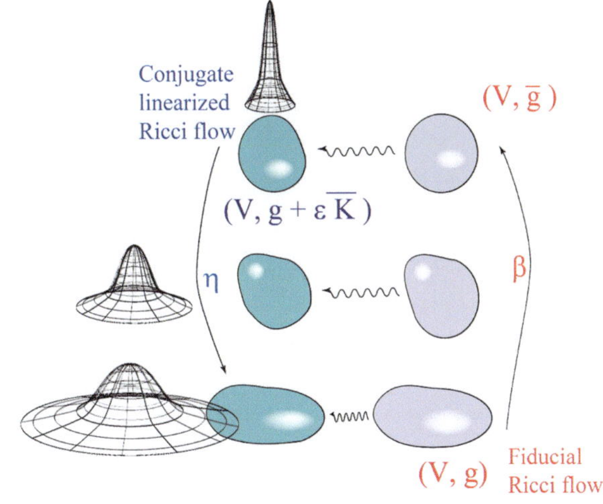

Fig. 4.34 The conjugate Hodge–DeRham–Lichnerowicz heat equation has a number of unexpected properties allowing a better control of the linearized Ricci flow. These properties are related to the fact that the Ricci curvature evolves according to the forward Lichnerowicz heat equation

$$\bigcirc_d \left(\delta_{g(\beta)}\, h(\beta)\right) = \delta_{g(\beta)}\left(\bigcirc_d h(\beta)\right) - 2\mathrm{R}^{ik}(\beta)\,\nabla_i h_{kl}(\beta)\,dx^l$$

$$(4.282)$$

$$-2\,h_{ik}(\beta)\left(\nabla_l\mathrm{R}^{ik}(\beta) - \nabla^k\mathrm{R}^i_l(\beta)\right)dx^l\,,$$

$$\bigcirc^*_d\left(\delta^*_{g(\beta)}\,v(\beta)\right) = \delta^*_{g(\beta)}\left(\bigcirc^*_d v(\beta)\right) - \big[v_a\nabla^k\mathrm{R}_{kb}(\beta) + v_b\nabla^k\mathrm{R}_{ka}(\beta)$$

$$(4.283)$$

$$+2\,v^k(\beta)\left(\nabla_a\mathrm{R}_{bk}(\beta) + \nabla_b\mathrm{R}_{ak}(\beta) - \nabla_k\mathrm{R}_{ab}(\beta)\right)\big]\,dx^a\otimes dx^b\,,$$

where $tr_{g(\beta)}$ and $^\sharp$ respectively denote the $g(\beta)$–dependent trace and the $g(\beta)$–raising operator along the fiducial flow.

Proof The relation (4.277), describing the evolution of the L^2 pairing between $\mathscr{T}_{(V^n,g(\beta))}\mathscr{M}et(V^n)$ and $C^\infty(V^n\times\mathbb{R},\otimes^2_S TV^n)$, immediately follows from adding and subtracting $H^{ab}\Delta_L h_{ab}$ to the integral below

$$\frac{d}{d\beta}\,\langle h(\beta),H(\beta)\rangle_{L^2(d\mu_{g(\beta)})}$$

$$(4.284)$$

$$= \int_{V^n}\left[h_{ab}\frac{\partial}{\partial\beta}H^{ab} + H^{ab}\frac{\partial}{\partial\beta}h_{ab} - Rh_{ab}H^{ab}\right]d\mu_{g(\beta)}\,,$$

and exploiting the fact that Δ_L is a self–adjoint operator with respect to $d\mu_{g(\beta)}$. The commutation relations (4.278) and (4.279) between the $g(\beta)$–dependent trace and the operators $\bigcirc_d$ and $\bigcirc^*_d$ are elementary consequences of the Ricci flow evolution. Similarly well–known, (see e.g. [23]), is the commutation rule (4.280). A direct computation shows that (4.282) is a consequence of the Weitzenböck formula, (see e.g. [17]),

$$\nabla^k\Delta_L S_{k\ell} = \Delta\,\nabla^k S_{k\ell} + S_{ka}\,\nabla_\ell R^{ka} - R^a_\ell\,\nabla^k S_{ka} - 2S_{ka}\,\nabla^k R^a_\ell\,,\qquad(4.285)$$

and of the Ricci flow rule

$$\frac{\partial}{\partial\beta}\,\nabla^k S_{k\ell} = g^{ik}\,\nabla_i\left(\frac{\partial}{\partial\beta}S_{k\ell}\right) + 2R^{ik}\nabla_i S_{k\ell} + S_{mi}\,\nabla_\ell R^{mi}\,,\qquad(4.286)$$

both valid for any symmetric bilinear form $S\in C^\infty(V^n\times\mathbb{R},\otimes^2_S T^*V^n)$. Explicitly we compute

$$\left(\frac{\partial}{\partial\beta} - \Delta_d\right)\nabla^k h_{k\ell} = \nabla^k\left(\frac{\partial}{\partial\beta}\,h_{k\ell}\right) + 2R^{ik}\nabla_i h_{k\ell}\qquad(4.287)$$

$$+ h_{mi} \nabla_\ell \mathrm{R}^{mi} - \Delta(\nabla^k h_{k\ell}) + \mathrm{R}^j_\ell \nabla^k h_{k\ell}$$

$$= \nabla^k \left(\frac{\partial}{\partial\beta} h_{k\ell} \right) + 2\mathrm{R}^{ik}\nabla_i h_{k\ell} + h_{mi} \nabla_\ell \mathrm{R}^{mi}$$

$$- \nabla^k (\Delta_L h_{k\ell}) + h_{ka}\nabla_\ell \mathrm{R}^{ka} - \mathrm{R}^a_\ell \nabla^k h_{ka} - 2 h_{ka} \nabla^k \mathrm{R}^a_\ell + \mathrm{R}^j_\ell \nabla^k h_{kj}$$

$$= \nabla^k \left(\frac{\partial}{\partial\beta} h_{k\ell} - \Delta_L h_{k\ell} \right) + 2\mathrm{R}^{ik}\nabla_i h_{k\ell} + 2h_{ik}(\nabla_\ell \mathrm{R}^{ik} - \nabla^k \mathrm{R}^i_\ell) \, ,$$

where in the first and in the fourth line we exploited (4.286) and (4.285), respectively. This proves (4.280). The basic relation (4.281) follows from a rather lengthy but otherwise straightforward computation. According to the definition of $\bigcirc^*_d$ we have

$$\bigcirc^*_d \left(\nabla_a H^{ab} \right) = -\frac{\partial}{\partial\beta} \left(g^{bl} \nabla_a H^a_l \right) - (\Delta_d - \mathrm{R}) \, \nabla_a H^{ab} \tag{4.288}$$

$$= -2\mathrm{R}^{bl}\nabla^a H_{al} - g^{bl}\frac{\partial}{\partial\beta}\nabla^a H_{al} - \Delta_g \nabla_a H^{ab} + \mathrm{R}^b_l \nabla_a H^{al} + \mathrm{R}\nabla_a H^{ab} \, .$$

By exploiting again (4.286) and (4.285), this latter expression reduces to

$$= -2\mathrm{R}^{bl}\nabla^a H_{al} - g^{bl}\nabla^a \left(\frac{\partial}{\partial\beta}H_{al} \right) - 2g^{bl}\mathrm{R}^{ia}\nabla_i H_{al} \tag{4.289}$$

$$- g^{bl}H_{ai}\nabla_l \mathrm{R}^{ai} - \Delta_g \nabla_a H^{ab} + \mathrm{R}^b_l \nabla_a H^{al} + \mathrm{R}\nabla_a H^{ab}$$

$$= -\mathrm{R}^{bl}\nabla^a H_{al} - g^{bl}\nabla^a \left(\frac{\partial}{\partial\beta} \left(g_{ac}\, g_{ld}H^{cd} \right) \right) - 2g^{bl}\mathrm{R}^{ia}\nabla_i H_{al}$$

$$- g^{bl}H_{ai}\nabla_l \mathrm{R}^{ai} - \Delta_g \nabla_a H^{ab} + \mathrm{R}\nabla_a H^{ab}$$

$$= -\mathrm{R}^{bl}\nabla^a H_{al} - g^{bl}\nabla^a \left[g_{ac}\, g_{ld}\frac{\partial}{\partial\beta}H^{cd} - 2\mathrm{R}_{ac}\, g_{ld}\, H^{cd} \right.$$

$$\left. - 2\mathrm{R}_{ld}\, g_{ac}\, H^{cd} \right] - 2g^{bl}\mathrm{R}^{ia}\nabla_i H_{al}$$

$$- g^{bl}H_{ai}\nabla_l \mathrm{R}^{ai} - \Delta_g \nabla_a H^{ab} + \mathrm{R}\nabla_a H^{ab}$$

$$= -\mathrm{R}^{bl}\nabla^a H_{al} + 2H^{cb} \nabla^a \mathrm{R}_{ac} + 2\mathrm{R}_{ac} \nabla^a H^{cb} + 2H^d_a \nabla^a \mathrm{R}^b_d$$

$$+ 2\mathrm{R}^b_d\nabla^a H^d_a - \nabla_c \left(\frac{\partial}{\partial\beta}H^{cb} \right) - 2\mathrm{R}^{ia}\nabla_i H^b_a - H^{mi}\nabla^b \mathrm{R}_{mi}$$

$$- \nabla_a \Delta_L H^{ab} + H_{ka}\nabla^b \mathrm{R}^{ka} - \mathrm{R}^a_b\nabla^k H_{ka} - 2H_{ka}\nabla^k \mathrm{R}^{ab} + \mathrm{R}\nabla_a H^{ab}$$

$$= \nabla_c \left[-\frac{\partial}{\partial\beta} - \Delta_L + \mathrm{R} \right] H^{cb} \, ,$$

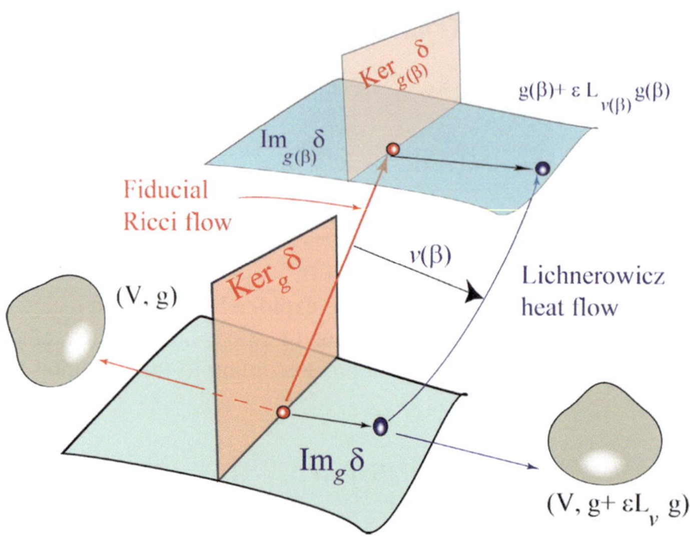

Fig. 4.35 The forward evolution along the linearized Ricci flow (in the HDRL representation) naturally preserves Im $\delta^*_{g(\beta)}$

where the last line follows from cancelling terms and by using the contracted Bianchi identity. This proves (4.281). Finally, (4.283) is a consequence of the known Ricci flow identities, (see e.g. [23]),

$$\nabla_i \left(\Delta v_j - R_j^k v_k \right) = \Delta_L \nabla_i v_j - v^k \left(\nabla_i R_{jk} + \nabla_j R_{ik} - \nabla_k R_{ij} \right) , \tag{4.290}$$

$$\frac{\partial}{\partial \beta} \nabla_i v_j = \nabla_i \left(\frac{\partial}{\partial \beta} v_j \right) + v^k \left(\nabla_i R_{jk} + \nabla_j R_{ik} - \nabla_k R_{ij} \right) , \tag{4.291}$$

which hold for any smooth β–dependent covector field $v(\beta)$. According to these we compute

$$\left(\frac{\partial}{\partial \beta} + \Delta_d - R \right) \left(\nabla_a v_b + \nabla_b v_a \right) \tag{4.292}$$

$$= \nabla_a \left(\frac{\partial}{\partial \beta} + \Delta_d - R \right) v_b + \nabla_b \left(\frac{\partial}{\partial \beta} + \Delta_d - R \right) v_a$$

$$+ 4 v^k \left(\nabla_a R_{bk} + \nabla_b R_{ak} - \nabla_k R_{ab} \right) + v_a \nabla^k R_{kb} + v_b \nabla^k R_{ka} ,$$

from which (4.283) follows.

From the commutation rule (4.280) we get a familiar property of the linearized Ricci flow which we state in the following result (Fig. 4.35).

Lemma 4.6 *If $\beta \mapsto v(\beta)$, $0 \leq \beta \leq \beta^*$, is a solution of $\bigcirc_d \, v(\beta) = 0$, then the induced flow*

$$\beta \mapsto \delta^*_{g(\beta)} \, v^\sharp(\beta) \in \mathcal{U}_{g(\beta)} \cap \mathcal{T}_{(V^n,g(\beta))} \mathcal{M}et(V^n) \times [0, \beta^*] \,, \qquad (4.293)$$

is a solution of the linearized Ricci flow $\bigcirc_d \, \delta^*_{g(\beta)} \, v^\sharp(\beta) = 0$.
This implies that the forward evolution along the linearized Ricci flow naturally preserves Im $\delta^*_{g(\beta)}$, and that initial data of the form $h(\beta = 0) = \delta^*_g \, v^\sharp\big|_{\beta=0}$ evolves in $\mathcal{U}_{g(\beta)} \times [0, \beta^*]$. Conversely, if $h(\beta) \in \ker \bigcirc_d \cap \mathcal{T}_{(V^n,g(\beta))} \mathcal{M}et(V^n)$, $0 \leq \beta \leq \beta^*$, is a solution of the linearized Ricci flow with $h(\beta = 0) \in \ker \delta_{g(\beta=0)}$, then in general $h(\beta) \notin \ker \delta_{g(\beta)}$ for $\beta > 0$, and $\beta \mapsto h(\beta)$ does not evolve in the affine slices $\mathcal{S}_{g(\beta)} \times [0, \beta^*]$.

Remark 4.3 If we decompose $h(\beta) \in \ker \bigcirc_d$ according to the $g(\beta)$–dependent Berger–Ebin splitting

$$h_{ab}(\beta) = h^T_{ab}(\beta) + L_{w^\sharp(\beta)} \, g_{ab}(\beta) \,, \quad \nabla^a \, h^T_{ab}(\beta) = 0 \,, \qquad (4.294)$$

we immediately get from $\bigcirc_d \, h_{ab}(\beta) = 0$, and the commutation rule (4.280), the relation

$$\bigcirc_d \, h^T_{ab}(\beta) = -2 \, \delta^*_{g(\beta)} \, (\bigcirc_d \, w(\beta))^\sharp \,. \qquad (4.295)$$

This directly shows that the non–trivial part $h^T_{ab}(\beta)$ of a solution of the linearized Ricci flow $h(\beta)$ dynamically generates $\mathrm{Diff}(V^n)$ reparametrizations, $\delta^*_{g(\beta)} \, (\bigcirc_d \, w(\beta))$, of the given fiducial flow. Thus, whereas elements in $\ker (\bigcirc_d) \cap C^\infty(V^n \times \mathbb{R}, T^* V^n)$ generate a natural evolution in $\mathcal{U}_{g(\beta)} \times [0, \beta^*]$, there is no natural way of preserving the subspace $\ker \delta_{g(\beta)}$ along the forward flow (4.260) if we do not impose strong restrictions on the underlying fiducial Ricci flow $\beta \mapsto g_{ab}(\beta)$ [4, 13, 47, 48, 87].
The situation is fully reversed if we consider the conjugated flow generated on $C^\infty(V^n \times \mathbb{R}, \otimes^2 T \, V^n)$ by $\bigcirc^*_d$, since in such a case the commutation relation (4.281) immediately implies the following lemma [17] (Fig. 4.36).

Lemma 4.7 *Let $\beta \mapsto g_{ab}(\beta)$ be a Ricci flow with bounded geometry on $V^n_\beta \times [0, \beta^*]$, $\beta^* < T_0$, and let $\eta \mapsto g_{ab}(\eta)$, $\eta \doteq \beta^* - \beta$, denote the corresponding backward Ricci flow on $V^n_\eta \times [0, \beta^*]$obtained by the time reversal $\beta \mapsto \eta \doteq \beta^* - \beta$. Then $\ker \delta_{g(\eta)}$ is an invariant subspace for $\bigcirc^*_d$ along $\eta \mapsto g_{ab}(\eta)$, i.e.,*

$$\bigcirc^*_d \left(\ker \delta_{g(\eta)} \right) \subset \ker \delta_{g(\eta)} \,, \qquad (4.296)$$

In particular, if $\eta \mapsto H(\eta)$ with $H(\eta = 0) \in \ker \delta_{g(\eta=0)}$ is a flow solution of the parabolic initial value problem $\bigcirc^*_d \, H(\eta) = 0$ on $V^n \times [0, \beta^*]$, i.e.

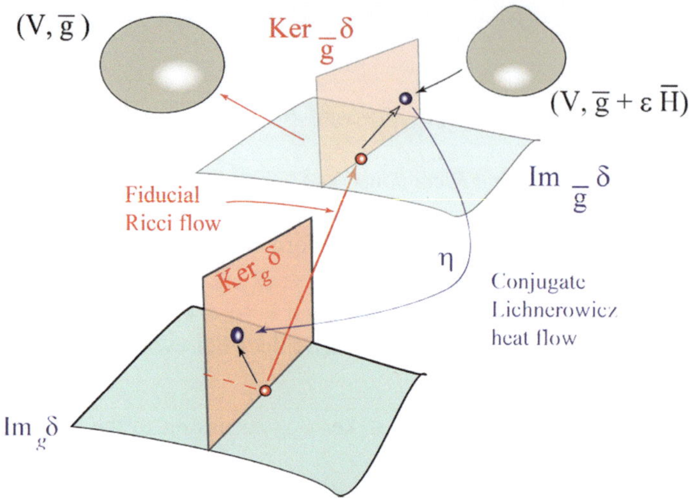

Fig. 4.36 The backward evolution along the conjugate linearized Ricci flow (in the HDRL representation) naturally preserves metric perturbations which are in ker $\delta_{\overline{g}}$

$$\frac{\partial}{\partial\eta} H^{ab} = \Delta_d H^{ab} - R H^{ab} \,,$$

$$H^{ab}(\eta = 0) = H^{ab} \in C^{\infty}(V^n, \otimes^2 T V^n) \cap \ker \delta_g \,,$$

(4.297)

then $\eta \mapsto H(\eta) \in \mathscr{S}_{g(\eta)} \cap \mathscr{T}_{(V^n, g(\eta))} \mathscr{M}et(V^n) \times [0, \beta^*]$.

Notice that, according to (4.277), the conjugate linearized Ricci flow $\eta \mapsto H^{ab}(\eta)$ is characterized by

$$\frac{d}{d\eta} \int_{V^n} h_{ab}(\eta) H^{ab}(\eta) d\mu_g(\eta) = 0 \,,$$

(4.298)

along any solution $\beta \mapsto h_{ab}(\beta)$ of the linearized Ricci flow $\bigcirc_d h(\beta) = 0$ on $(V^n, g(\beta)) \times [0, \beta^*]$, (with $\beta = \beta^* - \eta$). In particular, from the commutation relation (4.280), it follows that

$$\frac{d}{d\eta} \int_{V^n} \left(\delta^*_{g(\beta)} v^{\sharp}(\eta) \right)_{ab} H^{ab}(\eta) d\mu_g(\eta) = 0 \,,$$

(4.299)

for all $\beta \mapsto v(\beta)$, $\bigcirc_d v(\beta) = 0$. Surprisingly, these elementary properties directly imply the following strong geometric characterization of $\eta \mapsto H^{ab}(\eta)$:

Theorem 4.3 (Non-dissipative directions along the conjugate flow (see [17]))
Let $\eta \mapsto H^{ab}(\eta)$ *be a solution of the conjugate linearized Ricci flow (4.297) on* $V^n \times [0, \beta^*]$. *Then along the backward flow* $\eta \mapsto (g(\eta), H(\eta))$ *we have*

$$\frac{d}{d\eta} \int_{V^n} R_{ab}(\eta) H^{ab}(\eta) d\mu_g(\eta) = 0 \,, \qquad\qquad (4.300)$$

$$\frac{d}{d\eta} \int_{V^n} \left(g_{ab}(\eta) - 2\eta\, R_{ab}(\eta) \right) H^{ab}(\eta) d\mu_g(\eta) = 0 \,. \qquad\qquad (4.301)$$

Proof This is a direct consequence of the basic (and baffling) fact that along the Ricci flow $\beta \longmapsto g(\beta), \beta \in [0, \beta^*]$, the Ricci curvature $\mathrm{Ric}(g(\beta))$ evolves according to the forward Hodge-DeRham-Lichnerowicz heat equation

$$\frac{\partial}{\partial\beta} \mathrm{R}_{ij} = \Delta_L \mathrm{R}_{ij} = \Delta \mathrm{R}_{ij} + 2\mathrm{R}_{kijl}\mathrm{R}^{kl} - 2\mathrm{RR}_{ik}\mathrm{R}_j^k \,. \qquad\qquad (4.302)$$

In this case, (4.302) is obviously a non–linear parabolic PDE, nonetheless the conjugation condition (4.298) still holds and (4.301) immediately follows.

This result potentially plays a key role in the interaction between the Ricci flow and the renormalization group flow for $\mathrm{NL}\sigma\mathrm{M}$ since it characterizes the solutions $\eta \mapsto H(\eta)$ of the conjugate linearized Ricci flow $\bigcirc_d^* H(\eta) = 0$ as providing the directions in the space $\mathscr{M}et(V^n)$ which are non–dissipative in the $L^2(V^n, d\mu_{g(\beta)})$ sense along the Ricci flow evolution.

4.17 Perspectives

Real progress in obtaining viable extensions of the Ricci flow by exploiting the renormalization group analysis for $\mathrm{NL}\sigma\mathrm{M}$ along the lines described in the previous sections has been strongly hampered by the fact that the renormalization group flow is constructed perturbatively, under rather delicate geometrical and physical assumptions, often difficult to formalize, as we have seen. In particular, it is not clear how we should interpret the perturbative *embedding* of the Ricci flow in the renormalization group of non–linear σ models. This embedding provides a hierarchy of truncated geometric flows, generated by powers of the Riemann tensor, which are very difficult to handle and to analyze when the flows develop singularities (for a few remarkable attempts in this direction see paragraph 4.2 in [25] and the recent papers [26, 46]). Moreover, the validity of such hierarchical extension of the Ricci flow is biased by a *RG linearization stability problem* since it is not obvious in what sense a perturbative expansion approximates a full–fledged renormalization flow. These issues indicate that the idea of directly modeling Ricci flow extensions after the indications of perturbative quantum field theory is perhaps too naive in a geometric flow setting. For instance, the development of singularities, which plays a major role in Ricci flow theory, gives rise to phenomena which are not under control in a perturbative renormalization group. Hence, problems which are rather naturally addressed in Ricci flow theory become challenging issues in quantum field theory, ultimately questioning the very role that a renormalization group approach may have in extending Ricci flow.

Recently we have proposed [19] a way around these issues by showing that Wasserstein geometry and optimal transport offer an ideal framework for addressing the relation between Ricci flow and renormalization group from a novel point of view. The idea that Wasserstein geometry may play a central role also for renormalization group theory is not surprising in view of the recent work of Sturm, Lott–Villani, and Gigli–Mantegazza on Ricci flow in a metric space setting [41, 66, 67, 79, 85], nonetheless the explicit formulation of the non–linear σ model and of its relation with Ricci flow theory in terms of Wasserstein geometry is highly non trivial. It requires a delicate analysis of the geometry of dilatonic non–linear σ models in terms of maps between Riemannian surfaces (M, γ) and Riemannian measure spaces $(V^n, g, d\omega)$, where, as we have seen, the measure $d\omega$ describes the dilaton field. Several surprising phenomena appear in such a Wasserstein setting. The most important example is provided by a naturally induced scale–dependent renormalization map of the model [19]. This map is defined by the heat kernel embedding of $(V^n, g, d\omega)$ into the Wasserstein space of probability measures $(\text{Prob}(V^n, g), d_g^w)$ over (V^n, g), where d_g^w denotes the (quadratic) Wasserstein distance in $\text{Prob}(V^n, g)$. The geometrical deformation of $(V^n, g, d\omega)$ induced by such heat kernel embedding mimics a renormalization group flow for the corresponding dilatonic non–linear σ model, thus providing a mathematically well–defined prescription for a non–perturbative RG flow for the theory. It is possible to prove [19] that the RG scaling induced by the heat kernel embedding gives rise to a natural extension of the Hamilton–Perelman version of the Ricci flow with important monotonicity and gradient flow properties which are genuine extensions of the analogous properties in Ricci flow theory. The whole construction is based on a detailed analysis of the weighted heat kernel embedding and of its geometry in $(\text{Prob}(V^n, g), d_g^w)$. For further details we refer the interested reader to the original paper [19].

References

1. Alvarez-Gaume, L., Freedman, D.Z., Mukhi, S.: The background field method and the ultraviolet structure of supersymmetric nonlinear sigma model. Ann. Phys. **134**, 85–109 (1981)
2. Anderson, G., Chow, B.: A pinching estimate for solutions of the linearized Ricci flow system on 3–manifolds. Calc. Var. Partial Differ. Equ. **23**, 1–12 (2005)
3. Aubin, T.: Métriques riemanniennes et courbure. J. Differ. Geom. **4**(4), 383–519 (1970)
4. Avez, A.: Le Laplacien de Lichnerowicz sur les tenseurs. C. R. Acad. Sci. Paris. Ser. A Math. **284**, 1219–1220 (1977)
5. Bakas, I.: Geometric flows and (some of) their physical applications. In: AvH Conference Advances in Physics and Astrophysics of the 21st Century, Varna, 6–11 Sept 2005. hep-th/0511057
6. Bakas, I.: Renormalization group equations and geometric flows. arXiv:hep-th/0702034
7. Bakas, I., Sourdis, C.: Dirichlet sigma models and mean curvature flow. JHEP **0706**, 057 (2007). arXiv:0704.3985

8. Bourguignon, J.-P.: L'espace des structures riemanniennes d'une variété compacte: application á la courbure des produits de variétés: stratification par le groupe des isométries. Thesis–Paris VII (1974)

9. Bourguignon, J.-P.: Ricci curvature and Einstein metrics. In: Ferus, D., et al. (eds.) Global Differential Geometry and Global Analysis. Lecture Notes in Mathematics, vol. 838, pp. 42–63. Springer, Berlin (1981)

10. Bryant, R.L.: Gradient Kähler Ricci solitons. In: Géométrie différentielle, physique mathématique, mathématiques et société. I., Astérisque, vol. 321, pp. 51–97 (Spring, 2008). [math.DG/0407453]

11. Bryant, R.L.: Ricci flow solitons in dimension three with SO(3)-symmetries. Available at http://www.math.duke.edu/~bryant/3DRotSymRicciSolitons.pdf

12. Buser, P., Karcher, H.: Gromov's almost flat manifolds. Asterisque, vol. 81. Societe Mathematique de France, Paris (1981)

13. Buzzanca, C.: The Laplacian of Lichnerowicz on tensors. Boll. Un. Mat. Ital. B 3(6), 531–541 (1984)

14. Cao, H.-D.: Existence of gradient Kähler–Ricci solitons. In: Elliptic and parabolic methods in geometry, pp. 1–16. A. K. Peters, Wellesley (1996)

15. Cao, H.-D.: Geometry of Ricci solitons. Chin. Ann. Math. $27\mathbf{B}$(2), 121–142 (2006)

16. Cappelli, A., Friedan, D., Latorre, J.I.: c-theorem and spectral representation. Nucl. Phys. B $\mathbf{352}$, 616–670 (1991)

17. Carfora, M.: The conjugate linearized Ricci flow on closed 3-manifolds. Ann. Scuola Normale Superiore di Pisa (Cl. Sci. (5)) $\mathbf{VIII}$, 681–724 (2009). arXiv:0710.3342

18. Carfora, M.: Renormalization group and the Ricci flow. In: 150 Years of Riemann Hypothesis. Special Issue of Milan J. Math. $\mathbf{78}$, 319–353 (2010). arXiv:1001.3595

19. Carfora, M.: The Wasserstein geometry of non linear σ models and the Hamilton–Perelman Ricci flow. Rev. Math. Phys. $\mathbf{29}$, 1750001 (2017) [71 pages]. https://doi.org/10.1142/S0129055X17500015 (an extended version is available at arXiv: 1405.0827)

20. Carfora, M., Marzuoli, A.: Model geometries in the space of Riemannian structures and Hamilton's flow. Class. Quantum Grav. $\mathbf{5}$, 659–693 (1988)

21. Chang, S-Y.A., Gursky, M.J., Yang, P.: Conformal invariants associated to a measure. Proc. Natl. Acad. Sci. USA (PNAS) $\mathbf{103}$, 2535–2540 (2006)

22. Chow, B., Knopf, D.: The Ricci Flow: An Introduction. Mathematical Surveys and Monographs, vol. 110. American Mathematical Society, Providence (2004)

23. Chow, B., Lu, P., Ni, L.: Hamilton's Ricci Flow. Graduate Studies in Mathematics, vol. 77. American Mathematical Society, Providence (2007)

24. Chow, B., Chu, S.-C., Glickenstein, D., Guenther, C., Isenberg, J., Ivey, T., Knopf, D., Lu, P., Ni, L.: The Ricci Flow: Techniques and Applications: Part I: Geometric Aspects. Mathematical Surveys and Monographs, vol. 135. American Mathematical Society, Providence (2007)

25. Chow, B., Chu, S.-C., Glickenstein, D., Guenther, C., Isenberg, J., Ivey, T., Knopf, D., Lu, P., Ni, L.: The Ricci Flow: Techniques and Applications: PartIII: Geometric-Analytic Aspects. Mathematical Surveys and Monographs, vol. 163. American Mathematical Society, Providence (2010)

26. Cremaschi, L., Mantegazza, C.: Short-Time Existence of the Second Order Renormalization Group Flow in Dimension Three (2013). arXiv:1306.1721v1 [math.AP]

27. Derdzinski, A.: Compact Ricci solitons (2009, preprint)

28. DeTurck, D.: Deforming metrics in the direction of their Ricci tensor. J. Differ. Geom. $\mathbf{18}$, 157–162 (1983)

29. Douglas, M.R.: Spaces of Quantum Field Theories. Lectures at Erice (2009, unpublished)

30. Driver, B.K.: A Cameron-Martin type quasi-invariance theorem for Brownian motion on a compact manifold. J. Funct. Anal. $\mathbf{110}$, 272–376 (1992)

31. Eells, J., Fuglede, B.: Harmonic Maps Between Riemannian Polyhedra. Cambridge Tracts in Mathematics, vol. 142. Cambridge University Press, Cambridge (2001)

32. Eminenti, M., La Nave, G., Mantegazza, C.: Ricci solitons: the equation point of view. Manuscr. Math. $\mathbf{127}$(3), 345–367 (2008)

33. Feldman, M., Ilmanen, T., Knopf, D.: Rotationally symmetric shrinking and expanding gradient Kähler-Ricci solitons. J. Differ. Geom. **65**(2), 169–209 (2003)
34. Fradkin, E.S., Tseytlin, A.A.: Effective field theory from quantized strings. Phys. Lett. B **158**, 316 (1985)
35. Fradkin, E.S., Tseytlin, A.A.: Quantum string theory effective action. Nucl. Phys. B **261**, 1 (1985)
36. Friedan, D.: Nonlinear models in $2 + \epsilon$ dimensions. Ph.D. thesis (Berkeley). LBL-11517, UMI-81-13038, Aug 1980, 212pp.
37. Friedan, D.: Nonlinear models in $2 + \epsilon$ dimensions. Phys. Rev. Lett. **45**, 1057 (1980)
38. Friedan, D.: Nonlinear models in $2 + \epsilon$ dimensions. Ann. Phys. **163**, 318–419 (1985)
39. Gawedzki, K.: Conformal field theory. In: Deligne, P., Etingof, P., Freed, D.D., Jeffrey, L.C., Kazhdan, D., Morgan, J.W., Morrison, D.R., Witten, E. (eds.) Quantum Fields and Strings: A Course for Mathematicians. Institute for Advanced Study, vol. 2. AMS (1999)
40. Gegenberg, J., Suneeta, V.: The fixed points of RG flow with a Tachyon. JHEP **0609**, 045 (2006). arXiv:hep-th/0605230
41. Gigli, N., Mantegazza, C.: A flow tangent to the Ricci flow via heat kernels and mass transport (2012). arXiv:1208.5815v1
42. Gimre, K., Guenther, C., Isenberg, J.: Second-order renormalization group flow of three-dimensional homogeneous geometries (2012). arXiv:1205.6507v1 [math.DG]
43. Giveon, A., Halpern, M.B., Kiritsis, E.B., Obers, N.A.: Exact C-function and C-theorem on affine Virasoro space. Nucl. Phys. **B357**, 655–690 (1991)
44. Grigor'yan, A., Heat kernels on weighted manifolds and applications. In: The Ubiquitous Heat Kernel. Contemporary Mathematics, vol. 398, pp 93–191. American Mathematical Society, Providence (2006)
45. Gromov, M.: Metric Structures for Riemannian and Non-Riemannian Spaces. Progress in Mathematics, vol. 152. Birkhäuser, Boston (1999)
46. Guenther, C., Oliynyk, T.A.: Renormalization group flow for nonlinear sigma models. Lett. Math. Phys. **84**, 149–157 (2008). arXiv:0810.3954
47. Guenther, C., Isenberg, J., Knopf, D.: Linear stability of homogeneous Ricci solitons. arXiv:math.DG/0606793v4
48. Guenther, C., Isenberg, J., Knopf, D.: Stability of the Ricci flow at Ricci–flat metrics. Commun. Anal. Geom. **10**, 741–777 (2002)
49. Günther, M.: On the perturbation problem associated to isometric embeddings of Riemannian manifolds. Ann. Glob. Anal. Geom. **7**, 69–77 (1989)
50. Hamilton, R.S.: Three-manifolds with positive Ricci curvature. J. Diff. Geom. **17**, 255–306 (1982)
51. Hamilton, R.S.: The Ricci flow on surfaces. Contemp. Math. **71**, 237–261 (1988)
52. Hamilton, R.S.: The Formation of Singularities in the Ricci Flow. Surveys in Differential Geometry, vol. 2, pp. 7–136. International Press, Cambridge (1995)
53. Hamilton, R.S., Isenberg, J.: Quasi-convergence of Ricci flow for a class of metrics. Commun. Anal. Geom. **1**, 543–559 (1993)
54. Hélein, F.: Régularité des applications faiblement harmoniques entre una surface et une variété riemanienne. C.R. Acad. Sci. Paris **312**, 591–596 (1991)
55. Hélein, F., Wood, J.C.: Harmonic maps. In: Krupka, D., Saunders, D. (eds.) Handbook of Global Analysis. Elsevier, Amsterdam/Oxford (2007)
56. Honerkamp, J.: Chiral multiloops. Nucl. Phys. **B36**, 130–140 (1972)
57. Ivey, T.: Ricci solitons on compact three–manifolds. Diff. Geom. Appl. **3**, 301–307 (1993)
58. Jost, J.: Riemannian Geometry and Geometric Analysis, 2nd edn. Springer, Berlin/Heidelberg (1998)
59. Kiritis, E.: String Theory in a Nutshell. Princeton University Press, Princeton (2007)
60. Kiritsis, E., Nitti, F., Pimenta, L.S.: Exotic RG Flows from Holography (2016). arXiv:1611.05493

61. Koiso, N.: On rotationally symmetric Hamilton's equation for Kähler–Einstein metrics. In: Ochiai, T. (ed.) Recent Topics in Differential and Analytic Geometry. Advanced Studies in Pure Mathematics, vol. 18. Academic Press, Boston (1990)
62. Komargodski, Z., Schwimmer, A.: On renormalization group flows in four dimensions. J. High Energy Phys. **1112**, 099 (2011). arXiv:1107.3987
63. Korevaar, N.J., Schoen, R.M.: Sobolev spaces and harmonic maps for metric space targets. Commun. Anal. Geom. **1**, 561–659 (1993)
64. Leandre, R.: Stochastic Wess-Zumino-Novikov-Witten model on the torus. J. Math. Phys. **44**, 5530–5568 (2003)
65. Lott, J.: Renormalization group flow for general sigma models. Commun. Math. Phys. **107**, 165–176 (1986)
66. Lott, J.: Optimal transport and Ricci curvature for metric-measure spaces. In: Cheeger, J., Grove, K. (eds.) Metric and Comparison Geometry. Surveys in Differential Geometry, vol. 11, pp. 229–257. International Press, Somerville (2007)
67. Lott, J., Villani, C.: Ricci curvature for metric-measure spaces via optimal transport. Ann. Math. **169**, 903–991 (2009)
68. Morrey, C.B.: Multiple Integrals in the calculus of variations. Grundlehren der mathematischen Wissenschaften, vol. 130. Springer, New York (1966)
69. Moser, J.: On the volume elements on a manifold. Trans. Am. Math. Soc. **120**, 286–294 (1965)
70. Müller, R.: Differential Harnack Inequalities and the Ricci Flow. Series of Lecture in Mathematics. European Mathematical Society, Zurich (2006)
71. Oliynyk, T., Suneeta, V., Woolgar, E.: A gradient flow for worldsheet nonlinear sigma models. Nucl. Phys. **B739**, 441–458 (2006). hep-th/0510239
72. Perelman, G.: The entropy formula for the Ricci flow and its geometric applications. math.DG/0211159
73. Perelman, G.: Ricci flow with surgery on three-manifolds. math.DG/0303109
74. Perelman, G.: Finite extinction time for the solutions to the Ricci flow on certain three-manifolds. math.DG/0307245
75. Petersen, P., Wylie, W.: Rigidity of gradient Ricci solitons. Pac. Math. J. **241**, 329–345 (2009)
76. Polyakov, A.M.: Interaction of Goldstone particles in two dimensions. Applications to ferromagnets and massive Yang-Mills fields. Phys. Lett. **B59**, 79 (1975)
77. Schoen, R., Uhlenbeck, K.: The Dirichlet problem for harmonic maps. J. Differ. Geom. **18**, 253–268 (1983)
78. Schwartz, J.T.: Non Linear Functional Analysis. Gordon and Breach, New York (1969)
79. Sturm, K-T.: A curvature–dimension condition for metric measure spaces. C. R. Sci. Paris Ser. I **342**, 197–200 (2006)
80. Tao, T.: Poincaré's Legacies, Part II. American Mathematical Society, Providence (2009)
81. Taubes, C.H.: Constructions of measures and quantum field theories on mapping spaces. J. Differ. Geom. **70**, 23–58 (2005)
82. Tseytlin, A.A.: Conformal anomaly in two-dimensional sigma model on curved background and strings. Phys. Lett. **178B**, 34 (1986)
83. Tseytlin, A.A.: Sigma model Weyl invariance conditions and string equations of motion. Nucl. Phys. B **294**, 383 (1987)
84. Tseytlin, A.A.: On sigma model RG flow, "central charge" action and Perelman's entropy. Phys. Rev. **D75**, 064024 (2007). arXiv:hep-th/0612296
85. von Renesse, M.-K., Sturm, K-T.: Transport inequalities, gradient estimates, entropy and Ricci curvature. Commun. Pure Appl. Math. **58**, 923–940 (2005)
86. Weitsman, J.: Measures on Banach manifolds and supersymmetric quantum field theories. Commun. Math. Phys. **277**, 101–125 (2008)
87. Ye, R.: Ricci flow, Einstein metrics and space forms. Trans. Am. Math. Soc. **338**, 871–896 (1993)
88. Zamolodchikov, A.B.: Irreversibility of the flux of the renormalization group in a 2-D field theory. JEPT Lett **43**, 730 (1986)

Chapter 5
The Quantum Geometry of Polyhedral Surfaces: Variations on Strings and All That

This chapter is an introduction to the basic ideas of 2-dimensional quantum field theory and non-critical strings. This is classic material which nevertheless proves useful for illustrating the interplay between quantum field theory, the moduli space of Riemann surfaces, and the properties of polyhedral surfaces which are the *leitmotiv* of this LNP. At the root of this interplay lies 2D quantum gravity. It is well known that such a theory allows for two complementary descriptions: on the one hand we have a conformal field theory (CFT) living on a 2D world-sheet, a description that emphasizes the geometrical aspects of the Riemann surface associated with the world-sheet; while on the other, the theory can be formulated as a statistical critical field theory over the space of polyhedral surfaces (dynamical triangulations). We show that many properties of such 2D quantum gravity models are related to a geometrical mechanism which allows one to describe a polyhedral surface with N_0 vertices as a Riemann surface with N_0 punctures dressed with a field whose charges describe discretized curvatures (related to the deficit angles of the triangulation). Such a picture calls into play the (compactified) moduli space of genus g Riemann surfaces with N_0 punctures $\mathfrak{M}_{g;N_0}$, and allows one to prove that the partition function of 2D quantum gravity is directly related to computation of the Weil–Petersson volume of $\mathfrak{M}_{g;N_0}$. By exploiting the large N_0 asymptotics of such Weil–Petersson volumes, characterized by Manin and Zograf, it is then easy to relate the anomalous scaling properties of pure 2D quantum gravity, the KPZ exponent, to the Weil–Petersson volume of $\mathfrak{M}_{g;N_0}$. We also show that polyhedral surfaces provide a natural kinematical framework within which we can discuss open/closed string duality. A basic problem in such a setting is to provide an explanation of how open/closed duality is generated dynamically, and in particular how a closed surface is related to a corresponding open surface, with gauge-decorated boundaries, in such a way that the quantization of this correspondence leads to an open/closed duality. In particular, we show that from a closed polyhedral surface we naturally get an open hyperbolic surface with geodesic boundaries. This gives a geometrical mechanism describing the transition between closed and open surfaces. Such a

© Springer International Publishing AG 2017

M. Carfora, A. Marzuoli, *Quantum Triangulations*, Lecture Notes in Physics 942,
https://doi.org/10.1007/978-3-319-67937-2_5

correspondence is promoted to the corresponding moduli spaces: $\mathfrak{M}_{g;N_0} \times \mathbb{R}_+^N$, the moduli spaces of N_0-pointed closed Riemann surfaces of genus g whose marked points are decorated with the given set of conical angles, and $\mathfrak{M}_{g;N_0}(L) \times \mathbb{R}_+^{N_0}$, the moduli spaces of open Riemann surfaces of genus g with N_0 geodesic boundaries decorated by the corresponding lengths. Such a correspondence provides a nice kinematical setup for establishing an open/closed string duality, by exploiting the results by M. Mirzakhani on the relation between intersection theory over $\mathfrak{M}(g; N_0)$ and the geometry of hyperbolic surfaces with geodesic boundaries. The results in this chapter connect directly with many deep issues in 3D geometry, ultimately relating to the volume conjecture in hyperbolic geometry and to the role of knot invariants.

5.1   Introduction

As we have seen in Chap. 4, in the analysis of the quantum geometry of polyhedral surfaces a prominent role is played by the non–linear σ–model action

$$S[\gamma, \phi; \alpha] := \frac{1}{4\pi \, l_s^2} \int_M \left[tr_{\gamma(x)} \left(\phi^* \, g \right) \right] d\mu_\gamma, \tag{5.1}$$

where, for later convenience, we have expressed the coupling parameter a in terms of the *string length* $l_s := \sqrt{\frac{a}{4\pi}}$, and α stands for the metric coupling $\frac{g}{4\pi \, l_s^2}$. This very action and its deformations[28, 29], associated with the dilatonic $f(\phi)$ and a tachyonic[1] $U(\phi)$ coupling, e.g.

$$S[\gamma, \phi; g, f, U] = \frac{1}{4\pi \, l_s^2} \int_M \gamma^{\mu\nu} \partial_\mu \phi^i \partial_\nu \phi^j \, g_{ij} \, d\mu_\gamma + \frac{1}{4\pi \, l_s^2} \int_M (4\pi \, l_s^2) f(\phi) \, \mathscr{K}_\gamma \, d\mu_\gamma$$

$$+ \frac{1}{4\pi \, l_s^2} \int_M U(\phi) \, d\mu_\gamma,$$

play a basic role in (bosonic) string theory where (5.1) is known as *Polyakov* action. The connection between the quantum geometry of polyhedral surfaces and string theory is multifaceted, and in order to disclose it we need to discuss some further aspects of the quantum theory associated with (5.1).

---

[1]The concept of tachyon is slightly misleading in non–critical string theory, where low dimensions are sampled and the usual instability associated with tachyons in critical strings is not present. For details see Vol I, Chap. 9 of [57].

5.2   The Weyl Anomaly and Liouville Action

According to (4.103), the partition function over $Map(M, V^n)$ associated with (5.1) is provided by

$$Z[\gamma; \alpha] := \int_{\{Map(M, V^n)\}} D_\alpha[\phi]\, e^{-S[\gamma, \phi; \alpha]}. \tag{5.2}$$

String theory[2] extends this QFT picture by promoting also the two–dimensional metric γ over the surface M to a dynamical field, and the quantization of the action (5.1) is characterized by the functional integration of $Z[\gamma; \alpha]$ over the space of Riemannian structures $\mathcal{M}et(M)/\mathcal{D}iff(M)$ associated with a surface M_g of given genus[3] g, and then by summing over all topologies i.e.,

$$Z[\alpha] := \sum_{g=0}^{\infty} \int_{\mathcal{M}et(M_g)/\mathcal{D}iff(M_g)} D[\mathcal{M}/\mathcal{D}]_g \; Z[\gamma; \alpha], \tag{5.3}$$

where we have denoted by $D[\mathcal{M}/\mathcal{D}]_g$ a formal measure on $\mathcal{M}et(M_g)/\mathcal{D}iff(M_g)$.

The starting point for discussing the basic (perturbative) properties of (5.3), at least at fixed surface genus g, is the observation that the space of Riemannian metrics $\mathcal{M}et(M)$ on a 2-dimensional Riemannian manifold M, (working at fixed genus we drop the suffix g from notation), can be decomposed by means of a local slice for the action of the group of confeomorphisms $W(M) \ltimes \mathcal{D}iff(M)$, where $W(M) := C^\infty(M, \mathbb{R}_+)$ denotes the group of Weyl rescalings over M, and $\ltimes$ denotes the semidirect product. In particular, for $u \in C^\infty(M, \mathbb{R})$, there is a natural projection map

$$\pi : W(M) \ltimes \mathcal{D}iff(M) \;\rightarrow\; \mathcal{C}\mathcal{O}_{\widehat{\gamma}} \tag{5.4}$$

$$(e^u, \phi) \qquad \longmapsto \pi(e^u, \phi) := e^{2u}(\phi^* \widehat{\gamma}),$$

where $\mathcal{C}\mathcal{O}_{\widehat{\gamma}}$ is the $W(M) \ltimes \mathcal{D}iff(M)$–orbit in $\mathcal{M}et(M)$ of a given metric $\widehat{\gamma} \in \mathcal{M}et(M)$, and $\phi^* \widehat{\gamma}$ is the pull–back of $\widehat{\gamma}$ under $\phi \in \mathcal{D}iff(M)$. Note that

$$u \in C^\infty(M, \mathbb{R}) := \mathfrak{w}(M), \tag{5.5}$$

can be interpreted as an element of the Lie algebra $\mathfrak{w}(M)$ of $W(M)$.

---

[2]For the elementary notion of string theory we shall exploit we refer freely to the excellent presentation in [40, 57], and [38].

[3]There are clearly too many g's around, however no confusion should arise since it will be always clear from the context when we are dealing with the genus g of the surface M or with the running metric g of the target manifold V^n.

Let us also recall that the tangent space $\mathscr{T}_{(M,\gamma)}\mathscr{M}et(M)$ to $\mathscr{M}et(M)$ at (M,γ) is isomorphic to the space of (smooth) symmetric bilinear forms $C^\infty(M, \otimes^2 T^* M)$ over M, endowed with the pre–Hilbertian L^2 inner product

$$(U, V)_{L^2(M, d\mu_\gamma)} \doteq \int_M \langle U, V \rangle_\gamma d\mu_\gamma, \tag{5.6}$$

where $\langle U, V \rangle_\gamma$ is the pointwise γ–metric in $\otimes^2 T^*M$.

From (5.4) it follows that $\mathscr{T}_{(M,\gamma)}\mathscr{C}\mathscr{O}_{\widehat{\gamma}}$, the tangent space to the orbit $\mathscr{C}\mathscr{O}_\gamma$, is the image of the injective operator with closed range

$$P_1 : C^\infty(M, TM) \rightarrow C^\infty(M, \otimes^2 T^*M) \tag{5.7}$$

$$w \qquad \mapsto P_1(w) \doteq \pounds_w \gamma,$$

where $\pounds_w$ denotes the conformal Lie derivative along the vector field w

$$(\pounds_w \gamma)_{ab} := \nabla_a w_b + \nabla_b w_a - \nabla^c w_c. \tag{5.8}$$

Standard elliptic theory implies that the $L^2(M, d\mu_\gamma)$–orthogonal subspace to $Im\, P_1$ in $\mathscr{T}_{(M,\gamma)}\mathscr{M}et(M)$, is spanned by the kernel[4] of the L^2 adjoint $P_1^\dagger$ of P_1,

$$P_1^\dagger : C^\infty(M, \otimes^2 T^* M) \rightarrow C^\infty(M, T^* M) \tag{5.9}$$

$$h \qquad \mapsto P_1^\dagger h \doteq -2\,\gamma^{ij}\, \nabla_i h_{jk}\, dx^k.$$

This entails the well–known Berger–Ebin $L^2(M, d\mu_\gamma)$–orthogonal splitting [8, 21] (see Appendix A) of the tangent space $\mathscr{T}_{(M,\gamma)}\mathscr{M}et(M)$,

$$\mathscr{T}_{(M,\gamma)}\mathscr{M}et(M) \cong \left[\mathscr{T}_{(M,\gamma)}\mathscr{M}et(M) \cap Ker\, P_1^\dagger \right] \oplus Im\, P_1 \left[C^\infty(M, TM) \right] \tag{5.10}$$

$$\oplus\ Im\ tr_\gamma \left[C^\infty(M, \otimes^2 T^* M) \right],$$

according to which, for any given 2–tensor $h \in \mathscr{T}_{(M,\gamma)}\mathscr{M}et(M)$, we can write

$$h_{ab} = h_{ab}^{TT} + \pounds_w \gamma_{ab} + \left(tr_\gamma h \right) \gamma_{ab}, \tag{5.11}$$

where h_{ab}^{TT} denotes the *div*–free and trace–free part of h, ($\nabla^a h_{ab}^T = 0$, $tr_\gamma h^{TT} = 0$), and where the vector field w is characterized as the solution, (unique up to the conformal Killing vector fields of (M, γ)), of the elliptic PDE $P_1^\dagger P_1 w = P_1^\dagger h$. For a given $\gamma \in \mathscr{M}et(M)$, tensor fields $h_{ab}^{TT} \in \mathscr{T}_{(M,\gamma)}\mathscr{M}et(M)$ parametrize non–trivial infinitesimal deformations of the conformal class $[\gamma]$, whereas $\pounds_w \gamma_{ab}$ provide deformations tangent to the $\mathscr{D}iff(M) \ltimes W(M)$–orbit through $[\gamma]$.

---

[4]According to Riemann–Roch theorem dim $Ker\, P_1 - $ dim $Ker\, P_1^\dagger = 6 - 6g$. Since dim $Ker\, P_1 = 6$ for the sphere, $= 2$ for the torus, $= 0$ for surfaces with $g \geq 2$, it follows that in the case of surfaces dim $Ker\, P_1^\dagger$ is finite–dimensional, and is given by $= 0, 2, 6g - 6$ when $g = 0, 1, \geq 2$, respectively.

If we identify $\frac{\mathcal{M}et(M)}{\mathcal{D}iff(M)\ltimes W(M)}$ with the moduli space $\mathfrak{M}_g(M)$ of genus g Riemann surfaces, then the above properties of the action of $W(M)\ltimes\mathcal{D}iff(M)$ on $\mathcal{M}et(M)$ allow to parametrize locally a generic metric $\gamma\in\mathcal{M}et(M)$ by means of a diffeomorphism $\psi\in\mathcal{D}iff(M)$, a Weyl rescaling $f:=e^{2u}\in W(M)$, and $3g-3$ complex parameters $\{\mu_\kappa\}$ varying in $\mathfrak{M}_g(M)$, i.e.

$$\wp:\mathfrak{M}_g\times(W(M)\ltimes\mathcal{D}iff(M))\to\mathcal{M}et(M) \tag{5.12}$$

$$(\widehat{\gamma}_{\alpha\beta}(\mu_\kappa),e^{2u},\psi)\longmapsto\wp(\widehat{\gamma}_{\alpha\beta}(\mu_\kappa),e^{2u},\psi)\doteq\gamma_{\alpha\beta}=e^{2u}\left(\psi^*\widehat{\gamma}(\mu_\kappa)\right)_{\alpha\beta},$$

where $\widehat{\gamma}_{\alpha\beta}(\mu_\kappa)$ are the components of the reference metric on the surface M whose conformal class defines the point $\{\mu_\kappa\}$ in $\mathfrak{M}_g$ we are considering, (e.g., a constant curvature metric). Moreover, if $D[\gamma]$ denotes a formal measure on $\mathcal{M}et(M)$ then, its pull–back on $\mathfrak{M}_g(M)\times\mathcal{D}iff(M)\times W(M)$ under the slice map (5.12) can be expressed[5] as [15, 16]

$$\wp^*(D[\gamma])=\left[\det{}'\left(P_1^\dagger P_1\right)\right]^{\frac{1}{2}}_{\gamma=e^{2u}(\psi^*\widehat{\gamma}(\mu_\kappa))}d\omega_{WP}[\mu_k]\,D_{\widehat{\gamma}}[u]\,D[\psi], \tag{5.13}$$

where $d\omega_{WP}[\mu_k]$ denotes the Weil–Petersson measure on the Riemann moduli space $\mathfrak{M}_g(M)$, whereas $D[\psi]$ and $D_{\widehat{\gamma}}[u]$ denote formal functional measure over (the tangent spaces to) $\mathcal{D}iff(M)$ and $W(M)$, respectively. The Jacobian of the slice map (5.12), (the Faddev–Popov determinant), is represented in (5.13) by $\det{}'\left(P_1^\dagger P_1\right)$, the ζ-regularized determinant, (restricted to the non-zero modes), of the elliptic operator

$$P_1^\dagger P_1\;:C^\infty(M,TM)\longrightarrow C^\infty(M,TM) \tag{5.14}$$

$$w^a\longmapsto(P_1^\dagger P_1)^b_a\,w^a=-2\,\nabla_a\left(\nabla^a w^b+\nabla^b w^a-\gamma^{ab}\nabla_c w^c\right),$$

acting on vector fields.

The determinant $\det{}'\left(P_1^\dagger P_1\right)$ does not transform equivariantly under the conformal rescaling $\widehat{\gamma}\mapsto e^{2u}(\psi^*\widehat{\gamma}(\mu_\kappa))$, and we have [6, 15, 16]

$$\left[\det{}'\left(P_1^\dagger P_1\right)\right]^{\frac{1}{2}}_{\gamma=e^{2u}(\psi^*\widehat{\gamma}(\mu_\kappa))}d\omega_{WP}[\mu_k]\,D_{\widehat{\gamma}}[u]\,D[\psi] \tag{5.15}$$

---

[5]We are restricting for simplicity to the case of surface genus $g\geq 2$. In the case $g=0,1$, (5.13) acquires finite–dimensional determinants related to the presence of conformal Killing vectors. A careful analysis of the whole subject is presented in [15, 16].

$$= \left[\det{}' \left(P_1^\dagger P_1\right)\right]^{\frac{1}{2}}_{\widehat{\gamma}(\mu_\kappa)} \; e^{-26\,S_L(\widehat{\gamma},u)} \; d\omega_{WP}\,[\mu_k] \; D_{\widehat{\gamma}}\,[u]\;D\,[\psi]\,,$$

where

$$S_L(\widehat{\gamma},u) = \frac{1}{48\pi}\int_M d\mu_{\widehat{\gamma}} \left[\frac{1}{2}\widehat{\gamma}^{\,\mu\nu}\partial_\mu\,u\partial_\nu\,u + R(\widehat{\gamma})\,u\right] + \Gamma_0^2\int_M d\mu_{\widehat{\gamma}}\,e^{\beta u}, \qquad (5.16)$$

is the Liouville action,[6] (the parameters Γ_0 and β associated with the potential term $e^{\beta u}$ depend upon the procedure used for regularizing the functional measure (5.15)). We also have [16]

$$\left[\frac{\det{}'\,\Delta}{A(M)}\right]^{\frac{1}{2}}_{\gamma=e^{2u}(\psi*\widehat{\gamma}(\mu_\kappa))} = \left[\frac{\det{}'\,\Delta}{A(M)}\right]^{\frac{1}{2}}_{\widehat{\gamma}(\mu_\kappa)} \; e^{-\,S_L(\widehat{\gamma},u)}, \qquad (5.17)$$

where Δ and $A(M)$ respectively are the scalar Laplacian and the area of the surface M, (in the corresponding metrics γ and $\widehat{\gamma}$). If we assume that the target manifold V^n is flat, then these results, together with the standard characterization[7] of the Gaussian measure $D_\alpha[\phi]$ over $\{Map(M, V^n)\}$ [15, 58]

$$\int_{\{Map(M,V^n)\}} D_\alpha[\phi]\,e^{-\widetilde{\mathscr{S}}_\gamma[\phi;\alpha]} := \left[\frac{\det'\,\Delta_\gamma}{A_\gamma(M)}\right]^{-\frac{n}{2}}, \qquad (5.18)$$

provide the celebrated relation

$$Z[\gamma;\,\alpha]\left[\det{}'\left(P_1^\dagger P_1\right)\right]^{\frac{1}{2}}_{\gamma}\,d\omega_{WP}\,D_\gamma\,[u]\,D_\gamma\,[\psi]\Bigg|_{\gamma=e^{2u}(\psi*\widehat{\gamma})} \qquad (5.19)$$

$$= \left[\frac{\det{}'\,\Delta}{A(M)}\right]^{-\frac{n}{2}}_{\widehat{\gamma}}\left[\det{}'\left(P_1^\dagger P_1\right)\right]^{\frac{1}{2}}_{\widehat{\gamma}}\,e^{-(26-n)\,S_L(\widehat{\gamma},u)}\,d\omega_{WP}\;D_{\widehat{\gamma}}\,[u]\;D_{\widehat{\gamma}}\,[\psi]\,,$$

which, for bosonic matter $\{\phi\}$ with central charge $c_\phi = n = 26$ allows the decoupling of the Liouville field u from the path–integration, a property which characterizes critical (bosonic) string theory.

---

[6] $S_L(\widehat{\gamma},u)$ is actually the difference of the Liouville actions $S_L(\varphi)$ and $S_L(\varphi + u)$ respectively associated with the conformal metrics $\widehat{\gamma} = e^\varphi|dz|^2$ and $\gamma = e^{\varphi+u}|dz|^2$. The explicit characterization of $S_L(\varphi)$ is very delicate since e^φ is not a function but the $(1,1)$ component of the metric tensor. A thorough analysis of the subject, with the relevant references, is discussed in [64].

[7] For notational ease, we have chosen units for the fields ϕ^k such that $\frac{1}{4\pi\,l_s^2} = 1$.

5.3 Non–critical Strings and 2D Quantum Gravity

In 2D quantum gravity we are interested in keeping track of the Liouville mode since it describes the interaction of the conformal matter fields $\{\phi^k\}_{k=1}^n$ with two-dimensional gravity. According to the remarks of the previous section, this interaction is formally described by the partition function

$$\int_{\mathfrak{M}_g(M)} d\omega_{WP} \left[\frac{\det{}'\Delta}{A(M)}\right]_{\widehat{\gamma}}^{-\frac{n}{2}} \left[\det{}'\left(P_1^\dagger P_1\right)\right]_{\widehat{\gamma}}^{\frac{1}{2}} \int_{W(M)} e^{-(26-n)\,S_L(\widehat{\gamma},u)}\, D_{\widehat{\gamma}}[u]\,,$$

(5.20)

where we have dropped the inessential integration over $D_{\widehat{\gamma}}[\psi]$, (i.e., we have factored out the (∞) volume of the diffeomorphism group $\mathscr{D}iff(M)$). Note that formally (5.20) will be dominated by the classical configurations extremizing the Liouville action $S_L(\widehat{\gamma}, u)$ when the central charge $c_\phi = n$ of the matter fields $\{\phi^k\}_{k=1}^n$ tends to $-\infty$. The whole subject of *classical vs. quantum* Liouville theory is nicely discussed, with many illustrative examples, in the remarkable set of lectures [68]. Here, we stress those basic aspects which directly relate to the role of polyhedral surfaces in 2D quantum gravity.

The starting point of quantum Liouville theory is the observation that off criticality, (i.e. for $c_\phi = n \neq 26$), we need to look more carefully into the structure of the measure $D_\gamma[u]$ over the Liouville mode in (5.20). Formally, given a fiducial metric $\widehat{\gamma}$, the measure $D_{\gamma=e^{2u}(\psi*\widehat{\gamma})}[u]$ at $u \in W(M)$ is associated with the e^u–weighted L^2–inner product

$$(f, h)_{L^2(M, e^u d\mu_{\widehat{\gamma}})} := \int_M fh\, e^u\, d\mu_{\widehat{\gamma}} = \int_M fh\, d\mu_\gamma\,, \qquad f, h \in \mathscr{T}_u W(M), \qquad (5.21)$$

induced on $\mathscr{T}_u W(M)$ by the $L^2(M, d\mu_\gamma)$ pairing (5.6). It is notoriously difficult [17] to use the measure $D_{\gamma=e^{2u}(\psi*\widehat{\gamma})}[u]$ for generating the Feynman rules of the (quantum) Liouville theory defined by (5.19), a consequence of the fact that the inner product $(f, h)_{L^2(M, e^u d\mu_{\widehat{\gamma}})}$ is not invariant under the *translations*

$$u \longmapsto \widetilde{u} := u(x) - w(x),$$

(5.22)

in the Lie algebra $\mathfrak{w}(M)$, where $w(x)$ is an arbitrary given[8] function $\in C^\infty(M, \mathbb{R})$. To introduce a more manageable measure, let us consider the (exponential) mapping from the Lie algebra $\mathfrak{w}(M)$ to the group $W(M)$,

---

[8]However, (5.21) is invariant under the combined action of the above translation and of the conformal rescaling $\widehat{\gamma} \mapsto \widehat{\gamma}\, e^{2\,w(x)}$.

$$\pi_u : \mathfrak{w}(M) \longrightarrow W(M) \tag{5.23}$$

$$u \longmapsto e^u,$$

and notice that the inner product $(f,h)_{L^2(M,e^u d\mu_{\widehat{\gamma}})}$, on the tangent space $\mathscr{T}_{e^u} W(M)$ at e^u, pulls–back under such a map to the inner product on $\mathfrak{w}(M)$, (identified as the tangent space $\mathscr{T}_1 W(M)$ at the identity $e^u = 1$), defined by

$$\pi_u^* [(f,h)_{L^2(M,e^u d\mu_{\widehat{\gamma}})}] = \int_M fh \, d\mu_{\widehat{\gamma}} := (f,h)_{L^2(M,d\mu_{\widehat{\gamma}})}. \tag{5.24}$$

Let us denote by $\widetilde{D}_{\widehat{\gamma}}[u]$ the Gaussian measure over $\mathfrak{w}(M)$ associated with this latter pairing. Then, the pull–back, under the map (5.23), of the measure space $\left(W(M), D_{\gamma = e^{2u}(\psi * \widehat{\gamma})}[u] \right)$ to the measure space $\left(\mathfrak{w}(M) \simeq C^\infty(M, \mathbb{R}), \widetilde{D}_{\widehat{\gamma}}[u] \right)$, is characterized by a functional Jacobian,

$$\pi_u^* \left(D_{\gamma = e^{2u}(\psi * \widehat{\gamma})}[u] \right) = |Jac(\pi_u)| \, \widetilde{D}_{\widehat{\gamma}}[u]. \tag{5.25}$$

which, as argued at various level of mathematical rigor [11, 17, 19], is provided by a (renormalized) Liouville–type action of the form

$$S_L^{(ren)}(\widehat{\gamma},u) \longmapsto \frac{1}{4\pi} \int_M d\mu_{\widehat{\gamma}} \left[\widehat{\gamma}^{\mu\nu} \partial_\mu u \partial_\nu u + Q R(\widehat{\gamma}) u \right] + \Gamma^2 \int_M d\mu_{\widehat{\gamma}} e^{2bu}, \tag{5.26}$$

in term of parameters Q and b to be determined by requiring that the theory is independent from the (arbitrary) choice of the background metric $\widehat{\gamma}$ within a given conformal class.

The (formal) measure space $\left(\mathfrak{w}(M), e^{-S_L^{(ren)}(\widehat{\gamma},u)} \widetilde{D}_{\widehat{\gamma}}[u] \right)$ is the one typically used for discussing quantum Liouville theory, and we can rewrite the factorization of the measure in non-critical string theory, (see (5.19)), as

$$d\omega_{WP} \left[\frac{\det' \Delta}{A(M)} \right]_{\widehat{\gamma}}^{-\frac{n}{2}} \left[\det' \left(P_1^\dagger P_1 \right) \right]_{\widehat{\gamma}}^{\frac{1}{2}} e^{-S_L^{(ren)}(\widehat{\gamma},u)} \widetilde{D}_{\widehat{\gamma}}[u]. \tag{5.27}$$

To make sense of the modular integration in (5.27) we need to enforce in (5.27) the independence from the choice of a background metric $\widehat{\gamma}$, (within a given conformal class). To this end let us consider the mapping

$$\widetilde{\pi}_\xi : \mathfrak{w}(M) \times \mathscr{M}et(M) \longrightarrow \mathfrak{w}(M) \times \mathscr{M}et(M) \tag{5.28}$$

$$(\widetilde{u}, \widehat{\gamma}_{\mu\nu}) \longmapsto \left(u = \widetilde{u} - \frac{Q}{2}\xi, \widehat{\gamma}_{\mu\nu} = e^\xi \widetilde{\gamma}_{\mu\nu} \right),$$

where $\widetilde{\gamma}_{\mu\nu}$ is a new reference metric, $\xi \in C^\infty(M, \mathbb{R})$ is a smooth conformal factor, and $\widetilde{u}$ the ξ–translated u field in $\mathfrak{w}(M)$. As we shall see momentarily, it is straightforward to analyze the behavior of $S_L^{(ren)}(\hat{\gamma}, u)$ under the combined transformation (5.28). The only delicate term in (5.26) is represented by $\int_M d\mu \, \hat{\gamma} \, e^{2b u}$. Indeed, the exponential of a (path–integrated) field, $e^{2b u}$, requires a regularization which is metric–dependent and, on the reference $(M, \hat{\gamma})$, one should write $e^{2b u}$ more explicitly as

$$\left[e^{2b u} \right]_{\hat{\gamma}}, \tag{5.29}$$

with

$$\left[e^{2b u} \right]_{e^{\xi} \tilde{\gamma}} = e^{b^2 \xi} \left[e^{2b u} \right]_{\tilde{\gamma}}, \tag{5.30}$$

(see e.g. [68]). Thus, under the action of (5.28) we have that

$$\int_M d\mu_{\hat{\gamma}} \left[e^{2b u} \right]_{\hat{\gamma}} = \int_M d\mu_{\tilde{\gamma}} \, e^{(b^2 - bQ + 1)} \left[e^{2b \tilde{u}} \right]_{\tilde{\gamma}}, \tag{5.31}$$

and $\left[e^{2b u} \right]_{\hat{\gamma}}$ is a $(1, 1)$ tensor, (i.e. has conformal weight 1), iff

$$b^2 - b Q + 1 = 0 , \quad \Rightarrow \quad Q = b + b^{-1}. \tag{5.32}$$

By exploiting this latter constraint, the relation

$$R(\hat{\gamma}) \, d\mu_{\hat{\gamma}} = \left(R(\tilde{\gamma}) - \tilde{\Delta} \, \xi \right) d\mu_{\tilde{\gamma}}, \tag{5.33}$$

and the integration by parts formula

$$\int_M u \, \tilde{\Delta} \, \xi \, d\mu_{\tilde{\gamma}} = - \int_M \tilde{\gamma}^{\mu\nu} \partial_\mu u \, \partial_\nu \xi \, d\mu_{\tilde{\gamma}}, \tag{5.34}$$

where $\tilde{\Delta}$ denotes the Laplacian on $(M, \tilde{\gamma})$, it is straightforward to verify that under the combined transformation (5.28) we can write

$$S_L^{(ren)}(\hat{\gamma}, u) = S_L^{(ren)}(\tilde{\gamma}, \tilde{u}) - \frac{6 Q^2}{48 \pi} \int_M d\mu_{\tilde{\gamma}} \left[\frac{1}{2} \tilde{\gamma}^{\mu\nu} \partial_\mu \xi \partial_\nu \xi + R(\tilde{\gamma}) \, \xi \right]. \tag{5.35}$$

The structure of the Liouville action (5.16) suggests to add and subtract to this expression the quantity

$$\Gamma_0 \int_M d\mu_{\tilde{\gamma}} \left[e^{2b \tilde{u}} \right]_{\tilde{\gamma}}, \tag{5.36}$$

so that, if we shift the parameter Γ in (5.26) according to

$$\Gamma \longmapsto \Gamma - \Gamma_0, \tag{5.37}$$

we can rewrite (5.35) as

$$S_L^{(ren)}(\widehat{\gamma}, u) = S_L^{(ren)}(\widetilde{\gamma}, \widetilde{u}) - 6 Q^2 S_L(\widetilde{\gamma}, \widetilde{u}). \tag{5.38}$$

Moreover, since the measure $\widetilde{D}_{\widehat{\gamma}}[u]$ is formally invariant under the translation $u = \widetilde{u} - \frac{Q}{2}\xi$ in $\mathfrak{w}(M)$, its pull–back under $\widetilde{\pi}_\xi$ is provided by

$$\widetilde{\pi}_\xi^* \left(\widetilde{D}_{\widehat{\gamma}}[u] \right) = e^{S_L(\widetilde{\gamma},\,\widetilde{u})}\, \widetilde{D}_{\widetilde{\gamma}}[\widetilde{u}], \tag{5.39}$$

with the usual Jacobian, associated with the conformal anomaly of a free–field theory, provided by the Liouville action $S_L(\widetilde{\gamma}, \widetilde{u})$. Thus, we eventually get[9]

$$e^{-\widetilde{\pi}_\xi^*\left(S_L^{(ren)}(\widehat{\gamma},u)\right)}\, \widetilde{\pi}_\xi^* \left(\widetilde{D}_{\widehat{\gamma}}[u] \right) = e^{(1+6Q^2) S_L(\widetilde{\gamma},\,\widetilde{u})}\, e^{-S_L^{(ren)}(\widetilde{\gamma},\,\widetilde{u})}\, \widetilde{D}_{\widetilde{\gamma}}[\widetilde{u}], \tag{5.40}$$

and the pull–back of the full measure (5.27) under the mapping $\widetilde{\pi}_\xi$ is provided by

$$\widetilde{\pi}_\xi^* \left(d\omega_{WP} \left[\frac{\det{}'\Delta}{A(M)} \right]_{\widehat{\gamma}}^{-\frac{n}{2}} \left[\det{}' \left(P_1^\dagger P_1 \right) \right]_{\widehat{\gamma}}^{\frac{1}{2}}\, e^{-S_L^{(ren)}(\widehat{\gamma},u)}\, \widetilde{D}_{\widehat{\gamma}}[u] \right) \tag{5.41}$$

$$= d\omega_{WP} \left[\frac{\det'\Delta}{A(M)} \right]_{\widetilde{\gamma}}^{-\frac{n}{2}} \left[\det{}' \left(P_1^\dagger P_1 \right) \right]_{\widetilde{\gamma}}^{\frac{1}{2}}\, e^{[(1+6Q^2)-(26-n)] S_L(\widetilde{\gamma},\,\widetilde{u})}\, e^{-S_L^{(ren)}(\widetilde{\gamma},\,\widetilde{u})}\, \widetilde{D}_{\widetilde{\gamma}}[\widetilde{u}],$$

where, (in transforming $\widehat{\gamma}$ to $\widetilde{\gamma}$), we have exploited the corresponding conformal anomaly formula (5.19).

From (5.41) it immediately follows that the functional measure describing (quantum) Liouville theory is independent from the choice of the background metric $\widehat{\gamma}$ iff $[(1 + 6 Q^2) - (26 - n)] = 0$, i.e. if the parameter Q takes the value

$$Q = \sqrt{\frac{25 - n}{6}}. \tag{5.42}$$

Introducing this expression in (5.32) also yields the corresponding value(s) for b

$$b^{\pm} = \sqrt{\frac{25 - n}{24}} \pm \sqrt{\frac{1 - n}{24}}. \tag{5.43}$$

In order that the semiclassical limit holds,[10] we have to select the $-$ sign in (5.43)

---

[9]This shows that $S_L^{(ren)}(\gamma, u)$ characterizes a conformal field theory with central charge $c_u = 1 + 6 Q^2$, [38, 40, 53].

[10]We remind the reader that formally the classical limit corresponds to $n \searrow -\infty$, see the comments to the expression of the partition function (5.20).

$$b := \sqrt{\frac{25-n}{24}} - \sqrt{\frac{1-n}{24}}, \tag{5.44}$$

(we also have the reality condition restricting $c_\phi = n < 1$, however, for reasons that will be clear momentarily, we do not stress this restriction describing the *minimal gravity* realization of quantum Liouville theory [68]). Thus, we eventually get for (5.26) the expression

$$S_L^{(ren)}(\widehat{\gamma}, u) \tag{5.45}$$

$$= \frac{1}{4\pi \, l_s^2} \int_M d\mu_{\widehat{\gamma}} \left[\widehat{\gamma}^{\,\mu\nu} \partial_\mu (u \, l_s) \partial_\nu (u \, l_s) + l_s \sqrt{\frac{25-n}{6}} \, R(\widehat{\gamma}) \, (u \, l_s) \right]$$

$$+ \frac{T_0}{l_s^2} \int_M d\mu_{\widehat{\gamma}} \, e^{2 \, l_s^{-1} b \, (u \, l_s)},$$

where b is given by (5.44), and where we have reintroduced the string length l_s, and set $\Gamma^2 := \frac{T_0}{l_s^2}$ for later convenience.

5.4 A Spacetime Interpretation of the Liouville Mode

The structure of the action (5.46), with the presence of a kinetic term, suggests an intriguing interpretation of the role of the Liouville mode [10]. The idea is to promote the Liouville field u to an extra spacetime variable ϕ^{n+1} by setting

$$\phi^{n+1} := l_s \sqrt{\left(\frac{25-n}{6}\right)} \, u , \tag{5.46}$$

so as to rewrite the Liouville action as

$$S_L^{(ren)}(\widehat{\gamma}, u) = \frac{1}{4\pi \, l_s^2} \int_M \widehat{\gamma}^{\,\mu\nu} \left[\frac{6}{25-n} \partial_\mu \phi^{n+1} \partial_\nu \phi^{n+1} \right] d\mu_{\widehat{\gamma}} \tag{5.47}$$

$$+ \frac{1}{4\pi} \int_M l_s \, \phi^{n+1} R(\widehat{\gamma}) \, d\mu_{\widehat{\gamma}} + \frac{T_0}{l_s^2} \int_M e^{2b \sqrt{\frac{6}{25-n}} l_s^{-1} \phi^{n+1}} \, d\mu_{\widehat{\gamma}}.$$

At this point, let us make a step back, and consider the path integral

$$\int_{\{\mathfrak{w}(M)\}} \int_{\{Map(M,V^n)\}} \tilde{D}_{\widehat{\gamma}}[u] \, D_\alpha[\phi] \, e^{-\tilde{\mathscr{I}}_{\widehat{\gamma}}[\phi;\alpha] - S_L^{(ren)}(\widehat{\gamma},u)}. \tag{5.48}$$

Note that if, according to (5.46), we identify u with ϕ^{n+1} and introduce on $V^{n+1} \simeq V^n \times \mathbb{R}$ the product metric

$$g_{ab}\, d\phi^a \otimes d\phi^b := g_{ij}\, d\phi^i \otimes d\phi^j + \frac{6}{25 - n}\, d\phi^{n+1} \otimes d\phi^{n+1}, \tag{5.49}$$

with $a,\ b = 1, \ldots, n+1$, then we can write

$$\tilde{\mathscr{S}}_{\widehat{\gamma}}[\phi; \alpha] + S_L^{(ren)}(\widehat{\gamma}, u) = \mathscr{S}_{\widehat{\gamma}}[\phi; f(\phi), U(\phi)] \tag{5.50}$$

$$:= \frac{1}{4\pi\, l_s^2} \int_M \widehat{\gamma}^{\,\mu\nu} \left[\partial_\mu \phi^a \partial_\nu \phi^b\, g_{ab} \right] d\mu_{\widehat{\gamma}} + \int_M f(\phi)\, R(\widehat{\gamma})\, d\mu_{\widehat{\gamma}} +$$

$$+ \frac{1}{4\pi\, l_s^2} \int_M 4\pi\, T_0 \exp\left[4\pi \sqrt{\frac{6}{25 - n}} \left(\sqrt{\frac{25 - n}{6}} - \sqrt{\frac{1 - n}{6}} \right) f(\phi) \right] d\mu_{\widehat{\gamma}},$$

where we have introduced the linear dilaton field

$$f(\phi) := \frac{l_s}{4\pi}\, \phi^{n+1}, \tag{5.51}$$

and where

$$U(\phi) = 4\pi\, T_0 \exp\left[4\pi \sqrt{\frac{6}{25 - n}} \left(\sqrt{\frac{25 - n}{6}} - \sqrt{\frac{1 - n}{6}} \right) f(\phi) \right] \tag{5.52}$$

plays the role of a tachyonic field, (screening the strong coupling $f \nearrow \infty$ linear dilaton regime[11]). Thus, $\mathscr{S}_{\widehat{\gamma}}[\phi; f(\phi), U(\phi)]$ formally has the structure of a non–linear σ–model action in a flat $V^{n+1} \simeq V^n \times \mathbb{R}$ with a linear dilatonic coupling in the $n + 1$-direction, (see (5.2)), and a tachyonic term. To show that this formal structure is dynamical, let us remark that since $\mathfrak{w}(M) = C^\infty(M, \mathbb{R})$, we can identify $Map(M, V^n) \times \mathfrak{w}(M) \simeq Map(M, V^{n+1} := V^n \times \mathbb{R})$. Thus, under the natural projection

$$\pi : Map(M, V^{n+1} \simeq V^n \times \mathbb{R}) \longrightarrow Map(M, V^n) \times \mathfrak{w}(M) \tag{5.53}$$

$$\{\phi^a\}_{a=1}^{n+1} \longmapsto (\phi^k, \phi^{n+1} = l_s \sqrt{\left(\frac{25 - n}{6} \right)}\, u),$$

---

[11]When $\phi^{n+1} \nearrow \infty$ and $U(\phi)$ is real, $(n \leq 1)$, the term $\exp U(\phi)$ dominates the action $\mathscr{S}_{\widehat{\gamma}}[\phi; f(\phi), U(\phi)]$ which then becomes large and positive, suppressing, in the path integral over $Map(M, V^{n+1})$, the configurations for which $\phi^{n+1} \nearrow \infty$.

we can formally pull back the product measure $\tilde{D}_{\widehat{\gamma}}[u]\, D_\alpha[\phi]$ to a measure over $Map(M, V^{n+1} := V^n \times \mathbb{R})$,

$$D_\alpha^{(n+1)}[\phi] := \pi^*\left(\tilde{D}_{\widehat{\gamma}}[u]\, D_\alpha[\phi]\right), \tag{5.54}$$

so as to get

$$\int_{\{\mathfrak{w}(M)\}} \int_{\{Map(M,V^n)\}} \tilde{D}_{\widehat{\gamma}}[u]\, D_\alpha[\phi]\, e^{-\tilde{\mathscr{S}}_{\widehat{\gamma}}[\phi;\alpha] - S_L^{(ren)}(\widehat{\gamma},u)} \tag{5.55}$$

$$= \int_{\{Map(M,V^{n+1})\}} D_\alpha[\phi]\, e^{-\mathscr{S}_{\widehat{\gamma}}[\phi;f(\phi),U(\phi)]},$$

where now the measure $D_\alpha[\phi]$ runs over the map space $Map(M, V^{n+1})$. Thus, even if the original Liouville field theory does not decouple from the matter fields $\{\phi^k\}_{k=1}^n$, its properties can be exploited to generate an extra space dimension[12] which characterizes a CFT with central charge $c[\{\phi^a\}_{a=1}^{n+1}] = 26$, compensating the $\mathscr{D}iff(M)$ central charge $c_\psi = -26$ and yielding for an effective anomaly free theory. The price one has to pay, for such an effective field theory extension, is the presence of the linear dilaton $f(\phi)$, (whose presence can make the theory unreliable as $f \to \infty$), and of the attendant (screening) tachyonic field $U(\phi)$. Eventually, we can write the partition function (5.20) describing quantum Liouville theory as

$$\int_{\mathfrak{M}_g(M)} d\omega_{WP} \left[\frac{\det{}'\Delta}{A(M)}\right]_{\widehat{\gamma}}^{-\frac{n}{2}} \left[\det{}'\left(P_1^\dagger P_1\right)\right]_{\widehat{\gamma}}^{\frac{1}{2}} \int_{W(M)} e^{-(26-n)\,S_L(\widehat{\gamma},u)}\, D_{\widehat{\gamma}}[u],$$

$$\tag{5.56}$$

$$= \int_{\mathfrak{M}_g(M)} d\omega_{WP} \left[\det{}'\left(P_1^\dagger P_1\right)\right]_{\widehat{\gamma}}^{\frac{1}{2}} \int_{\{Map(M,V^{n+1})\}} D_\alpha[\phi]\, e^{-\mathscr{S}_{\widehat{\gamma}}[\phi;f(\phi),U(\phi)]}.$$

5.5 KPZ Scaling

As we have seen in Sect. 4.12 the effective action governs the low energy limit of the metric and dilaton couplings as *spacetime fields* obeying Einstein type equations.[13] This implies that a spacetime interpretation of the Liouville field goes much deeper

---

[12]The extra dimension is actually time–like if $n > 26$.

[13]The effective action (4.167) is written in the so called *string frame*. By a conformal transformation it is possible to move to the *Einstein frame* where (4.167) takes a manifest Einstein–Hilbert structure; see e.g. [40, 53] for details. The picture becomes more complex with the presence of

than a formal rewriting of the action (5.50). In particular, since quantum Liouville field theory governed by (5.50) appears as a weak–coupling (conformally invariant) fixed point of the renormalization group, a natural question concerns the existence of other (possibly, strong coupling) fixed points and of a renormalization group interpolation among such field configurations. Connected with this problem are the sample properties of the quantum Liouville field: How do the fields ϕ^a probe the spacetime geometry of the manifold V^{n+1}? To appreciate the nature of the issue under discussion is perhaps worthwhile making a parallel with diffusive motion by using brownian paths, where free motion distance is proportional to the square root of time, (rather than to time, as in ballistic motion). Thus, free diffusive motion, as described by Wiener measure, is rather inefficient in sampling space,[14] (however, this is an inefficiency which pays off in the long run, since what is lost in distance is regained in a better control, via diffusion, of the ambient geometry!). Similarly, the quantum Liouville measure associated with (5.50) describes a sort of diffusive motion of a random surface around a point $\in V^{n+1}$, (this point can be identified with the background field $\phi^a_{(0)}$ associated with the constant map dominating the path integral in the weak coupling regime). To what an extent this diffusion process is an efficient sampling[15] of V^{n+1} naturally brings to the fore the Knizhnik–Polyakov–Zamolodchikov susceptibility exponents related to the quantum Liouville action (5.50).

In order to discuss the standard Liouville field theory derivation of the KPZ susceptibility, it is convenient to redefine the dilatonic field $f(\phi)$ according to

$$f(\phi) \longmapsto f'(\phi) := 4\pi \sqrt{\frac{6}{25-n}} f(\phi) = \sqrt{\frac{6}{25-n}} \, l_s \, \phi^{n+1}, \qquad (5.57)$$

and (5.50) as

$$\mathscr{S}_{\widehat{\gamma}}[\phi; f'(\phi), T_0] \qquad (5.58)$$

$$:= \frac{1}{4\pi \, l_s^2} \int_M \widehat{\gamma}^{\mu\nu} \left[\partial_\mu \phi^a \partial_\nu \phi^b \, g_{ab} \right] d\mu_{\widehat{\gamma}} + \frac{Q}{4\pi} \int_M f'(\phi) \, R(\widehat{\gamma}) \, d\mu_{\widehat{\gamma}} +$$

$$+ \frac{T_0}{l_s^2} \int_M \exp \left[\left(\sqrt{\frac{25-n}{6}} - \sqrt{\frac{1-n}{6}} \right) f'(\phi) \right] d\mu_{\widehat{\gamma}},$$

---

the tachyonic coupling $U(\phi)$, and in general the implementation of conformal invariance just at leading order is not believed to be sufficient for a reliable effective field theory description.

[14] This remark on diffusive motion is nicely stressed and discussed by W. G. Faris in [27].

[15] One can well argue that such a sampling process characterizes V^{n+1} in a neighborhood of the given point.

where $Q = \sqrt{\frac{25-n}{6}}$. With these notational remarks along the way, if we consider the fixed–area partition function at given genus g

$$Z_g[A] := \int_{\mathfrak{M}_g(M)} d\omega_{WP} \int_{Map(M,V^{n+1})} \mathcal{D}_\alpha[\phi]\, e^{-\mathscr{S}_\gamma[\phi;f',T_0]} \tag{5.59}$$

$$\delta\left(A - \int_M \exp\left[\left(\sqrt{\frac{25-n}{6}} - \sqrt{\frac{1-n}{6}}\right)f'\right] d\mu_{\widehat{\gamma}}\right),$$

associated with (5.58), where $\delta(\ldots)$ denotes the Dirac function, then we get [11, 19, 41] the relation

$$Z_g[A] = Z_g[1]\, A^{\frac{(1-g)}{12}[n-25-\sqrt{(25-n)(1-n)}]-1}, \tag{5.60}$$

characterizing the scaling, with surface area A, of $Z_g[A]$. This is a particular case of the celebrated Knizhnik–Polyakov–Zamolodchikov (KPZ) relations, [11, 19, 41], and it is considered a characteristic signature of the interaction between matter and 2D quantum gravity. In particular, if we define the associated susceptibility[16] exponent Γ_{KPZ} via the asymptotic relation

$$Z_g[A] \sim_{A>>1} A^{\Gamma_{KPZ}-3}, \tag{5.61}$$

then we get

$$\Gamma_{KPZ} = \frac{(1-g)}{12}\left(n - 25 - \sqrt{(25-n)(1-n)}\right) + 2. \tag{5.62}$$

A heuristic derivation of (5.60) is elementary, in sharp contrast with the difficulties in providing both a mathematical characterization of KPZ scaling, not to mention its rigorous proof. As for the heuristics, let us consider the one–parameter (ϵ) family of constant shifts in the fields $\{\phi^a\}_{a=1}^{n+1}$ defined by

$$\phi^a \longmapsto \phi_\epsilon^a := \phi^a + \epsilon\, l_s\, Q\left(\sqrt{\frac{25-n}{6}} - \sqrt{\frac{1-n}{6}}\right)^{-1}, \qquad 0 \le \epsilon \le \ln A. \tag{5.63}$$

Note that this induces a corresponding shift in the dilatonic field $f'(\phi)$ given by

$$f' \longmapsto f_\epsilon := f' + \epsilon\left(\sqrt{\frac{25-n}{6}} - \sqrt{\frac{1-n}{6}}\right)^{-1}, \qquad 0 \le \epsilon \le \ln A. \tag{5.64}$$

---

[16]Our definition of Γ_{KPZ} is concocted in such a way to explicitly keep track of the genus g. Another standard definition of the *string susceptibility exponent* Γ_{string} is via the asymptotic scaling $Z_g[A] \sim A^{(\Gamma_{string}-2)(1-g)-1}$. The two critical exponents are related by $\Gamma_{string} = (\Gamma_{KPZ} - 2g)/(1-g)$.

Under the action of (5.63), we immediately compute

$$\mathscr{S}_{\widehat{\gamma}}(\phi,f_\epsilon; T_0) = \mathscr{S}_{\widehat{\gamma}}(\phi,f'; e^\epsilon T_0) + \epsilon\, Q \left(\sqrt{\frac{25-n}{6}} - \sqrt{\frac{1-n}{6}} \right)^{-1} \chi(M),$$

(5.65)

where we have exploited the Gauss–Bonnet theorem $\int_M d\mu_{\widehat{\gamma}} R(\widehat{\gamma}) = 4\pi\,\chi(M)$. Similarly, we compute

$$\delta\left(A - \int_M \exp\left[\left(\sqrt{\frac{25-n}{6}} - \sqrt{\frac{1-n}{6}} \right) f_\epsilon \right] d\mu_{\widehat{\gamma}} \right)$$

(5.66)

$$= e^{-\epsilon}\delta\left(e^{-\epsilon} A - \int_M \exp\left[\left(\sqrt{\frac{25-n}{6}} - \sqrt{\frac{1-n}{6}} \right) f' \right] d\mu_{\widehat{\gamma}} \right),$$

which follows from $\delta(a\,x) = |a|^{-1}\,\delta(x)$. Since the formal path measure $D[\phi]$ over $Map(M, V^{n+1})$ in (5.59) is invariant under a shift, the above transformation laws imply that

$$Z_g[A_\epsilon] := \int_{\mathfrak{M}_g(M)} d\omega_{WP} \int_{Map(M,V^{n+1})} D[\phi]\, e^{-\mathscr{S}_{\widehat{\gamma}}(\phi,f_\epsilon; T_0)}$$

(5.67)

$$\delta\left(A - \int_M \exp\left[\left(\sqrt{\frac{25-n}{6}} - \sqrt{\frac{1-n}{6}} \right) f_\epsilon \right] d\mu_{\widehat{\gamma}} \right),$$

$$= e^{-\epsilon\left(\varrho\left(\sqrt{\frac{25-n}{6}} - \sqrt{\frac{1-n}{6}} \right)^{-1} \chi(M)+1 \right)} Z_g[e^{-\epsilon} A],$$

where we have set

$$Z_g[e^{-\epsilon} A] := \int_{\mathfrak{M}_g(M)} d\omega_{WP} \int_{Map(M,V^{n+1})} D[\phi]\, e^{-\mathscr{S}_{\widehat{\gamma}}(\phi,f'; e^\epsilon T_0)}$$

(5.68)

$$\delta\left(e^{-\epsilon} A - \int_M \exp\left[\left(\sqrt{\frac{25-n}{6}} - \sqrt{\frac{1-n}{6}} \right) f' \right] d\mu_{\widehat{\gamma}} \right).$$

Since we are considering the constant shift (5.64) at a fixed area, we have $Z_g[A_\epsilon] = Z_g[A]$, and (5.68) implies the scaling law

$$Z_g[A] = e^{-\epsilon\left(\varrho\left(\sqrt{\frac{25-n}{6}} - \sqrt{\frac{1-n}{6}} \right)^{-1} \chi(M)+1 \right)} Z_g[e^{-\epsilon} A].$$

(5.69)

In particular, if we set $\epsilon = \ln A$, then we get

$$Z_g[A] = A^{-\left(\varrho\left(\sqrt{\frac{25-n}{6}}-\sqrt{\frac{1-n}{6}}\right)^{-1}\chi(M)+1\right)} Z_g[1], \tag{5.70}$$

which immediately yields (5.60).

More generally, (5.60) extends to the coupling between conformal matter of central charge c_m and 2D quantum gravity according to

$$Z_g[A;\, c_m] = Z_g[1;\, c_m]\, A^{\frac{(1-g)}{12}\left[c_m-25-\sqrt{(25-c_m)(1-c_m)}\right]-1}, \tag{5.71}$$

with the attendant susceptibility exponent

$$\Gamma_{KPZ}[c_m] := \frac{(1-g)}{12}\left(c_m - 25 - \sqrt{(25-c_m)(1-c_m)}\right) + 2. \tag{5.72}$$

The above heuristic argument leading to the scaling ansatz (5.71) is believed to be reliable for c_m small, a fact reflected in the expression for $\gamma_{string}[c_m]$ which is ambiguous for $c_m > 1$. From the point of view of non–critical string theory, we are entering a region of strong coupling regime between conformal matter and gravity, the *screening* effect of the tachyonic potential $U(\phi)$ in (5.55) is no longer active, and the analysis of the coupling between matter and quantum gravity requires the understanding of the quantum dynamics of Liouville theory. We have already stressed that, in contrast with the rather effective heuristic arguments leading to (5.60) and its generalization, it has been extremely hard to provide a fully rigorous proof of KPZ scaling. Only recently, and in the very general (and mathematically compelling) setting of the theory of random fields, B. Duplantier and S. Sheffield [20], and in a remarkable paper, R. Rhodes and V. Vargas [59], have been able to address rigorously a probabilistic derivation of the KPZ scaling in Liouville quantum gravity. Another interesting analysis has been proposed by F. David and M. Bauer [12] by exploiting a heat kernel technique. All this has recently evolved in a rigorous definition, by probabilistic techniques, of the full (and non perturbative) quantum Liouville theory. The interested reader is urged to look at the beautiful *Liouville Quantum Gravity on the Riemann sphere* by F. David, A. Kupiainen, R. Rhodes, and V. Vargas [14]. As stressed in the preface to this second edition of our LNP, it is a pity not to be able to review these important recent advances in Liouville theory, but our limited knowledge of the subtle probabilistic techniques used in these approaches would not do justice to the authors.

A few comments are in order as to the nature of the difficulties in deriving the KPZ susceptibility in a *triangulation* setting (in our opinion still an interesting problem–see the paragraph 5.3 *Relation with discretized* 2d *quantum gravity* in [14]). Let us start by observing that for pure gravity, i.e. for $n = 0$, (5.62) reduces to the well-known value

$$\Gamma_{KPZ} = \frac{5g-1}{2}. \tag{5.73}$$

which shows that the scaling properties of the surfaces sampled by the functional measure (5.56) have a two–fold origin: one is associated with the Liouville mode itself, the other is generated by the coupling of the Liouville mode with conformal matter. Already the scaling property of the pure gravity case appear a little bit intriguing and one may wonder as to its geometrical origin. This has its roots in the formal definition (5.16) of the Liouville action which, if taken at face value, has a few drawbacks from the mathematical point of view. These latter have been clearly emphasized by P. Zograf and L. Takhtadzhyan [70], (see also [64]). Their point is that if $\{(U_{(\alpha)}, z_{(\alpha)})\}_{\alpha \in I}$ is a local covering of M with coordinate charts $(U_{(\alpha)}, z_{(\alpha)})$ and transition functions $f_{\alpha\beta} : z_{(\beta)}(U_{(\alpha)} \cap U_{(\beta)}) \rightarrow z_{(\alpha)}(U_{(\alpha)} \cap U_{(\beta)})$, then the conformal factors $\{\varphi_{(\alpha)}\}$ in the corresponding local parametrization of the metric $\gamma|_{(\alpha)} = e^{\varphi_{(\alpha)}}|dz_{(\alpha)}|^2$ are not scalar functions, but rather the components of a metric tensor field, thus transforming according to

$$\varphi_{(\beta)} \;=\; \varphi_{(\alpha)} \circ f_{\alpha\beta} \;+\; \ln \left|f'_{\alpha\beta}\right|^2. \tag{5.74}$$

This remark directly implies that Dirichlet type integrals of the form

$$S_L[M, \{(U_{(\alpha)}, z_{(\alpha)})\}] \;:=\; \frac{\sqrt{-1}}{2} \int_M \left(|\partial_z\varphi|^2 + e^\varphi\right) dz \wedge d\bar{z}, \tag{5.75}$$

(hence also (5.16) and (5.58)), cannot be defined only in terms of the Riemann surface $\left(M, \{(U_{(\alpha)}, z_{(\alpha)})\}\right)$ as the integral of the local 2–form [64]

$$\omega|_{(\alpha)} \;:=\; \left(\left|\partial_{z_{(\alpha)}}\varphi_{(\alpha)}\right|^2 + e^{\varphi_{(\alpha)}}\right) dz_{(\alpha)} \wedge d\bar{z}_{(\alpha)}, \tag{5.76}$$

since the term $\left|\partial_{z_{(\alpha)}}\varphi_{(\alpha)}\right|^2$ does not transform equivariantly under (5.74). The simplest decoration of the Riemann surface $\left(M, \{(U_{(\alpha)}, z_{(\alpha)})\}\right)$ that allows for a coherent definition of the Liouville action is provided by the insertion of punctures in M, say N_0 to conform to our notation, and this allows to define the Liouville action as $\sqrt{-1}/2 \int_{M\backslash N_0} \omega$ by introducing a suitable regularization at the punctures [64, 70]. A basic property of such a regularized Liouville action $S_L^{reg}[M \setminus N_0]$ is that $-S_L^{reg}[M \setminus N_0]$ appears as the Kähler potential of the Weil–Petersson metric on the Teichmüller space $\mathfrak{T}_{(g,N_0)}(M)$ of N_0 punctured Riemann surfaces [64, 70], i.e.

$$\bar{\partial}\, \partial \left(- S_L^{reg}[M \setminus N_0, \{(U_{(\alpha)}, z_{(\alpha)})\}]\right) \;=\; 2\sqrt{-1}\,\omega_{wp}. \tag{5.77}$$

Thus, in the weak coupling limit (around a stationary value ϕ_0 of the field ϕ), we can write,[17] (up to the constant factor $S_L^{reg}[M \setminus N_0]|_{\phi_0}$),

---

[17]A formulation of Quantum Liouville theory which takes into care the subtleties of the definition of the Liouville action is discussed in [64].

$$- S_L^{reg}[M \setminus N_0] \simeq \sqrt{-1}\, \omega_{wp} + \dots . \qquad (5.78)$$

As a consequence of these remarks, we may tentatively identify (5.59) with

$$Z_g[A] \propto \int_{\mathfrak{M}_{(g,N_0)}(M)} e^{\omega_{WP}} = Vol\left[\mathfrak{M}_{(g,N_0)}(M)\right], \qquad (5.79)$$

which would indicate a modular origin to the pure gravity KPZ exponent. As we have seen, punctured Riemann surfaces can be rather directly put into correspondence with polyhedral surfaces, and it does not come as a surprise that the above rough heuristics can be put on a firm basis in the polyhedral setting.

5.6   2D QG and Polyhedral Surfaces: General Remarks

From a mathematical point of view, and in line with the renormalization group strategy described in the previous sections, one may well argue that 2D quantum gravity reduces to the study of a critical filtration of decorated probability measures

$$\{Z[\gamma, \{\phi(x_i)\}; \alpha]\, D_\alpha[\gamma]\}_{\alpha_1} \preceq \dots \{Z[\gamma, \{\phi(x_i)\}; \alpha]\, D_\alpha[\gamma]\}_{\alpha_k} \preceq \dots , \qquad (5.80)$$

on the space $\mathscr{M}et\,(M)/\mathscr{D}iff\,(M)$, where $Z[\gamma, \{\phi(x_i)\}; \alpha]$ is a matter field correlation function over $Map(M, V^n)$, (see (4.103)), and where criticality is related to the onset of local scale invariance driven by a set of tunable parameters α. At criticality, averaging over metrics with respect to $\{Z[\gamma, \{\phi(x_i)\}; \alpha]\, D_\alpha[\gamma]\}_{\alpha_{crit}}$, eliminates scale dependence in the QFT defined over $Map(M, V^n)$, and the random fields $\{\phi(x_i)\}$ will describe a distribution of geometrical objects whose scaling dimensions, related to the KPZ susceptibility exponents, provide a rather direct evidence of a quantum regime in the interaction between matter fields and geometry.

A natural way for addressing the study of such critical families of probability measures over $\mathscr{M}et\,(M)/\mathscr{D}iff\,(M)$ emerges naturally when we approximate the set of Riemannian surfaces with polyhedral metrics. In particular, equilateral polyhedral surfaces (Dynamical Triangulations) provide one of the most powerful technique for analyzing two-dimensional quantum gravity in regimes which are not accessible to the standard field-theoretic formalism. This is basically due to the circumstance that in such a discretized setting the filtration (5.80) reduces to a suitably constrained enumeration of distinct triangulations admitted by a surface of given topology, (see e.g., [3] for a review). At criticality, (corresponding to tuning the number of vertices $N_0 \nearrow \infty$ and the edge–length $l \searrow 0$), one recovers a scaling measure, describing the gravitational dressing of conformal operators in the continuum theory. It has been argued, mainly as a consequence of a massive numerical evidence, that such a (counting) measure automatically accounts for the anomalous scaling properties of the measure $D[\gamma]$ and $D_\gamma[u]$ governing the continuum path-quantization of 2D gravity, (see (5.13)). However, the geometrical origin of such a property is rather elusive and it is not clear how the counting for dynamical triangulations factorizes, so to speak, in terms of a discrete analogous of a moduli space measure $D[\mu_k]$ and of a Liouville measure $e^{S_L(\widehat{\gamma},u)} D_\gamma[u]$ over the conformal degrees of freedom of the theory.

The geometrical explanation of the behavior of the dynamical triangulation measure is not obvious if we only consider polyhedral surfaces as a sort of approximating net in the space of Riemannian structures $\frac{\mathcal{M}et(M)}{\mathcal{D}iff(M)}$. Even if such heuristics is mathematically motivated by the density of equilateral polyhedral surfaces in Teichmüller space, as implied by *Belyĭ*'s theorem, (see Theorem 3.4 and [66]), this does not provide a detailed hint to the rationale which connects triangulation counting to the quantum gravity measure. The effectiveness of dynamical triangulation is also related to the (old[18]) matrix models. These latter provide natural generating functions for counting the distinct dual ribbon graphs associated with polyhedral surfaces, (possibly decorated with matter fields), and they have a life of their own[19] in different guises both in string theory, integrable models, as well as in moduli space theory where they have played a fundamental role [37, 43]. Notwithstanding the success achieved on a mathematical and physical level, a deep understanding of their *raison d'être* is still unclear. On the physical side, in particular from the point of view of strings, the basic role of (decorated) geometrical discretization lies in the observation that large N matrix field theory has (under rather general hypotheses) the structure of a string theory. This was first observed by 't Hooft for $SU(N)$ gauge theories [62], hinting to a connection which nowadays has its roots in open and closed strings duality. On the mathematical side the explanation of the successful role of polyhedral surfaces is deeply connected with the geometrical mechanism we have been discussing in the previous chapters, and which allows to describe a polyhedral surface with N_0 vertices as a N_0-pointed Riemann surface dressed with a Liouville field whose charges describe localized curvatures. As we have seen, the Riemann moduli spaces for closed and open surfaces, $\overline{\mathfrak{M}}_{g,N_0}$ and $\overline{\mathfrak{M}}_{g,N_0}(L)$, are naturally called into play in such a representation. This suggests again that a form of open closed/duality is at work here, too. The origin of the interplay between strings, 2D quantum gravity, and polyhedral surfaces, (with the attendant old and new matrix models), lies indeed in the distinct cellularization of moduli spaces which allow to provide *good cellular decomposition* of moduli space in the form either of ribbon graph [37, 51, 61], or light–cone coordinates[20] [31], or Penner hyperbolic parametrization [55, 56]. This

---

[18]As compared with the Matrix theory describing flat 11–dimensional M–theory in the discrete light–cone quantization–see [40] for a review and relevant references.

[19]An excellent review is provided by [18].

[20]The light–cone cellular decomposition of the N–pointed Teichmüller space arises from the structure theory of abelian differential of the third kind. As in the case of quadratic differentials associated to ribbon graphs, also here we get a graph structure yielding for a cellular decomposition which descends to Riemann moduli space and exhibits certain computational advantages with respect to the ribbon graph cellularization [52].

is a basic issue in string field theory[21] that we will not pursue here, but which nicely blends together the mathematical and the physical rationale on the role of polyhedral surfaces (and matrix models) alluded before.

To conclude these preliminary remarks, it is worthwhile stressing that a rationale explaining the common combinatorial threads among moduli space, matrix models, string theory, and integrable models has recently emerged in the work of Eynard and Orantin, and in the striking results of M. Mirzakhani. Building on ideas familiar in matrix models (and string field theory [26]), Eynard and Orantin [22, 24, 25] have introduced a deep framework for *Topological Recursion* [23]. In a nutshell, if $v_{g,N}$ is a geometrical quantity defined over the moduli space $\mathfrak{M}_{g,n}$ of n–pointed Riemann surfaces of genus g, (under the usual stability assumptions on (n,g)), then their theory establishes recursive formulas of the form, (here we are following [9])

$$
v_{g,n} = f_1(v_{g,n-1}) + f_2(v_{g-1,n+1}) + \sum_{g_1+g_2=g;\, n_1+n_2=n-1} f_3(v_{g_1,n_1+1}, v_{g_1,n_2+1}),
$$

$$(5.81)$$

where the summation runs over all possible partitions of g and $n-1$ such that $2g_i - 1 + n_i > 0$, $i = 1, 2$, and where f_1, f_2, are linear operators and f_3 is bilinear. The strategy underlying these relations is the familiar one of constructing structured objects, $(v_{g,n})$, over moduli space, out of simpler pieces, $(v_{g,n-1}$ and $v_{g-1,n+1})$, defined in terms of buiding blocks of Riemann surfaces, (typically pair of pants or thrice–punctured spheres). Intersection numbers in Witten–Kontsevich theory [43], metrical ribbon graphs, matrix models, afford relevant examples obeying recursion relations of the type (5.81), which thus appear as the common combinatorial framework underlying these theories. Also the remarkable recursion formula discovered, in the context of hyperbolic geometry, by M. Mirzakhani [46, 47] falls within this class. A deep geometric connection between these two approaches is discussed at length in a series of remarkable papers by Mulase and Safnuk [50], to which we refer for further details.

Even if the analysis of the work of Eynard and Orantin is beyond the scope of these lecture notes, we shall make use of Mirzakhani's results. We start by connecting 2D quantum gravity to the orbifold integral representation (3.104) of the symplectic volume of $\overline{\mathfrak{M}}_{g,N_0}$, provided by Theorem 3.6. Then, by exploiting the large N_0 asymptotics of the Weil-Petersson volume of $\overline{\mathfrak{M}}_{g,N_0}$ recently discussed by Manin and Zograf [44, 69], we show that the anomalous scaling properties of pure gravity is only due to the modular degrees of freedom of $\overline{\mathfrak{M}}_{g,N_0}$, as suggested by (5.79).

---

[21]For a nice and clear presentation see [38].

5.7   The Moduli Space $\overline{\mathfrak{M}}_{g,N_0}$ and 2D Quantum Gravity

As we have recalled in the previous sections, the non–critical string partition function, at fixed genus, is provided by giving meaning to the formal path integral

$$Z\left[\alpha\right] := \int_{\mathscr{M}et(M)/\mathscr{D}iff(M)} D[\mathscr{M}/\mathscr{D}] \int_{\{Map(M,V^n)\}} D_\alpha[\phi]\, e^{-\mathscr{S}_\gamma[\phi;\alpha]}, \qquad (5.82)$$

(see (5.2) and (5.3)), in particular by factorizing a Gaussian measure on the (tangent) space of Riemannian metrics $\mathscr{M}et(M)$ into a moduli space measure, a (non–Gaussian) Weil measure, and a Gaussian measure on the tangent space to $\mathscr{D}iff(M)$. In such a picture, pure 2D quantum gravity is characterized by eliminating the degrees of freedom represented by $\phi \in Map(M, V^n)$ by setting the central charge $c = n \equiv 0$. According to the QFT heuristics we have discussed before, this procedure still leaves a residual Liouville mode $\phi^{n+1} \propto$ the dilaton field f, which is responsible for a non-trivial KPZ scaling characterized by a susceptibility exponent $\Gamma_{KPZ} = (5g - 1)/2$, (see (5.73)). To address the analysis of (5.82) from a more geometrical point of view, we can associate with it the filtration (5.80) obtained by replacing the Riemannian surfaces (M, γ) with corresponding polyhedral surfaces (T_l, M) with N_0 vertices, and discuss how the behavior of the associated sequence of partition functions behave for large N_0. If we set $c = n \equiv 0$, then at fixed genus g and at fixed area A we can replace (5.82) with an infinite sequence of finite statistical sums

$$Z\left[\alpha\right]_{n=0} \longmapsto \left\{Z_{g,N_0}[\alpha]\right\} \qquad (5.83)$$

whose generic term Z_{g,N_0}, at fixed N_0, heuristically has the form

$$Z_{g,N_0}[\alpha] := \text{``} \sum_{Polyhedral\ surfaces} \text{''}\, e^{-\alpha_0 N_0 - \beta\, \chi(M)}, \qquad (5.84)$$

where $\alpha := (\alpha_0, \beta)$ are running coupling constants, $\chi(M) = 2 - 2g$ is the Euler characteristic of the surface M, and the summation is over all polyhedral surfaces $POL_{g,N_0}(A)$ of genus g, with N_0 vertices and given area A. The summation is between quotes because we have to decide how to define it, yet. There are two alternatives: (i) Since (5.84) is a regularization we may well decide to pick a suitable finite set of representative polyhedral surfaces in $POL_{g,N_0}(A)$. Dynamical triangulations are quite a convenient choice, owing to their density properties for large N_0 and their simple geometry in discussing the asymptotic behavior of (5.84); (ii) The other alternative is to consider the full space $POL_{g,N_0}(A)$. This is apparently a more complicated choice since we need to replace the sum in (5.84) with an integral, with the attendant difficult problem of selecting the appropriate measure[22] for $POL_{g,N_0}(A)$. As discussed in Chap. 3 we have a good control over $POL_{g,N_0}(A)$ and on its natural (pre)symplectic measure, thus here we take this second choice. In particular, since the term $e^{-\alpha_0 N_0 - \beta \chi(M)}$, at fixed g and N_0, is constant over $POL_{g,N_0}(A)$, we can regularize $Z[\alpha]_{n=0}$ according to

Definition 5.1 The regularizing filtration associated with the pure gravity partition function $Z[\alpha]_{n=0}$ is provided by the sequence $\{Z_{g,N_0}[\alpha]\}$, $N_0 \in \mathbb{N}$, defined by

$$Z_{g,N_0}[\alpha] := Vol\left[POL_{g,N_0}(A)\right] e^{-\alpha_0 N_0 - \beta \chi(M)}, \qquad (5.85)$$

where $Vol\left[POL_{g,N_0}(A)\right]$ is the (pre)-symplectic volume of $POL_{g,N_0}(A)$ defined by Theorem 3.5, and where $\alpha := (\alpha_0, \beta)$ are running coupling constants.
We shall see momentarily that this characterization indeed leads to the correct scaling properties for pure 2D quantum gravity.

Since the computation of the volume of $POL_{g,N_0}(A)$ involves an orbifold integration we have, as in the case of Theorem 3.5, an explicit representation of (5.85) in terms of a finite summation over combinatorially distinct triangulations

$$Z_{g,N_0}[\alpha] = e^{-\alpha_0 N_0 - \beta \chi(M)} \qquad (5.86)$$

$$\times \sum_{[T] \in \mathscr{T}_{g,N_0}} \frac{1}{|Aut(T)|} \int_{\Delta_\Theta} d\mu_E^{N_0-1} \int_{\left.\overline{\mathscr{T}}_{g,N_0}^{met}(M;\Theta)\right|_{[T]}} \exp\left(\omega_{WP}^\Theta\right)$$

$$= \frac{\sqrt{N_0}\,[2\pi\,(N_0 + 2g - 2)]^{(N_0-1)}}{(N_0 - 1)!\,\sqrt{2^{N_0-1}}} e^{-\alpha_0 N_0 - \beta \chi(M)}\, Vol_{WP}\left[\mathfrak{M}_{g,N_0}\right],$$

where the notation is that of Theorem 3.5. In particular, the sum runs over the finite set of equivalence classes $[T]$ of distinct triangulations (T, M) in $\mathscr{T}_{g,N_0}$, and $\left.\overline{\mathscr{T}}_{g,N_0}^{met}(M;\Theta)\right|_{[T]}$ denotes the set of stable polyhedral surfaces in $\overline{\mathscr{T}}_{g,N_0}^{met}(M;\Theta)$ whose incidence is in the equivalence class $[T]$ defined by (T, M). With these preliminary remarks along the way we have

Theorem 5.1 *For N_0 sufficiently large, the generic term $Z_{g,N_0}[\alpha]$ in the filtration* (5.85) *scales with N_0 according to*

$$Z_{g,N_0}[\alpha] \approx \frac{\sqrt{N_0}\,[2\pi\,(N_0 + 2g - 2)]^{(N_0-1)}}{(N_0 - 1)!\,\sqrt{2^{N_0-1}}} \times \qquad (5.87)$$

---

[22]The situation is apparently similar to what happens in standard two–dimensional Regge calculus. There, however, a poor understanding of the correct measure to use over (a badly selected part of) $POL_{g,N_0}(A)$ has hampered the use of Regge calculus for regularizing 2D quantum gravity.

$$\times (N_0 + 1)^{\frac{5g-7}{2}} \, C^{-N_0} \left(B_g + \sum_{k=1}^{\infty} \frac{B_{g,k}}{(N_0 + 1)^k} \right) \, e^{-\alpha_0 \, N_0 - \beta \, \chi(M)},$$

where $C = -\frac{1}{2} j_0 \frac{d}{dz} J_0(z)|_{z=j_0}$, *($J_0(z)$ the Bessel function, j_0 its first positive zero); (note that $C \simeq 0.625....$), and where the parameters B_g and $B_{g,k}$ depends only from the genus g.*

Proof The proof is a direct consequence of the large N_0 asymptotics of $Vol_{W-P}(\overline{\mathfrak{M}}_{g,N_0})$ discussed by Manin and Zograf [44, 69]. They obtained

$$Vol_{W-P}(\overline{\mathfrak{M}}_{g,N_0}) = \pi^{2(3g-3+N_0)} \times \tag{5.88}$$

$$\times (N_0 + 1)^{\frac{5g-7}{2}} \, C^{-N_0} \left(B_g + \sum_{k=1}^{\infty} \frac{B_{g,k}}{(N_0 + 1)^k} \right),$$

where $C = -\frac{1}{2} j_0 \frac{d}{dz} J_0(z)|_{z=j_0} \simeq 0.625....$, *with* $J_0(z)$ *the Bessel function, and* j_0 *its first positive zero. The genus dependent parameters* B_g *are explicitly given* [44] *by*

$$\begin{cases} B_0 = \dfrac{1}{A^{1/2} \Gamma(-\frac{1}{2}) C^{1/2}}, & B_1 = \dfrac{1}{48}, \\[2ex] B_g = \dfrac{A^{\frac{g-1}{2}}}{2^{2g-2}(3g-3)! \, \Gamma(\frac{5g-5}{2}) C^{\frac{5g-5}{2}}} \left\langle \tau_2^{3g-3} \right\rangle, & g \geq 2 \end{cases} \tag{5.89}$$

where $A \doteq -j_0^{-1} J_0'(j_0)$, *and* $\left\langle \tau_2^{3g-3} \right\rangle$ *is a Kontsevich-Witten intersection number,* (the coefficients $B_{g,k}$ can be computed similarly-see [44] for details). By inserting this asymptotics in (5.86) we get the stated result. ∎
From (5.87) we immediately get the

Lemma 5.1 *In the large N_0 limit $Z_{g,N_0}[\alpha]$ scales with N_0 according to*

$$Z_{g,N_0}[\alpha] \approx B_g \, \pi^{(6g-6)} \, e^{\mu_0 \, N_0} \, N_0^{\frac{5g-1}{2}-3} \left(1 + O(\frac{1}{N_0}) \right) e^{-\alpha_0 \, N_0 - \beta \, \chi(M)}, \tag{5.90}$$

where $e^{\mu_0} := \frac{\sqrt{2} \, e \, \pi^3}{C} > 1$, *and where C is the constant appearing in (5.88).*

Proof From Stirling's formula in the form

$$k! \simeq \sqrt{2\pi} \, (k+1)^{k+\frac{1}{2}} \, e^{-k-1} \left[1 + O\left(\frac{1}{k} \right) \right] \quad k \to \infty, \tag{5.91}$$

(which is accurate also for small k), we get the estimate

$$\frac{\sqrt{N_0} \, [2\pi \, (N_0 + 2g - 2)]^{(N_0-1)}}{(N_0 - 1)! \, \sqrt{2^{N_0-1}}} \approx \left(\frac{2\pi \, e}{\sqrt{2}} \right)^{N_0}. \tag{5.92}$$

By inserting this into (5.87) and gathering terms we get the stated asymptotic. Thus, (5.90) provides the correct genus-g pure gravity critical exponent

$$\Gamma_{KPZ} = \frac{5g-1}{2}.$$
(5.93)

The nature of the asymptotic (5.90) also shows that Γ_{KPZ} has a modular origin, arising from the cardinality of the cell decomposition of $\overline{\mathfrak{M}}_{g,N_0}$. To go deeper into such a remark, let us recall the expression for the large $N_0(T)$ asymptotics of the distinct triangulation counting in (5.86). This latter asymptotics can be obtained from purely combinatorial (and matrix theory) arguments, (see [3] for the relevant references starting from the classical paper (for the $g = 0$ case) by W. J. Tutte [65], see also [7]), to the effect that

$$Card\left[\mathscr{T}_{g,N_0}\right] \sim \frac{16c_g}{3\sqrt{2\pi}} \cdot e^{v_0 N_0(T)} N_0(T)^{\frac{5g-7}{2}} \left(1 + O(\frac{1}{N_0})\right),$$
(5.94)

where c_g is a numerical constant depending only on the genus g, and $e^{v_0} = (108\sqrt{3})$ is a (non-universal) parameter depending on the set of triangulations considered.[23] Through a comparison of the asymptotics (5.90) and (5.94), we easily get that the integral over the local cells $\overline{\mathscr{T}}_{g,N_0}^{met}(M;\Theta)\Big|_{[T]}$ in (5.86)

$$\int_{\Delta_\Theta} d\mu_E^{N_0-1} \int_{\overline{\mathscr{T}}_{g,N_0}^{met}(M;\Theta)\Big|_{[T]}} \exp\left(\omega_{WP}^\Theta\right),$$
(5.95)

is proportional to A^{N_0-1} for some constant A and does not contribute to the KPZ scaling.

5.8 Polyhedral Liouville Action and KPZ Scaling

A natural question concerns the possibility of extending the above analysis to the more general case of random polyhedral surfaces supporting a $Map(M, V^n)$ QFT with $n \neq 0$. To begin with, let us remark that in such a setting the path integral over $Map(M, V^n)$ is typically discretized according to

$$\int_{\{Map(M,V^n)\}} D_\alpha[\phi]\, e^{-\mathscr{S}_\gamma[\phi;\alpha]}$$
(5.96)

---

[23]Here we deal with generalized triangulations, barycentrically dual to trivalent graphs; in the case of regular triangulations in place of $108\sqrt{3}$ we would get $e^{v_0} = (\frac{4^4}{3^3})$. Also note that the parameter c_g does not play any relevant role in 2D quantum gravity.

$$\simeq e^{-\alpha_0 N_0 - \beta \chi(M)} \int \prod_{k=1}^{N_0} D\phi_k \, \exp\left(-\sum_{ij}(\phi_i - \phi_j)^2 / 2\, l_s^2\right),$$

where the sum is over the edge–connected vertices $\sigma^0(i)$ and $\sigma^0(j)$ of (T_l, M), (see e.g. [3]). However, here we take a different perspective and avoid regularizing the action $\mathscr{S}_\gamma[\phi; \alpha]$ via a discretized Laplacian. Indeed, if we consider the generic polyhedral surface $(T_l, M) \in \mathscr{T}_{g,N_0}^{met}(M; \Theta)$ of given area $A_\gamma(M)$ as a N_0–pointed Riemann surface $((M, N_0), \gamma)$ with conical singularities, then it is still profitable (and fully rigorous) to consider the Gaussian measure $D_\alpha[\phi]$ over $\{Map(M, V^n)\}$ defined by [15, 58]

$$\int_{\{Map(M,V^n)\}} D_\alpha[\phi] \, e^{-\tilde{\mathscr{S}}_\gamma[\phi;\alpha]} := \left[\frac{\det' \Delta((M, N_0), \gamma)}{A_\gamma(M)}\right]^{-\frac{n}{2}}, \tag{5.97}$$

where $\tilde{\mathscr{S}}_\gamma[\phi; \alpha]$ is the harmonic map action (5.1), and $\det' \Delta((M, N_0), \gamma)$ denotes the ζ–function regularized determinant for the scalar Laplacian on $((M, N_0), \gamma)$. Moreover, the discretization (5.96) does not take into account the basic fact that even in the polyhedral setting we still have a large residual action of the diffeomorphism group $\mathscr{D}iff(M, N_0)$ that plays an important role in the theory. In particular, together with the scalar Laplacian $\Delta((M, N_0), \gamma)$ we need to consider the regularized determinant of the vector Laplacian

$$P_1^\dagger P_1 \; : C^\infty((M, N_0), TM) \longrightarrow C^\infty((M, N_0), TM) \tag{5.98}$$

$$w^a \longmapsto (P_1^\dagger P_1)_a^b \, w^a = -2\, \nabla_a\left(\nabla^a w^b + \nabla^b w^a - \gamma^{ab} \nabla_c w^c\right),$$

(see (5.14)), acting on smooth vector fields w^a that vanish at the N_0 vertices of the polyhedral surface (T_l, M).

This strategy in discussing the interplay between the geometry of polyhedral surfaces and two-dimensional Quantum Field Theory is justified by the fact that recently the determinant $\det' \Delta((M, N_0), \gamma)$ has been fully characterized by A. Kokotov[24] in a very elegant way [42].

For the convenience of the reader we have provided a rather complete introduction to Kokotov's results in Appendix C, and in such a framework we start by examining how (5.97) behaves under conformal transformations among polyhedral surfaces. To this end, following [42], we consider for $\eta \in [0, 1]$ two distinct families of polyhedral surfaces with different areas $A_{(1)} \neq A_{(2)}$,

---

[24]In this respect, the situation is here quite simpler than that described in the delicate and prescient analysis of the measure issue in Regge calculus addressed in a series of paper by P. Menotti and P.P. Peirano, (see [45] and references therein).

$$\eta \mapsto (T_{(1)}, M)_\eta \in POL_{g, N_0}(A_{(1)}) \tag{5.99}$$

$$\eta \mapsto (T_{(2)}, M)_\eta \in POL_{g, \widehat{N}_0}(A_{(2)}), \tag{5.100}$$

and with a corresponding different set of marked vertices

$$\{p_k(\eta)\}_{k=1}^{N_0} := \{\sigma^0_{(1)}(k, ; \eta)\}_{k=1}^{N_0}, \tag{5.101}$$

$$\{q_h(\eta)\}_{h=1}^{\widehat{N}_0} := \{\sigma^0_{(2)}(h, \eta)\}_{h=1}^{\widehat{N}_0}, \tag{5.102}$$

(note that generally $N_0 \neq \widehat{N}_0$). We assume that these two vertex sets are disjoint for all $\eta \in [0, 1]$, and that they support distinct η–independent conical singularities $\{\Theta_{(1)}(k)\}$ and $\{\Theta_{(2)}(h)\}$. We also assume that $(T_{(1)}, M)_\eta$ and $(T_{(2)}, M)_\eta$, $\eta \in [0, 1]$ define the same (η–independent) conformal structure

$$((M, N_0), \mathscr{C}^{(1)}_{sg}) \simeq ((M, \widehat{N}_0), \mathscr{C}^{(2)}_{sg}) \simeq (M, \mathscr{C}). \tag{5.103}$$

According to (2.70) the conical metric $ds^2_{T_{(1)}}$ around the generic conical point $p_k(\eta)$ is given, in term of a local conformal parameter $z_{(1)}(k, \eta)$, (with $p_k(\eta) \mapsto z_{(1)}(k, \eta) = 0$), by

$$ds_{T_{(1)}}(k)^2 := \frac{[L(k)]^2}{4\pi^2 \left|z_{(1)}(k, \eta)\right|^2} \left|z_{(1)}(k, \eta)\right|^{2\left(\frac{\Theta_{(1)}(k)}{2\pi}\right)} \left|d\, z_{(1)}(k, \eta)\right|^2, \tag{5.104}$$

whereas the conical metric $ds^2_{T_{(2)}}$ around the generic conical point $q_h(\eta)$ is given, in term of a local conformal parameter $z_{(2)}(h, \eta)$, (with $q_h(\eta) \mapsto z_{(2)}(h, \eta) = 0$), by

$$ds^2_{T_{(2)}}(h) := \frac{[L'(h)]^2}{4\pi^2 \left|z_{(2)}(h, \eta)\right|^2} \left|z_{(2)}(h, \eta)\right|^{2\left(\frac{\Theta_{(2)}(h)}{2\pi}\right)} \left|d\, z_{(2)}(h, \eta)\right|^2. \tag{5.105}$$

Since the points $\{q_h(\eta)\}_{h=1}^{\widehat{N}_0} \in M$, supporting the conical singularities of the metric $ds^2_{T_{(2)}}$, are disjoint from the conical set $\{p_k(\eta)\}_{k=1}^{N_0}$, they are *regular points as seen by* $(T_{(1)}, M)$. To exploit such a remark we need to explicitly identify the vertices $\{q_h\}_{h=1}^{\widehat{N}_0}$ with a corresponding set of smooth points of $(T_{(1)}, M)$. It is clear that such an identification is defined modulo the action of the diffeomorphism group $\mathscr{D}iff(M, N_0)$ preserving setwise the N_0 vertices $\{p_k\}_{k=1}^{N_0}$ of $(T_{(1)}, M)$, (see Appendix B). Thus, we assume that there is a (smooth) embedding

$$\widehat{\iota} : (T_{(1)}, M) \hookrightarrow (T_{(2)}, M), \tag{5.106}$$

injecting the N_0 vertices $\{p_k\}_{k=1}^{N_0}$ of $(T_{(1)}, M)$ into $(T_{(2)}, M)$ in such a way that $\widehat{\iota}(p_i) \notin \{q_h\}_{h=1}^{\widehat{N_0}} \; \forall p_i \in \{p_k\}_{k=1}^{N_0}$. Let $U_k \subset (T_{(1)}, M)$ be a neighborhood of the generic vertex $p_k \in (T_{(1)}, M)$ such that $\{(U_k, z_{(1)(k)})\}_{k=1}^{N_0}$ is a covering of $(T_{(1)}, M)$ with local conformal parameter neighborhoods, and let $\widehat{\iota}_k := \widehat{\iota}|_{U_k}$ denote the restriction of the embedding $\widehat{\iota}$ to such U_k. According to the above remarks, if we take the pull–back, under $\widehat{\iota}$, of the conical metric $ds_{T_{(2)}}^2$ of $(T_{(2)}, M) \cap \widehat{\iota}_k$, then we can assume [42] that there are smooth functions $\mathfrak{f}_{T_{(2)}}[T_{(1)}, z_{(1)}(k, \; \eta)]$ of the local conformal parameter $z_{(1)}(k, \; \eta)$ such that, in a neighborhood of the vertices $\{p_k(\eta)\}_{k=1}^{N_0} \in (T_{(1)}, M)_\eta$, the pull–backed metric $\widehat{\iota}^* \left(ds_{T_{(2)}}^2 \right)$ can be written as

$$\widehat{\iota}^* \left(ds_{T_{(2)}}^2 \right)\Big|_{p_k(\eta)} := \left| \mathfrak{f}_{T_{(2)}}[T_{(1)}, z_{(1)}(k, \; \eta)] \right|^2 \left| d\, z_{(1)}(k, \; \eta) \right|^2. \tag{5.107}$$

Similarly, if we interchange the role of $(T_{(1)}, M)_\eta$ and $(T_{(2)}, M)_\eta$, then there is an embedding

$$\iota \; : \; (T_{(2)}, M) \; \hookrightarrow \; (T_{(1)}, M), \tag{5.108}$$

and smooth functions of the local conformal parameter $z_{(2)}(h, \; \eta)$,

$$\mathfrak{f}_{T_{(1)}}[T_{(2)}, \; z_{(2)}(h, \; \eta)], \tag{5.109}$$

such that in a neighborhood of the points $\{q_h(\eta)\}_{h=1}^{\widehat{N_0}} \in M$ the pull–backed metric $\iota_h^* \left(ds_{T_{(1)}}^2 \right)$ takes the form

$$\iota_h^* \left(ds_{T_{(1)}}^2 \right)\Big|_{q_h(\eta)} := \left| \mathfrak{f}_{T_{(1)}}[T_{(2)}, \; z_{(2)}(h, \; \eta)] \right|^2 \left| d\, z_{(2)}(h, \; \eta) \right|^2. \tag{5.110}$$

With these notational remarks along the way, we have the following basic result by A. Kokotov [42], (see also Appendix C),

Theorem 5.2 *Let* $\det' \Delta(T_{(1)})$ *and* $\det' \Delta(T_{(2)})$ *respectively denote the ζ–regularized determinants of the Laplacian associated with the conical metrics* $ds_{T_{(1)}}^2$ *and* $ds_{T_{(2)}}^2$. *Then, there is a constant* C_{12} *independent of* $\eta \in [0, 1]$ *such that*

$$\frac{\det' \; \Delta(T_{(1)})}{\det' \; \Delta(T_{(2)})} = C_{12} \; \frac{A_{(1)}}{A_{(2)}} \; \frac{\prod_{h=1}^{\widehat{N_0}} \left| \mathfrak{f}_{T_{(1)}}(T_{(2)}, h) \right|^{\frac{1}{6}\left(\frac{\Theta_{(2)}(h)}{2\pi} - 1 \right)}}{\prod_{k=1}^{N_0} \left| \mathfrak{f}_{T_{(2)}}(T_{(1)}, k) \right|^{\frac{1}{6}\left(\frac{\Theta_{(1)}(k)}{2\pi} - 1 \right)}}, \tag{5.111}$$

where[25] we have set

$$\mathfrak{f}_{T_{(1)}}(T_{(2)},\, h) := f_{T_{(1)}}[T_{(2)},\, z_{(2)}(h,\, \eta) = 0], \tag{5.112}$$

and

$$\mathfrak{f}_{T_{(2)}}(T_{(1)},\, k) := f_{T_{(2)}}[T_{(1)},\, z_{(1)}(k,\, \eta) = 0]. \tag{5.113}$$

It is important to stress that the assignment of these Liouville fields (5.116), and (5.117) cannot be arbitrary. Indeed, if we consider three polyhedral surfaces $(T_{(\alpha)}, M) \in POL_{g, N_0(\alpha)}(A_{(\alpha)})$, $\alpha = 1, 2, 3$, within the same conformal class, and define [42]

$$Q(\alpha) := \frac{\det' \Delta_{(\alpha)}}{A_{(\alpha)}}, \tag{5.114}$$

then by considering the product of ratios $\prod_\alpha (Q(\alpha)/Q(\alpha + 1))$, (with $\alpha + 1 = 1$ for $\alpha = 3$), one gets a basic result of A. Kokotov [42], consequence of the Weil reciprocity law, that we rewrite in our setting as the cocycle condition

$$\prod_{\alpha=1}^{3} \prod_{k_\alpha=1}^{N_0(\alpha)} \left[\frac{|\mathfrak{f}_{T_{(\alpha+2)}}(T_{(\alpha)},\, k_\alpha)|}{|\mathfrak{f}_{T_{(\alpha+1)}}(T_{(\alpha)},\, k_\alpha)|} \right]^{\frac{1}{6}\left(\frac{\Theta^{(\alpha)}(k_\alpha)}{2\pi} - 1 \right)} = 1, \tag{5.115}$$

where α is, as usual, defined   mod 3. Note that the $\mathfrak{f}$–ratios in the above expression are independent from the local conformal parameter z chosen.

We can profitably rewrite (5.111) in a more symmetric form:

Lemma 5.2 *Let*

$$\mathfrak{u}_{T_{(2)}}(T_{(1)},\, k) := \{2 \ln |\mathfrak{f}_{T_{(2)}}(T_{(1)},\, k)|\}, \tag{5.116}$$

denote the discretized Liouville fields on $(T_{(1)}, M)$ describing the polyhedral surface $(T_{(2)}, M)$ around the vertices $\{p_k\}$ of $(T_{(1)}, M)$. Similarly, let

$$\mathfrak{u}_{T_{(1)}}(T_{(2)},\, h) := \{2 \ln |\mathfrak{f}_{T_{(1)}}(T_{(2)},\, h)|\}, \tag{5.117}$$

be the discretized Liouville fields on $(T_{(2)}, M)$ locally describing $(T_{(1)}, M)$ around the vertices $\{q_h\}$ of $(T_{(2)}, M)$. Then we have

---

$$\frac{\det' \Delta(T_{(1)})}{A_{(1)}} \exp \frac{1}{12} \left[\sum_{k=1}^{N_0} \left(\frac{\Theta_{(1)}(k)}{2\pi} - 1 \right) \mathrm{u}_{T_{(2)}}(T_{(1)}, k) \right] \qquad (5.118)$$

$$= C_{12} \frac{\det' \Delta(T_{(2)})}{A_{(2)}} \exp \frac{1}{12} \left[\sum_{h=1}^{\widehat{N_0}} \left(\frac{\Theta_{(2)}(h)}{2\pi} - 1 \right) \mathrm{u}_{T_{(1)}}(T_{(2)}, h) \right].$$

Note that the exponential terms in (5.118) are the natural counterpart, on the polyhedral surfaces $(T_{(1)}, M)$ and $(T_{(2)}, M)$, of the dilaton coupling $\int_M f\, R(\widehat{\gamma})\, d\mu_{\widehat{\gamma}}$.

As shown in [42], the ratio on the right hand side of (5.111) corresponds, for a polyhedral surface, to the standard Liouville action

$$\frac{\det' \Delta_{(1)}}{\det' \Delta_{(2)}} = C_{12}\, \frac{A_{(1)}}{A_{(2)}} \exp \left\{ \frac{\sqrt{-1}}{6\,\pi} \int_M \ln \frac{\rho_{(2)}}{\rho_{(1)}}\, \partial_z \partial_{\bar{z}} \ln \left(\rho_{(1)}\, \rho_{(2)} \right) dz \wedge d\bar{z} \right\},$$

$$(5.119)$$

associated with two smooth, conformally related, metrics $ds^2_{(1)} = \rho^{-2}_{(1)}(z, \bar{z})\, dz \otimes d\bar{z}$ and $ds^2_{(2)} = \rho^{-2}_{(2)}(z, \bar{z})\, dz \otimes d\bar{z}$. More explicitly, by comparing (5.111) and (5.118) with (5.17) we formally have the correspondence

Lemma 5.3

$$S_L(\widehat{\gamma}, u) \Longrightarrow S_L^{(Pol)} \left(T_{(1)}, T_{(2)} \right) := \frac{1}{24} \left[\sum_{k=1}^{N_0} \left(\frac{\Theta_{(1)}(k)}{2\pi} - 1 \right) \mathrm{u}_{T_{(2)}}(T_{(1)}, k) \right.$$

$$\left. - \sum_{h=1}^{\widehat{N_0}} \left(\frac{\Theta_{(2)}(h)}{2\pi} - 1 \right) \mathrm{u}_{T_{(1)}}(T_{(2)}, h) - 24 \ln C_{12} \right].$$

$$(5.120)$$

Kokotov's analysis naturally extends to the regularized determinant of the vector Laplacian (5.98) on the (ordered[26]) pair of polyhedral surfaces $(T_{(1)}, M)$ and $(T_{(2)}, M)$, and we have

Theorem 5.3

$$\left[\det' \left(P_1^{\dagger} P_1 \right)_{T_{(1)}} \right]^{\frac{1}{2}} = \left[\det' \left(P_1^{\dagger} P_1 \right)_{T_{(2)}} \right]^{\frac{1}{2}} e^{-26\, S_L^{(Pol)} \left(T_{(1)}, T_{(2)} \right)}, \qquad (5.121)$$

---

[26]The ordering is important for characterizing the diffeomorphism group relevant to the problem: $\mathscr{D}iff(M, N_0)$ if we are injecting $(T_{(1)}, M)$ into $(T_{(2)}, M)$ so as to consider the neighborhoods of the vertices $\{q_h\} \in (T_{(2)}, M)$ as (conformally) smooth as seen by $(T_{(1)}, M)$, whereas $\mathscr{D}iff(M, \widehat{N_0})$ is the appropriate group when we inject $(T_{(2)}, M)$ into $(T_{(1)}, M)$.

or equivalently

$$\left[\det{}'\left(P_1^\dagger P_1\right)_{T_{(1)}}\right]^{\frac{1}{2}} \exp \frac{26}{24}\left[\sum_{k=1}^{N_0}\left(\frac{\Theta_{(1)}(k)}{2\pi}-1\right)\, \mathrm{u}_{T_{(2)}}(T_{(1)},k)\right] \qquad (5.122)$$

$$= C_{12}^{26}\left[\det{}'\left(P_1^\dagger P_1\right)_{T_{(2)}}\right]^{\frac{1}{2}} \exp \frac{26}{24}\left[\sum_{h=1}^{\widehat{N_0}}\left(\frac{\Theta_{(2)}(h)}{2\pi}-1\right)\, \mathrm{u}_{T_{(1)}}(T_{(2)},h)\right].$$

Proof As stressed above, it is not difficult to directly check that Kokotov's technique extends to the heat kernel characterization of $\det{}'\left(P_1^\dagger P_1\right)$ on N_0–pointed Riemann surfaces (M, N_0), (for this latter see e.g. [16] sections F and L), decorated with conical singularities. $\blacksquare$

If we fix a *reference* polyhedral surface $(\widehat{T}, M)$ and compare it with the whole set of polyhedral surfaces $(T_l, M) \in POL_{g,N_0}(A)$ we can specialize the above results to the

Lemma 5.4 *Let* $(\widehat{T}, M) \in POL_{g,\widehat{N}_0}(\widehat{A})$ *be a given polyhedral surface of area* $\widehat{A}$, *vertex set* $\{q_i\}_{i=1}^{\widehat{N}_0}$, *conical angles* $\{\widehat{\theta}(i)\}_{i=1}^{\widehat{N}_0}$ *and local metric*

$$ds_{\widehat{T}}^2(h) = \frac{\widehat{L}^2(h)}{4\pi^2}\,|\widehat{z}_h|^{2\left(\frac{\widehat{\theta}(h)}{2\pi}-1\right)}\,|d\widehat{z}_h|^2. \qquad (5.123)$$

Let us denote by $(T_l, M) \in POL_{g,N_0}(A)$ *the generic polyhedral surface of area* A *and vertex set* $\{p_k\}_{k=1}^{N_0}$ *supporting conical angles* $\{\theta(k)\}_{k=1}^{N_0}$. *We assume that* $A \neq \widehat{A}$, $\widehat{N}_0 \geq N_0$, *and that for any* (T_l, M) *there exists a corresponding embedding*

$$\widehat{\iota}_T : (T_l, M) \hookrightarrow (\widehat{T}, M), \qquad (5.124)$$

injecting the vertices $\{p_k\}_{k=1}^{N_0}$ *of* (T_l, M) *into smooth points of* $(\widehat{T}, M)$, *i.e.* $\widehat{\iota}_T(p_k) \notin \{q_i\}_{i=1}^{\widehat{N}_0}$. *We let* $\{(U_k, z_k)\} \subset (T_l, M)$ *be local coordinate neighborhoods of the vertices* $\{p_k\} \in (T_l, M)$, *and for a vertex* $q_h \in \widehat{\iota}_T(U_k)$ *we denote by*

$$\zeta_h(k) := \widehat{\iota}_T^{-1}(q_h)\big|_{U_k}, \qquad (5.125)$$

the corresponding value of the conformal parameter z_k *in* U_k. *Then, for any* (T_l, M) *there exists a corresponding set of smooth Liouville fields*

$$u_{\widehat{T}}(T, z_k) := \ln\left[\frac{\widehat{L}^2(h)}{4\pi^2}\,|z_k-\zeta_h(k)|^{2\left(\frac{\widehat{\theta}(h)}{2\pi}-1\right)}\right], \qquad (5.126)$$

such that

$$\widehat{\iota}_T^*\big|_{U_k}\left(ds_{\widehat{T}}^2\right) = e^{u_{\widehat{T}}(T,z_k)}\,|dz(k)|^2\,, \qquad k=1,\ldots,N_0\,, \qquad (5.127)$$

provides the (smooth) conformal structure locally describing $(\widehat{T}, M)$ around sufficiently small neighborhoods $\{\tilde{U}_k\}$ of the vertices $\{p_k\}$ of (T_l, M).

Proof Since the points $\widehat{\iota}_T^{-1}\left(\{q_h(\eta)\}_{h=1}^{\widehat{N_0}}\right)$ are disjoint from the vertices $\{p_k(\eta)\}_{k=1}^{N_0}$ of (T_l, M), by restricting the neighborhood U_k to a suitably smaller coordinate neighborhood $(\tilde{U}_k, z_k)$ of the generic vertex p_k, (e.g. by requiring that $|z_k| < |\zeta_h(k)|$ for any $\zeta_h(k) \in \widehat{\iota}_T^{-1}(q_h)$), we immediately get the expression of the smooth Liouville fields (5.126) in terms of which we can explicitly write the parametrization (5.107) for $(\widehat{T}, M)$. ∎

As (T_l, M) varies in $POL_{g, N_0}(A)$ we can characterize the area of M in the pulled–back metric $\widehat{\iota}_T^*\left(ds_{\widehat{T}}^2\right)$ according to

$$\widehat{A}(T) = \sum_{k=1}^{N_0} \int_{U_k} \varphi_k(z_k)\, \widehat{\iota}_T^*(d\widehat{\mu})|_{U_k}\,, \tag{5.128}$$

where

$$\widehat{\iota}_T^*(d\widehat{\mu})|_{U_k} = \frac{\sqrt{-1}}{2}\, e^{u_{\widehat{T}}(T, z_k)}\, dz_k \wedge d\bar{z}_k, \tag{5.129}$$

is the pull–back of the Riemannian measure on $(\widehat{T}, M)$ associated with the metric (5.123), and $\{\varphi_k\}_{k=1}^{N_0}$ is a partition of unity subordinated to the covering $\{U_k\}_{k=1}^{N_0}$ of (T_l, M). In general such an area $\widehat{A}(T)$ will not be equal to the area $\widehat{A}$ of the given polyhedral surface $(\widehat{T}, M)$ since, under the vertex injection $\widehat{\iota}_T : (T_l, M) \hookrightarrow (\widehat{T}, M)$, the image $\widehat{\iota}_T(T_l, M)$ may fold, intersect, or accumulate around a subset of vertices of $(\widehat{T}, M)$. To somewhat characterize such a behavior let us start by observing that since around p_k the corresponding Liouville field $u_{\widehat{T}}(T, z_k)$ is smooth we can exploit the mean value theorem, and introduce a parameter $\beta_k(T) \in \mathbb{R}$ such that, for any given coordinate neighborhood U_k,

$$\int_{\overline{U}_k} \varphi_k(z_k)\, \widehat{\iota}_T^*(d\widehat{\mu})|_{U_k} = e^{\beta_k(T)\, u_{\widehat{T}}(T, k)}, \tag{5.130}$$

where

$$u_{\widehat{T}}(T, k) := u_{\widehat{T}}(T, z_k = 0) = \ln\left[\frac{\widehat{L}^2(h)}{4\pi^2}\, |\zeta_h(k)|^{2\left(\frac{\widehat{\theta(h)}}{2\pi} - 1\right)}\right]. \tag{5.131}$$

Geometrically, $\beta_k(T)$ represents the logarithm of the area of the *Liouville vertex cell*

$$\left(\overline{U}_k, \widehat{\iota}_T^*\left(ds_{\widehat{T}}^2\right)\right), \tag{5.132}$$

describing the given $(\widehat{T}, M)$ around the p_k–vertex of (T_l, M). We can take the average of the $\{\beta_k(T)\}_{k=0}^{N_0}$ over the N_0 vertexes of (T_l, M) so as to obtain

$$\beta(T) := N_0^{-1} \sum_{k=1}^{N_0} \beta_k(T), \tag{5.133}$$

providing the average (of the logarithm of the) vertex area associated with the description of $(\widehat{T}, M)$ with respect to (T_l, M). Eventually, we are naturally led to average this quantity over the whole set of polyhedral surfaces (T_l, M) in $POL_{g, N_0}(A)$ and characterize the *Liouville free vertex area*[27] associated with $\left(POL_{g, N_0}(A), \widehat{\imath}\right)$ according to

$$\beta(\imath) := Vol\left[POL_{g, N_0}(A)\right]^{-1} \int_{\overline{POL_{g, N_0}(A)}} \beta(T) \, d\mu_E^{N_0-1} \exp\left(\omega_{WP}\right), \tag{5.134}$$

(see Definition 3.1 and the accompanying remarks for the definition of $POL_{g, N_0}(A)$ averages). Since for large values of N_0 the integral over $POL_{g, N_0}(A)$ may be dominated by degenerate triangulations, it is not obvious that the parameter $\beta(\imath)$ so defined is, for $N_0 \nearrow \infty$, always well–behaved, and, (owing to the presence of the logarithm), it may be even appropriate to analytically continue it in the complex domain. In any case, it characterizes the effective area of $(\widehat{T}, M)$ as described by the vertex Liouville fields $\{u_{\widehat{T}}(T, k)\}_{k=1}^{N_0}$ as they fluctuate over the triangulations in $POL_{g, N_0}(A)$. Thus, we introduce the following

Definition 5.2 (Effective Area) The effective area of $(\widehat{T}, M)$ generated by the Liouville fields $\{u_{\widehat{T}}(T, k)\}_{k=1}^{N_0}$ over a polyhedral surface $(T_l, M) \in POL_{g, N_0}(A)$ is provided by

$$\widehat{A}_{eff}(T) := \sum_{k=1}^{N_0} e^{\beta(\widehat{\imath}) \, u_{\widehat{T}}(T, k)}, \tag{5.135}$$

where $\beta(\imath)$ is the Liouville free vertex area associated with $\left(POL_{g, N_0}(A), \imath\right)$ defined by (5.134).

Armed with these preparatory results and remarks, let us envisage now the situation where the *given* polyhedral surface $(\widehat{T}, M)$ is generated out of random immersions $\phi : (T_l, M) \hookrightarrow (V^n, g_{ab} = \delta_{ab})$ of the polyhedral surfaces $(T, M) \in POL_{g, N_0}(A)$ in the (flat) ambient manifold (V^n, δ), (δ denoting the Euclidean metric), *viz.* $(\widehat{T}, M) := \phi\left((T_l, M)\right)$ with

$$ds_{\widehat{T}}^2 = \phi^*(\delta). \tag{5.136}$$

In such a case, we can naturally identify the injection $\widehat{\imath}_T\left(\{p_k\}^{N_0}\right)$ of the vertex set $\{p_k\}^{N_0}$ of (T_l, M) in the given $(\widehat{T}, M)$ with the image $\phi\left(\{p_h\}^{N_0}\right)$. For N_0 finite, and in full analogy with the field theoretic analysis described in the previous sections, the corresponding random Liouville fields $u_{\widehat{T}}(T, z_k)$ on (T_l, M) can be promoted to an immersion variable

$$\phi^{n+1} := u_{\widehat{T}}(T, z_k), \tag{5.137}$$

so as to consider the given $(\widehat{T}, M)$ as being actually generated by random immersions of the polyhedral surfaces $(T, M) \in POL_{g, N_0}(A)$ in the (flat) ambient manifold (V^{n+1}, δ). Also note that in such a setting the parameter $\beta(\imath)$, defined by (5.134), providing the Liouville free vertex area associated with the pair $\left(POL_{g, N_0}(A), \widehat{\imath}\right)$ is now a functional of the embedding variables $\{\phi\}$, and in order to emphasize such a dependence we correspondingly write $\beta(\imath) \equiv \beta_n$.

With the configuration just described we can naturally associate the functional

$$(\widehat{T}, M) \longmapsto Z_{g, N_0, A}\left[(\widehat{T}, M); \widehat{A}_{eff}\right] := \tag{5.138}$$

$$:= \int_{\overline{POL_{g, N_0}(A)}} \left(\left[\frac{\det' \Delta_{(T)}}{A_{(T)}}\right]^{-\frac{(n+1)}{2}} \left[\det'\left(P_1^\dagger P_1\right)_{T_{(1)}}\right]^{\frac{1}{2}}\right.$$

$$\times e^{\frac{[(n+1)-26]}{24}\left[\sum_{k=1}^{N_0}\left(\frac{\Theta^{(1)}(k)}{2\pi}-1\right)u_{\widehat{T}}(T, k)\right]}$$

$$\left. \times \delta\left(\widehat{A}_{eff} - \sum_{h=1}^{N_0} e^{\beta_n u_{\widehat{T}}(T, h)}\right)\right) d\mu_E \exp\left(\omega_{WP}\right),$$

which, normalized with respect to the volume of $POL_{g, N_0}(A)$, can indeed be interpreted as the probability of generating the given $(\widehat{T}, M)$, with $\widehat{N}_0$ vertices and with a given effective area $\widehat{A}_{eff}$, out of random immersions of the polyhedral surfaces $(T, M) \in POL_{g, N_0}(A)$ in a flat Euclidean space $V^{(n+1)}$. Note that in (5.138) we have promoted the Liouville field ϕ^{n+1}, locally described by $u_{\widehat{T}}(T, z_k)$, to a free dynamical field by introducing a further determinant factor $\det' \Delta_{(T)}/A_{(T)}$ associated with the free–field scalar action $\mathscr{S}_\gamma[\phi^{n+1}]$, so that

$$\left[\frac{\det' \Delta_{(T)}}{A_{(T)}}\right]^{-\frac{(n)}{2}} \Rightarrow \left[\frac{\det' \Delta_{(T)}}{A_{(T)}}\right]^{-\frac{(n+1)}{2}}. \tag{5.139}$$

Even if averaged over $POL_{g,N_0}(A)$, the functional $Z_{g,N_0,A}[(\widehat{T},M);\widehat{A}_{eff}]$ exhibits a non–trivial scale dependence. Explicitly, let us consider a one–parameter constant shift of the Liouville field $u_{\widehat{T}}(T,z_k)$ according to

$$u_{\widehat{T}}(T,\,k) \longmapsto u_{\widehat{T}}^{\epsilon}(T,\,k) + \epsilon\,\beta_n^{-1}\,, \qquad 0 \le \epsilon \le \ln\widehat{A}_{eff}. \tag{5.140}$$

Under this shift we compute

$$Z_{g,N_0,A}[(\widehat{T},M);\widehat{A}_{eff}^{\epsilon}] :=$$

$$:= \int_{\overline{POL_{g,N_0}(A)}} \left(\left[\frac{\det{}'\,\Delta_{(T)}}{A_{(T)}} \right]^{-\frac{n+1}{2}} \left[\det{}'\left(P_1^{\dagger}P_1\right)_{T_{(1)}} \right]^{\frac{1}{2}} \right.$$

$$\times e^{\frac{(n-25)}{24}\left[\sum_{k=1}^{N_0} \left(\frac{\Theta^{(1)}(k)}{2\pi} - 1 \right) u_{\widehat{T}}^{\epsilon}(T,k) \right]}$$

$$\left. \times \delta\left(\widehat{A}_{eff} - \sum_{h=1}^{N_0} e^{\beta_n\,u_{\widehat{T}}^{\epsilon}(T,h)} \right) \right)\, d\mu_E\, \exp\left(\omega_{WP}\right)$$

$$= e^{-\epsilon\left(\frac{(n-25)}{24}\,\beta_n^{-1}\,\chi(M)\,+1 \right)}\, Z_{g,N_0,A}[(\widehat{T},M);\,e^{-\epsilon}\,\widehat{A}_{eff}^{\epsilon}], \tag{5.141}$$

where

$$Z_{g,N_0,A}[(\widehat{T},M);\,e^{-\epsilon}\,\widehat{A}_{eff}] := \tag{5.142}$$

$$:= \int_{\overline{POL_{g,N_0}(A)}} \left(\left[\frac{\det{}'\,\Delta_{(T)}}{A_{(T)}} \right]^{-\frac{n+1}{2}} \left[\det{}'\left(P_1^{\dagger}P_1\right)_{T_{(1)}} \right]^{\frac{1}{2}} \right.$$

$$\times e^{\frac{(n-25)}{24}\left[\sum_{k=1}^{N_0} \left(\frac{\Theta^{(1)}(k)}{2\pi} - 1 \right) u_{\widehat{T}}(T,k) \right]}$$

$$\left. \times \delta\left(e^{-\epsilon}\,\widehat{A}_{eff} - \sum_{h=1}^{N_0} e^{\beta_n\,u_{\widehat{T}}(T,h)} \right) \right)\, d\mu_E\, \exp\left(\omega_{WP}\right).$$

Setting $\epsilon = \ln\widehat{A}_{eff}$, we get from (5.141) the area scaling law

$$Z_{g,N_0,A}[(\widehat{T},M);\,e^{-\epsilon}\,\widehat{A}_{eff}] = A^{-\left(\frac{(n-25)}{24}\,\beta_n^{-1}\,\chi(M)\,+1 \right)}\, Z_{g,N_0,A}[(\widehat{T},M);\,1]. \tag{5.143}$$

Note that this is consistent with the KPZ scaling exponent Γ_{KPZ} given by (5.62) if the parameter β_n assumes the value

$$\beta_n = \frac{n-25}{12}\frac{b^2}{b^2+1},\tag{5.144}$$

where b is provided by

$$b := \sqrt{\frac{25-n}{24}} - \sqrt{\frac{1-n}{24}},\tag{5.145}$$

(see (5.44)).

The above analysis, while not being yet a proof of KPZ scaling for the coupling of conformal matter interacting with quantum polyhedral surfaces, clearly goes a long way in showing the pivotal role of Kokotov's discretized Liouville action in addressing such a delicate question. And we expect that the techniques described may led to a deeper understanding of the anomalous scaling of conformal matter interacting with 2D gravity.

5.9   Polyhedral Surfaces and Open/Closed String Duality

In a rather general sense, polyhedral surfaces provide also a natural kinematical framework within which we can discuss open/closed string duality.[28] Indeed, advances in our understanding of Open/Closed string duality have provided a number of paradigmatical connections between Riemann moduli space theory, piecewise-linear geometry, and the study of the gauge/gravity correspondence. These connections have a two-fold origin. On the mathematical side they are deeply related to the fact that moduli space admits natural polyhedral decompositions which are in a one-to-one correspondence with classes of suitably decorated graphs. On the physical side they are consequence of the observation that these very decorated (Feynman) graphs parametrize consistently the quantum dynamics of conformal and gauge fields.

A basic problem in any such a setting is to provide an explanation of how open/closed duality is dynamically generated. In particular how a closed surface is related to a corresponding open surface, with gauge-decorated boundaries, in such a way that the quantization of such a correspondence leads to a open/closed duality. Typically, the natural candidate for such a mapping is Strebel's theorem which, as we have been recalling, allows to reconstruct a closed N-pointed Riemann surfaces M of genus g out of the datum of a the quadratic differential associated

---

[28]This is a vast subject with thousands of relevant papers. A nice selection, among those emphasizing the connection with combinatorial aspects, is provided by [1, 2, 13, 30, 32–36].

with a ribbon graph [49, 61]. Looked at face value, ribbon graphs are open Riemann surfaces which one closes by inserting punctured discs, (so generating semi-infinite cylindrical ends). The dynamics of gauge fields decorating the boundaries of the ribbon graph is naturally framed within the context of boundary conformal field theory (BCFT) which indeed plays an essential role in the onset of a open/closed duality regime. The rationale of such a role of BCFT is to be seen in the fact that BCFT is based on algebraic structures parametrized by the moduli space $\mathfrak{M}(g; N)$. This parametrization is deeply connected with Strebel's theorem in the sense that it is consistent with the operation of sewing together any two ribbon graphs (open surfaces) with (gauge-decorated) boundaries, provided that we match the complex structure and the decoration in the overlap and keep track of which puncture is ingoing and which is outgoing. In such a setting a BCFT leads to a natural algebra, over the decorated cell decomposition of Riemann moduli space, which can be related to the algebra of physical space of states of the theory and to their boundary dynamics. It is fair to say that in such a sense BCFT realizes open/closed duality as the quantization of a gauge-decorated Strebel's mapping.

Are ribbon graphs, with the attendant BCFT techniques, the only key for addressing the combinatorial aspects of Open/Closed String Duality? In Chap. 3, while discussing the symplectic geometry of the space of polyhedral surfaces, we have been naturally led to the Penner and Thurston (see e.g. [55, 63]) parametrization of moduli space. This does not emphasize the role of conformal geometry, but rather exploits the parametrization of the moduli space in terms of hyperbolic surfaces, and one generates a combinatorial decomposition of Riemann moduli space via the geometry of surface geodesics. As we have seen, one of the advantages of the hyperbolic point of view is that one has a subtle interplay between the hyperbolic geometry of open surfaces and the singular Euclidean structure associated with closed polyhedral surface. In particular we have shown that, in assembling a surface with ideal hyperbolic triangles out of the glueing pattern of a closed polyhedral surface (T_l, M), we get an open surface with boundary $\Omega(T_l, M)$, where the boundary lengths are naturally associated with the conical angles of (T_l, M). This gives a geometrical mechanism describing the transition between closed and open surfaces which, in a dynamical sense, is more interesting than Strebel's construction. As we have proven, the correspondence between closed polyhedral surfaces and open hyperbolic surface is easily promoted to the corresponding moduli spaces: $\mathfrak{M}_{g;N_0} \times \mathbb{R}_+^N$ the moduli spaces of N_0-pointed closed Riemann surfaces of genus g whose marked points are decorated with the given set of conical angles, and $\mathfrak{M}_{g;N_0}(L) \times \mathbb{R}_+^{N_0}$ the moduli spaces of open Riemann surfaces of genus g with N_0 geodesic boundaries decorated by the corresponding lengths. This provides a nice kinematical set up for establishing a open/closed string duality once the appropriate field decoration is activated.[29]

---

[29]Different geometrical aspects of the role of the hyperbolic point of view in open/closed string duality have been discussed also by R. Kaufmann and R. Penner [39].

We start by recalling that, according to Theorem 3.5, if the spaces $POL_{g,N_0}(A)$ and $POL_{g,N_0}(\Theta, A)$ denote the set of polyhedral structures in $POL_{g,N_0}(M)$ with given polyhedral area A and with a given sequence of conical angles $\{\Theta(k)\}_{k=1}^{N_0}$, then the symplectic volumes of the Riemann moduli spaces $\mathfrak{M}_{g,N_0}$ and $\mathfrak{M}_{g,N_0}(L)$ induce natural volume forms on $POL_{g,N_0}(A)$ and $POL_{g,N_0}(\Theta, A)$, and we have

$$Vol\left[POL_{g,N_0}(A)\right] = \frac{\sqrt{N_0}\,[2\pi\,(N_0 + 2g - 2)]^{(N_0-1)}}{(N_0-1)!\,\sqrt{2^{N_0-1}}}\,Vol_{WP}\left[\mathfrak{M}_{g,N_0}\right], \qquad (5.146)$$

$$Vol\left[POL_{g,N_0}(\Theta, A)\right] = Vol_{WP}\left[\mathfrak{M}_{g,N_0}(L)\right], \qquad (5.147)$$

where the boundary lengths vector L is given by

$$L = \left(\left\{L_k = \left|\ln\frac{\Theta(k)}{2\pi}\right|\right\}_{k=1}^{N_0}\right). \qquad (5.148)$$

We can actually compute the dependence of $Vol_{WP}\left(\mathfrak{M}_{g,N_0}(L)\right)$ from the boundary lengths $L = (L_1, \ldots, L_{N_0})$ by exploiting a striking result due to M. Mirzakhani [46, 47]

Theorem 5.4 (Maryam Mirzakhani (2003)) *The Weil-Petersson volume $Vol_{WP}\left(\mathfrak{M}_{g,N_0}(L)\right)$ is a polynomial in $L_1, \ldots, L_{N_0}$*

$$Vol_{WP}\left(\mathscr{M}_{g,N_0}(L)\right) = \sum_{\substack{(\alpha_1,\ldots,\alpha_{N_0})\in(\mathbb{Z}_{\geq 0})^{N_0} \\ |\alpha|\leq 3g-3+N_0}} C_{\alpha_1\ldots\alpha_{N_0}}\,L_1^{2\alpha_1}\ldots L_{N_0}^{2\alpha_{N_0}}, \qquad (5.149)$$

where $|\alpha| = \sum_{i=1}^{N_0}\alpha_i$ and where the coefficients $C_{\alpha_1\ldots\alpha_{N_0}} > 0$ are (recursively determined) numbers of the form

$$C_{\alpha_1\ldots\alpha_{N_0}} = \pi^{6g-6+2N_0-2|\alpha|}\cdot q \qquad (5.150)$$

for rationals $q \in \mathbb{Q}$.
Moreover Mirzakhani [46, 47] is also able to express $C_{\alpha_1\ldots\alpha_{N_0}}$ in terms of the intersection numbers $< \tau_{\alpha_1}\ldots\tau_{\alpha_{N_0}} >$ [67] of the tautological line bundles $\mathscr{L}(i)$ over $\overline{\mathfrak{M}}_{g,N_0}$ according to

$$C_{\alpha_1\ldots\alpha_{N_0}} = \frac{2^{m(g,N_0)|\alpha|}}{2^{|\alpha|}\prod_{i=1}^{N_0}\alpha_i!(3g-3+N_0-|\alpha|)!} < \tau_{\alpha_1}\ldots\tau_{\alpha_{N_0}} >, \qquad (5.151)$$

$$< \tau_{\alpha_1}\ldots\tau_{\alpha_{N_0}} > \doteq \int_{\overline{\mathscr{M}}_{g,N_0}} \psi_1^{\alpha_1}\ldots\psi_{N_0}^{\alpha_{N_0}}\cdot\omega_{WP}^{3g-3+N_0-|\alpha|}, \qquad (5.152)$$

where ψ_i is the first Chern class of $\mathscr{L}(i)$, and where $m(g, N_0) \doteq \delta_{g,1}\delta_{N_0,1}$. Note in particular that the constant term $C_{0\ldots0}$ of the polynomial $Vol_{WP}\left(\mathfrak{M}_{g,N_0}(L)\right)$ is the volume of $\overline{\mathfrak{M}}_{g,N_0}$ i.e.,

$$C_{0\ldots0} = Vol_{WP}\left(\overline{\mathfrak{M}}_{g,N_0}\right) = \int_{\overline{\mathfrak{M}}_{g,N_0}} \frac{\omega_{WP}^{3g-3+N_0(T)}}{(3g-3+N_0(T))!}. \tag{5.153}$$

We use this basic result of Mirzakhani by exploiting the identification (5.147) between $Vol_{WP}\left[\mathfrak{M}_{g,N_0}(L)\right]$ and $Vol\left[POL_{g,N_0}(\Theta, A)\right]$. We also need the polyhedral characterization of first Chern class ψ_i of $\mathscr{L}(i)$ provided by Theorems 1.5 and 2.5, according to which we have

$$\omega_k := \sum_{1 \leq \alpha < \beta \leq q(k)-1} d\left(\frac{\theta_\alpha(k)}{\Theta(k)}\right) \wedge d\left(\frac{\theta_\beta(k)}{\Theta(k)}\right) = -\pi_k^*\left[\Upsilon^*\left(c_1(\mathscr{L}_k)\right)\right],$$

$$\tag{5.154}$$

(see (2.92)), where π_k is the polyhedral map defined by (1.96). Under such a dictionary, we can interpret Mirzakhani's result as expressing a natural kinematical duality between the modular geometry of open Riemann surfaces and the quantum geometry of closed polyhedral surfaces. Explicitly we have

Theorem 5.5 *The symplectic volume of* $Vol\left[POL_{g,N_0}(\Theta, A)\right]$ *is provided by*

$$Vol\left[POL_{g,N_0}(\Theta, A)\right] = \sum_{\substack{(\alpha_1,\ldots,\alpha_{N_0})\in(\mathbb{Z}_{\geq0})^{N_0} \\ |\alpha|\leq 3g-3+N_0}} C_{\alpha_1\ldots\alpha_{N_0}} \left|\ln\frac{\Theta(1)}{2\pi}\right|^{2\alpha_1} \cdots \left|\ln\frac{\Theta(N_0)}{2\pi}\right|^{2\alpha_{N_0}},$$

$$\tag{5.155}$$

where

$$C_{\alpha_1\ldots\alpha_{N_0}} = \frac{2^{m(g,N_0)|\alpha|}(-1)^{|\alpha|}}{2^{|\alpha|}\prod_{i=1}^{N_0}\alpha_i!(3g-3+N_0-|\alpha|)!} \tag{5.156}$$

$$\times \int_{\overline{POL}_{g,N_0}(\Theta,A)} \omega_1^{\alpha_1}\ldots\omega_{N_0}^{\alpha_{N_0}}\cdot\left(\omega_{WP}^\Theta\right)^{3g-3+N_0-|\alpha|}.$$

In particular, If $\overline{\mathcal{T}}_{g,N_0}^{met}(M;\Theta)$ *denotes the set of stable polyhedral surfaces locally modeling* $POL_{g,N_0}(\Theta, A)$, *then*

$$\int_{\overline{\mathcal{M}}_{g,N_0}} \psi_1^{\alpha_1}\ldots\psi_{N_0}^{\alpha_{N_0}}\cdot\omega_{WP}^{3g-3+N_0-|\alpha|} \tag{5.157}$$

$$= \sum_{[T] \in \mathscr{T}_{g,N_0}} \frac{(-1)^{|\alpha|}}{|Aut(T)|} \int_{\overline{\mathscr{T}}^{met}_{g,N_0}(M;\Theta)\big|_{[T]}} \omega_1^{\alpha_1} \dots \omega_{N_0}^{\alpha_{N_0}} \cdot \left(\omega_{WP}^{\Theta}\right)^{3g-3+N_0-|\alpha|}.$$

Proof The expression for $Vol\left[POL_{g,N_0}(\Theta, A)\right]$ follows from a direct substitution of the length vector $L := \{L_k\}$ given by (5.148) into Mirzakhani's theorem, and from the bijective mapping between the space of stable polyhedral structures $\overline{POL}_{g,N_0}(\Theta, A)$ and the moduli space $\overline{\mathscr{M}}_{g,N_0}(L)$. By parametrizing the orbifold integration over $\overline{\mathscr{M}}_{g,N_0}(L)$ with the integration over the orbicells $\overline{\mathscr{T}}^{met}_{g,N_0}(M;\Theta)\big|_{[T]}$ provides (5.157). ∎

The identification (5.157) shows that intersection theory over the moduli space $\overline{\mathscr{M}}_{g,N_0}$ has an explicit realization in terms of polyhedral surfaces, and that in particular it is generated by summing over distinct polyhedral surfaces with a given distribution of conical angles. A further manifestation of the fact that the Witten–Kontsevich model corresponds to a topological sector of 2D quantum gravity, not dynamically coupled to curvature.

In order to discuss at a deeper level the duality between polyhedral and hyperbolic surfaces implied by the above theorem, let us go back to the sky mapping we defined in Sect. 3.3. There, we considered the decoration of each triangle $\sigma^2(k, h, j)$ and of each vertex $\sigma^0(i)$ of a polyhedral surface $(T_l, M) \in \mathscr{T}^{met}_{g,N_0}(M)$ with a corresponding null vector, $\sigma^2(k, h, j) \mapsto \overrightarrow{\xi}(k, h, j)$, $\sigma^0(i) \longmapsto \overrightarrow{\xi}(i)$, belonging to the future light cone $\mathbb{L}^+$. The vector $\overrightarrow{\xi}(k, h, j)$ defines a *visual* horosphere $\Sigma_\infty(k, h, j)$ where an observer $\mathscr{O}_\infty$ in a neighborhood of $\{\infty\} \in \mathbb{H}^{3,+}_{up}$ describes a Euclidean triangle $\sigma^2(k, h, j) \in (T_l, M)$ resulting from the projection of a hyperbolic triangle $\sigma^2_{hyp}(k, h, j)$ living in the $\mathbb{R}^2 \times \{0\}$ portion of the boundary of $\mathbb{H}^{3,+}_{up}$. Such a projection takes place along the $\mathbb{H}^{3,+}_{up}$ geodesics defined by the vertex null vectors $\{\overrightarrow{\xi}(k)\}$. Thus, the horosphere $\Sigma_\infty(k, h, j)$ represents a (local) *screen* and the pair $(\Sigma_\infty(k, h, j), \Sigma_k)$ characterizes the *visual incoming direction* from which the observer describes the boundary component[30] $\partial\Omega_k$. The vertex null vectors $\{\overrightarrow{\xi}(k)\}$ can be interpreted as fields on the hyperbolic surface Ω with preassigned Dirichlet conditions on the distinct $\partial\Omega_k$. Since it would be very difficult to work explicitly with these fields, let us consider instead the corresponding λ-lengths $\lambda(\Sigma_\infty, \Sigma_k)$ associated with the vertical null geodesic connecting $v^0(0) \simeq \infty$ with the generic boundary component $\partial\Omega_k$ of the hyperbolic surface Ω. According to formula (3.10), governing the distance scaling in hyperbolic three-geometry, as we move the screen Σ_∞, $\beta \to \Sigma_\infty(\beta)$, $0 \leq \beta \leq \infty$, along these vertical geodesics, we experience a rescaling of the lengths of the geodesic boundaries $\{L_k\}$ given by

$$\left\{e^{\delta(\Sigma_\infty(\beta), \Sigma_k)} L_k\right\}_{k=1}^{N_0}, \tag{5.158}$$

---

[30]Recall that these boundary components correspond, under hyperbolic completion, to the ideal vertices $\sigma^0_{hyp}(k) := v^0(k)$.

where $\delta(\Sigma_\infty(\beta), \Sigma_k)$ is the signed hyperbolic distance between the respective horosphere, (recall that $\lambda(\Sigma_\infty, \Sigma_k) = \sqrt{e^{\delta(\Sigma_\infty, \Sigma_k)}}$, see (3.8)). We can consider the Weil–Petersson volume $Vol_{WP}\left(\mathfrak{M}_{g,N_0}(L_\delta)\right)$ associated with the moduli space of such boundary rescaled hyperbolic surfaces. According to Mirzakhani's theorem this volume can be expressed as

$$Vol_{WP}\left(\mathfrak{M}_{g,N_0}(e^{\delta(\Sigma_\infty(\beta),\Sigma_k)}\,L_k)\right) = \frac{2^{m(g,N_0)|\alpha|}}{(3g-3+N_0-|\alpha|)!} \times \tag{5.159}$$

$$\times \sum_{\substack{(\alpha_1,\ldots,\alpha_{N_0})\in(\mathbb{Z}_{\geq 0})^{N_0} \\ |\alpha|\leq 3g-3+N_0}} \int_{\overline{\mathcal{M}}_{g,N_0}} \prod_{i=1}^{N_0} \frac{L_i^{2\alpha_i}\, e^{\alpha_i\,\delta(\Sigma_\infty(\beta),\Sigma_i)}}{2^{|\alpha|}\,\alpha_i!}\, \psi_i^{\alpha_i}\, \omega_{WP}^{3g-3+N_0-|\alpha|}.$$

Let us consider the scaling regime in which the β–varying signed hyperbolic distances $\delta(\Sigma_\infty(\beta), \Sigma_k)$ increases according to

$$\delta(\Sigma_\infty, \Sigma_k) \to \delta(\Sigma_\infty(\beta), \Sigma_k) := \beta\,\delta(\Sigma_\infty, \Sigma_k), \tag{5.160}$$

and simultaneously the corresponding boundary lengths shrinks

$$L_k \to L_k(\beta) := \beta^{-1}\,L_k\,, \qquad 0 \leq \beta \leq \infty\,, \tag{5.161}$$

in such a way that

$$L_k(\beta)e^{\delta(\Sigma_\infty(\beta),\Sigma_k)}, \tag{5.162}$$

remains constant. In other words, we are sampling hyperbolic surfaces which are far away from the Euclidean projection screen, and projecting there as polyhedral surfaces (T_l, M) with conical angles $\{\Theta(k)\}$ such that

$$\left|\ln\frac{\Theta(k)}{2\pi}\right| = \lim_{\beta\nearrow\infty} L_k(\beta)e^{\delta(\Sigma_\infty(\beta),\Sigma_k)}. \tag{5.163}$$

Under such scaling regime, we indeed get from (5.159) and theorem 5.5

$$Vol_{WP}\left(\mathfrak{M}_{g,N_0}(e^{\delta(\Sigma_\infty(\beta),\Sigma_k)}\,L_k(\beta))\right) = Vol\left[POL_{g,N_0}(\Theta,A)\right]. \tag{5.164}$$

The open/closed duality we have been discussing so far has aspects that naturally involve three–dimensional hyperbolic manifolds. One may wonder whether or not such a geometry is just a static arena or an active player. Thus, in the next section, we go back to geometric considerations for addressing such an issue. This will provide a suitable framework for introducing a set up related with Chern–Simons theory, a subject that will be discussed in more detail in the next two chapters.

5.10   Glimpses of Hyperbolic 3-Manifolds and of Their Volume

The open/closed duality between polyhedral surfaces and hyperbolic surfaces with geodesic boundaries can be naturally extended to three-dimensional hyperbolic cone-manifolds. Recall that to a polyhedral surface (T_l, M) we can associate either the marked horosphere (Σ_k, z_k) or, equivalently, the (unique) geodesic $\gamma(k, \infty)$ in $\mathbb{H}^{3,+}_{up}$ connecting the vertex $v^0(k)$ with the vertex at ∞ of an ideal tetrahedron $\sigma^3_{hyp}(\infty, k, h, j)$ in $\mathbb{H}^{3,+}_{up}$. In particular, to any two adjacent triangles sharing a common edge, say $\sigma^2(k, h, j)$ and $\sigma^2(k, j, l)$, correspond pairwise adjacent tetrahedra, $\sigma^3_{hyp}(\infty, k, h, j)$ and $\sigma^3_{hyp}(\infty, k, j, l)$, that can be glued along the isometric faces $\sigma^2_{hyp}(j, k, \infty)$ and $\sigma^2_{hyp}(\infty, k, j)$. Each face-pairing is realized by an isometry of $\mathbb{H}^{3,+}_{up}$

$$f_{jk} : \sigma^2_{hyp}(j, k, \infty) \longrightarrow \sigma^2_{hyp}(\infty, k, j) \tag{5.165}$$

which reverses orientation (so as to have orientability of the resulting complex). In this way, by pairwise glueing the $q(k)$ ideal tetrahedra $\left\{ \sigma^3_{hyp}(\infty, k, h_\alpha, h_{\alpha+1}) \right\}$, associated with the corresponding Euclidean triangles $\sigma^2(k, h_\alpha, h_{\alpha+1})$, we generate a polytope

$$P^3(k) \doteq \sqcup^{q(k)}_{\alpha=1} \sigma^3_{hyp}(\infty, k, h_\alpha, h_{\alpha+1}) \Big/ \{f_{h_\alpha k}\} \tag{5.166}$$

with a conical singularity along the core geodesic $\gamma(k, \infty)$. Explicitly, let us denote by $\psi_{\cdot,\cdot}$ the dihedral angles associated with the edges $\sigma^1_{hyp}(\cdot, \cdot)$ of this polytope. From the relations between the dihedral angles of each hyperbolic tetrahedron $\sigma^3_{hyp}(\infty, k, h, j)$ and the vertex angles of the corresponding Euclidean triangle $\sigma^2(k, h, j)$ it easily follows that

$$\psi_{\infty, h} = \theta_{hjk} + \theta_{kjl}, \tag{5.167}$$

$$\psi_{kj} = \theta_{khj} + \theta_{klj},$$

$$\psi_{hj} = \theta_{hkj},$$

$$\psi_{\infty, k} = \sum_{\alpha=1}^{q(k)} \theta_{\alpha, k, \alpha+1} = \Theta(k).$$

Note in particular that the conical defect $\Theta(k)$ at the vertex $\sigma^0(k) \in Star[\sigma^0(k)]$ propagates as a conical defect along the core geodesic $\gamma(k, \infty)$ of $\mathbb{H}^{3,+}_{up}$. It follows that $P^3(k)$ has a non-complete hyperbolic metric and that the singularity on $\gamma(k, \infty)$ is conical with angle $\Theta(k)$. In order to endow $P^3(k)$ with a hyperbolic structure, let $\tilde{P}_\gamma(k)$ denote the universal cover in $\mathbb{H}^{3,+}_{up}$ of $P^3(k)$, with the core geodesic $\gamma(k, \infty)$ removed. $\tilde{P}_\gamma(k)$ carries a natural hyperbolic structure and the holonomy representation of its fundamental group, $\pi_1(\tilde{P}_\gamma(k)) = \mathbb{Z}$, is generated by an isometry of $\tilde{P}_\gamma(k) \subset \mathbb{H}^{3,+}_{up}$ of the form

$$\rho(k) : \pi_1(\tilde{P}_\gamma(k)) \longrightarrow Isom\left(\mathbb{H}^{3,+}_{up}\right) \tag{5.168}$$

$$(c_s, s) \longmapsto \left[a(k) \begin{pmatrix} e^{i\phi(s)} & 0 \\ 0 & e^{-i\phi(s)} \end{pmatrix}, s \right],$$

where $a(k) > 1$ and $s \mapsto c_s$, $0 \leq s < \infty$ is closed curve winding around the link of $\sigma^0(k)$ in $Star\left[\sigma^0(k)\right]$ with $\phi(s = 2\pi) = \Theta(k)$. Since $a(k) > 1$, the isometry is hyperbolic (fixing the point $v^0(k)$ and ∞ in $\mathbb{H}^{3,+}_{up}$). For simplicity, let us identify $v^0(k)$ with the origin of $\mathbb{H}^{3,+}_{up}$. The horosphere Σ_k intersects $\sqcup \sigma^3_{hyp}(\infty, k, h_\alpha, h_{\alpha+1})$ along a sequence of offset horocycle segments $\left\{ F^{h_\alpha}_k \right\}$ such that

$$d_{\mathbb{H}^3}(F^{h_1}_k, \widehat{F}^{h_{q(k)}}_k) = \left| \ln \frac{\sum_{\alpha=1}^{q(k)} \theta_{\alpha+1,k,\alpha}}{2\pi} \right|. \tag{5.169}$$

Similarly the concentric horosphere $^*\Sigma_k$ defined by $z = a(k)\, z_k$, $a(k) \geq 1$, intersects the $\sqcup \sigma^3_{hyp}(\infty, k, h_\alpha, h_{\alpha+1})$ along a sequence of horocycle segments $\left\{ ^*F^{h_\alpha}_k \right\}$ such that $d_{\mathbb{H}^3}(^*F^{h_1}_k, ^*\widehat{F}^{h_{q(k)}}_k) = \left| \ln \frac{\Theta(k)}{2\pi} \right|$. Let us consider the rectangular parallelepiped labeled by the segments $\left(F^{h_1}_k, \widehat{F}^{h_{q(k)}}_k; ^*F^{h_1}_k, ^*\widehat{F}^{h_{q(k)}}_k \right)$. A straightforward application of (3.10) provides the following relations between the (hyperbolic) lengths of the sides of this parallelepiped

$$\left| ^*\widehat{F}^{h_{q(k)}}_k \right| = e^{-d_{\mathbb{H}^3}(\Sigma_k, ^*\Sigma_k)} \left| \widehat{F}^{h_{q(k)}}_k \right|, \tag{5.170}$$

$$\left| ^*F^{h_1}_k \right| = e^{-d_{\mathbb{H}^3}(\Sigma_k, ^*\Sigma_k)} \left| F^{h_1}_k \right|,$$

where $|\ldots|$ denotes the length of the corresponding horocycle segment. Since

$$d_{\mathbb{H}^3}(\Sigma_k, ^*\Sigma_k) = 2 \tanh^{-1} \frac{a(k) - 1}{a(k) + 1} = \ln a(k), \tag{5.171}$$

we get

$$\left| ^*\widehat{F}^{h_{q(k)}}_k \right| = a(k)^{-1} \left| \widehat{F}^{h_{q(k)}}_k \right|, \quad \left| ^*F^{h_1}_k \right| = a(k)^{-1} \left| F^{h_1}_k \right|. \tag{5.172}$$

Moreover, from (3.51) we have

$$\left| \widehat{F}^{h_{q(k)}}_k \right| = e^{\left| \ln \frac{\Theta(k)}{2\pi} \right|} \left| F^{h_1}_k \right|, \quad \left| ^*\widehat{F}^{h_{q(k)}}_k \right| = e^{\left| \ln \frac{\Theta(k)}{2\pi} \right|} \left| ^*F^{h_1}_k \right|. \tag{5.173}$$

By comparing these expressions, it follows that we can match the length of horocycle segment $*\widehat{F}_k^{h_{q(k)}}$ with the length of the segment $F_k^{h_1}$ if we choose the parameter $a(k)$ according to

$$a(k) = e^{\left| \ln \frac{\Theta(k)}{2\pi} \right|}. \tag{5.174}$$

Such a matching condition allows, under the action of $\rho(k)\left(\pi_1\left(\tilde{P}_\gamma(k)\right)\right)$, an (offset) identification between opposite faces of $\left(F_k^{h_1}, \widehat{F}_k^{h_{q(k)}}; *F_k^{h_1}, *\widehat{F}_k^{h_{q(k)}}\right)$, and consequently we can choose this rectangular parallelepiped as a fundamental domain for the action of the holonomy representation $\rho(k)$. The resulting developing map describes $\tilde{P}_\gamma(k)$ as an incomplete manifold and $P_{hyp}^3(k) \doteq \tilde{P}_\gamma(k) \setminus \rho(k)$ is topologically equivalent to a solid torus $\mathbb{S}^1 \times B^2$, (B^2 being the meridianal 2-dimensional disc) with the central geodesic missing. Note that such a geodesic can be naturally identified with the geodesic boundary component $\partial\Omega_k$ of the open hyperbolic surface Ω. In order to get an intuitive picture of what happens, observe that the identification polytope $P^3(k)$, cut by the horosphere Σ_∞, is topologically a solid cylinder sliced by the faces of the component tetrahedra. If we remove a tube of small (infinitesimal) width around the central geodesic $\gamma(k,\infty)$ we get a topological solid torus sliced into parallelepipeds, with a thin and long tubular hole associated with the removed geodesic. The isometry (5.168) twists up this solid torus with a shearing motion, like a 3-dimensional photographic diaphragm. Adjacent parallelepipeds slide one over the other tilting up, while the central tube correspondingly winds up accumulating towards an horizontal $\mathbb{S}^1$.

We can formally extend this geometric analysis to the whole polyhedral surface (T_l, M) by forming the support space (for a compatible hyperbolic structure)

$$V \doteq \sqcup_{\sigma_{hyp}^2(l,m,\infty)}^{N_1(T)} \sigma_{hyp}^3(\infty, k, h, j) \Big/ \{f_{lm}\}, \tag{5.175}$$

(the number of hyperbolic faces to be paired is equal to the number $N_1(T)$ of edges in $|T_l| \to M$). Note that the link of the vertex at ∞ in V is

$$link\ [\infty] \doteq \bigcup_{\sigma_{hyp}^1(l,m)}^{N_1(T)} \sigma_{hyp}^2(k, h, j), \tag{5.176}$$

where the glueing along the edges $\left\{\sigma_{hyp}^1(l,m)\right\}$ is modelled after the polyhedral surface (T_l, M). If this latter has genus g, then from the Euler and Dehn-Sommerville relations

$$N_0(T) - N_1(T) + N_2(T) = 2 - 2g, \tag{5.177}$$

$$2N_1(T) = 3N_2(T),$$

we get that the support space V has

$$N_2(T) = 2N_0(T) + 4g - 4 \geq N_0(T) + g \tag{5.178}$$

ideal tetrahedra with $N_0(T)$ vertices associated with its boundary components ∂V. As we have seen in Sect. 3.5, the edge-glueing of $\{\sigma^2_{hyp}(k, h, j)\}$ gives rise to an incomplete hyperbolic surface and consequently also V cannot support, as it stands, a complete hyperbolic structure. To take care of this, we start by removing from V an open (horospherical) neighborhood of the vertices. In this way, each tetrahedron $\sigma^3_{hyp}(\infty, k, h, j)$ becomes a octahedron with four (Euclidean) triangular faces (in the same similarity class which defines the given tetrahedron), and four (hyperbolic) exagonal faces. Note that the boundary of the removed open neighborhood of ∞ is triangulated by Euclidean triangles and it reproduces $|T_l| \to M$. Note also that the removed neighborhoods cut out an open disk D_k around each vertex $v^0(k)$ in ∂V. Next, we remove from V also an open neighborhood of the geodesics $\{\gamma(k, \infty)\}_{k=1}^{N_0(T)}$. In this way we get from the support space V a handlebody H_V. Topologically, H_V is $[0, 1] \times \Omega$, where Ω is the surface with boundary $(\sqcup \partial \Omega_k)$ associated with the hyperbolic completion of $\sqcup \sigma^2_{hyp}(k, h, j)$. The handlebody H_V plays here the role of the polytope $\tilde{P}_\gamma(k)$ introduced in connection with the support space (5.166). By identifying the bottom $\Omega_0 \simeq \partial H_V|_0$ and top $\Omega_1 \simeq \partial H_V|_1$ copies of the surface Ω by means of the appropriate orientation reversing boundary homeomorphism $h : \partial H_V|_0 \to \partial H_V|_1$, with $h(\partial \Omega_k|_0) = -\partial \Omega_k|_1$, we get the support space

$$V\left(\{\Theta(k)\}_{k=1}^{N_0(T)}\right) \setminus K \doteq H_V \setminus \sim^h \tag{5.179}$$

($V \setminus K$, for notational ease), where K is the knot-link generated in H_V by the action of the identification homeomorphism h on the boundaries connecting the tubes associated with the removed core geodesics $\{\gamma(k, \infty)\}_{k=1}^{N_0(T)}$. It is not yet obvious that $V \setminus K$ admits a complete hyperbolic structure. First, we have been rather cavalier on the delicate issue concerning orientation in glueing the ideal tetrahedra, (for semi-simplicial triangulations problems connected with orientability of the hyperbolic complexes obtained upon face-identifications can be rather serious and we may end up in a ideal triangulation which may actually not define a manifold). Moreover, around the removed geodesics $\{\gamma(k, \infty)\}_{k=1}^{N_0(T)}$ the geometry is conical, and in order to establish completeness for the hyperbolic structure we have to discuss how hyperbolic Dehn filling can be extended to cone manifolds. These are delicate issues which, to the best of our knowledge, do not have answers that can be easily given in general terms. Notwithstanding the technical difficulties in characterizing complete hyperbolic structures on $V \setminus K$, their existence, when established, implies a number of important consequences which bear relevance to our analysis.

First of all, if the support space $V \setminus K$ generated by (T_l, M), is indeed a three-dimensional hyperbolic manifold $V_{hyp}(\{\Theta(k)\}) \setminus K$, then we can easily compute its hyperbolic volume in terms of the conical angles $(\{\Theta(k)\}_{k=1}^{N_0(T)})$. As a matter of fact we can associate to any triangle $\sigma^2(k,h,j)$ of $|T_l| \to M$ the volume $Vol[\sigma^3_{hyp}]$ of the corresponding ideal tetrahedron σ^3_{hyp}. According to Milnor's formula, (see e.g. [5], for a very informative analysis), such a volume can be expressed in terms of the Lobachevsky functions $\mathscr{L}(\theta_{jkh})$, $\mathscr{L}(\theta_{khj})$, and $\mathscr{L}(\theta_{hjk})$ of the respective vertex angles of $\sigma^2(k,h,j)$, where

$$\mathscr{L}(\theta_{jkh}) \doteq -\int_0^{\theta_{jkh}} \ln|2\sin x|\,dx. \tag{5.180}$$

In our setting, this translates into the mapping

$$\sigma^2(k,h,j) \longmapsto Vol\left[\sigma^3_{hyp}(\infty,k,h,j)\right] = \tag{5.181}$$

$$= \mathscr{L}(\theta_{j\,kh}) + \mathscr{L}(\theta_{khj}) + \mathscr{L}(\theta_{hj\,k}),$$

which is well-defined since, due to the symmetries of the dihedral angles, the valu-tation of $Vol\left[\sigma^3_{hyp}(\infty,k,h,j)\right]$ is independent from which vertex of the tetrahedron is actually mapped to ∞, (see [5], prop. C.2.8). Thus, we can compute the volume of the three-dimensional hyperbolic manifold $V_{hyp} \setminus K$ as

$$Vol\left[V_{hyp}\left(\{\Theta(k)\}_{k=1}^{N_0(T)}\right) \setminus K\right] = \tag{5.182}$$

$$\sum_{\{\sigma^2(k,h_\alpha,h_{\alpha+1})\}}^{N_2(T)} \left[\mathscr{L}(\theta_{\alpha+1,k,\alpha}) + \mathscr{L}(\theta_{k,\alpha,\alpha+1}) + \mathscr{L}(\theta_{\alpha,\alpha+1,k})\right],$$

where the summation extends over all triangles $\sigma^2(k,h_\alpha,h_{\alpha+1})$ in the Regge triangulated surface $|T_l| \to M$. Equivalently, in terms of the complex moduli $\zeta_{\alpha+1,k,\alpha}$ of the triangles $\sigma^2(k,h_\alpha,h_{\alpha+1})$, we get

$$Vol\left[V_{hyp}\left(\{\Theta(k)\}_{k=1}^{N_0(T)}\right) \setminus K\right] = \tag{5.183}$$

$$\sum_{\{\sigma^2(k,h_\alpha,h_{\alpha+1})\}}^{N_2(T)} \left[\mathscr{L}(\arg\zeta_{\alpha+1,k,\alpha}) + \mathscr{L}(\arg\zeta_{k,\alpha,\alpha+1}) + \mathscr{L}(\arg\zeta_{\alpha,\alpha+1,k})\right].$$

It is worthwhile to remark that if one computes the Hessian of $Vol\left[V_{hyp}\right]$ with respect the angular variables $\{\theta_{\alpha+1,k,\alpha}\}$ of the generic triangle $\sigma^2(k,h_\alpha,h_{\alpha+1})$ one gets

$$H_{k,k} \doteq \frac{\partial^2}{\partial\,\theta_{\alpha+1,k,\alpha}^2} Vol\left[V_{hyp}\left(\{\Theta(k)\}_{k=1}^{N_0(T)}\right)\right] = -\cot\theta_{\alpha+1,k,\alpha} = \tag{5.184}$$

$$= \frac{l^2(h_{\alpha+1}, k) + l^2(k, h_\alpha) - l^2(h_\alpha, h_{\alpha+1})}{4\,\Delta(\alpha+1,\,k,\,\alpha)}\,,$$

$$H_{\alpha,\alpha} \doteq \frac{\partial^2}{\partial\,\theta_{k,\alpha,\alpha+1}{}^2}\,Vol\left[V_{hyp}\left(\{\Theta(k)\}_{k=1}^{N_0(T)}\right)\right] = -\cot\theta_{k,\alpha,\alpha+1} =$$

$$= \frac{l^2(k, h_\alpha) + l^2(h_\alpha, h_{\alpha+1}) - l^2(h_{\alpha+1}, k)}{4\,\Delta(k,\alpha,\,\alpha+1)}\,,$$

$$H_{\alpha+1,\alpha+1} \doteq \frac{\partial^2}{\partial\,\theta_{\alpha,\alpha+1,k}{}^2}\,Vol\left[V_{hyp}\left(\{\Theta(k)\}_{k=1}^{N_0(T)}\right)\right] = -\cot\theta_{\alpha,\alpha+1,k} =$$

$$= \frac{l^2(h_\alpha, h_{\alpha+1}) + l^2(h_{\alpha+1}, k) - l^2(k, h_\alpha)}{4\,\Delta(\alpha,\,\alpha+1,k)}\,,$$

where $\Delta \doteq \Delta(\alpha+1,\,k,\,\alpha)$ denotes, up to cyclic permutation, the Euclidean area of the triangle $\sigma^2(k, h_\alpha, h_{\alpha+1})$. From (5.184) we get

$$l^2(h_{\alpha+1}, k) = 2\,\Delta(H_{\alpha+1,\alpha+1} + H_{k,k}), \tag{5.185}$$

$$l^2(k, h_\alpha) = 2\,\Delta\,(H_{k,k} + H_{\alpha,\alpha})\,,$$

$$l^2(h_\alpha, h_{\alpha+1}) = 2\,\Delta(H_{\alpha,\alpha} + H_{\alpha+1,\alpha+1}),$$

which provide sign conditions on H_{lm}. Actually, it is relatively easy ([60]) to show that the restriction of the Hessian of $Vol\left[V_{hyp}\setminus K\right]$ to the local Euclidean structure on each $\sigma^2(k, h_\alpha, h_{\alpha+1})$ is negative-definite. This latter remark implies that (minus) the Hessian of the hyperbolic volume can be used as a natural quadratic form, on the space of deformations of the Euclidean structures associated with polyhedral surfaces, which naturally pairs with the Weil-Petersson measure. It is also clear that formally the hyperbolic volume (5.182) does not require the existence of a complete hyperbolic structure on the support space $V\setminus K$, and we may well associate the function (5.182) to $V\setminus K$. However, the existence of a complete hyperbolic structure implies that such a volume function is a topological invariant by Mostow rigidity. Moreover, one can formulate the so-called volume conjecture, (see [54] for a review), which, in our setting, may be phrased by stating that if K is not a split link and $J_n(K; t)$ is its colored Jones polynomial associated with the n-dimensional irreducible representation of $sl_2(\mathbb{C})$, then

$$2\pi\,\lim_{n\to\infty}\frac{\ln\left|J_n(K; \exp\left[\frac{2\pi i}{n}\right])\right|}{n} = Vol\left[V_{hyp}\left(\{\Theta(k)\}_{k=1}^{N_0(T)}\right)\setminus K\right] \tag{5.186}$$

(in the standard formulation of the volume conjecture the role of the support space $V(\{\Theta(k)\}_{k=1}^{N_0(T)})$ is played by $\mathbb{S}^3$, and one assumes that the complement $\mathbb{S}^3\setminus K$ of the link K admits a (complete) hyperbolic structure). $J_n(K; t)$ is defined through the n-dimensional irreducible representations of the quantum group $U_q(sl(2, \mathbb{C}))$.

For some hyperbolic knots in $\mathbb{S}^3$, in particular for the figure eight knot (and for torus links, which are non-hyperbolic and yield 0 on the right member of (5.186)), the conjecture has been proved.[31] This connection between knot polynomials and hyperbolic volume has been actually promoted to be part of a more general conjecture relating the asymptotics of the colored Jones polynomials to the Chern-Simons invariant

$$2\pi i \cdot \lim_{n \to \infty} \frac{\ln J_n(K; \exp\left[\frac{2\pi i}{n}\right])}{n} = CS\left[V_{hyp}\big/K\right] + i\, Vol\left[V_{hyp}\big/K\right] \qquad (5.187)$$

and

$$\lim_{n \to \infty} \frac{J_{n+1}(K; \exp\left[\frac{2\pi i}{n}\right])}{J_n(K; \exp\left[\frac{2\pi i}{n}\right])} = \exp\left(\frac{1}{2\pi i}\left(CS\left[V_{hyp}\big/K\right] + i\, Vol\left[V_{hyp}\big/K\right]\right)\right)$$

$$(5.188)$$

where again we have formally referred all quantities to $V_{hyp}\big/K$, in particular $CS\left[V_{hyp}\big/K\right]$ is the Chern-Simons invariant of the connection defined by the hyperbolic metric on $V_{hyp}\big/K$. It should be clear that these statements have a status quite more conjectural then the original ones owing to the conical nature of $V_{hyp}\big/K$, nonetheless they are reasonable in view of the holographic principle. Recall that a geometrical version of *classical* holography is familiar in hyperbolic geometry as the Ahlfors-Bers theorem which applies to hyperbolic manifolds V containing a compact subset determining a conformal structure on the boundary at ∞ of V. In such a case the geometry of V is uniquely determined by such induced conformal structure at ∞. It should be clear from its very set-up that our approach to closed/open duality is, geometricaly speaking, holographic in nature. Roughly speaking it is akin to a simplicial version of Ahlfors-Bers theorem, (for a serious analysis of this issue for conical hyperbolic manifolds see [48]).

References

1. Aharony, O., Komargodski, Z., Razamat, S.S.: On the worldsheet theories of strings dual to free large N gauge theories. JHEP **0605**, 16 (2006) [arXiv:hep-th/06020226]
2. Akhmedov, E.T.: Expansion in Feynman graphs as simplicial string theory. JETP Lett. **80**, 218 (2004) [Pisma Zh. Eksp. Teor. Fiz. **80**, 247 (2004)] [arXiv:hep-th/0407018]
3. Ambjørn, J., Durhuus, B., Jonsson, T.: Quantum geometry. Cambridge Monograph on Mathematical Physics. Cambridge University Press, Cambridge/New York (1997)
4. Baseilhac, S., Benedetti, R.: QHI, 3-manifolds scissors congruence classes and the volume conjecture. In: Ohtsuki, T., et al. (eds.) Invariants of Knots and 3-Manifolds. Geometry & Topology Monographs, vol. 4, pp. 13–28. University of Warwick, Coventry (2002) [arXiv:math.GT/0211053]

---

[31] See [4] for a deep analysis and the relevant references.

5. Benedetti, R., Petronio, C.: Lectures on Hyperbolic Geometry. Universitext. Springer, New York (1992)
6. Bost, J.B., Jolicoeur, T.: A holomorphy property and the critical dimension in string theory from an index theorem. Phys. Lett. B **174**, 273–276 (1986)
7. Brézin, E., Itzykson, C., Parisi, G., Zuber, J.B.: Planar diagrams. Commun. Math. Phys. **59**, 25–51 (1978)
8. Cantor, M.: Elliptic operators and the decomposition of tensor fields. Bull. Am. Math. Soc. **5**, 235–262 (1981)
9. Chapman, K.M., Mulase, M., Safnuk, B.: The Kontsevich constants for the volume of the moduli of curves and topological recursion. arXiv:1009.2055 [math.AG]
10. Das, S.R., Naik, S., Wadia, S.R.: Quantization of the Liouville mode and string theory. Mod. Phys. Lett. **A4**, 1033–1041 (1989)
11. David, F.: Conformal field theories coupled to 2D gravity in the conformal gauge. Mod. Phys. Lett. A **3**, 1651–1656 (1988)
12. David, F., Bauer, M.: Another derivation of the geometrical KPZ relations. J. Stat. Mech. **3**, P03004 (2009). arXiv:0810.2858
13. David, J.R., Gopakumar, R.: From spacetime to worldsheet: four point correlators. arXiv:hep-th/0606078
14. David, F., Kupiainen, A., Rhodes, R., Vargas, V.: Liouville quantum gravity on the Riemann sphere. Commun. Math. Phys. **342**, 869–907 (2016)
15. D'Hoker, E.: Lectures on strings, IASSNS-HEP-97/72
16. D'Hoker, E., Phong, D.H.: The geometry of string perturbation theory. Rev. Mod. Phys. **60**(4), 917–1065 (1988)
17. D'Hoker, E., Kurzepa, P.S.: 2-D quantum gravity and Liouville theory. Mod. Phys. Lett. **A5**, 1411–1422 (1990)
18. Di Francesco, P.: 2D Quantum Gravity, Matrix Models and Graph Combinatorics. Lectures given at the summer school Applications of Random Matrices in Physics, Les Houches, June 2004. arXiv:math-ph/0406013v2
19. Distler, J., Kaway, H.: Conformal field theory and 2D quantum gravity. Nucl. Phys. B **321**, 509–527 (1989)
20. Duplantier, B., Sheffield, S.: Liouville quantum gravity and KPZ. Invent. Math. (2008, to appear on). arXiv:0808.1560
21. Ebin, D.: The manifolds of Riemannian metrics, Global analysis. Proc. Sympos. Pure Math. **15**, 11–40 (1968)
22. Eynard, B.: Recursion between Mumford volumes of moduli spaces. arXiv:0706.4403[math-ph]
23. Eynard, B.: Counting Surfaces: CRM Aisenstadt Chair Lectures. Progress in Mathematical Physics, vol. 70. Springer, Basel (2016)
24. Eynard, B., Orantin, N.: Invariants of algebraic curves and topological expansion. Commun. Number Theory Phys. **1**, 347–452 (2007)
25. Eynard, B., Orantin, N.: Weil–Petersson volume of moduli spaces, Mirzhakhani's recursion and matrix models. arXiv:0705.3600[math-ph]
26. Eynard, B., Orantin, N.: Geometrical interpretation of the topological recursion, and integrable string theory. arXiv:0911.5096[math-ph]
27. Faris, W.G. (ed.): Diffusion, Quantum Theory, and Radically Elementary Mathematics. Mathematical Notes, vol. 47. Princeton University Press, Princeton/Oxford (2006)
28. Fradkin, E.S., Tseytlin, A.A.: Effective field theory from quantized strings. Phys. Lett. B **158**, 316 (1985)
29. Fradkin, E.S., Tseytlin, A.A.: Quantum string theory effective action. Nucl. Phys. B **261**, 1 (1985)
30. Gaiotto, D., Rastelli, L.: A paradigm of open/closed duality: Liouville D-branes and the Kontsevich model. JHEP **0507**, 053 (2005) [arXiv:hep-th/0312196]
31. Giddings, S.B., Wolpert, S.A.: A triangulation of moduli space from light-cone string theory. Commun. Math. Phys. **109**, 177–190 (1987)

32. Gopakumar, R.: From free fields to adS. Phys. Rev. D **70**, 025009 (2004) [arXiv:hep-th/0308184]
33. Gopakumar, R.: From free fields to ads. II. Phys. Rev. D **70** (2004) 025010 [arXiv:hep-th/0402063]
34. Gopakumar, R.: Free field theory as a string theory? C. R. Phys. **5**, 1111 (2004) [arXiv:hep-th/0409233]
35. Gopakumar, R.: From free fields to adS. III. Phys. Rev. D **72**, 066008 (2005) [arXiv:hep-th/0504229]
36. Gopakumar, R., Vafa, C.: Adv. Theor. Math. Phys. **3**, 1415 (1999) [hep-th/9811131]
37. Harer, J.L., Zagier, D.: The Euler characteristic of the moduli space of curves. Invent. Math. **85**, 457–485 (1986). See also: The cohomology of the moduli spaces of curves. In: Harer, J.L. (ed.) Theory of Moduli, Montecatini Terme, 1985. Lecture Notes in Mathematics, vol. 1337, pp. 138–221. Springer, Berlin (1988)
38. Kaku, M.: Strings, Conformal Fields, and M-Theory, 2nd ed. Springer, New York (1999)
39. Kaufmann, R., Penner, R.C.: Closed/open string diagrammatics. arXiv:math.GT/0603485
40. Kiritis, E.: String Theory in a Nutshell. Princeton University Press, Princeton (2007)
41. Knizhnik, V.G., Polyakov, A.M., Zamolodchikov, A.B.: Fractal structure of 2D quantum gravity. Mod. Phys. Lett. A **3**, 819–826 (1988)
42. Kokotov, A.: Compact polyhedral surfaces of an arbitrary genus and determinant of Laplacian. arXiv:0906.0717 (math.DG)
43. Kontsevitch, M.: Intersection theory on the moduli space of curves and the matrix airy functions. Commun. Math. Phys. **147**, 1–23 (1992)
44. Manin, Y.I., Zograf, P.: Invertible cohomological filed theories and Weil-Petersson volumes. Annales de l' Institute Fourier **50**, 519–535 (2000)
45. Menotti, P., Peirano, P.P.: Diffeomorphism invariant measure for finite dimensional geometries. Nucl. Phys. **B488**, 719–734 (1997). arXiv:hep-th/9607071v1
46. Mirzakhani, M.: Simple geodesics and Weil–Petersson volumes of moduli spaces of bordered Riemann surfaces. Invent. Math. **167**, 179–222 (2007)
47. Mirzakhani, M.: Weil–Petersson volumes and intersection theory on the moduli spaces of curves. J. Am. Math. Soc. **20**, 1–23 (2007)
48. Moroianu, S., Schlenker, J.-M.: Quasi-Fuchsian manifolds with particles. arXiv:math.DG/0603441
49. Mulase, M., Penkava, M.: Ribbon graphs, quadratic differentials on Riemann surfaces, and algebraic curves defined over $\overline{\mathbb{Q}}$. Asian J. Math. **2**, 875–920 (1998) [math-ph/9811024 v2]
50. Mulase, M., Safnuk, B.: Mirzakhani's recursion relations, Virasoro constraints and the KdV hierarchy. Indian J. Math. **50**, 189–228 (2008)
51. Mumford, D.: Towards an enumerative geometry of the moduli space of curves. In: Selected Papers on the Classification of Varieties and Moduli Spaces, pp. 235–292. Springer, New York (2004)
52. Nakamura, S.: A calculation of the orbifold Euler number of the moduli space of curves by a new cell decomposition of the Teichmüller space. Tokyo J. Math. **23**, 87–100 (2000)
53. Nakayama, Y.: Liouville field theory – a decade after the revolution. Int. J. Mod. Phys. **A19**, 2771–2930 (2004). arXiv:hep-th/0402009
54. Ohtsuki, T. (ed.): Problems on invariants of knots and 3-manifolds. In: Kohno, T., Le, T., Murakami, J., Roberts, J., Turaev, V. (eds.) Invariants of Knots and 3-Manifolds. Geometry and Topology Monographs, vol. 4, pp. 377. Mathematics Institute, University of Warwick, Coventry (2002)
55. Penner, R.C.: The decorated Teichmüller space of punctured surfaces. Commun. Math. Phys. **113**, 299–339 (1987)
56. Penner, R.C.: Perturbation series and the moduli space of Riemann surfaces. J. Differ. Geom. **27**, 35–53 (1988)
57. Polchinski, J.: String Theory, vols. I and II. Cambridge University Press, Cambridge (1998)
58. Polyakov, A.M.: Quantum geometry of bosonic strings. Phys. Lett. B **103**, 207–210 (1981)

59. Rhodes, R., Vargas, V.: KPZ formula for log-infinitely divisible multi-fractal random measures (2008). ESAIM, P& S **15**, 358–371 (2011)
60. Rivin, T.: Euclidean structures on simplicial surfaces and hyperbolic volume. Ann. Math. **139**, 553–580 (1994)
61. Strebel, K.: Quadratic Differentials. Springer, Berlin (1984)
62. 't Hooft, G.: A planar diagram theory for strong interactions. Nucl. Phys. **B 72**, 461Ű-470 (1974)
63. Thurston, W.P.: Three-dimensional geometry and topology, vol. 1. In: Levy, S. (ed.) Princeton Mathematical Series, vol. 35. Princeton University Press, Princeton (1997)
64. Takthajan, L.A., Teo, L.-P.: Quantum Liouville theory in the background field formalism I. Compact Riemannian surfaces. Commun. Math. Phys. **268**, 135–197 (2006)
65. Tutte, W.J.: A census of planar triangulations. Can. J. Math. **14**, 21–38 (1962)
66. Voevodskii, V.A., Shabat, G.B.: Equilateral triangulations of Riemann surfaces, and curves over algebraic number fields. Sov. Math. Dokl. **39**, 38 (1989)
67. Witten, E.: Two dimensional gravity and intersection theory on moduli space. Surv. Diff. Geom. **1**, 243 (1991)
68. Zamolodchikov, A., Zamolodchikov, A.: Lectures on Liouville Theory and Matrix Models. http://qft.itp.ac.ru/ZZ.pdf
69. Zograf, P.G.: Weil-Petersson volumes of moduli spaces of curves and the genus expansion in two dimensional gravity. math.AG/9811026
70. Zograf, P.G., Takhtadzhyan, L.A.: On Liouville's equation, accessory parameters, and the geometry of Teichmuller space for Riemann surfaces of genus 0. Math. USRR Sbornik **60**, 143–161 (1988)

Chapter 6
State Sum Models and Observables

From a historical viewpoint the Ponzano–Regge asymptotic formula for the $6j$ symbol of the group $SU(2)$ [52], together with Penrose's original idea of combinatorial spacetime out of coupling of angular momenta – or *spin networks* – [51], is the precursor of the discretized approaches to 3–dimensional Euclidean quantum gravity collectively referred to as 'state sum models' after the 1992 paper by Turaev and Viro [69]. The prominent role here is played by the *colored* tetrahedron encoding the tetrahedral symmetry of the $6j$ symbol – reminiscent of the Platonic solid shown in the reproduction of Fig. 6.1 – and recognized in the semiclassical limit as a *geometric* 3–simplex whose edge lengths are irreps labels from the representation ring of either $SU(2)$ or its universal enveloping algebra $\mathscr{U}_q(sl(2))$ with deformation parameter $q = $ root of unity.

From a historical viewpoint the Ponzano–Regge asymptotic formula for the $6j$ symbol of the group $SU(2)$ [52], together with Penrose's original idea of combinatorial spacetime out of coupling of angular momenta – or *spin networks* – [51], is the precursor of the discretized approaches to 3–dimensional Euclidean quantum gravity collectively referred to as 'state sum models' after the 1992 paper by Turaev and Viro [69]. The prominent role here is played by the *colored* tetrahedron encoding the tetrahedral symmetry of the $6j$ symbol – reminiscent of the Platonic solid shown in the reproduction of Fig. 6.1 – and recognized in the semiclassical limit as a *geometric* 3–simplex whose edge lengths are irreps labels from the representation ring of either $SU(2)$ or its universal enveloping algebra $\mathscr{U}_q(sl(2))$ with deformation parameter $q = $ root of unity.

We start with a review of some basic facts on Ponzano–Regge and Turaev–Viro state sum models for closed 3–manifolds and relationships of the latter with the generating functional of $SU(2)$ Chern–Simons–Witten (CSW) Topological Quantum Field Theory (TQFT). It is worth noting that there exist very informative and more complete reviews dealing with such topics in connection with geometric topology, discretized quantum gravity models, TQFT and associated 2D Conformal Field Theories [2, 20, 21, 26, 29, 30, 36, 38, 45, 56, 68], just to mention a few.

© Springer International Publishing AG 2017

M. Carfora, A. Marzuoli, *Quantum Triangulations*, Lecture Notes in Physics 942,

https://doi.org/10.1007/978-3-319-67937-2_6

Fig. 6.1 An ancient view to a
tetrahedron

Our treatment relies basically on the original formulations given in [15, 52], topic 9 and [69] respectively, thus avoiding the algebraic language of unitary modular tensor categories [45, 68] and focusing on some basic geometric and field–theoretic aspects.

The main body of the chapter is devoted to address a number of generalizations and extensions of the state sum functionals – some of which developed by the authors in collaboration with Gaspare Carbone – also in view of applications to topological quantum computing in the next chapter, thus conveying the idea that the two pictures might actually merge together [19].

6.1 The Wigner 6j Symbol and the Tetrahedron

Given three angular momentum operators $\mathbf{J}_1, \mathbf{J}_2, \mathbf{J}_3$ – associated with three kinematically independent quantum systems – the Wigner–coupled Hilbert space of the composite system is an eigenstate of the total angular momentum

$$\mathbf{J}_1 + \mathbf{J}_2 + \mathbf{J}_3 \doteq \mathbf{J} \tag{6.1}$$

and of its projection J_z along the quantization axis. The degeneracy can be completely removed by considering binary coupling schemes such as $(\mathbf{J}_1 + \mathbf{J}_2) + \mathbf{J}_3$ and $\mathbf{J}_1 + (\mathbf{J}_2 + \mathbf{J}_3)$, and by introducing intermediate angular momentum operators defined by

$$(\mathbf{J}_1 + \mathbf{J}_2) = \mathbf{J}_{12}; \ \ \mathbf{J}_{12} + \mathbf{J}_3 = \mathbf{J} \tag{6.2}$$

and

$$(\mathbf{J}_2 + \mathbf{J}_3) = \mathbf{J}_{23}; \ \ \mathbf{J}_1 + \mathbf{J}_{23} = \mathbf{J}, \tag{6.3}$$

respectively. In Dirac notation the simultaneous eigenspaces of the two complete sets of commuting operators are spanned by basis vectors

$$|j_1 j_2 j_{12} j_3; jm\rangle \quad \text{and} \quad |j_1 j_2 j_3 j_{23}; jm\rangle, \tag{6.4}$$

where j_1, j_2, j_3 denote eigenvalues of the corresponding operators, j is the eigenvalue of $\mathbf{J}$ and m is the total magnetic quantum number with range $-j \leq m \leq j$ in integer steps. Note that j_1, j_2, j_3 run over $\{0, \frac{1}{2}, 1, \frac{3}{2}, 2, \dots\}$ (labels of $SU(2)$ irreducible representations), while $|j_1 - j_2| \leq j_{12} \leq j_1 + j_2$ and $|j_2 - j_3| \leq j_{23} \leq j_2 + j_3$ (all quantum numbers are in $\hbar$ units).

The Wigner $6j$ symbol expresses the transformation between the two schemes (6.2) and (6.3), namely

$$|j_1 j_2 j_{12} j_3; jm\rangle = \sum_{j_{23}} [(2j_{12} + 1)(2j_{23} + 1)]^{1/2} \begin{Bmatrix} j_1 & j_2 & j_{12} \\ j_3 & j & j_{23} \end{Bmatrix} |j_1 j_2 j_3 j_{23}; jm\rangle \tag{6.5}$$

apart from a phase factor,[1] and where $(2j_{12} + 1)$ and $(2j_{23} + 1)$ are the dimensions of the irreps labeled by j_{12} and j_{23}, respectively. It follows that the square of the $6j$ symbol[2] represents the probability that a system prepared in a state of the coupling scheme (6.2), where j_1, j_2, j_3, j_{12}, j have definite magnitudes, will be measured to be in a state of the coupling scheme (6.3).

The $6j$ symbol can be written as sums of products of four Clebsch–Gordan coefficients or Wigner $3j$ symbols. The relations between the $6j$ and the latter is given explicitly by (see e.g. [70])

$$\begin{Bmatrix} a & b & c \\ d & e & f \end{Bmatrix} = \sum (-)^{\Phi} \begin{pmatrix} a & b & c \\ \alpha & \beta & -\gamma \end{pmatrix} \begin{pmatrix} a & e & f \\ \alpha & \epsilon & -\varphi \end{pmatrix} \begin{pmatrix} d & b & f \\ -\delta & \beta & \varphi \end{pmatrix} \begin{pmatrix} d & e & c \\ \delta & -\epsilon & \gamma \end{pmatrix}, \tag{6.6}$$

where $\Phi = d + e + f + \delta + \epsilon + \varphi$. Here Latin letters stand for j–type labels (integer or half–integers non–negative numbers) while Greek letters denote the associated magnetic quantum numbers (each varying in integer steps between $-j$ and j, $j \in \{a, b, c, d, e, f\}$). The sum is over all possible values of $\alpha, \beta, \gamma, \delta, \epsilon, \varphi$ with only three summation indices being independent.

On the basis of the above decomposition the $6j$ symbol is invariant under any permutation of its columns or under interchange the upper and lower arguments in each of any two columns. These algebraic relations involve $3! \times 4 = 24$ different $6j$ with the same value and are referred to as 'classical' as opposite to 'Regge' symmetries to be discussed below.

---

[1] Actually the above expression should contain the Racah W–coefficient $W(j_1 j_2 j_3 j; j_{12} j_{23})$ which differs from the $6j$ by the factor $(-)^{j_1 + j_2 + j_3 + j}$.

[2] According to Condon–Shortely conventions adopted here, the $6j$ is a real orthogonal matrix, and the same holds true for Clebsch–Gordan and Wigner coefficients.

The *6j* symbol is naturally endowed with a *tetrahedral symmetry*, by noticing first that each *3j* (or Clebsch–Gordan) coefficient vanishes unless its *j*–type entries satisfy the triangular condition, namely $|b - c| \leq a \leq b + c$, etc.. This suggests that each of the four *3j*'s in (6.6) can be associated with either a 3–valent vertex or a triangle. Accordingly, there are two graphical representation of the *6j* exhibiting its symmetry properties. Here we adopt the three–dimensional picture introduced in [52], rather than Yutsis' 'dual' representation as a complete graph on four vertices [75]. Then the *6j* is thought of as a real solid tetrahedron T with edge lengths $\ell_1 = a + \frac{1}{2}, \ell_2 = b + \frac{1}{2}, \ldots, \ell_6 = f + \frac{1}{2}$ in $\hbar$ units[3] and triangular faces associated with the triads (abc), (aef), (dbf), (dec). This implies in particular that the quantities $q_1 = a + b + c, q_2 = a + e + f, q_3 = b + d + f, q_4 = c + d + e$ (sums of the edge lengths of each face), $p_1 = a + b + d + e, p_2 = a + c + d + f, p_3 = b + c + e + f$ are all integer with $p_h \geq q_k$ ($h = 1, 2, 3, k = 1, 2, 3, 4$). The conditions addressed so far are in general sufficient to guarantee the existence of a non–vanishing *6j* symbol, but they are not enough to ensure the existence of a geometric tetrahedron T, topologically a 3–disk bounded by a 2–sphere in Euclidean 3–space, as Fig. 6.1 is aimed to suggest. More precisely, T exists in this sense if (*and only if*, see the discussion in the introduction of [52]) its square volume $V(T)^2 \equiv V^2$, evaluated by means of the Cayley–Menger determinant, is positive.

6.1.1 The Racah Polynomial and Algebraic Identities for the 6j Symbol

The generalized hypergeometric series, denoted by $_pF_q$, is defined on p real or complex numerator parameters $a_1, a_2, \ldots, a_p$, q real or complex denominator parameters $b_1, b_2, \ldots, b_q$ and a single variable z by

$$
_pF_q \begin{pmatrix} a_1 \ldots a_p \\ \\ b_1 \ldots b_q \end{pmatrix}; z \Bigg) = \sum_{n=0}^{\infty} \frac{(a_1)_n \cdots (a_p)_n}{(b_1)_n \cdots (b_p)_n} \frac{z^n}{n!}, \tag{6.7}
$$

where $(a)_n = a(a+1)(a+2) \cdots (a+n-1)$ denotes a rising factorial with $(a)_0 = 1$. If one of the numerator parameter is a negative integer, as actually happens in the following formula, the series terminates and the function is a polynomial in z.

The key expression for relating the *6j* symbol to hypergeometric functions is the Racah sum rule (see e.g. [15], topic 11 and [70], Ch. 9 also for the original references). Then the expression of the so–called *Racah polynomial* in terms of the $_4F_3$ hypergeometric function evaluated at $z = 1$ reads

---

[3] The $\frac{1}{2}$–shift is crucial in the analysis developed in [52]: for high quantum numbers the length $[j(j + 1)]^{1/2}$ of an angular momentum vector is close to $j + \frac{1}{2}$, see the semiclassical analysis given in Sect. 6.1.2 below.

$$\begin{Bmatrix} a & b & d \\ c & f & e \end{Bmatrix} = \Delta(abe)\,\Delta(cde)\,\Delta(acf)\,\Delta(bdf)\,(-)^{\beta_1}(\beta_1+1)!$$

$$\times \frac{{}_4F_3\left(\begin{matrix} \alpha_1-\beta_1 & \alpha_2-\beta_1 & \alpha_3-\beta_1 & \alpha_4-\beta_1 \\ -\beta_1-1 & \beta_2-\beta_1+1 & \beta_3-\beta_1+1 \end{matrix} ; 1\right)}{(\beta_2-\beta_1)!(\beta_3-\beta_1)!(\beta_1-\alpha_1)!(\beta_1-\alpha_2)!(\beta_1-\beta_3)!(\beta_1-\alpha_4)!}\,, \qquad (6.8)$$

where

$$\beta_1 = \min(a+b+c+d; a+d+e+f; b+c+e+f)$$

and the parameters β_2, β_3 are identified in either way with the pair remaining in the 3–tuple $(a+b+c+d; a+d+e+f; b+c+e+f)$ after deleting β_1. The four α's may be identified with any permutation of $(a+b+e; c+d+e; a+c+f; b+d+f)$. Finally, the Δ–factors in front of ${}_4F_3$ are defined, for any triad (abc) as

$$\Delta(abc) = \left[\frac{(a+b-c)!(a-b+c)!(-a+b+c)!}{(a+b+c+1)!}\right]^{1/2}$$

Such a seemly complicated notation is indeed the most convenient for the purpose of listing a number of analytical and algebraic properties of the Wigner $6j$ symbol.

- The Racah polynomial is placed at the top of the Askey hierarchy including all of hypergeometric orthogonal polynomials of one (discrete or continuous) variable [9]. Thus most commonly encountered families of special functions in quantum mechanics can be obtained from the Racah polynomial by applying suitable limiting procedures, as recently reviewed in [53]. Moreover, such an unified scheme provides in a straightforward way the algebraic 'defining relations' of the Wigner $6j$ symbol. In the standard notation adopted in the rest of this chapter such relations are:

  the Biedenharn–Elliott identity $(R = a+b+c+d+e+f+p+q+r)$:

$$\sum_x (-)^{R+x}\,(2x+1)\begin{Bmatrix} a & b & x \\ c & d & p \end{Bmatrix}\begin{Bmatrix} c & d & x \\ e & f & q \end{Bmatrix}\begin{Bmatrix} e & f & x \\ b & a & r \end{Bmatrix}$$

$$= \begin{Bmatrix} p & q & r \\ e & a & d \end{Bmatrix}\begin{Bmatrix} p & q & r \\ f & b & c \end{Bmatrix}; \qquad (6.9)$$

  the orthogonality relation (δ is the Kronecker delta)

$$\sum_x (2x+1)\begin{Bmatrix} a & b & x \\ c & d & p \end{Bmatrix}\begin{Bmatrix} c & d & x \\ a & b & q \end{Bmatrix} = \frac{\delta_{pq}}{(2p+1)}, \qquad (6.10)$$

where the summation label x runs, for any fixed values of the other entries, over a finite range owing to triangular inequalities which must hold for each triad in the $6j$'s.

- Given the Racah polynomial as in (6.8), the unexpected new symmetry of the $6j$ symbol discovered in 1958 by Regge [54] (see also [14, 70]) is recognized as a 'trivial' set of permutations on the parameters α, β that leaves $_4F_3$ invariant. Combining the Regge symmetry and the 'classical' ones, one get a total number of 144 algebraic symmetries for the $6j$. Note in passing that implications of Regge symmetry on the geometry of the quantum tetrahedron, discussed first in [59], should deserve further investigations.

- The Askey hierarchy of orthogonal polynomials can be extended to a q-hierarchy [9], on top of which the q-analog of the $_4F_3$ polynomial stands. Accordingly, the q-analog of the Wigner $6j$ symbol might be consistently introduced (actually for q real between 0 and 1), but we postpone this issue to Sect. 6.2.2 where the q-$6j$ is recovered as a basic ingredient in the construction of the representation ring of the ($q = $ root of unity)–analog of $SU(2)$ (crf. [13, 42] for accounts on the theory of q–special functions and its relations with q–tensor algebra).

- The $6j$ symbol satisfies a 3–term recursion relation in one variable [70] which derives directly from the Biedenharn–Elliott identity (6.9). It can be recasted in the form of a second order difference equation, which in turn becomes a differential equation for a continuous variable in the semiclassical limit analyzed within the WKB framework, see [43, 52, 63, 64], [15], topic 9. This remark has to do with a deep property of hypergeometric polynomials in the Askey scheme [9, 42]: the defining difference (differential) equation and the recursion relation are 'dual' to each other but the $6j$ happens to be 'self–dual' so that the two viewpoints can be used equivalently.

6.1.2 Ponzano–Regge Asymptotic Formula

According to Bohr's correspondence principle, classical concepts become increasingly valid in the regime where quantum numbers are large. In handling with angular momenta variables measured in units of $\hbar$, the classical limit $\hbar \to 0$ implies that, for finite angular momenta, both the j–quantum numbers and the magnetic ones are much bigger than one. For what concerns pure angular momentum states as those introduced in (6.4), when approaching classical limit all the components of the operators $\mathbf{J}$'s are confined to narrower ranges around specific values. Thus geometric concepts typical of the semiclassical vector model arise naturally and the corresponding physical quantities have to be thought as averaged out. Angular momentum functions such as Wigner rotation matrices and $3j$ symbols, as well as $6j$ symbols, admit well defined 'asymptotic limits', whose absolute squares (probabilities) correspond to classical limits of the related physical quantities (cfr. [15], topic 9 for a survey).

The (positive frequency part of) asymptotic formula for the $6j$ symbol proposed by Ponzano and Regge in [52] reads

$$\begin{Bmatrix} a & b & d \\ c & f & e \end{Bmatrix} \sim \frac{1}{\sqrt{24\pi V}} \exp \left\{ i \left(\sum_{r=1}^{6} \ell_r \theta_r + \frac{\pi}{4} \right) \right\} \tag{6.11}$$

where the limit is taken for all entries $\gg 1$ (recall that $\hbar = 1$) and $\ell_r \equiv j_r + 1/2$ with $\{j_r\} = \{a, b, c, d, e, f\}$, *cfr.* Footnote 3. V is the Euclidean volume of the tetrahedron T, θ_r is the angle between the outer normals to the faces which share the edge ℓ_r and the formula is valid in the the classically allowed region determined by the requirement $V^2 > 0$.

From a purely quantum mechanical viewpoint, the above probability amplitude has the form of a semiclassical (wave) function since the factor $1/\sqrt{24\pi V}$ is slowly varying with respect to the spin variables while the exponential is a rapidly oscillating dynamical phase (*cfr.* Wigner's semiclassical estimate for the probability, namely $\begin{Bmatrix} a & b & d \\ c & f & e \end{Bmatrix}^2 \sim 1/12\pi\, V$). It is worth noting that, quite independently from the remarks to be addressed below, the literature concerning more or less 'rigorous' derivations of (6.11) – and of its counterpart in the classically forbidden region – spreads over decades, starting from [15, 43, 63, 64], topic 9, up to [59], just to mention a few.

Coming to Feynman's path sum interpretation of quantum mechanics, the argument of the exponential in (6.11) can be recognized as a classical action involving pairs of canonical variables (angular momenta and their conjugate angles). Such an interpretation, already sketched in [52], has been improved recently by resorting to rigorous multidimensional WKB analysis and geometric quantization methods [4, 6].

However, the most intriguing and far–reaching physical interpretation of the asymptotic formula (6.11) stems from the role it plays as a semiclassical limit of a path–sum over all quantum fluctuations to be associated with the simplest 3–dimensional discrete 'spacetime', an Euclidean tetrahedron T. In fact the argument in the phase reproduces the classical Regge action $S_R(\ell)$ for T [55] since in the present case $(d - 2)$ simplices σ_i^1 are the edges of T and $\mathrm{Vol}(\sigma_i^1) \equiv \ell_i$ are the associated lengths. On such a crucial remark relies the so–called Ponzano–Regge state sum for discretized gravity in dimension 3 addressed in the next section.

In this monography we do not address the issue of asymptotic expansions of more general $SU(2)$ recoupling coefficients or $3nj$ symbols (see [15], topic 12 for the necessary background material). We just mention the results of [31], where the asymptotic formula for the $9j$ is worked out on a rigorous basis, and a general scheme for dealing systematically with (partial) asymptotics of $3nj$ symbols proposed in [3]. From the viewpoint of special function theory (Sect. 6.1.1) such coefficients are related to hypergeometric polynomials in more than one discrete variable (the $9j$ is actually in two variables [42]) and the analysis of the associated recursion relations and differential equations in the large–j limit would be interesting also in view of applications in quantum chemistry [5].

6.2   State Sum Functionals for Closed 3–Manifolds

6.2.1   Ponzano–Regge State Sum and Semiclassical Euclidean Gravity

With a slight change of notations with respect to the previous chapters, denote by $\mathscr{T}^3(j) \to M^3$ a particular triangulation of a (not necessarily oriented) closed 3–dimensional simplicial PL manifold M^3 of fixed topology obtained by assigning $SU(2)$ spin variables, or 'colors', $\{j\}$ to the edges of $\mathscr{T}^3$. The assignment satisfies a number of conditions, better illustrated if we introduce the *state functional* associated with $\mathscr{T}^3(j)$, namely

$$\mathbf{Z}[\mathscr{T}^3(j) \to M^3; L] = \Lambda(L)^{-N_0} \prod_{A=1}^{N_1} (-1)^{2j_A} \mathsf{w}_A \prod_{B=1}^{N_3} \phi_B \begin{Bmatrix} j_1 & j_2 & j_3 \\ j_4 & j_5 & j_6 \end{Bmatrix}_B \tag{6.12}$$

where N_0, N_1, N_3 are the number of vertices, edges and tetrahedra in $\mathscr{T}^3(j)$, $\Lambda(L) = 4L^3/3C$ (L is a fixed length and C an arbitrary constant), $\mathsf{w}_A \doteq (2j_A + 1)$ are the dimensions of irreps of $SU(2)$ which weigh the edges, $\phi_B = (-1)^{\sum_{p=1}^{6} j_p}$ and $\{:::\}_B$ are $6j$ symbols to be associated with the tetrahedra of the triangulation. The Ponzano–Regge *state sum* is obtained by summing over triangulations corresponding to all assignments of spin variables $\{j\}$ bounded by the cut–off L

$$\mathbf{Z}_{PR}[M^3] = \lim_{L \to \infty} \sum_{\{j\} \leq L} \mathbf{Z}[\mathscr{T}^3(j) \to M^3; L], \tag{6.13}$$

where the cut–off is formally removed by taking the limit in front of the sum (a regularization procedure that does not provide necessarily a finite result, see the next section). From a historical viewpoint (6.13) was recognized, see e.g. [47], as the first example of a consistent discretized 'path integral'[4] for simplicial quantum gravity, the approach founded in the early 1980's by R. Williams, H. Hamber, J. Hartle and others (see the review [72] also for a complete list of references up to 1992). Further developments of these ideas, in particular the 'dynamical triangulations' setting, have been extensively addressed up to the end of the 1990's and we refer to the Lecture Notes co-authored with Jan Ambjørn [1] for a comprehensive review and list of references.

---

[4]By consistency we mean that the discretized counterpart of the functional measure in the Euclidean path integral would be proportional to $\prod (2j + 1)dj$, according to the identification between 'colors' and edge lengths. Triangular (tetrahedral) inequalities, quite difficult to be implemented within a purely simplicial PL background, are automatically fulfilled since the $6j$ symbol vanishes whenever a constraint of this kind is violated, *cfr.* the introductory part of this section and further remarks on regularized functional measures in Sect. 6.2.2.

It is not easy to review in short the huge number of implications and further improvements of Ponzano–Regge state sum functional (6.13), as well as its deep and somehow unexpected relationships with so many different topics in quantum field theory and in pure mathematics. In particular we have to leave aside the analysis of the prominent role of Ponzano–Regge model in the Loop Quantum Gravity approach, see e.g. [27, 56, 62] and references therein.

We are going to focus in the rest of this section on the issue of topological invariance of Ponzano–Regge state sum. As already noted in [52] and reviewed for instance in [21], the state sum $\mathbf{Z}_{PR}[M^3]$ (if finite) is a combinatorial invariant of the manifold M^3 because its value is independent of the particular triangulation $\mathscr{T}^3(j)$ used to evaluate it. Such feature can be read from a field–theoretic viewpoint as related to the behavior of the discretized path sum not only under refinements of the mesh but also under compositions associated to different dissections on the underlying spacetime (see Sect. 6.3). The 'combinatorial moves' that leave (6.13) invariant are expressed algebraically in terms of the Biedenharn-Elliott identity (6.9), representing the moves (2 *tetrahedra*) $\leftrightarrow$ (3 *tetrahedra*)

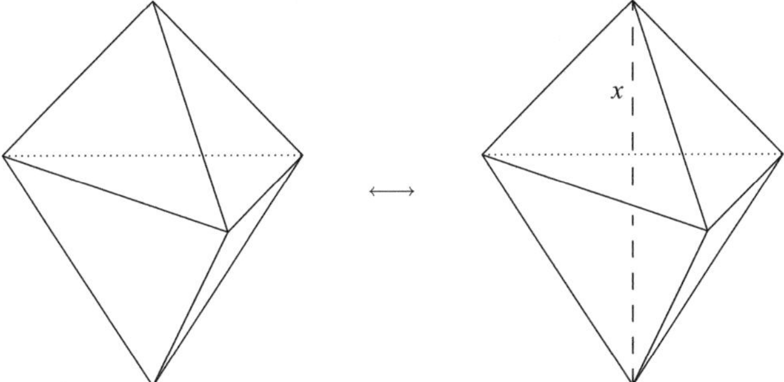

and of both the Biedenharn–Elliott identity and the orthogonality conditions (6.10) for $6j$ symbols, which represents the so–called Alexander move together its inverse, namely (1 *tetrahedra*) $\leftrightarrow$ (4 *tetrahedra*)

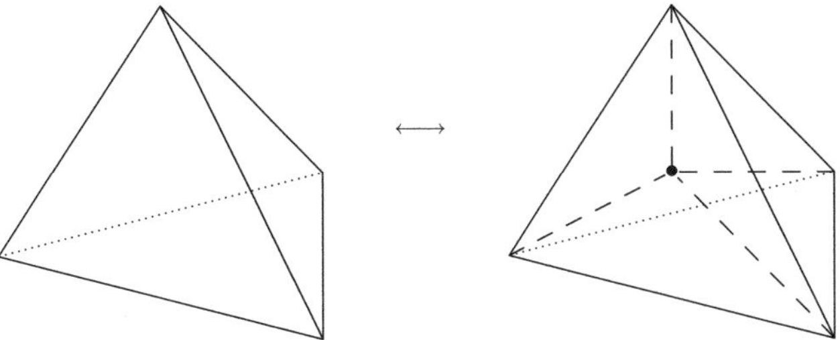

Then, in view of the remarks in Sect. 6.1.1, these algebraic identities play the twofold role of 'defining relations' for the $6j$ and 'combinatorial constraints' on local arrangements of colored tetrahedra in 3–dimensional triangulations. Moreover, borrowing the language of (braided) tensor categories [33], the Biedenharn–Elliott identity is recognized as the 'pentagon' constraint for the representation ring of $SU(2)$, while the so–called 'hexagon' constraints would correspond to the Racah identity for the $6j$ symbol, whose combinatorial content is related to braiding (a morphism which is actually trivial in the $SU(2)$ case), see Sect. 6.4 below.

In Appendix A of [1] we gave a comprehensive account of results from (d-dimensional) geometric topology concerning relations between PL–homeomorphic closed PL manifolds and equivalence of their simplicial dissections under families of combinatorial moves named 'bistellar' after Pachner's paper [48]. Without entering into much details, we are going to review just a few basic definitions with the aim of setting up a shorthand notation for combinatorial moves to be extended in Sect. 6.3 in dealing with manifolds with boundary.

Two closed PL manifolds M_1^d and M_2^d are PL–homeomorphic if and only if there exists a map $f : M_1^d \to M_2^d$ which is both a homeomorphism and a simplicial isomorphism, $M_1^d \simeq M_2^d$. Given two triangulations $T_1^d \to |T_1^d| \simeq M_1^d$ and $T_2^d \to |T_2^d| \simeq M_2^d$, Pachner [48] proved that they are PL–homeomorphic if and only if T_1^d can be transformed into T_2^d by a finite sequence of d–dimensional bistellar moves (and their inverse moves), formally

$$|T_1^d| \simeq |T_2^d| \quad \Leftrightarrow \quad T_1^d \propto_{\text{bst}} T_2^d. \tag{6.14}$$

The classification of the allowed moves is based on the observation that each move involves a finite arrangement of connected d–simplices, topologically a d–dimensional disk bounded by a $(d-1)$–dimensional PL sphere, which must be kept fixed. Each local operation of this kind is characterized in terms of a pair of sub–simplices, σ and τ, whose dimensionalities satisfy $\dim(\sigma) + \dim(\tau) = d$.

If n_d is the number of d–simplices $\in T^d$ involved in a bistellar operation, then

$$\left\{ [n_d \to (d+1) - (n_d - 1)]_{\text{bst}}^{\mathbf{d}}, n_d = 1, 2, \ldots, d+1 \right\} \tag{6.15}$$

is a convenient notation for the set admissible moves in dimension d. For instance the case $n_d = 1$ corresponds to the move $[1 \to (d+1)]_{\text{bst}}^d$ – one d–simplex is transformed into $(d+1)$ d–simplices with the same boundary configuration – and it can be easily recognized as the Alexander (sometimes called 'barycentric') move in any dimension d, obtained by putting one new vertex inside one d-simplex and joining it to the original vertices. Its inverse move is represented by $[(d+1) \to 1]_{\text{bst}}^d$, and both of them satisfy the constraint $\dim(\sigma) + \dim(\tau) = d$ since the sub–simplices involved are indeed one 0–simplex and one d–simplex. According to the above notation, the admissible bistellar moves in $d = 3$, depicted in the Figures above, can be listed as

$$[1 \leftrightarrow 4]_{\text{bst}}^{3} \quad \text{and} \quad [2 \leftrightarrow 3]_{\text{bst}}^{3}, \tag{6.16}$$

where a double arrow denotes the move and its inverse. Similarly in $d = 2$ we have

$$[1 \leftrightarrow 3]_{\text{bst}}^{2} \quad \text{and} \quad [2 \to 2]_{\text{bst}}^{2}, \tag{6.17}$$

which stand for the Alexander move (and its inverse) and the 'flip' move acting on a pair of triangles joined along an edge (*cfr.* the role of the latter in dealing with polyhedral surfaces in Chap. 1, and notice that here operations are blind with respect to their effective metric realization).

6.2.2 Turaev–Viro Quantum Invariant

In the seminal paper [69] a 'regularized' version of (6.13) based on the representation theory of the quantum deformation of the universal enveloping algebra of $sl(2, \mathbb{C})$ (the complexified counterpart of the algebra $su(2)$) denoted by $\mathcal{U}_q\,(sl(2))$, was shown to provide be a well–defined (finite) *quantum invariant* for closed 3–dimensional PL manifolds. Its expression (for a triangulation with N_0 vertices, N_1 edges and N_3 tetrahedra) reads

$$\mathbf{Z}_{TV}\,[M^3; q] \;=\; \sum_{\{j\}} \omega^{-2N_0} \prod_{A=1}^{N_1} \omega_A^2 \prod_{B=1}^{N_3} \left| \begin{matrix} j_1 & j_2 & j_3 \\ j_4 & j_5 & j_6 \end{matrix} \right|_B , \tag{6.18}$$

where the summation is over all $\{j\}$ labeling highest weight irreps of $\mathcal{U}_q\,(sl(2))$ for $q = \exp\{2\pi i/r\}$ and $j \in \{0, 1/2, 1, \ldots, (r-2)/2\}$. The weights in the state sum are

$$\omega^2 = \frac{-2r}{(q^{1/2} - q^{-1/2})^2}$$

$$\omega_j = (-)^{2j}\,[2j+1]^{1/2} \tag{6.19}$$

where a q–integer $[n]$ is defined as

$$[n] := [n]_q = \frac{q^{n/2} - q^{-n/2}}{q^{1/2} - q^{-1/2}} . \tag{6.20}$$

Finally each tetrahedron is weighted by

$$\left| \begin{matrix} j_1 & j_2 & j_3 \\ j_4 & j_5 & j_6 \end{matrix} \right| = (\sqrt{-1})^{-2\sum j_i} \left\{ \begin{matrix} j_1 & j_2 & j_3 \\ j_4 & j_5 & j_6 \end{matrix} \right\}_q , \tag{6.21}$$

where on the right–hand side there appears the q-6j symbols of $\mathcal{U}_q\,(sl(2))$ whose entries are the colors $j_i, i = 1, 2, \ldots, 6$. The proof that (6.18) for each value of q is an invariant of the PL structure was made in [69] by resorting to a sophisticated analysis that we are not going to review here. What suffices to say is that a prominent role is played by the q–analog of Biedenharn–Elliott identity (6.9) together with the properties of the representation ring of $\mathcal{U}_q\,(sl(2))$, which ensure that the state sum is actually finite owing to the relations $\sum_k \omega_k^2 = \omega_i^2\,\omega_l^2$ and $\omega^2 = \omega_l^{-2} \sum_{k,i} \omega_k^2\,\omega_i^2$ (where (k, i, l) belong to a triad satisfying triangular inequalities). However (6.18) should be also an extension of Ponzano–Regge state sum (6.13) as a physical model as far as $\mathbf{Z}_{TV}\,[M^3; q = 1] = \mathbf{Z}_{PR}\,[M^3]$ and in the limit $q \to 1$ ($r \to \infty$): $[2j+1] \approx (2j+1) + O(r^{-2})$ and $\omega^2 \approx r^3/2\pi^2\,(1 + O(r^{-2}))$.

The issue of the $(q \to 1)$–limit of $\mathbf{Z}_{TV}$ has been discussed soon after the discovery of this quantum invariant [41] and later on by a number of authors.[5] The correct physical interpretation has been recognized by going through the 'one–loop' expansion of TV state sum (as well as PR state sum) to quantum gravity in the first–order formalism [73], where the classical action is of Chern–Simons–type for the gauge group $SU(2)$ with (or without) a cosmological constant (see [2, 20] for accounts and original references).

Actually most developments of Turaev–Viro model have taken place in low–dimensional geometric topology with the birth of an entirely new branch dealing with quantum invariants and their 'perturbative' or finite–type counterparts (see the monography [45] and the rich bibliography therein).

Both the physical interpretation and most mathematical applications of quantum invariants rely on the connection between the state sum (6.18) *at a fixed value* of the root of unity q and a Chern–Simons–Witten generating functional addressed in the next section (we postpone to Sect. 6.4.3 a few remarks on the strictly related Reshetikhin–Turaev invariant [57] obtained by presenting the 3–manifold as the complement of a knot in the 3–sphere).

6.2.3   Chern–Simons–Witten Generating Functional and Turaev–Viro Invariant

Let M^3 represent in this section a closed *oriented* smooth 3–manifold and recall that, once given a C^∞ atlas, M^3 belongs to a unique PL–type. The state sum (6.18) can be shown to be equal to the square of the modulus of the Chern–Simons–Witten (Reshetikhin–Turaev) invariant [57, 74], which in turn represents the quantum generating functional of an $SU(2)$ Chern–Simons topological field theory [20, 26, 30, 38]. For any fixed root of unity q one has

$$\mathbf{Z}_{TV}\,[M^3;q] \;\longleftrightarrow\; \mathbf{Z}_{CSW}\,[M^3;k]\,\mathbf{Z}_{CSW}\,[\bar{M}^3;k] \equiv |\mathbf{Z}_{CSW}\,[M^3;k]|^2, \qquad (6.22)$$

where the 'level' k of the Chern–Simons functional is related to the deformation parameter $q = \exp\{2\pi i/r\}$ by identifying it with r and $\mathbf{Z}_{CSW}\,[\bar{M}^3;k]$ denotes the invariant evaluated for M^3 with the opposite orientation. The proof of this equivalence has been provided by Turaev himself [68], by Walker [71], by Roberts [58], Beliakova and Durhuhs [12] and by Kauffman and Lins [37] by resorting to different methodologies (see also the review [45]). Without entering into details on the proof, we conclude this section by reviewing in brief the original construction of the 3–manifold invariant $\mathbf{Z}_{CSW}\,[M^3;k]$.

---

[5] Actually the geometric content of the q-$6j$ symbol comes out in such a perturbative (not quite 'semiclassical') limit, and interestingly its emerging geometry is spherical at q a root of unity and hyperbolic in case of q real positive [65].

Topological Quantum Field Theories (TQFT) are particular types of gauge theories, namely field theories quantized through the (Euclidean) path integral prescription starting from a classical Yang–Mills action defined on a suitable d–dimensional space(time). TQFT are characterized by partition functions and observables (correlation functions) which depend only on the global features of the space on which these theories live, namely independent of any metric which may be used to define the underlying classical theory. The geometrical and topological generating functionals and correlation functions of such theories are computable by standard techniques in quantum field theory and provide novel representations of certain global invariants (for d-manifolds and for particular submanifols embedded in the ambient space).

Focusing on the 3–dimensional case denote by Σ_1 and Σ_2 a pair of disjoint 2–dimensional manifolds and by M^3 a 3–dimensional manifold with boundary $\partial M^3 = \Sigma_1 \cup \Sigma_2$ (all manifolds here are Riemannian, compact, smooth and oriented and $\cup$ is the disjoint union). According to the axiomatization given by Atiyah [10] a unitary 3d quantum field theory corresponds to the assignment of

(i) finite dimensional Hilbert spaces (endowed with non–degenerate bilinear forms) $\mathscr{H}_{\Sigma_1}$ and $\mathscr{H}_{\Sigma_2}$ to Σ_1 and Σ_2, respectively;

(ii) a map ('functor') connecting such Hilbert spaces

$$\mathscr{H}_{\Sigma_1} \xrightarrow{\ \mathbf{Z}\,[M^3]\ } \mathscr{H}_{\Sigma_2} \tag{6.23}$$

where M^3 is a cobordism, namely it interpolates between Σ_1 (incoming boundary) and Σ_2 (outgoing boundary). Without entering into details concerning a few more axioms (diffeomorphism invariance, factorization etc.) we just recall that unitarity implies that

(iii) if $\bar{\Sigma}$ denotes the surface Σ with the opposite orientation, then $\mathscr{H}_{\bar{\Sigma}} = \mathscr{H}_{\Sigma}^*$, where $*$ stands for complex conjugation;

(iv) the functors (6.23) are unitary and $\mathbf{Z}[\bar{M}^3] = \mathbf{Z}^*[M^3]$, where $\bar{M}^3$ denotes the manifold with the opposite orientation.

In Chern–Simons theory (discussed here for the gauge group $SU(2)$ but generalizable to any $SU(N)$) the functor (6.23) is the partition function $\mathbf{Z}_{CSW}$ associated with the classical action

$$S_{CS}(A) = \int_{M^3} \mathrm{Tr}\left(A \wedge dA + \frac{2}{3}A \wedge A \wedge A\right) \tag{6.24}$$

written for the case $\partial M^3 = \emptyset$. Here A is a connection on the trivial $SU(2)$–bundle over M^3, namely it is a skew–Hermitian matrix of Lie algebra–valued 1–forms with trace 0, $A \in \Omega(M^3, su(2))$, d is the exterior differential and the wedge products $\wedge$ of differential forms and traces are combined with matrix multiplication.

The partition function is obtained formally by integrating the exponential of the classical action (6.24) over the space $A/\mathscr{G}$ of equivalence classes of (flat) connections modulo gauge transformations

$$\mathbf{Z}_{CSW}[M^3; k] = \int_{A/\mathcal{G}} [DA] \exp\left\{\frac{ik}{4\pi} S_{CS}(A)\right\}, \tag{6.25}$$

where $\hbar = 1$ and the coupling constant k, the level of the theory, is constrained to be a positive integer by the quantization procedure. $A/\mathcal{G}$ is the quotient of $\Omega(M^3, su(2))$ by the action of the Lie group $\mathcal{G}$ of maps $g : M^3 \to G$ given by

$$g \cdot A = g^{-1} A g + g^{-1} dg; \quad A \in A/\mathcal{G}. \tag{6.26}$$

In the field–theoretic setting of path integrals, 'observables' of the theory are quantities which must be gauge–invariant and invariant under diffeomorphisms of the ambient manifols. These properties are shared by Wilson loop operators $\mathscr{W}_k(K)$, namely holonomies of the connection 1–form evaluated on closed (knotted) curves K in M^3, formally

$$\mathscr{W}_k(K) = \frac{\int_{A/\mathcal{G}} [DA] \, e^{\frac{ik}{4\pi} S_{CS}(A)} \mathrm{Tr}\,(\mathrm{hol}\,K)}{\int_{A/\mathcal{G}} [DA] \, e^{\frac{ik}{4\pi} S_{CS}(A)}} \tag{6.27}$$

where $\mathrm{hol}\,K \doteq \mathscr{P} \exp \int_K A$, $\mathscr{P}$ is the path ordering and the trace is over Lie–algebra indices. Recall that, once chosen the adjoint representation of the Lie algebra of the group $SU(2)$ to express the holonomies, (6.27) represents the Jones polynomial of the knot K [34, 74]. Both the partition function $\mathbf{Z}_{CSW}$ and Jones–type polynomial invariants of knots will be discussed in the next chapter in connection with (quantum) computational questions.

Remark The extension of (6.25) to the case $\partial M^3 \neq \emptyset$ requires modifications of the classical action (6.24) by suitable Wess–Zumino–Witten (WZW) terms providing non–trivial quantum degrees of freedom on the boundary surfaces, see [20] for a review. In [7] direct correspondence between *2d* (boundary) Regge triangulations and punctured Riemann surfaces has been established, thus providing a novel explicit characterization of the WZW model on triangulated surfaces on any genus at a fixed level $\ell = k + 2$.

6.3   State Sum Functionals for 3–Manifolds with Boundary

Given the 'topological' nature of Turaev–Viro (Ponzano–Regge) state sums in case of closed 3–manifolds , whenever a 2–dimensional boundary occurs in M^3, giving rise to a simplicial PL–pair (M^3, Σ), things might change. The reason for dealing with such extended state sums – not simply addressing this issue on the basis of the correspondence (6.22) with CSW functional and following for instance the approach quoted in the previous remark – is related to the possibility of addressing in a straightforward way 2d discretized (lattice) models on the boundary which are useful in applications, see the central part and the final remark of Sect. 6.3.3 below. Moreover, in view of the difficulties in dealing with heuristic path integral definitions of the invariants given in (6.25) and (6.27), the statistical sum approach certainly provides a more effective computational scheme, *cfr.* Sect. 6.4.

6.3.1   *Turaev–Viro Invariant with a Fixed Boundary Triangulation*

The Turaev–Viro invariant (6.18) can be easily generalized to a simplicial pair $(M^3, \partial M^3 = \mathcal{T}^2(j'))$, where the 2–dimensional boundary triangulation is kept fixed [68]. More precisely, given the assignments of weights to the interior of a particular colored triangulation $\mathcal{T}^3(j) \rightarrow M^3$ as in Sect. 6.2.2, consider the triangulation it induces on the boundary, $\partial \mathcal{T}^3(j') := \mathcal{T}^2(j')$, where the collective variable j' denotes the subset of colorings labeling boundary edges (and from now on $\{j\}$ are the admissible colorings of $\text{Int}(\mathcal{T}^3(j))$). Then the state sum

$$
\mathbf{Z}_{TV}\left[(M^3, \mathcal{T}^2(j')); q\right] = \sum_{\{j\}} \omega^{-2N_0 + n_0} \prod_{A=1}^{N_1} \omega_A^2 \prod_{A'=1}^{n_1} \omega_{A'} \prod_{B=1}^{N_3} \left| \begin{matrix} j_1 & j_2 & j_3 \\ j_4 & j_5 & j_6 \end{matrix} \right|_B ,
$$

(6.28)

where n_0, n_1 are the numbers of vertices and edges in $\mathcal{T}^2(j')$, is a quantum invariant depending explicitly on the fixed boundary coloring $\{j'\}$. The weight of each boundary vertex is ω and a boundary edge labeled by j' contributes with $\omega_{j'}$ given in (6.19). Then these assignments are compatible with the correct behavior of Turaev–Viro partition functions under composition (corresponding to invariance under arbitrary dissections of 3-triangulations). We postpone to Sect. 6.4.1 the illustration of the original axiomatic formulation of such kind of statistical sums in terms of *initial data sets*.

An extension to 3–manifold with boundary can be made for Ponzano–Regge state sum (6.13) as well by setting $q = 1$ simply in (6.28) and replacing each ω^2–factor with $\Lambda(L)$.

6.3.2   *Ponzano–Regge State Sum for a Pair $(M^3, \partial M^3)$*

In this section we are going to extend the Ponzano–Regge state sum of Sect. 6.2.1 to the case of a compact simplicial PL–pair $(M^3, \partial M^3)$ where the boundary triangulation is not kept fixed [17]. The resulting functional (up to regularization) is shown to be an invariant by resorting to a general set of combinatorial moves (bistellar operations and shellings).

This result has been generalized to d–dimensional simplicial PL–pairs $(M^d, \partial M^d)$ in [18], thus providing an inductive procedure to built up an entire hierarchy of state sum invariants to be identified with discretized versions of generating functionals of topological quantum field theory of the BF type [36] (see a few more remarks at the end of this section).

Given a 3–dimensional simplicial pair $(M^3, \partial M^3)$ consider the *colored* triangulation

$$(\mathscr{T}^3(j), \partial \mathscr{T}^3(j', m)) \longrightarrow (M^3, \partial M^3) \tag{6.29}$$

representing an admissible assignment of both spin variables to the collection of the edges $((d-2)$–simplices) in $(\mathscr{T}^3, \partial \mathscr{T}^3)$ and of momentum projections to the subset of edges lying in $\partial \mathscr{T}^3$. The collective variable $j \equiv \{j_A\}$, $A = 1, 2, \ldots, N_1$, denotes all the spin variables, n_1' of which are associated with the edges in the boundary (for each A: $j_A = 0, 1/2, 1, 3/2, \ldots$ in $\hbar$ units). Notice that the last subset is labeled both by $j' \equiv \{j_C'\}$, $C = 1, 2, \ldots, n_1'$, and by $m \equiv \{m_C\}$, where m_C is the projection of j_C' along the fixed reference axis (of course, for each m, $-j \leq m \leq j$ in integer steps). The consistency in the assignment of the j, j', m variables is ensured by requiring that

- each 3-simplex σ_B^3, $(B = 1, 2, \ldots, N_3)$, in $(T^3, \partial T^3)$ must be associated, apart from a phase factor, with a $6j$ symbol of $SU(2)$, namely

$$\sigma_B^3 \longleftrightarrow (-1)^{\sum_{p=1}^6 j_p} \begin{Bmatrix} j_1 & j_2 & j_3 \\ j_4 & j_5 & j_6 \end{Bmatrix}_B ; \tag{6.30}$$

- each 2-simplex σ_D^2, $D = 1, 2, \ldots, n_2'$ in ∂T^3 must be associated with a Wigner $3j$ symbol of $SU(2)$ according to

$$\sigma_D^2 \longleftrightarrow (-1)^{(\sum_{s=1}^3 m_s)/2} \begin{pmatrix} j_1' & j_2' & j_3' \\ m_1 & m_2 & -m_3 \end{pmatrix}_D . \tag{6.31}$$

Then the following state sum can be defined

$$\mathbf{Z}_{PR}[(M^3, \partial M^3)]$$

$$= \lim_{L \to \infty} \sum_{\{j, j', m \leq L\}} \mathbf{Z}\left[(\mathscr{T}^3(j), \partial \mathscr{T}^3(j', m)) \to (M^3, \partial M^3); L\right], \tag{6.32}$$

where

$$\mathbf{Z}\left[(\mathscr{T}^3(j),\partial\mathscr{T}^3(j',m))\to(M^3,\partial M^3);L\right]$$

$$= \Lambda(L)^{-N_0}\prod_{A=1}^{N_1}(-1)^{2j_A}(2j_A+1)\prod_{B=1}^{N_3}(-1)^{\sum_{p=1}^{6}j_p}\begin{Bmatrix}j_1 & j_2 & j_3\\ j_4 & j_5 & j_6\end{Bmatrix}_B \qquad (6.33)$$

$$\times \prod_{D=1}^{n_2'}(-1)^{(\sum_{s=1}^{3}m_s)/2}\begin{pmatrix}j_1' & j_2' & j_3'\\ m_1 & m_2 & -m_3\end{pmatrix}_D.$$

N_0, N_1, N_3 denote respectively the total number of vertices, edges and tetrahedra in $(\mathscr{T}^3(j),\partial\mathscr{T}^3(j',m))$, while n_2' is the number of 2-simplices lying in $\partial T^3(j',m)$. Note that there appears a factor $\Lambda(L)^{-1}$ for each vertex in $(\mathscr{T}^3(j),\partial\mathscr{T}^3(j',m))$, with $\Lambda(L)\equiv 4L^3/3C$, C an arbitrary constant. The state sum given in (6.32) and (6.33) when $\partial M^3=\emptyset$ reduces to the usual Ponzano–Regge partition function for a closed manifold M^3 given in (6.13).

The state functional (6.33) is clearly invariant under bistellar moves [48] performed in the interior of $\mathscr{T}^3(j)$ by the same argument employed in the remark at the end of Sect. 6.2.1 for the closed case.

The presence of a $\partial M^3\neq\emptyset$ calls into play other topological transformations introduced by Pachner in dealing with compact d-dimensional PL–pairs, the *elementary shellings* [49, 50]. This kind of operation involves the cancellation of one d-simplex at a time in a given triangulation $(T^d,\partial T^d)\to(M^d,\partial M^d)$ of a compact PL-pair of dimension d. In order to be deleted, the d-simplex must have some of its $(d-1)$–dimensional faces lying in the boundary ∂T^d. Moreover, for each elementary shelling there exists an inverse move which corresponds to the attachment of a new d–simplex to a suitable component in ∂T^d. It is possible to enumerate these moves by setting

$$\left\{[n_{d-1}\to d-(n_{d-1}-1)]_{\mathrm{sh}}^{\mathbf{d}},n_{d-1}=1,2,\ldots,d\right\}, \qquad (6.34)$$

$$\left\{[n_{d-1}\to d-(n_{d-1}-1)]_{\mathrm{insh}}^{\mathbf{d}},n_{d-1}=1,2,\ldots,d\right\}, \qquad (6.35)$$

where n_{d-1} represents the number of $(d-1)$–simplices (belonging to a single d–simplex) involved in an elementary shelling or inverse shelling, respectively. In $d=3$ the admissible shellings are (here the double arrow is just a shorthand notation, not representing an 'inverse' move)

$$[1\leftrightarrow 3]_{\mathrm{sh}}^{3}\quad\text{and}\quad[2\to 2]_{\mathrm{sh}}^{3}. \qquad (6.36)$$

It is worth noting the close similarity of the above notations with those used for 2–dimensional bistellar moves in (6.17), namely $[1\leftrightarrow 3]_{\mathrm{bst}}^{2}$ and $[2\to 2]_{\mathrm{bst}}^{2}$: actually there is a 1–1 correspondence between $[\cdot]_{\mathrm{bst}}^{2}$ and either $[\cdot]_{\mathrm{sh}}^{3}$ or $[\cdot]_{\mathrm{insh}}^{3}$, reflecting natural projection mappings as depicted in the figures below.

In [17] algebraic identities representing the three types of elementary shellings have been established (see also [1]). The first identity represents the move $[2\to 2]_{\mathrm{sh}}^{3}$,

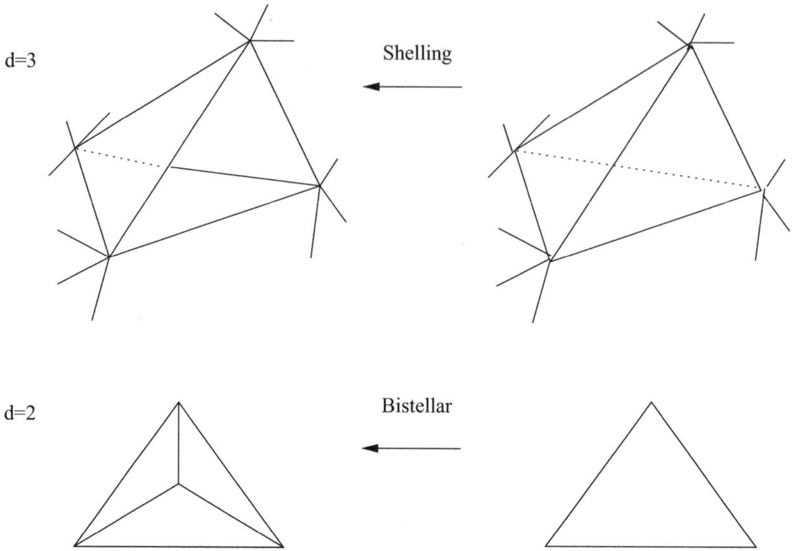

Fig. 6.2 Top: the shelling $[2 \to 2]^3_{sh}$ where a tetrahedron with two faces lying on the boundary disappears while the previously hidden two faces come to the surface. Bottom: the associated bistellar (flip) move $[2 \to 2]^2_{bst}$ can be seen as the 2–dimensional projection of the shelling

the topological content of which is drawn on the top of Fig. 6.2 while its expression reads

$$
\sum_{c\gamma} (2c+1)(-)^{2c-\gamma} \begin{pmatrix} a & b & c \\ \alpha & \beta & \gamma \end{pmatrix} \begin{pmatrix} c & r & p \\ -\gamma & \rho & \psi \end{pmatrix} (-1)^{\Phi} \begin{Bmatrix} a & b & c \\ r & p & q \end{Bmatrix}
$$
$$
= (-)^{-2\rho} \sum_{\kappa} (-1)^{-\kappa} \begin{pmatrix} p & a & q \\ \psi & \alpha & -\kappa \end{pmatrix} \begin{pmatrix} q & b & r \\ \kappa & \beta & -\rho \end{pmatrix} ,
$$

$$(6.37)$$

where Latin letters $a, b, c, r, p, q, \ldots$ denote angular momentum variables, Greek letters $\alpha, \beta, \gamma, \rho, \psi, \kappa, \ldots$ are the corresponding momentum projections and $\Phi \equiv a + b + c + r + p + q$. Notice that from now on we agree that all j–variables in $3j$ symbols are associated with edges lying in $\partial \mathcal{T}^3$, while j–arguments of the $6j$ may belong to either $\partial \mathcal{T}^3$ (if they have a counterpart in the nearby $3j$) or $\mathrm{Int}\,(\mathcal{T}^3)$.

The other identities $[1 \leftrightarrow 3]^3_{sh}$ can be derived (up to regularization factors) from (6.37) and from both the orthogonality conditions for the $6j$ symbols (6.10) and the unitarity conditions for the $3j$ symbols

$$
\sum_{\psi\kappa} (2a+1)(-)^{p-\psi+q-\kappa} \begin{pmatrix} a & p & q \\ -\alpha & \psi & \kappa \end{pmatrix} \begin{pmatrix} a' & p & q \\ \alpha' & -\psi & -\kappa \end{pmatrix} = (-)^{a+\alpha}\, \delta_{aa'}\, \delta_{\alpha\alpha'}
$$

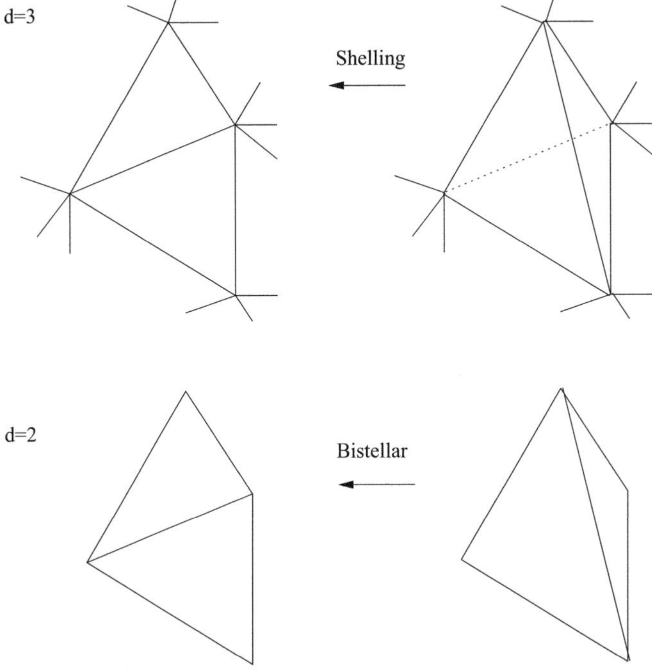

Fig. 6.3 Top: the shelling $[1 \rightarrow 3]^3_{sh}$ where a tetrahedron with one faces lying on the boundary disappears while the previously hidden three faces come to the surface. Bottom: the associated bistellar move $[1 \rightarrow 3]^2_{bst}$

$$\sum_{q\kappa}(2q+1)(-)^{q-\kappa}\begin{pmatrix} a & b & q \\ -\alpha & -\beta & \kappa \end{pmatrix}\begin{pmatrix} q & a & b \\ -\kappa & \alpha' & \beta' \end{pmatrix} = (-)^{a+\alpha+b+\beta}\,\delta_{\alpha\alpha'}\,\delta_{\beta\beta'}.$$

(6.38)

The shelling $[1 \rightarrow 3]^3_{sh}$ is sketched on the top of Fig. 6.3 and the corresponding identity is given

$$\begin{pmatrix} a & b & c \\ \alpha & \beta & \gamma \end{pmatrix}(-1)^{\Phi}\begin{Bmatrix} a & b & c \\ r & p & q \end{Bmatrix}$$

$$= \sum_{\kappa\psi\rho}(-1)^{-\psi-\kappa-\rho}\begin{pmatrix} p & a & q \\ \psi & \alpha & -\kappa \end{pmatrix}\begin{pmatrix} q & b & r \\ \kappa & \beta & -\rho \end{pmatrix}\begin{pmatrix} r & c & p \\ \rho & \gamma & -\psi \end{pmatrix}.$$

(6.39)

Finally, the shelling $[3 \rightarrow 1]^3_{sh}$ is depicted on the top of Fig. 6.4 and the associated identity reads

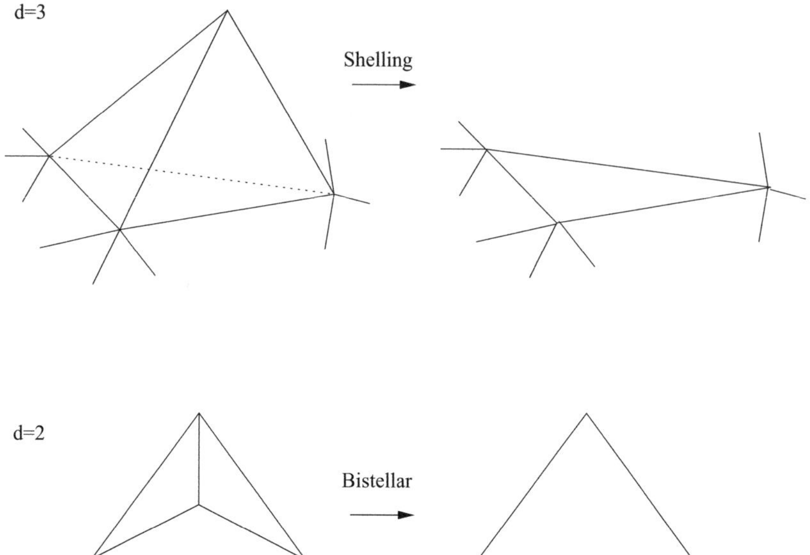

Fig. 6.4 Top: the shelling $[3 \to 1]^3_{sh}$ where a tetrahedron with three faces lying on the boundary disappears while the previously hidden face comes to the surface. Bottom: the associated bistellar move $[3 \to 1]^2_{bst}$

$$\Lambda(L)^{-1} \sum_{q\kappa',p\psi',r\rho'} (-1)^{-\psi'-\kappa'-\rho'}(-1)^{2(p+q+r)}(2p+1)(2r+1)(2q+1)$$

$$\times \begin{pmatrix} a & p & q \\ \alpha & -\psi' & \kappa' \end{pmatrix} \begin{pmatrix} b & q & r \\ \beta & -\kappa' & \rho' \end{pmatrix} \begin{pmatrix} c & r & p \\ \gamma & -\rho' & \psi' \end{pmatrix} (-1)^{\Phi} \begin{Bmatrix} a & b & c \\ r & p & q \end{Bmatrix}$$

$$= \begin{pmatrix} b & a & c \\ \beta & \alpha & \gamma \end{pmatrix}$$

$$(6.40)$$

where $\Lambda(L)$ is defined as in (6.33). Notice that each of the above identities we can be suitably modified by resorting to orthogonality relations in order to reproduce inverse shellings (6.35), namely the operations consisting in the attachment of one 3-simplex to the suitable component(s) in $\partial\mathcal{T}^3$.

Comparing the above identities representing the elementary shellings and their inverse moves with the expression given in (6.33), we see that the state sum $\mathbf{Z}_{PR}\,[(M^3, \partial M^3)]$ in (6.32) is formally invariant under both (a finite number of) bistellar moves in the interior of M^3, and (a finite number of) elementary boundary operations. Then, according to [50] we conclude that (6.32) (up to regularization) is indeed an invariant of the PL–structure of the pair $(M^3, \partial M^3)$.

6.3.3   Q–Extension, Induced State Sums and D–Dimensional Hierachies

The Ponzano–Regge–type state sum (6.32) and (6.33) for $(M^3, \partial M^3)$ can be extended in terms of the representation ring of the quantum enveloping algebra $\mathcal{U}_q(sl(2))$, q a root of unity. According to the standard conventions given in (6.19) of Sect. 6.2.2, $(-1)^{2j}(2j+1)$ are replaced by ω_j^2, while each $\Lambda(L)^{-1}$ becomes ω^{-2}. The $6j$ symbols weighting tetrahedra are replaced by the $q - 6j$ given in (6.21) and each Wigner symbol $3j$ of $SU(2)$ is replaced by its (normalized) q-analog. Recall from [40, 44] that the relation between the quantum Clebsch–Gordan coefficient $(j_1 m_1 j_2 m_2 | j_3 m_3)_q$ and the q-$3j$ symbol is given by

$$(j_1 m_1 j_2 m_2 | j_3 m_3)_q = (-1)^{j_1-j_2+m_3}([2j_3+1])^{1/2} \begin{pmatrix} j_1 & j_2 & j_3 \\ m_1 & m_2 & -m_3 \end{pmatrix}_q, \qquad (6.41)$$

where, as usual, an m variable runs in integer steps between $-j$ and $+j$, and the classical expression is recovered when $q = 1$. The symmetry properties of the q-$3j$ symbol read

$$\begin{pmatrix} j_1 & j_2 & j_3 \\ m_1 & m_2 & -m_3 \end{pmatrix}_q = (-1)^{j_1+j_2+j_3} \begin{pmatrix} j_2 & j_1 & j_3 \\ m_2 & m_1 & -m_3 \end{pmatrix}_{1/q}$$

$$\begin{pmatrix} j_1 & j_2 & j_3 \\ m_1 & m_2 & -m_3 \end{pmatrix}_q = (-1)^{j_1+j_2+j_3} q^{-m_1/2} \begin{pmatrix} j_1 & j_3 & j_2 \\ m_1 & m_3 & -m_2 \end{pmatrix}_{1/q} \qquad (6.42)$$

$$\begin{pmatrix} j_1 & j_2 & j_3 \\ m_1 & m_2 & -m_3 \end{pmatrix}_q = (-1)^{j_1+j_2+j_3} \begin{pmatrix} j_1 & j_2 & j_3 \\ -m_1 & -m_2 & m_3 \end{pmatrix}_q.$$

Thus we define the *normalized* $q - 3j$ symbols, for deformation parameters q and $1/q$ respectively, according to

$$\begin{bmatrix} j_1 & j_2 & j_3 \\ m_1 & m_2 & -m_3 \end{bmatrix}_q \doteq q^{(m_1-m_2)/6} \begin{pmatrix} j_1 & j_2 & j_3 \\ m_1 & m_2 & -m_3 \end{pmatrix}_q$$

$$\begin{bmatrix} j_1 & j_2 & j_3 \\ m_1 & m_2 & -m_3 \end{bmatrix}_{1/q} \doteq q^{(m_2-m_1)/6} \begin{pmatrix} j_1 & j_2 & j_3 \\ m_1 & m_2 & -m_3 \end{pmatrix}_{1/q}. \qquad (6.43)$$

The orthogonality relation involving the normalized symbols (the analog of the second formula in (6.38) used in handling identities representing elementary shellings) reads

$$\sum_{jm} \omega_j^2 (-)^{\phi} q^{(m_2-m_1)/3} \begin{bmatrix} j_1 & j_2 & j \\ m_1 & m_2 & -m \end{bmatrix}_q \begin{bmatrix} j_2 & j_1 & j \\ -m_2' & -m_1' & m \end{bmatrix}_q = \delta_{m_1 m_1'}\, \delta_{m_2 m_2'}, \qquad (6.44)$$

where $\phi = m_1 + m_2 + m_3$.

The (invariant) state sum introduced in the previous section induces a well defined state sum on triangulated 2–dimensional PL–manifolds (to be thought of as oriented for definiteness). This task is accomplished by dropping out the contributions of tetrahedra and noting that the structure of local arrangements of 2-simplices in the state sum (6.32) is naturally encoded in (6.37), (6.39), and (6.40). Then a state functional for a 2–dimensional colored triangulation of a closed *PL*-manifold M^2

$$\mathscr{T}^2(j; m, m') \longrightarrow M^2 \tag{6.45}$$

can be consistently defined by requiring that

- each 2-simplex $\sigma^2 \in \mathscr{T}^2$ is associated with the product of two Wigner symbols (a *double* $3j$ symbol for short)

$$\sigma^2 \longleftrightarrow (-1)^{\sum_{s=1}^{3}(m_s+m'_s)/2} \begin{pmatrix} j_1 & j_2 & j_3 \\ m_1 & m_2 & -m_3 \end{pmatrix} \begin{pmatrix} j_1 & j_2 & j_3 \\ m'_1 & m'_2 & -m'_3 \end{pmatrix}, \tag{6.46}$$

where $\{m_s\}$ and $\{m'_s\}$ are two different sets of momentum projections associated with the same angular momentum variables $\{j_s\}$, $-j \leq m_s, m'_s \leq j \ \forall s = 1, 2, 3$. Then the state sum reads

$$\mathbf{Z}\left[\mathscr{T}^2(j; m, m') \to M^2; L\right]$$

$$= \Lambda(L)^{-N_0} \prod_{A=1}^{N_1} (2j_A + 1)(-1)^{2j_A}(-1)^{-m_A-m'_A} \tag{6.47}$$

$$\times \prod_{B=1}^{N_2} \begin{pmatrix} j_1 & j_2 & j_3 \\ m_1 & m_2 & -m_3 \end{pmatrix}_B \begin{pmatrix} j_1 & j_2 & j_3 \\ m'_1 & m'_2 & -m'_3 \end{pmatrix}_B,$$

where N_0, N_1, N_2 are the numbers of vertices, edges and triangles in $\mathscr{T}^2$, respectively. Summing over all of the admissible assignments of $\{j; m, m'\}$ we get

$$\mathbf{Z}[M^2] = \lim_{L \to \infty} \sum_{\{j; m, m' \leq L\}} \mathbf{Z}[\mathscr{T}^2(j; m, m') \to M^2; L], \tag{6.48}$$

where the regularization is formally carried out according to the usual prescription.

According to Pachner theorem for closed PL–manifolds the invariance of (6.48) is ensured by improving bistellar moves in $d = 2$, given symbolically by $[2 \to 1]^2_{bst}$ $[3 \leftrightarrow 1]^2_{bst}$ and and depicted at the bottom of Figs. 6.2 and 6.3, 6.4 respectively. Algebraic identities encoding these moves, where the basic symbols are double $3j$, are given explicitly in [17, 18]. The state sum given in (6.47) – as well as its the q-counterpart – is shown to be algebraic–compatible with the bistellar moves, thus providing a well defined quantum invariant given by

$$\mathbf{Z}[M^2; q] = \omega^2 \, \omega^{-2\chi(M^2)}, \tag{6.49}$$

where ω^2 is defined in (6.19) and $\chi(M^2)$ is the Euler number of the closed oriented surface M^2. Then the 2–dimensional state sum naturally induced by the 3–dimensional model with a non empty boundary is not trivial, being given in terms of the only significant topological invariant for closed oriented surfaces.

Actually the scheme outlined above is aimed to provide an inductive procedure for generating (state sum) invariants for closed PL–manifolds in contiguous dimensions, namely $\mathbf{Z}[M^{d-1}] \to \mathbf{Z}[M^d]$ as discussed at length in [18]. Without entering into details here, the general philosophy underlying the construction of a bottom–up hierarchy of invariants out of $SU(2)$ or $\mathscr{U}_q(sl(2))$–colored triangulations (at a fixed root of unity) can be summarized as follows

1. Given a (suitable defined) invariant state sum $\mathbf{Z}[M^{d-1}]$ for a closed $(d-1)$–dimensional PL–manifold M^{d-1}, one extends it to a state sum for a colored pair $(\mathscr{T}^d, \partial\mathscr{T}^d \equiv \mathscr{T}^{d-1})$. This is achieved by assembling in a suitable way the *square roots* of the symbols associated with the fundamental blocks in $\mathbf{Z}[M^{d-1}]$ in order to pick up the $3nj$ symbol to be associated with the d–dimensional simplex (the dimension of the $SU(2)$–labeled (or q-colored) $(d-2)$–simplices is kept fixed when passing from $\mathscr{T}^{d-1}$ to $\mathscr{T}^d \supset \mathscr{T}^{d-1}$).
2. The state sum for $(\mathscr{T}^d, \partial\mathscr{T}^d)$ gives rise to a PL–invariant $Z[(M^d, \partial M^d)]$ owing in particular to its invariance under elementary shellings of Pachner [50] (the algebraic identities associated with such moves in $d = 3$ are given in the previous section, while for higher dimensions the diagrammatical method is more suitable).
3. From the expression of $Z[(M^d, \partial M^d)]$ it is possible to extract a state functional for a *closed* colored triangulation $\mathscr{T}^d$. The proof of its PL–invariance relies now on the algebrization in any dimension d of the *bistellar moves*. The inductive procedure turns out to be consistent with known results in dimension 4 and provides for each d a PL–invariant $Z[M^d]$ where each d–simplex in the colored triangulation $\mathscr{T}^d$ is represented by a $\{3(d-2)(d+1)/2\}j$ recoupling coefficient of $SU(2)$ (or by its q-analog).
4. By construction all of these state sum functionals turn out to be well behaved under composition (a quite obvious property in case of state sums with fixed colored triangulations on boundaries, see Sect. 6.3.1). This is achieved by observing that simplices of maximal dimension in a boundary component are associated with 'double' symbols which, once glued to their counterparts to get a closed triangulation, are going to disappear owing to orthogonality conditions (recall that both j–type and m–type colors have to be summed up).

It is worth noting that the invariant in the closed $d = 4$ case, the so–called $15j$ model introduced originally in [23, 46] is not 'new' in the sense that it was recognized to be related to other known topological invariants, the signature and the Euler character of the manifold [58]. The other invariants in $d > 4$ seem 'new' but it is not yet clear if they could be of some interest in physical applications.

As a matter of fact all state sum invariants discussed so far share a nice interpretation in the field–theoretic setting, being interpreted as discretized counterparts of Topological Quantum Field Theories of the BF–type. Recall that, unlike Chern–Simons–Witten partition functions, BF theories (with or without a cosmological constant) [22, 38] can be defined in any spacetime dimension. In particular the generating function of a BF theory with a cosmological constant in $d = 3$ corresponds to a 'double' CSW functional, thus improving the correspondence between Chern–Simons–Witten and Turaev–Viro models discussed in Sect. 6.2.3 above.

The $15j$–model mentioned above has been radically generalized first by Barrett and Crane [11], and in the next few years in a number of papers dealing with the spin foam model (the 4–dimensional extension of the 'spin network' setting of loop quantum gravity based on the Plebanski formulation of gravity as a constrained BF theory), see e.g. [24, 62] and references therein.

Remark As a final application of the Ponzano–Regge state sum $\mathbf{Z}_{PR}[(M^3, \partial M^3)]$ of Sect. 6.3.2 we mention the results we found as co–authors of [8]. Unlike the case examined in Sect. 6.3.3, where the 2–dimensional boundary surface is completely decoupled from the 'bulk' 3–manifold and the resulting boundary theory is purely topological, it is possible to perform an holographic projection starting from a suitably chosen state functional $\mathbf{Z}_{PR}[(\mathscr{T}^3, \partial \mathscr{T}^3); L]$. The main feature of the procedure is to provide a semiclassical partition function in the bulk, while the state functional on the boundary triangulation in the decoupling limit remains quantum and acquires non–trivial degrees of freedom. This result represents the first attempt of dealing with 't Hooft holographic principle [66] within an ab initio discretized quantum gravity model.

6.4   Observables in the Turaev–Viro Environment

In Sect. 6.2.3 the relation between the Turaev–Viro and the Witten–Reshetikhin–Turaev approaches has been discussed very briefly in connection with the correspondence (6.22) between their generating functionals. In this section we are going to present an unifying scheme aimed to address observables, particularly 3-valent graphs and links invariants, in a TV background theory. Soon after the seminal paper by Turaev and Viro [69], Viro himself [67, 68] and others [12, 25, 35, 71] dealt with colored (fat) graphs and links and calculations of specific invariants were carried out there as well as in [16, 32]. It is worth noting that in the smooth category of unitary topological quantum field theories it has been shown that observables – vacuum expectation values of Wilson loop operators [30, 38, 74] as given in the formal expression (6.27) – are basically the same in the Chern–Simons and in a suitable BF setting, see e.g. [22, 38]. Also in view such a relationship between a BF theory with cosmological constant and the Turaev–Viro state sum we do not expect to find out any essentially new link invariants in the colored–triangulations setting. However, the method we are going to illustrate with the final scope of giving explicit, diagrammatic expressions for observables is quite instructive by itself and gives us the chance of filling a few gaps left aside in previous sections.

6.4.1 Turaev–Viro Quantum Initial Data

Following [67] with some minor changes of notation, we start by defining an abstract *initial data set* . Let K be a commutative ring with unit and K^* the subgroup of its invertible elements. Given a finite set I, choose two functions $I \to K^*$ by setting $i \to \omega_i$ and $i \to q_i$ for each $i \in I$ and fix a distinguished element $\omega \in K^*$. Assume that in I^3 there exists an unordered set *Adm* of *admissible triples*. An ordered 6-tuple $(i, j, k, l, m, n,) \in I^6$ is admissible if the triples (i, j, k), (k, l, m), (m, n, i), (j, l, n) are in *Adm*. Then

$$(i, j, k, l, m, n,) \quad \to \quad \begin{vmatrix} i & j & k \\ l & m & n \end{vmatrix} \tag{6.50}$$

represents the formal assignment of an element of K to each admissible 6-tuple. If the 6-tuple is not admissible then its symbol is equal to zero. The symbol $\left| \begin{smallmatrix} ::: \end{smallmatrix} \right|$ actually shares the same 24 'classical' symmetries of the $6j$ symbol of $SU(2)$ described in the introductory part of Sect. 6.1, but at this level there is no reference to any representation ring. The initial data are called *irreducible* if, for any $i, j \in I$, there exists a sequence $l_1, l_2, \ldots, l_r$ with $l_1 = i, l_r = j$ such that the triple (l_t, l_{t+1}, l_{t+2}) is in *Adm* for all $t = 1, 2, \ldots, r - 2$.

The above initial data satisfy the algebraic conditions listed below, where, in agreement with the definitions given in (6.19) and (6.20)

$$\omega_j^2 = [2j + 1] \ \ \forall \, j \in I \tag{6.51}$$

is the quantum dimension of the *irrep* associated with j.

I. $\forall \ j_1, j_2, j_3, j_4, j_5, j_6 \in I$ such that the triples (j_1, j_3, j_4), (j_2, j_4, j_5) (j_1, j_3, j_6), (j_2, j_5, j_6) are in *Adm*

$$\sum_{j \in I} \omega_j^2 \begin{vmatrix} j_2 & j_1 & j \\ j_3 & j_5 & j_4 \end{vmatrix} \begin{vmatrix} j_3 & j_4 & j_6 \\ j_2 & j_5 & j \end{vmatrix} = \delta_{j_4 j_6} . \tag{6.52}$$

II. $\forall \ a, b, c, e, f, j_1, j_2, j_3, j_{23} \in I$ such that the 6-tuples $(j_{23}, a, e, j_1, f, b)$ and $(j_3, j_2, j_{23}, b, f, c)$ are admissible

$$\sum_{d \in I} \omega_d^2 \begin{vmatrix} j_2 & a & d \\ j_1 & c & b \end{vmatrix} \begin{vmatrix} j_3 & d & e \\ j_1 & f & c \end{vmatrix} \begin{vmatrix} j_3 & j_2 & j_{23} \\ a & e & d \end{vmatrix} = \begin{vmatrix} j_{23} & a & e \\ j_1 & f & b \end{vmatrix} \begin{vmatrix} j_3 & j_2 & j_{23} \\ b & f & c \end{vmatrix} . \tag{6.53}$$

III. For any $i \in I$

$$\omega^2 = \omega_j^{-2} \sum_{\substack{k, l \in I \\ (j, k, l) \in Adm}} \omega_k^2 \, \omega_l^2 . \tag{6.54}$$

Conditions I.–III. have been used originally in [69] in the construction of the Turaev–Viro invariant for closed manifolds (see Sect. 6.2.2 above). Actually I. and II. axiomatize the orthogonality relation and the Biedenharn–Elliott identity for (classical) $6j$ symbols given in (6.10) and (6.9), respectively. Condition III. does not have a counterpart in the representation ring of $SU(2)$ (and is obviously strictly related to the finiteness property of the Turaev–Viro invariant). Irreducible initial data satisfying condition I. are shown to be independent of the choice of $j \in I$, so that in this case condition III. actually selects a unique distinguished element $\omega^2 \in K^*$, to be looked at as part of the initial data.

The issue of link invariants will call into play the functions $q_i \in K^*$ ($i \in I$) which turn out to enter the analog of the Racah identity for classical $6j$ symbols

IV. For any $j_1, j_2, j_{12}, j_3, j, j_{23} \in I$

$$\sum_{j_{13} \in I} \omega_{j_{13}}^2 \, q_{j_{13}} \begin{vmatrix} j_3 & j_1 & j_{13} \\ j_2 & j & j_{12} \end{vmatrix} \begin{vmatrix} j_2 & j_3 & j_{23} \\ j_1 & j & j_{13} \end{vmatrix} = \frac{q_j \, q_{j_1} \, q_{j_2} \, q_{j_3}}{q_{j_{12}} \, q_{j_{23}}} \begin{vmatrix} j_1 & j_2 & j_{12} \\ j_3 & j & j_{23} \end{vmatrix}. \tag{6.55}$$

In [67] a few consequences of conditions I.–IV. were proven. In particular

$$\sum_{g \in I} \omega_g^2 \, (q_b \, q_d \, q_f \, q_g)^\varepsilon \begin{vmatrix} j_2 & a & g \\ j_1 & c & b \end{vmatrix} \begin{vmatrix} j_3 & g & e \\ j_1 & d & c \end{vmatrix} \begin{vmatrix} j_3 & a & f \\ j_2 & e & g \end{vmatrix}$$

$$= \sum_{h \in I} \omega_h^2 \, (q_a \, q_c \, q_e \, q_h)^\varepsilon \begin{vmatrix} j_3 & b & h \\ j_2 & d & c \end{vmatrix} \begin{vmatrix} j_3 & a & f \\ j_1 & h & b \end{vmatrix} \begin{vmatrix} j_2 & f & e \\ j_1 & d & h \end{vmatrix}, \tag{6.56}$$

where $\varepsilon = \pm 1$, represents the counterpart of Yang–Baxter relation written in terms of q-$6j$ symbols (and can be actually recognized as an identity satisfied by the q-analog of the $9j$ symbol, see [44]), while

$$\sum_{j \in I} \omega_j^2 \, (q_a \, q_j \, q_b^{-2})^\varepsilon \begin{vmatrix} i & b & j \\ i & b & a \end{vmatrix} = (q_i)^{2\varepsilon}. \tag{6.57}$$

can be recognized as the Markov relation for R–matrices written in terms of q-$6j$ symbols. The diagrammatic counterparts of the identities (6.55) and (6.57) will be given in the next section, where their role in the proof of invariance of state sums associated with colored fat graphs will become clear.

The notion of irreducible *quantum* initial data comes out when the axiomatization is modeled on the representation ring of $\mathcal{U}_q(sl(2))$, $q = \exp\{2\pi i/r\}$, with the finite set I given by $\{0, 1/2, 1, \ldots, (r-2)/2\}$ for $r \geq 3$ and ground ring $\mathbb{C}$ (see Sect. 6.2.2). Then the identifications made in (6.19), the definition of q-integer (6.20) and the relation between $| ::: |$ and the Racah–Wigner symbol $\{:::\}_q$ in (6.21) hold true. Note that the q-$6j$ can be defined as a hypergeometric polynomial (at the top of the q-Askey hierarchy) through an expression formally similar to the Racah sum rule given in (6.8), see e.g. [57] or [9].

To complete the quantum data, the functions q_j are defined according to

$$q_j = e^{\pi i \left\{ j - \frac{j(j+1)}{r} \right\}} \tag{6.58}$$

with $i = \sqrt{-1}$.

In Sect. 6.3.1 the Turaev–Viro state sum for a pair $(M^3, \partial M^3 = \mathcal{T}^2(j'))$, where $\mathcal{T}^2(j')$ is a fixed triangulation on the boundary surface. Here we rewrite the same expression by adopting the more concise notation of [67]. Thus M stands for a compact triangulated 3–manifold and a *coloring* (*I*–coloring) μ is the assignment to each edge of M of an element of I such that: (i) for every 2–simplex $\subset M$ the colors of its three edges form an admissible triple $\in Adm$; (ii) each 3–simplex T is associated to the element $|T|_\mu = | \text{:::} |$ as in (6.50). The set of colorings is denoted by $Col\,(M)$ and for quantum initial data the root of unity q is kept fixed (and omitted in the following).

In the presence of a non empty boundary ∂M we agree that the set of colorings is naturally extended to $Col\,(M, \partial M)$ by taking the triangulation of ∂M induced by any given one of M. The state functional of the colored triangulated pair reads

$$|M, \partial M|_\mu = \omega^{-2\alpha+\beta} \prod_e \omega^2_{\mu(e)} \prod_{e'} \omega_{\mu(e')} \prod_T |T|_\mu, \tag{6.59}$$

where α and β are the number of vertices in M and ∂M respectively, e and e' run over edges in the interior of M and in ∂M respectively and the last product is over all tetrahedra in $(M, \partial M)$. The Turaev–Viro state sum is

$$|M, \partial M| = \sum_{\mu \in Col\,(M,\partial M)} |M, \partial M|_\mu. \tag{6.60}$$

The state functional (6.59) can be modified to deal with selected subsets lying in ∂M. If F is a certain union of components in ∂M then the state functional

$$|M, F \subset \partial M|_\mu = \omega^{-2\alpha+\beta'} \prod_e \omega^2_{\mu(e)} \prod_{e''} \omega_{\mu(e'')} \prod_T |T|_\mu := |M, F|_\mu, \tag{6.61}$$

once summed over μ, provides a relative invariant which will enter the expression of other state sums addressed in the next section. In the above expression α is the number of vertices in M, β' is the number of vertices in $\partial M \backslash F$, e are edges in M not in ∂M, e'' runs over edges in $\partial M \backslash F$ and T runs over over all tetrahedra in $(M, \partial M)$.

Finally, the Turaev–Viro invariant (6.60) can be slightly modified in order to emphasize the role of the colorings of the boundary (components). If $\lambda \in Col\,(\partial M)$ is an admissible coloring of a fixed 2–dimensional triangulation of ∂M, then it is always possible to extend it to a (not unique) triangulation in the interior of M. The state sum

$$< M, \partial M \mid \lambda > \; \dot{=} \sum_{\substack{\mu \,\in\, Col\,(M,\,\partial M) \\ \mu|_{\partial M} \,=\, \lambda}} |M, \partial M|_{\mu} \qquad (6.62)$$

is a topological invariant of $(M, \partial M)$ which does not depend on the extension of the triangulation but of course the colorings μ of the interior of M are constrained to be compatible with the given coloring λ of ∂M. The above expression complies with (6.28) of Sect. 6.3.1.

6.4.2   State Sum Invariants of Colored Fat Graphs in 3–Manifolds

As stated in the introduction of [67], the axiomatic approach developed there (and reviewed here) is aimed to provide a consistent and rigorous 3-dimensional interpretation of the Jones polynomial for knots (and its generalizations) not resorting to Witten's path integral formalism (Sect. 6.2.3, see expression (6.27)). The basic ingredients used by Turaev in constructing invariants of fat colored graphs in generic 3–manifold – graphs which can be specialized to framed colored links in the 3–sphere S^3 – are the (quantum, irreducible) initial data of the previous section, namely q-$6j$ symbols and the functions ω, ω_j, q_j taking values in the ground ring $\mathbb{C}$. Actually the invariants are defined in terms of the exterior (complement) of the graphs (links) by purely geometric–combinatorial tools, thus leaving aside the quantum group machinery and associated R–matrix representations [39, 40, 57].

From now on let N denote a compact, either closed or with a non–empty boundary ∂N, not necessarily oriented smooth 3–manifold which can be equipped with a colored triangulation as done for a pair $(M, \partial M)$ in the previous section. The most general state model will be written in the shorthand form $< N, \Gamma | \psi >$, with Γ a fat, 3-valent colored graph smoothly embedded in the interior of N and ψ an ordinary 3-valent colored graph lying in general position with respect to Γ. This sort of bra–ket notation, already used in (6.62) above, is somehow reminiscent of the proper use of $< \mid >, < \mid \mid >$ in quantum mechanics and quantum field theory. For instance, as shown in Sect. 6.4.2.3 below, the ket $|\psi >$ might be interpreted as a vector quantity to be associated with the set of colorings $\{\lambda\}$ of a 2–dimensional surface embedded in N. The construction of $< N, \Gamma | \psi >$ proceeds through a number of quite challenging steps, each providing specialized state sums which are interesting by themselves.

Let us remind a few basic definitions about colored 3-valent graphs (denoted by small Greek letters $\phi, \psi, \gamma, \dots$) and their fat counterparts (denoted by $\Gamma, \Gamma', \dots$). A finite graph ψ with unoriented edges and possibly with loops is 3-valent if the number of edges incident on any vertex is either two or three. An I–coloring of ψ is the assignment of a color to each edge in such a way that any triple of edges stemming from a 3-valent vertex is colored with an admissible triple of colors, while any pair of edges incident to a 2-valent vertex has the same color. Starting from a 3-valent colored graph, a 3-valent fat graph Γ (smoothly embedded in the ambient manifold N) is the extension to small disks and to narrow bands of the vertices and edges of a 3-valent graph, respectively. Γ inherits a coloring from its *core* $c(\Gamma)$, an ordinary 3-valent colored graph, and $Col(\Gamma)$ denotes the set of admissible coloring of $c(\Gamma)$. Note that fat graphs are not to be confused with ribbon graphs since vertices of the latter are endowed with enriched structures, see Chap. 2. The topological union of disks and bands of Γ is the *surface* of Γ, $\mathscr{S}(\Gamma) \subset N$, a 2-dimensional compact submanifold with a non–empty boundary. A fat graph with an oriented surface is called oriented. Two fat graphs $\Gamma, \Gamma' \subset N$ are *ambient isotopic* if they can be smoothly deformed into each other in the class of fat graphs in N. Coloring–preserving isotopies of fat graphs are defined in the obvious way.

Remark Recall that a link L in N is a finite collection of disjoint unoriented circles smoothly embedded in N. A coloring of an m-component link is the assignment of a color from I to each component. A framed link is a link endowed with a non singular normal vector field (a framing). Every m-component framed link L in N gives rise to a fat graph $\Gamma_L \subset N$ consisting of m vertices and m edges (loops), where the edges are topological annuli (for N oriented) and annuli and Möbius bands (for N not oriented) and the position of the vertex (disk) on each annulus is chosen arbitrarily. Each coloring of L induces a coloring of Γ_L by assigning the same color to an edge and to the corresponding component of L. Thus there exists a natural injective map $L \to \Gamma_L$ from the set of (isotopy classes of) colored framed links into the set of (isotopy classes of) colored 3-valent fat graphs in N. Whatever invariant of colored fat graphs will be defined, it can be specialized in a straightforward way to colored, framed links.

6.4.2.1 State Models Associated with Graphs in a Cylinder $F \times [-1, 1]$

The first state model is defined by assuming that the ambient manifold is a cylinder $F \times [-1, 1]$, where F is a compact and oriented surface. A fat, 3-valent graph Γ is embedded in the cylinder and its surface inherits the orientation of F. A pair of (possibly intersecting) colored 3-valent graphs, ϕ and ψ, are embedded in F too. Here F is going to play the same role as a 2-disk in $\mathbb{R}^3$ with respect to projection of knots or graphs, giving rise to 2-dimensional configurations or *diagrams*. A diagram D of a (generic) graph is obtained by an immersion in F with only double transverse crossings. An additional rule must be applied: at each crossings, one of the two edges is cut so that the other overcrosses it. When a fat graph Γ is considered, one more convention is necessary: when embedded in $F \times [-1, 1]$ its vertices (disks) keep on lying on F, while the thickened edges which overcross are pushed up into $F \times (0, 1]$ and the bands which undercross are tied down into $F \times [-1, 0)$. The choice of solving a transverse crossing in either under – or over–passing is arbitrary in dealing with graphs (contrary to what happens when projecting a knot living in $\mathbb{R}^3$ onto a plane) but it can be shown that the final state model does not depend on this choice.

Actually it can be proven that two graph diagrams on F give rise to isotopic fat graphs in $F \times [-1, 1]$ if and only if they are related, as projected diagrams $D(c(\Gamma))$, $D(c(\Gamma'))$ (c is the core of the fat graph), by local Reidemeister–type moves. We do not insist on this issue here since these moves are encoded into the algebraic prescriptions given below for the construction of the state model $<\phi|\ \Gamma\ |\psi> \ \in \mathbb{C}$.

The Graph Σ

The initial configuration on F includes the colored 3-valent graphs ϕ, ψ and the diagram $D(c(\Gamma)) \equiv D(\Gamma)$ obtained as explained above and endowed with under- and over–crossings. By 2-dimensional ambient isotopies these three structures can be arranged in general position, so that all intersections in $\phi \cup \psi \cup D(\Gamma)$ are double transverse crossings of edges. In order to construct an overall graph diagram, let $D(\phi)$ be a diagram of ϕ obtained by requiring that it lies everywhere over the rest of the configuration. Similarly, a diagram $D(\psi)$ associated with ψ is required to lie everywhere under the other two diagrams. Denote the union $D(\phi) \cup D(\psi) \cup D(\Gamma)$ by σ. From σ, forgetting over – and under–crossing information, a new graph

$$\Sigma = \phi \cup \psi \cup D(\Gamma) \subset F \tag{6.63}$$

is obtained, the number of vertices of which is the sum of the number of the (3-valent) vertices of ϕ, ψ, Γ and the number of (4-valent) vertices generated by joining back crossing edges from the diagram σ. Note that, since each edge of Σ is contained in an edge of either ϕ or ψ or Γ, Σ inherits a consistent coloring from the colorings of its components.

Classification of Vertices of Σ

The graph Σ has five types of vertices

  (i)   the 2-valent vertices of ϕ, ψ, Γ;
 (ii)   the 3-valent vertices of ϕ, ψ, Γ;
(iii)   the crossing points of ϕ with ψ;
 (vi)   the crossing points of $D(\Gamma)$ with ϕ and ψ;
  (v)   the self–crossing points of $D(\Gamma)$.

Area Coloring of $F \setminus \Sigma$

The graph Σ provides a subdivision of the compact surface F into a finite number of domains. In particular, any connected component of $F \setminus \Sigma$ is a *region* of $D(\Gamma)$ with respect to ϕ and ψ. An area coloring η is an arbitrary mapping from the set of regions into the set of colors I which is admissible if, for each edge $e \in \Sigma$, the color of e together and the η-colors of the two regions adjacent to e combine in an admissible triple. The admissible area (and edge) coloring are denoted by $Adm\,(D(\Gamma))$, a shorthand notation for $Adm\,(D(\Gamma);\phi,\psi)$.

Assignments of Weights to Vertices and Regions

The collection of vertices of Σ and regions in $F\setminus\Sigma$ is going to be decorated with weights, denoted by $|\cdot|_\eta$ with $\eta\in Adm\,(D(\Gamma))$, suitably chosen in the (quantum) initial data of Sect. 6.4.1.

The weight of a region y is given by

$$|\,y\,|_\eta \;=\; \omega_{\eta(y)}^{2\chi(y)} \tag{6.64}$$

where $\omega_{\eta(y)}^2$ is the quantum dimension associated with the value of the given color (see (6.51)) and χ is the Euler characteristic of the region y (for y a 2-disk $\chi(y)=1$).

Weights $|\,v\,|_\eta$ for vertices v in Σ depend on the classification given above. Thus

(i) a 2-valent vertex has a trivial weight

$$|\,v\,|_\eta \;=\; 1\,;$$

(ii) if (i,j,k) is the triple of admissible colors assigned to edges incident on the 3-valent vertex v, let (l,m,n) denote the η–colors of the opposite regions.

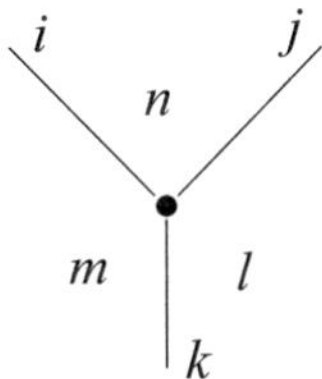

The weight of v is

$$|\,v\,|_\eta \;=\; \begin{vmatrix} i & j & k \\ l & m & n \end{vmatrix}, \tag{6.65}$$

where $|\,{:}{:}{:}\,|$ is the q-$6j$ symbol (6.50).

Types (iii)–(v) refer to 4-valent vertices (crossing points of the underlying graph diagrams). If v is any such vertex with two branches, let l be the color of the upper branch and i the color of the lower one (recall that ϕ lies over $\psi\cup D(\Gamma)$ and ψ under $\phi\cup D(\Gamma)$). Then the four regions surrounding v are η–colored with j,k,l,m, where the orientation is by convention anticlockwise and derives from the orientation of the underlying F.

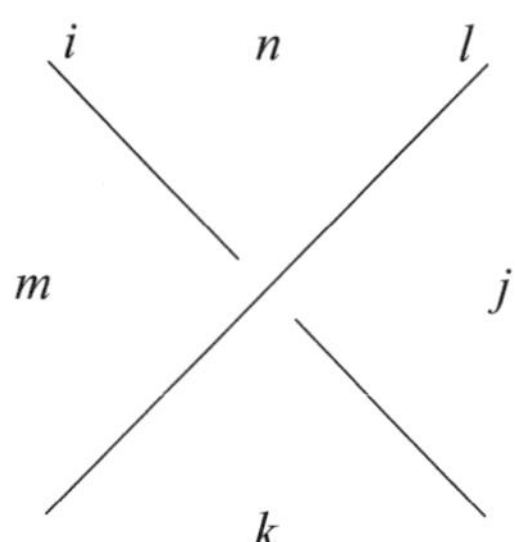

In particular

(iii) if v is a crossing of ϕ with ψ its weight is

$$
\mid v \mid_\eta \; = \; q_k^{1/2} \, q_n^{1/2} \, q_j^{-1/2} \, q_m^{-1/2} \left| \begin{matrix} i & j & k \\ l & m & n \end{matrix} \right| , \tag{6.66}
$$

where the q-factors are defined in (6.58);

(iv) if v is a crossing of $D(\Gamma)$ with ϕ or ψ

$$
\mid v \mid_\eta \; = \; q_k^{1/2} \, q_n^{1/2} \, q_j^{-1/2} \, q_m^{-1/2} \left| \begin{matrix} i & j & k \\ l & m & n \end{matrix} \right| ; \tag{6.67}
$$

(v) if v is a self–crossing point of $D(\Gamma)$

$$
\mid v \mid_\eta \; = \; q_k \, q_n \, q_j^{-1} \, q_m^{-1} \left| \begin{matrix} i & j & k \\ l & m & n \end{matrix} \right| . \tag{6.68}
$$

The State Model $< \phi \mid \Gamma \mid \psi >$

For each η–coloring in $Adm \, (D(\Gamma); \phi, \psi)$, define the functional

$$
< \phi \mid D(\Gamma) \mid \psi >_\eta \; := \; \prod_y \mid y \mid_\eta \, \prod_v \mid v \mid_\eta . \tag{6.69}
$$

Then the state sum

$$
< \phi \mid \Gamma \mid \psi > \; = \sum_{\eta \, \in Adm(D(\Gamma))} < \phi \mid D(\Gamma) \mid \psi >_\eta \; \in \mathbb{C} \tag{6.70}
$$

for any choice of ϕ, ψ, Γ is invariant under ambient isotopies of ϕ, ψ in F and isotopies of the fat graph Γ in $F \times [-1, 1]$.

The proof is given in Th. 4.1, 4.2 of [67] and relies on a mixing of combinatorial and algebraic tools. Let us illustrate just a few of them in connection with invariance with respect to 2–dimensional ambient isotopies, encoded in Reidemeister–type colored moves performed on the graph diagram $\sigma = D(\phi) \cup D(\psi) \cup D(\Gamma)$.

The first Reidemeister move is the transformation of a single strand colored with $i \in I$, bounding two regions with η-colors a, b, into a self–crossing strand, where the new internal region is colored with j.

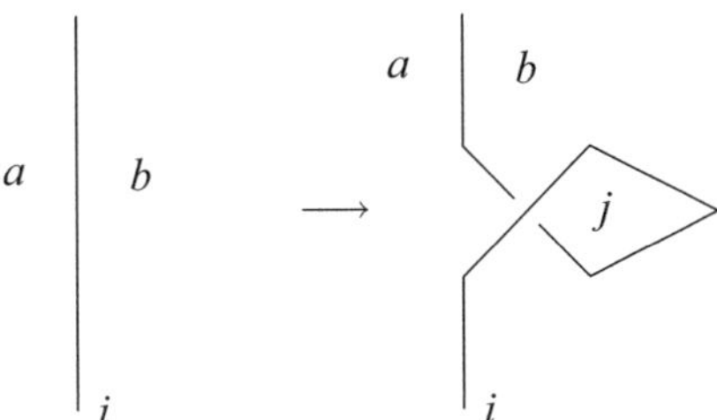

The algebraic counterpart of this move is given by the identity (6.57) (the analog of the Markov relation for R–matrix), rewritten here for convenience

$$\sum_{j \in I} \omega_j^2 \, (q_a \, q_j \, q_b^{-2})^{\varepsilon} \begin{vmatrix} i & b & j \\ i & b & a \end{vmatrix} = (q_i)^{2\varepsilon}.$$

The second Reidemeister moves has to be applied to a colored 3-valent vertex configuration (case ii above), where the lower leg is twisted with the two upper ones kept fixed, giving rise to a new internal region colored with j_{13}.

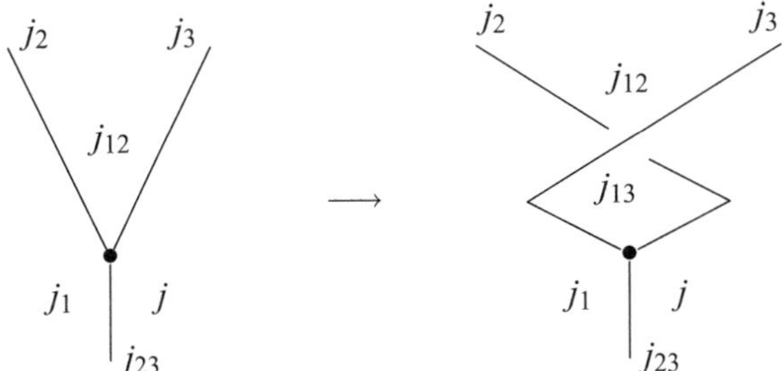

The algebraic counterpart of this move is the Racah identity (6.55), namely

$$\sum_{j_{13} \in I} \omega_{j_{13}}^2 \, q_{j_{13}} \begin{vmatrix} j_3 & j_1 & j_{13} \\ j_2 & j & j_{12} \end{vmatrix} \begin{vmatrix} j_2 & j_3 & j_{23} \\ j_1 & j & j_{13} \end{vmatrix} = \frac{q_j \, q_{j_1} \, q_{j_2} \, q_{j_3}}{q_{j_{12}} \, q_{j_{23}}} \begin{vmatrix} j_1 & j_2 & j_{12} \\ j_3 & j & j_{23} \end{vmatrix}.$$

The third Reidemeister move involve an exchange transformation between two arrangements containing three crossings, with the internal regions colored by g and h, respectively.

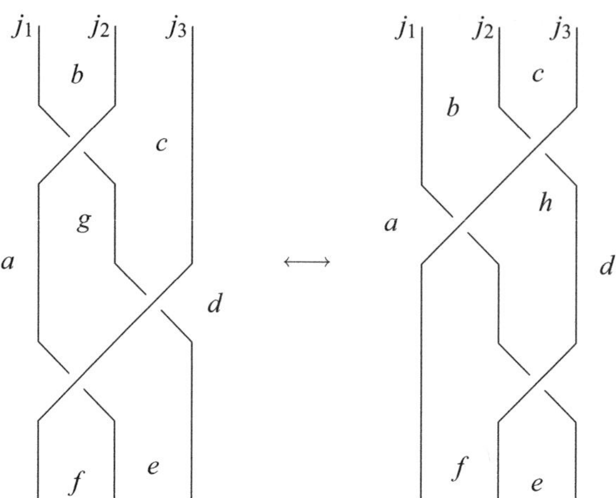

The algebraic counterpart is (6.56), representing the Yang–Baxter relation in terms of q-$6j$ symbols and weights of type q and ω, rewritten here for convenience

$$\sum_{g\in I} \omega_g^2 \left(q_b\, q_d\, q_f\, q_g\right)^\varepsilon \begin{vmatrix} j_2 & a & g \\ j_1 & c & b \end{vmatrix} \begin{vmatrix} j_3 & g & e \\ j_1 & d & c \end{vmatrix} \begin{vmatrix} j_3 & a & f \\ j_2 & e & g \end{vmatrix}$$

$$= \sum_{h\in I} \omega_h^2 \left(q_a\, q_c\, q_e\, q_h\right)^\varepsilon \begin{vmatrix} j_3 & b & h \\ j_2 & d & c \end{vmatrix} \begin{vmatrix} j_3 & a & f \\ j_1 & h & b \end{vmatrix} \begin{vmatrix} j_2 & f & e \\ j_1 & d & h \end{vmatrix}.$$

It is worth stressing that, in proving invariance of the above state model, all of the conditions I.–IV. on initial data have been employed. In dealing with Turaev–Viro state models for 3-manifolds (Sects. 6.2.2, 6.3.1, and 6.4.1) the explicit proof of invariance (through Pachner bistellar moves) is based on conditions I.–III., the Racah identity being putatively associated with a move (2 tetrahedra $\to$ 1 tetrahedron) not permitted on PL triangulated manifolds [61]. On the other hand, the structures introduced here are (diagrams of) colored graphs embedded in a surface F and information about the embeddings is encoded in the extra η-colorings to be assigned consistently to the 2–dimensional connected components of $F\backslash\Sigma$. Note finally that the state sum invariant does not contain any weight to be associated with the 3-dimensional cylinder: this is due to the fact that here the ambient manifold is topologically a product of F by a closed interval. In the next section, where Γ is going to be embedded into a non–trivial 3–manifold N, the corresponding state model will contain also a contribution from the complement of Γ in N.

Specific Examples

The invariants $< \phi\,|\,\Gamma\,|\,\psi >$ in (6.70) are non trivial even when ϕ, ψ are empty graphs, denoted by $\emptyset$. In particular, if F is a 2-disk (and thus $F\times[-1,1]$ an ordinary cylinder in $\mathbb{R}^3$) then

$$< \emptyset \,|\, \Gamma \,|\, \emptyset > = \sum_{\eta \,\in Adm(\phi,\psi)} < \phi \,|\emptyset|\, \psi >_{\eta} \tag{6.71}$$

is a generalization of the Jones polynomial of colored framed links L to colored fat graphs Γ in $\mathbb{R}^3$ (*cfr.* the remark at the end of the introductory part of Sect. 6.4.2).

A second case comes about when ϕ and ψ are dual graphs of 2–dimensional colored triangulations. This situation is easily obtained by assuming that F is a compact triangulated surface with a fixed coloring λ. Then the dual graph of the 1-skeleton of the triangulation, denoted here by γ_F, is obtained as usual by taking the barycenter of each triangle and connecting it to the barycenters of the three contiguous triangles by dual edges $\{e^*\}$. Each dual edge e^* inherits the color λ from the corresponding transverse edge e in the triangulation. The colored dual graph γ_F^{λ} is of course 3-valent and can take the role of ϕ, ψ. Thus in the interior of the simplicial cylinder $F \times [-1, 1]$ – whose bases are endowed with the same colored triangulation – it is possible to insert a fat colored graph Γ and carry out the procedure described above (in particular the η-coloring of regions of $D(\Gamma) \subset F$) to get the ambient isotopy invariant

$$< \gamma_F^{\lambda} \,|\, \Gamma \,|\, \gamma_F^{\lambda} > \;=\; \sum_{\eta \,\in Adm(D(\Gamma))} < \gamma_F^{\lambda} \,|D(\Gamma)|\, \gamma_F^{\lambda} >_{\eta} \,. \tag{6.72}$$

6.4.2.2   State Models for Colored Fat Graphs in 3–Manifolds with Triangulated Boundary

Here N is a compact, non necessarily oriented 3–manifold with a triangulated boundary ∂N and Γ an finite I–colored 3-valent fat graph in the interior of N. The surface of the fat graph $\mathscr{S}(\Gamma)$ must be orientable, and an orientation is chosen. The role of the surface F in the previous section is played here by the surface $F^{\Gamma} = \partial U$, where U is a closed tubular neighborhood of Γ in N. More precisely, F^{Γ} inherits its orientation from $\mathscr{S}(\Gamma)$ and from the choice of a normal vector field directed outwards, while U is an oriented handlebody made of 3–balls (regular neighborhoods of the 2–disks of Γ) and solid cylinders (regular neighborhoods of the bands of Γ). The fat graph Γ can be slightly shifted by isotopy to get a copy Γ' lying on F^{Γ}, the boundary of the handlebody. Each of the solid cylinder of the latter can be equipped with a meridian disk transverse to a portion of the band of Γ'. The boundary of each meridian disk is a circle $\subset F^{\Gamma}$, to be thought of as a graph with one edge and one vertex (chosen arbitrarily). The meridian circles (actually generators of the fundamental group of the handlebody U) are thus simple disjoint loops denoted by $\psi_1, \psi_2, \ldots, \psi_m$, where m is the number of edges in Γ. Given an ordered sequence of colors $J \equiv (j_1, j_2, \ldots, j_m) \in I^m$ denote by

$$\psi_J = \psi_1 \cup \psi_2 \cdots \cup \psi_m \,; \quad \omega_J = \prod_{k=1}^{m} \omega_{j_k} \tag{6.73}$$

the resulting colored graph and the associated weight, respectively.

If *Int* (U) is the interior of the closed handlebody U, define

$$N \setminus Int\,(U) := M \;\Rightarrow\; \partial M = \partial N \cup F^{\Gamma}, \tag{6.74}$$

so that the 3–manifold M is the closure of the complement of the handlebody U in the ambient manifold N. Recall that ∂N has been assumed to be triangulated, and fix now a coloring $\lambda \in Col\,(\partial N)$. As discussed in connection with the state model (6.62) in Sect. 6.4.1, the given triangulation in ∂N can be extended to the whole N and then induced by restriction also on M and ∂M. For each $\lambda \in Col\,(\partial N)$ define a relative state model of the pair (N, Γ) according to

$$< N, \Gamma \,|\, \lambda > := \omega^{2-2s} \sum_{\substack{\mu \,\in\, Col(M) \\ \mu|_{\partial N} \,=\, \lambda}} \sum_{J \in I^m} \omega_J \, |M, F^{\Gamma}|_{\mu} \; < \gamma_{F\Gamma}^{\mu_{F\Gamma}} \,|\, \Gamma' \,|\psi_J >, \tag{6.75}$$

where s is the number of vertices of Γ; Γ' is the copy of Γ lying on F^{Γ}; $|M, F^{\Gamma}|_{\mu}$ is the relative invariant (6.62); ω_J is defined in (6.73) and $\gamma_{F\Gamma}^{\mu_{F\Gamma}}$ is the dual graph of the 1–skeleton of the triangulation of F^{Γ} with $\mu_{F\Gamma}$ the restriction of μ to F^{Γ}.

As anticipated in the motivations at the beginning of Sect. 6.4.2, the evaluation of the invariant relies on a factorization, in which the terms $|M, F^{\Gamma}|_{\mu}$ weights the 3-dimensional triangulation of the complement of the tubular neighborhood of the fat graph and the $<\; |\; |\; >$ terms are calculated diagrammatically by means of expressions like (6.70) which are essentially 2–dimensional.

In Th. 5.1 and 5.2 of [67] the above state model is shown to be a topological invariant of the triple N, Γ, λ so that, in particular, it does not depend on the choice of a triangulation of M extending the given one in ∂N. At the same time (6.75) is an ambient isotopy invariant of Γ (the proofs are omitted).

Properties and Specific Examples

The condition $\partial N = \emptyset$ provides particular instances of the state sum denoted by

$$< N, \Gamma > \omega^{2-2s} \sum_{\mu \in Col(M)} \sum_{J \in I^m} \omega_J \, |M, F^{\Gamma}|_{\mu} \; < \gamma_{F\Gamma}^{\mu_{F\Gamma}} \,|\, \Gamma' \,|\psi_J >, \tag{6.76}$$

where, according to (6.74), ∂M is given by F^{Γ} and colored triangulations of the latter are induced now from those of $M \subset N$. Ambient isotopy invariants of colored framed links in closed (oriented) 3-manifolds are included in this kind of state sums. Actually in section 6 of [67] it is proven that, for any framed link L in S^3, the associated $< S^3, \Gamma_L >$ satisfies the standard skein relations of Jones'–type polynomials.

Another property of the (general) invariant is its multiplicativity with respect to a dissection of N into two compact 3-manifolds N_1, N_2 with $N_1 \cap N_2 = G$, a closed triangulated surface. If Γ_1 is the portion of Γ lying in N_1 and Γ_2 the portion in N_2 then the following splitting property holds

$$< N, \Gamma \,|\, \lambda > \,=\, \omega^{-2} \sum_{\nu \in Col\,(G)} < N, \Gamma_1 \,|\, \lambda_1 \cup \nu > \cdot < N, \Gamma_2 \,|\, \lambda_2 \cup \nu >, \qquad (6.77)$$

where λ is the coloring of ∂N and λ_i the reduction of λ to $\partial N_i \backslash G$.

Note finally that all state models considered so far, as well as those defined in the next section, are determined (and evaluated up to q-factors) by the colored core $c(\Gamma)$ of the fat graph Γ and by the orientation of its regular neighborhood. Thus all these invariants are well defined also for embedded (ordinary) colored 3-valent graphs (and colored links).

6.4.2.3   State Models for (N, Γ) Relatively to ϕ

A further generalization of the invariant of Sect. 6.4.2.2 arises when the structure of the set of colorings of a triangulated closed surface is taken into account. As shown in [69], the colorings of a triangulated closed surface G generate a vector space $Q(G)$. Denoting by $|\lambda >$ the generator in $Q(G)$ associated with $\lambda \in Col\,(G)$, each colored 3-valent graph $\phi \subset G$ gives rise to an element $|\phi > \in Q(G)$ defined as

$$|\phi > := \sum_{\lambda \in Col\,(G)} \omega_\lambda < \gamma_G^\lambda \,|\, \emptyset \,|\, \phi > \,|\, \lambda >, \qquad (6.78)$$

where $\emptyset$ is the empty (fat) graph in $G \times [-1, 1]$, γ_G^λ is the dual graph of the 1–skeleton of the triangulation of G colored with λ and

$$\omega_\lambda \,=\, \omega^{-\beta} \prod_{e'} \omega_{\lambda(e')} \qquad (6.79)$$

with β = number of vertices of G and $e' \in$ edges of G. The element $|\phi > \in Q(G)$ does not depend on the choice of the triangulation (up to isomorphisms) and in particular $|\gamma_G^\lambda > \,=\, \omega_\lambda^{-1} |\lambda >$.

If G is the boundary of a compact 3-manifold N and Γ a fat graph in the interior of N (as in the previous section), then for any colored 3-valent *oriented* graph ϕ on G the state sum

$$< N, \Gamma \,|\, \phi > \,=\, \sum_{\lambda \in Col\,(G)} \omega_\lambda < \gamma_G^\lambda \,|\, \emptyset \,|\, \phi > < N, \Gamma \,|\, \lambda > \qquad (6.80)$$

is a relative invariant of (N, Γ) with respect to ϕ which can be shown to be independent of the choice of the triangulation of G. This state sum is the most general in the class under consideration in [67] and contains all the previous ones as particular cases (e.g. if $\phi = \emptyset$ the absolute invariant $< N, \Gamma >$ of (6.76) is recovered, while in case the ambient manifold is a cylinder the invariants of Sect. 6.4.2.1 are obtained).

6.4.3 *Heegard Splitting Version of State Models for Closed Oriented 3-Manifolds*

Recall that any smooth, closed, oriented and connected 3–manifold can be generated by surgery operations performed on tubular neighborhoods of knots or links in the 3–sphere S^3 [60]. Let for instance K denote such a knot in S^3. A *Dehn surgery* amounts to remove first the interior *Int* (U) of a regular tubular neighborhood U of K. Then the two manifolds

$$S^3 \setminus Int\,(U) \equiv V \;\;\text{and}\;\; U \tag{6.81}$$

are to be looked at as distinct spaces (they are two handlebodies). Their boundaries

$$\partial(S^3 \setminus Int\,(U)) \;\;\text{and}\;\; \partial U \equiv F \tag{6.82}$$

are topologically tori. The next step consists in gluing back U and $S^3 \setminus Int\,(U)$ by an assigned orientation–preserving gluing homeomorphism

$$\mathfrak{h} : \partial U \to \partial(S^3 \setminus Int\,(U)). \tag{6.83}$$

K and $\mathfrak{h}$ define in a complete way the decomposition

$$N := (S^3 \setminus Int\,(U)) \cup_\mathfrak{h} U = V \cup_\mathfrak{h} U. \tag{6.84}$$

Actually the splitting depends on the homotopy class of $\mathfrak{h}(\psi)$, where ψ is a meridian loop on V (generator of the fundamental group of the handlebody). Two smooth 3–manifolds N, N' generated by different surgery operations are homeomorphic iff they are related by a finite sequence of Kirby–Rolfsen moves [60].

Consider now an N as defined in the last equality of (6.84). It is called a *presentation* of N as a Heegard splitting. (Note however that such a presentation can be actually achieved for any closed oriented smooth 3–manifold, not necessarily generated from a Dehn surgery operation on the 3–sphere as above). In any case the oriented surface F (of genus g) bounding the two handlebodies is called an Heegard surface. Let $\{\phi_1, \phi_2, \ldots, \phi_g\}$ and $\{\psi_1, \psi_2, \ldots, \psi_g\}$ be the boundaries of two systems of meridian disks on V and U, respectively. The two collections of disjoint loops on F, together with F itself, represent a *Heegard diagram* of N. (The splitting can be extended to a manifold with a non empty boundary ∂N where V is a handlebody and U is such that its boundary is $\partial N \cup F$, but then the loops ψ must be selected in a different manner).

According to the spirit of the construction of the state sums in Sect. 6.4.2.2 (considered here for $\partial N = \emptyset$), the two systems of loops are treated as graphs on F and colored by setting

$$\{\phi_1, \phi_2, \ldots, \phi_g\} \to H = \{h_1, h_2, \ldots, h_g\} \in I^g \tag{6.85}$$

$$\{\psi_1, \psi_2, \ldots, \psi_g\} \to J = \{j_1, j_2, \ldots, j_g\} \in I^g. \tag{6.86}$$

Consider an arbitrary colored 3-valent fat graph $\Gamma \subset N$ and deform it into a Γ' lying in the cylindrical neighborhood $F \times [-1, 1]$. Then the state model

$$< N, \Gamma > = \omega^{-2} \sum_{\substack{J \in I^g \\ H \in I^g}} \prod_{i=1}^{g} \omega_{j_i}^2 \prod_{k=1}^{g} \omega_{h_k}^2 \; < \phi_H \,|\, \Gamma' \,|\, \psi_J >, \qquad (6.87)$$

as the notation suggests, has the same value of the invariant (6.76) defined in Sect. 6.4.2.2 (for $\partial N = \emptyset$ and N oriented). Thus the present version of $< N, \Gamma >$ is a topological invariant of the pair (N, Γ) and in particular an ambient isotopy invariant of the fat graph Γ (see Th. A.1 in [67]).

The evaluation of (6.87), apart from ω–factors, is reduced to the calculation of elements of the type of $< \phi \,|\, \Gamma \,|\, \psi >$ as done in Sect. 6.4.2.1, while the evaluation of (6.76) includes the intrinsically 3–dimensional relative invariant $|M, F^\Gamma|_\mu$ defined in (6.61). It is not surprising that the prominent role is played here by a diagrammatic, essentially 2–dimensional computation, where the fat graph Γ is shifted to F and the generating loops of the handlebodies lie on the Heegard surface F by definition. (Note that, in comparing (6.76) of Sect. 6.4.2.2 and (6.87) above, some caution is in order. In the first state sum the triangulated handlebody M is associated with Γ itself, F being its triangulated boundary with dual graph $\gamma_{F\Gamma}$, and $\{\psi_J\}$ represent the generating loops of F itself. In the second state sum the collections $\{\phi_H\}$ and $\{\psi_J\}$, independent of Γ, are associated with the *surgery link* and its Heegard surface F bounding the two handlebodies, hidden by the notation.)

Note finally that (6.87), in case Γ is the empty fat graph

$$< N, \emptyset > \equiv < N > = \omega^{-2} \sum_{\substack{J \in I^g \\ H \in I^g}} \prod_{i=1}^{g} \omega_{j_i}^2 \prod_{k=1}^{g} \omega_{h_k}^2 \; < \phi_H \,|\, \emptyset \,|\, \psi_J >, \qquad (6.88)$$

is forced to give the value of the Turaev–Viro state sum for any N closed and oriented. It is worth noting that in the present case the diagrams to be associated with this kind of configuration of closed loops on the Heegard surface give rise exclusively to area colorings and vertices of type (iii) in the classification given in Sect. 6.4.2.1. The weights associated with these vertices are q-$6j$ symbols with extra q_i weights, but, on applying to concrete cases the identities (6.55), (6.56) or (6.57), the q_i weights are indeed absorbed so that the explicit form of (6.88) is formally recovered to be the same as the Turaev–Viro original invariant (with only ω^2, ω_i^2 and q-$6j$). Of course the meaning of the individual q-$6j$ is different in the two cases and the sets of moves which improve the invariance of the two state sums must be chosen accordingly.

Explicit calculations of the present version of the Turaev–Viro invariant have been carried out for classes of oriented 3–manifolds (such as lens spaces) [16, 32] for which $< \phi_H \mid \emptyset \mid \psi_J >$ can be put in the form

$$\prod_i < \phi_{k_i} \mid \emptyset \mid \psi_{j_i} > .$$

The resulting invariants can be written as suitable q-$3nj$ symbols of the second kind, see [70] for their definitions.

More crucially, such a Heegard–splitting version of the Turaev–Viro invariant can be compared directly with the Witten–Reshetikhin–Turaev generating functional for an oriented 3-manifold presented by using Dehn surgery along the tubular neighborhood of a link L in S^3 [57]. Since we do not give here details on the latter construction (see also [39]), we just want to remark that the correspondence (6.22) stated in Sect. 6.2.3 can be recasted by using $< N >$ as defined above instead of $\mathbf{Z}_{TV}[N; q]$ obtained by presenting N through colored triangulations (recall that every state model in Sect. 6.4 gives an invariant for each fixed value of the root of unity q).

References

1. Ambjørn, J., Carfora, M., Marzuoli, A.: The Geometry of Dynamical Triangulations. Lecture Notes in Physics Monographs, vol. 50. Springer, Berlin (1997)
2. Ambjørn, J., Durhuus, B., Jonsson, T.: Quantum Geometry. Cambridge University Press, Cambridge (1997)
3. Anderson, R.W., Aquilanti, V., Marzuoli, A.: 3nj morphogenesis and asymptotic disentangling. J. Phys. Chem. A **113**, 15106–15117 (2009)
4. Aquilanti, V., Haggard, H.M., Littlejohn, R.G., Yu, L.: Semiclassical analysis of Wigner 3j-symbol. J. Phys. A Math. Theor. **40**, 5637–5674 (2007)
5. Aquilanti, V., Bitencourt, A.P.C., da S. Ferreira, C., Marzuoli, A., Ragni, M.: Combinatorics of angular momentum recoupling theory: spin networks, their asymptotics and applications. Theor. Chem. Acc. **123**, 237–247 (2009)
6. Aquilanti, V., Haggard, H.M., Littlejohn, R.G., Poppe, S., Yu, L.: Asymptotics of the Wigner 6j symbol in a 4j model. Preprint (2010)
7. Arcioni, G., Carfora, M., Dappiaggi, C., Marzuoli, A.: The WZW model on random Regge triangulations. J. Geom. Phys. **52**, 137–173 (2004)
8. Arcioni, G., Carfora, M., Marzuoli, A., O' Loughin, M.: Implementing holographic projections in Ponzano–Regge gravity. Nucl. Phys. B **619**, 690–708 (2001)
9. Askey, R.: Ortogonal Polynomials and Special Functions, Society for Industrial and Applied Mathematics. Philadelphia (1975);
   R. Koekoek, R.F. Swarttouw, The Askey-Scheme of Hypergeometric Orthogonal Polynomials and Its Q-Analogue. Technische Universiteit Delft, Delft (1998). http://aw.twi.tudelft.nl/~koekoek/askey/
10. Atiyah, M.F.: The Geometry and Physics of Knots. Cambridge University Press, Cambridge (1990)
11. Barrett, J.W., Crane, L.: Relativistic spin networks and quantum gravity. J. Math. Phys. **39**, 3296–3302 (1998)
12. Beliakova, A., Durhuus, B.: Topological quantum field theory and invariants of graphs for quantum groups. Commun. Math. Phys. **167**, 395–429 (1995)
13. Biedenharn, L.C., Lohe, M.A.: Quantum Group Symmetry and Q-tensor Algebra. World Scientific, Singapore (1995)

14. Biedenharn, L.C., Louck, J.D.: Angular momentum in quantum physics: theory and applications. In: Rota G.-C. (ed.) Encyclopedia of Mathematics and Its Applications, vol. 8. Addison–Wesley Publications Co., Reading (1981)
15. Biedenharn, L.C., Louck, J.D.: The Racah–Wigner algebra in quantum theory. In: Rota G.-C. (ed.) Encyclopedia of Mathematics and Its Applications, vol. 9. Addison–Wesley Publications Co., Reading (1981)
16. Carbone, G.: Turaev–Viro invariant and 3nj symbols. J. Math. Phys. **41**, 3068–3084 (2000)
17. Carbone, G., Carfora, M., Marzuoli, A.: Wigner symbols and combinatorial invariants of three–manifolds with boundary. Commun. Math. Phys. **212**, 571–590 (2000)
18. Carbone, G., Carfora, M., Marzuoli, A.: Hierarchies of invariant spin models. Nucl. Phys. B **595**, 654–688 (2001)
19. Carfora, M., Marzuoli, A., Rasetti, M.: Quantum tetrahedra. J. Phys. Chem. A **113**, 15376–15383 (2009)
20. Carlip, S.: Quantum Gravity in 2+1 Dimensions. Cambridge University Press, Cambridge (1998)
21. Carter, J.S., Flath, D.E., Saito, M.: The Classical and Quantum 6j-Symbol. Princeton University Press, Princeton (1995)
22. Cattaneo, A.S., Cotta–Ramusino, P., Frölich, J., Martellini, M.: Topological BF theories in 3 and 4 dimensions. J. Math. Phys. **36**, 6137–6160 (1995)
23. Crane, L., Kauffman, L.H., Yetter, D.N.: State sum invariants of 4 manifolds. arXiv:hep–th/9409167
24. De Pietri, R., Freidel, L., Krasnov, K., Rovelli, C.: Barrett–Crane model from a Boulatov–Ooguri field theory over a homogeneous space. Nucl. Phys. B **574**, 785–806 (2000)
25. Durhuus, B., Jakobsen, H.P., Nest, R.: Topological quantum field theories from generalized 6j-symbols. Rev. Math. Phys. **5**, 1–67 (1993)
26. Freed, D.S.: Remarks on Chern–Simons theory. Bull. Am. Math. Soc. **46**, 221–254 (2009)
27. Freidel, L., Krasnov, K., Livine, E.R.: Holomorphic factorization for a quantum tetrahedron. Commun. Math. Phys. **297**, 45–93 (2010)
28. Freyd, P., Yetter, D., Hoste, J., Lickorish, W.B.R., Millett, K., Ocneanu, A.: A new polynomial invariant of knots and links. Bull. Am. Math. Soc. **12**, 239–246 (1985)
29. Gomez C., Ruiz–Altaba M., Sierra, G.: Quantum Group in Two–Dimensional Physics. Cambridge University Press, Cambridge (1996)
30. Guadagnini, E.: The link invariants of the Chern–Simons field theory. W. de Gruyter, Berlin/Boston (1993)
31. Haggard, H.M., Littlejohn, R.G.: Asymptotics of the Wigner 9j symbol. Class. Quant. Grav. **27**, 135010 (2010)
32. Ionicioiu, R., Williams, R.M.: Lens spaces and handlebodies in 3D quantum gravity. Class. Quant. Grav. **15**, 3469–3477 (1998)
33. Joyal, A., Street, R.: Braided tensor categories. Adv. Math. **102**, 20–78 (1993)
34. Jones, V.F.R.: A polynomial invariant for knots via von Neumann algebras. Bull. Am. Math. Soc. **12**, 103–111 (1985)
35. Karowski, M., Schrader, R.: A combinatorial approach to topological quantum field theories and invariants of graphs. Commun. Math. Phys. **167**, 355–402 (1993)
36. Kauffman, L.: Knots and Physics. World Scientific, Singapore (2001)
37. Kauffman, L., Lins, S.: Temperley–Lieb recoupling theory and invariants of 3-Manifolds. Princeton University Press, Princeton (1994)
38. Kaul, R.,K., Govindarajan, T. R., P. Ramadevi, P.: Schwarz type topological quantum field theories. In: Encyclopedia of Mathematical Physics. Elsevier (2005) (eprint hep–th/0504100)
39. Kirby, R., Melvin, P.: The 3-manifold invariant of Witten and Reshetikhin–Turaev for $sl(2, \mathbb{C})$. Invent. Math. **105**, 437–545 (1991)
40. Kirillov, A.N., Reshetikhin, N.Y.: In: Kac, V.G. (ed.) Infinite Dimensional Lie Algebras and Groups. Advanced Series in Mathematical Physics, vol. 7, pp. 285–339. World Scientific, Singapore (1988)

41. Mizoguchi, S., Tada, T.: 3-dimensional gravity from the Turaev–Viro invariant. Phys. Rev. Lett. **68**, 1795–1798 (1992)
42. Nikiforov, A.F., Suslov, S.K., Uvarov, V.B.: Classical Orthogonal Polynomials of a Discrete Variable. Springer, Berlin/New York (1991)
43. Neville, D.: A technique for solving recurrence relations approximately and its application to the 3-j and 6-J symbols. J. Math. Phys. **12**, 2438–2453 (1971)
44. Nomura, M.: Relations for Clebsch–Gordan and Racah coefficients in $su_q(2)$ and Yang–Baxter equations. J. Math. Phys. **30**, 2397–2405 (1989)
45. Ohtsuki T. (ed.): Problems on invariants of knots and 3–manifolds, RIMS geometry and topology monographs, vol. 4 (eprint arXiv: math.GT/0406190)
46. Ooguri, H.: Topological lattice models in four dimensions. Mod. Phys. Lett. A **7**, 2799–2810 (1992)
47. Ooguri, H.: Schwinger–Dyson equation in three-dimensional simplicial quantum gravity. Prog. Theor. Phys. **89**, 1–22 (1993)
48. Pachner, U.: Ein Henkel Theorem für geschlossene semilineare Mannigfaltigkeiten [A handle decomposition theorem for closed semilinear manifolds]. Result. Math. **12**, 386–394 (1987)
49. Pachner, U.: Shelling of simplicial balls and P.L. manifolds with boundary. Discr. Math. **81**, 37–47 (1990)
50. Pachner, U.: Homeomorphic manifolds are equivalent by elementary shellings. Eur. J. Comb. **12**, 129–145 (1991)
51. Penrose, R.: Angular momentum: an approach to combinatorial space–time. In: Bastin, T. (ed.) Quantum Theory and Beyond, pp. 151–180. Cambridge University Press, Cambridge (1971)
52. Ponzano, G., Regge, T.: Semiclassical limit of racah coefficients. In: Bloch F. et al. (eds.) Spectroscopic and Group Theoretical Methods in Physics, pp. 1–58. North–Holland, Amsterdam (1968)
53. Ragni, M., Bitencourt, A.P.C., da S. Ferreira, C. Aquilanti, V., Anderson, R.W., Littlejohn, R.G.: Exact computation and asymptotic approximations of 6j symbols: illustration of their semiclassical limits. Int. J. Quant. Chem. **110**, 731–742 (2009)
54. Regge, T.: Symmetry properties of Racah's coefficients. Nuovo Cimento **11**, 116–117 (1958)
55. Regge, T.: General relativity without coordinates. Nuovo Cimento **19**, 558–571 (1961)
56. Regge, T., Williams, R.M.: Discrete structures in gravity. J. Math. Phys. **41**, 3964–3984 (2000)
57. Reshetikhin, N., Turaev, V.G.: Invariants of 3-manifolds via link polynomials and quantum groups. Invent. Math. **103**, 547–597 (1991)
58. Roberts, J.D.: Skein theory and Turaev–Viro invariants. Topology **34**, 771–787 (1995)
59. Roberts, J.D.: Classical 6j-symbols and the tetrahedron. Geom. Topol. **3**, 21–66 (1999)
60. Rolfsen, D.: Knots and Links. Publish or Perish, Inc., Berkeley (1976)
61. Rourke, C.P., Sanderson, B.J.: Introduction to Piecewise–Linear Topology. Springer, Berlin (1972)
62. Rovelli, C.: Quantum Gravity. Cambridge University Press, Cambridge (2004)
63. Schulten, K., Gordon, R.G.: Exact recursive evaluation of 3j- and 6j-coefficients for quantum mechanical coupling of angular momenta. J. Math. Phys. **16**, 1961–1970 (1975)
64. Schulten, K., Gordon, R.G.: Semiclassical approximations to 3j- and 6j-coefficients for quantum mechanical coupling of angular momenta. J. Math. Phys. **16**, 1971–1988 (1975)
65. Taylor, Y.U., Woodward, C.T.: 6j symbols for $U_q(sl_2)$ and non–Euclidean tetrahedra. Sel. Math. New Ser. **11**, 539–571 (2005)
66. 't Hooft, G.: The scattering matrix approach for the quantum black hole, an overview. Int. J. Mod. Phys. **A11**, 4623–4688 (1996)
67. Turaev, V.G.: Quantum invariants of links and 3-valent graphs in 3-manifolds. Publ. Math. IHES **77**, 121–171 (1993)
68. Turaev, V.G.: Quantum Invariants of Knots and 3-manifolds. W. de Gruyter, Berlin (1994)
69. Turaev, V.G., Viro, O.Y.: State sum invariants and quantum 6j symbols. Topology **31**, 865–902 (1992)
70. Varshalovich, D.A., Moskalev, A.N., Khersonskii, V.K.: Quantum Theory of Angular Momentum. World Scientific, Singapore/Philadelphia (1988)

71. Walker, K.: On witten's 3-manifolds invariant. Preprint (1991). (An extended version dated 2001 is available on the web)
72. Williams, R.M., Tuckey, P.A.: Regge calculus: a bibliography and brief review. Class. Quant. Grav. **9**, 1409–1422 (1992)
73. Witten, E.: (2+1)-dimensional gravity as an exactly soluble system. Nucl. Phys. B **311**, 49–78 (1988/89)
74. Witten, E.: Quantum field theory and the Jones polynomial. Commun. Math. Phys. **121**, 351–399 (1989)
75. Yutsis, A.P., Levinson, I.B., Vanagas, V.V.: The Mathematical Apparatus of the Theory of Angular Momentum. Israel Program for Scientific Translations Ltd (1962)

Chapter 7
Combinatorial Framework for Topological Quantum Computing

In this chapter we describe a combinatorial framework for topological quantum computation, and illustrate a number of algorithmic questions in knot theory and in the theory of finitely presented groups, focusing in particular on the braid group. This list of problems give us the chance of defining (classical) complexity classes of algorithms by resorting to specific examples and not in a purely abstract way. In particular the algorithmic questions concerning the Jones polynomial are discussed and the basic definition of 'colored' Jones polynomials is given within an algebraic context. We address efficient quantum algorithms for the (approximate) evaluation of colored Jones polynomials and 3–manifold invariants, stressing the strong mutual connections between quantum geometry and quantum computing.

7.1 Introduction

Unlike perturbatively renormalizable quantum field theory – representing the basic tool in the standard model in particle physics, where the physically measurable quantities are obtained as finite limits of infinite series in the physical coupling constant – both $SU(2)$ Witten–Reshetikhin–Turaev (WRT) and Turaev–Viro models are actually 'solvable' for each fixed value of the coupling constant k or, equivalently, of the deformation parameter q. Such finiteness property is manifest in the case of the Turaev–Viro state sum (Sect. 6.2.2 of the previous chapter) and associated colored link invariants (Sect. 6.4 of the previous chapter) which are expressed as summations of a finite number of terms. In the WRT framework (Sect. 6.2.3 of the previous chapter) solvability relies on the 'heuristic' path integral quantization prescription of 3-dimensional topological quantum field theories (TQFT) and reflects the existence of an algebraic symmetry stemming from braid group representations and associated (quantum) Yang–Baxter equations.

© Springer International Publishing AG 2017

M. Carfora, A. Marzuoli, *Quantum Triangulations*, Lecture Notes in Physics 942,

https://doi.org/10.1007/978-3-319-67937-2_7

Looking in particular at the Jones polynomial [28, 55] as the prototype of observables in a WRT environment, an 'effective' procedure for calculating it can be carried out on applying recursively the so–called skein relations to a given knot diagram, see e.g. [32]. The question of the 'efficiency' of such an algorithm can be loosely formulated by asking how computer resources needed for computation should grow as the size of the input (measured here by the number of crossing of the knot diagrams) increases. Focusing on the time required to complete the calculation, and referring to the classical (Turing) model of computation, an algorithm is said to be polynomial (belonging to the complexity class denoted by **P**) if its time function grows at most as a polynomial in the size of the input. Algorithmic procedures in the class **P** are considered 'efficient', while other types of algorithmic behaviors, classified in various complexity classes (see below for precise definitions), are referred to as 'difficult' or even 'intractable'. It is a classic result in complexity theory that the calculation of the Jones polynomial on a classical computer is an intractable problem [27].

With the above remarks in mind, the issue of effective computability of the basic functionals and observables of quantum WRT addressed in this chapter, can be looked at as related to solvability/finiteness of the underlying theory. Turning the argument upside down, the search for new efficient quantum algorithms for processing 'invariant quantities' – characterizing suitably decorated geometrical objects or combinatorial patterns – represents an original and possibly very fruitful approach for improving the underlying physical models, typically formulated in $d = 2, 3$ space dimensions, with respect to their (yet unknown) solvability properties. On the other hand, the search for efficient computational protocols for topological invariants has been recognized in the last decade as a major achievement for quantum information theory [7, 14], also in connections with the search for anyonic quantum computing machines [10, 38].

The first section of this chapter is a short review of the combinatorial framework for topological quantum computation proposed by Mario Rasetti and one of the author. Note that we are not going to review specifically foundations of topological quantum computation, referring the reader to [10, 33, 48] for informative accounts.

The second section is aimed to illustrate a number of algorithmic questions in knot theory and in the theory of finitely presented groups, focusing in particular on the braid group. This list of problems give us the chance of defining (classical) complexity classes of algorithms by resorting to specific examples and not in a purely abstract way.

In Sect. 3 algorithmic questions concerning the Jones polynomial are discussed and the basic definition of 'colored' Jones polynomials is given within an algebraic context (this choice, further expanded in Sect. 4.1 makes this chapter quite indepen-dent from the field–theoretic setting of Sect. 2.3 of the previous chapter).

Efficient quantum algorithms for the (approximate) evaluation of colored Jones polynomials and 3–manifold invariants are addressed in Sect. 4. The construction actually bears on the interplay of three different contexts:

1. a topological context, where the problem is well–posed and makes it possible
   to recast the initial instance from the topological language of knot theory to the
   algebraic language of braid group theory;
2. a field theoretic context, where tools from $3d$ quantum Chern–Simons–Witten
   (CSW) and associated $2d$ conformal field theories provide unitary representations of the braid group;
3. a quantum information context, where specific features of the spin network
   framework for quantum computation are used to efficiently solve the original
   problem formulated in a field–theoretic language.

Finally, Sect. 5 deals with implications and improvements of the results of
Sect. 4 enlightening mutual connections between quantum geometry and quantum
computing.

7.2   The Spin Network Quantum Simulator

The model for universal quantum computation proposed in [44, 45], the 'spin
network simulator', is based on the (re)coupling theory of $SU(2)$ angular momenta
as formulated in the basic texts [3, 4] on the quantum theory of angular momentum
and the Racah–Wigner algebra, respectively, (*crf.* Sect. 1 of the previous chapter).
At the first glance the spin network simulator can be thought of as a non–Boolean
generalization of the Boolean *quantum circuit model*[1] [46], with finite–dimensional,
binary coupled computational Hilbert spaces associated with N mutually commuting angular momentum operators and unitary gates expressed in terms of:

(i) recoupling coefficients ($3nj$ symbols) between inequivalent binary coupling
     schemes of $N = (n + 1)$ $SU(2)$–angular momentum variables (j–gates);
(ii) Wigner rotations in the eigenspace of the total angular momentum $\mathbf{J}$ (M–gates)
     (that however will not be taken into account in what follows, see Sect. 3.2 of
     [45] for details).

---

[1]Recall that this scheme is the quantum version of the classical Boolean circuit in which strings
written in the basic binary alphabet $(0, 1)$ are replaced by collections of 'qubits', namely quantum
states in $(\mathbb{C}^2)^{\otimes N}$, and the gates are unitary transformations that can be expressed, similarly to
what happens in the classical case, as suitable sequences of elementary gates associated with the
Boolean logic operations *and, or, not.*

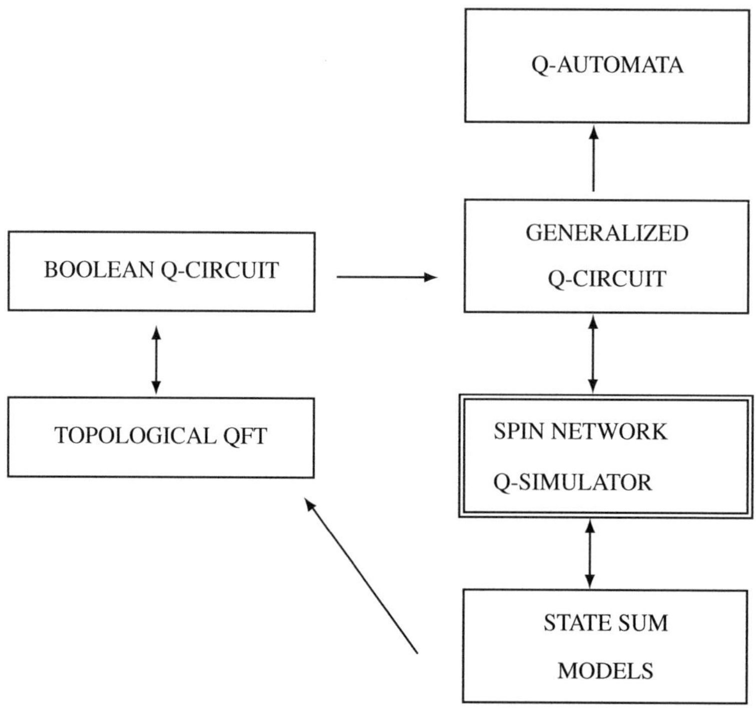

In the diagram we try to summarize various aspects of the spin network simulator together with its relationships with other models for quantum computation, in the light of underlying physical models discussed in the previous chapter.

On the left–hand portion of the diagram the standard Boolean quantum circuit is connected with a double arrow to the so–called topological approach to quantum computing proposed in [14] and developed in [15–17, 17]. Such an approach resorts to the Witten–Reshetikhin–Turaev (WRT) quantum functionals and associated unitary representations of the braid group of an $SU(2)$ Chern–Simons topological field theory (Sect. 2.3 of the previous chapter). It was shown in [16] that $SU(2)$ CSW functors at the fifth root of unity, whose domain is restricted to collections of 'topological' qubits (disks with three marked points) on which suitable unitary representations of the braid groups $\mathbf{B}_3$ and $\mathbf{B}_3 \times \mathbf{B}_3$ act, do reproduce the standard elementary gates of the quantum circuit model. The Boolean case is in turn connected one–way to the box of the generalized Q–circuit because it is a particular case of the latter when all N angular momenta are $\frac{1}{2}$–spins, see section 3.2 of [45].

On the right–hand column, the double arrows stemming from the box of the Q–simulator relate it to its reference models: from the viewpoint of quantum information theory it is a generalized Q–circuit, as already noted before, while its physical setting can be assimilated to state sum models discussed in the previous chapter.

The upper arrow is to be meant as generating, from the general spin networks computational scheme, families of *finite–state quantum automata* able to process a number of specific algorithmic problems, as will be shown in Sect. 4 below.

Besides the features described above, the kinematic structure of the Q–spin network complies with the requisites of an universal Q–simulator as defined by Feynman in [13], namely

- *locality*, reflected in the binary bracketing structure of the computational Hilbert spaces, which bears on the existence of poly–local, two–body interactions;
- *discreteness of the computational space*, reflected in the combinatorial structure of the (re)coupling theory of $SU(2)$ angular momenta [3, 4, 52, 58];
- *discreteness of time*, given by the possibility of selecting controlled, step–by–step applications of sequences of unitary operations for the generation of (any) process of computation;
- *universality*, guaranteed by the property that any unitary transformation operating on binary coupled Hilbert spaces (given in terms of $SU(2)$ $3nj$ symbols) can be reconstructed by taking a finite sequence of Racah–Wigner transforms each implemented by a $6j$ symbol as shown in [4], topic 12.

Then the Wigner $6j$ symbol, the *quantum tetrahedron* [8], plays a prominent role also in the spin network Q-simulator scheme, where it is recognized as the 'elementary' unitary operation, from which any algorithmic procedure can be built up. The role of the algebraic identities satisfied by the $6j$'s (Sect. 1 of the previous chapter) in the present context is analyzed at length in [45] (Sect. 4.2 and Appendix A).

Remark In dealing with specific algorithmic questions, a warning is however in order. The complexity class of any classical [quantum] algorithm is defined with respect to a *standard* classical [quantum] model of computation. A review of basic definitions in classical complexity theory, as framed within the standard, Turing model of computation (equivalent to the Boolean circuit model), is given in the following section. As for *quantum complexity classes*, let us anticipate here that a quantum algorithm for solving a given computational problem is *efficient* if it belongs to the complexity class **BQP**, the class of problems that can be solved in polynomial time by a Q–circuit with a fixed, bounded error, as functions of the *size* of a typical input. In most examples the size of the input is measured by the length of the string of qubits necessary to encode the generic sample of the algorithmic problem, as happens with the binary representation of an integer number, e.g. in calculations aimed to factorizing it in prime factors [46]. Once taken as reference model the Boolean Q–circuit, what would be necessary to verify is that a $6j$ symbol with generic entries can be efficiently (polynomially) processed by a suitably designed Q–circuit. This issue is discussed at length in Sect. 4.3 below. ∎

7.3   Knots, Braids and Complexity Classes

The analysis of algorithmic questions presented in this section and in the next one (based on [21]) is carried out with a twofold scope. The first one is about the necessity of defining classical and quantum complexity classes of computational problems or algorithms according to standard classifications [19, 46] while a few basic definitions in topological knot theory, together with its relation with the braid group and its representations, are needed to handle polynomial invariants of knots in Sect. 4.

A *knot K* is provided by a continuous embedding of the circle S^1 (the 1–dimensional sphere) into the Euclidean 3–space $\mathbb{R}^3$ or, equivalently, into the 3–sphere $S^3 \doteq \mathbb{R}^3 \cup \{\infty\}$. A *link L* is the embedding of the disjoint union of M circles, $\cup_{i=1}^{S} (S^1)_i$ into $\mathbb{R}^3$ or S^3, namely a finite collection of knots referred to as the components of L and denoted by $\{L_i\}_{i=1,2,\dots,S}$. Since each circle can be naturally endowed with an orientation, we can introduce naturally *oriented* knots (links).

Referring for simplicity to the unoriented case, two knots K_1 and K_2 are said to be *equivalent*, $K_1 \sim K_2$, if and only if they are *ambient isotopic*. An isotopy can be thought of as a continuous deformation of the shape of, say, $K_2 \subset \mathbb{R}^3$ which makes K_2 identical to K_1 without cutting and gluing back the 'closed string' K_2.

The *planar diagram*, or simply the *diagram*, of a knot K is the projection of K on a plane $\mathbb{R}^2 \subset \mathbb{R}^3$, in such a way that no point belongs to the projection of three segments, namely the singular points in the diagram are only transverse double points. Such a projection, together with over – and under–passing information at the crossing points – depicted in figures by breaks in the under–passing segments – is denoted by $D(K)$; a *link diagram* $D(L)$ is defined similarly. In what follows we shall sometimes identify the symbols K [L] with $D(K)$ [$D(L)$], although we can obviously associate with a same knot (link) an infinity of planar diagrams. Examples of diagrams are depicted in Fig. 7.1.

Fig. 7.1 Planar diagrams: the trefoil knot (top) and the Borromean link (bottom)

The first question on algorithmic complexity – related to the still unsolved problem in knot theory, namely a complete classification of (isotopy classes of) knots – is stated as

> *Problem 0. Give an effective algorithm for establishing when two knots or links are equivalent.*

Note that 'effective' is not to be confused with efficient: the latter adjective is deserved to algorithmic problems which can be solved by a Turing in a number of steps growing polynomially with the size of the input (this is actually the definition of the complexity class **P**). 'Effective' refers more loosely to a procedure that can be carried out systematically on each instance of a given computational problem.

Dealing with knots, the first issue concerns the choice of numerical quantities that can encode their topological structure. The number of crossings of a knot (diagram) is clearly a good indicator of the 'complexity' of the knot. Indeed, Tait in late 1800 initiated a program aimed to classifying systematically knots in terms of the number of crossings (see [5, 6, 40, 51] for exhaustive accounts on knot theory and for references to both older papers and knots tables).

Since a knot K with crossing number $\kappa(K)$ can be represented by planar diagrams with crossing numbers $c(D(K))$ for each $c(D(K)) > \kappa(K)$, the first issue is the search for procedures aimed to simplify as much as possible the diagrams of a knot K to get a $D'(K)$ with $c(D'(K)) = \kappa(K)$, which, from now on, denotes the 'minimum' crossing number. Reidemeister theorem (see e.g. [5]) gives the answer to this basic question.

Equivalence of knots (Reidemeister moves). Given any pair of planar diagrams D, D' of the same knot (or link), there exists a finite sequence of diagrams

$$D = D_1 \rightarrow D_2 \rightarrow \cdots \rightarrow D_k = D' \tag{7.1}$$

such that any D_{i+1} in the sequence is obtained from D_i on applying one of the Reidemeister moves (I, II, III) depicted in Fig. 7.2.

The procedure of Reidemeister theorem applies to subsets of link diagrams localized inside disks belonging to the plane where the diagram lives, and can be used in principle to compare pairs of arbitrarily chosen knot diagrams also in view of *Problem 0*. However, notwithstanding the recursively numerable character of the implementation of the Reidemeister moves with respect to the quite unfeasible notion of ambient isotopy, such moves can be hardly formalized and encoded into effective algorithms, basically because of their purely topological character. As we shall see, transformations on link diagrams can be consistently framed within a group–theoretic setting by exploiting their connection with braids, the new Markov 'moves' being reformulated in terms of algebraic operations.

In general, a *link invariant* is a map

$$L \longrightarrow f(L) \tag{7.2}$$

Fig. 7.2 The three
Reidemeister moves acting on
local configurations of link
diagrams

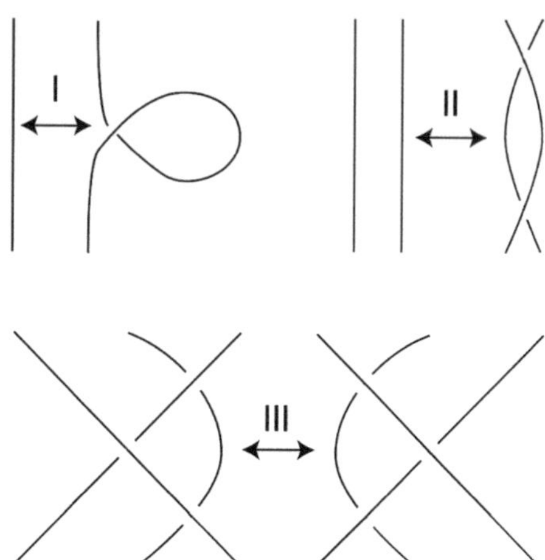

where the quantity $f(L)$ depends only on the type of the link, namely takes different
values on inequivalent links. Switching to link diagrams, we keep on using the same
notation as in (7.2), but now it is sufficient to verify that $f(L)$ ($\equiv f(D(L))$) does not
change under applications of Reidemeister moves I, II, III.

We have already met a numerical invariant, namely the (minimum) crossing
number κ. It is a natural number which takes the value 0 for the trivial knot
represented as an unknotted circle. Other invariants taking values in $\mathbb{Z}$ can be defined
for oriented link diagrams, where each crossing is marked by ± 1 according to some
fixed convention. For instance, the *writhe* $w(D(L))$ of a diagram D of an oriented
link L is the summation of the signs of the crossings of D, namely

$$w(D(L)) = \sum_{p} \epsilon_p, \qquad\qquad (7.3)$$

where the sum runs over the crossing points $\{p\}$ and $\epsilon_p = +1$ if the (directed) knot
path shows an overpass at the crossing point p, $\epsilon_p = -1$ for an underpass. Note
however that both the crossing number and the writhe do change under Reidmeister
move of type I, but are invariant under the moves II and III: this property defines a
restricted kind of isotopy, commonly referred to as *regular isotopy*. The notion of
regular isotopy is very useful because, by eliminating the move I, we do not really
lose any information about the topology of the link. Moreover, the evaluation of
crossing numbers and writhes can be carried out efficiently by a simple inspection
of the diagrams.

Over the years, mathematicians have provided a number of knot invariants,
by resorting to topological, combinatorial and algebraic methods. Nevertheless,
we do not have yet a complete invariant (nor a complete set of invariants) able
to characterize the topological type of each knot and to distinguish among all
possible inequivalent knots, recall *Problem 0*. As a matter of fact, the most effective
invariants have an group–theoretic origin, being framed in the notion of braid groups
and their representation theory. The relation between link and braids is governed by
Alexander theorem (see e.g. [6], Sect. 2).

Braids from links (Alexander theorem). Every knot or link in $S^3 = \mathbb{R}^3 \cup \{\infty\}$ can be represented as a closed braid, although not in a unique way.

The *Artin braid group* $\mathbf{B}_n$, whose elements are (open) braids β, is a finitely presented group on n 'standard' generators $\{\sigma_1, \sigma_2, \ldots, \sigma_{n-1}\}$ plus the identity element e, which satisfy the relations

$$\sigma_i \sigma_j = \sigma_j \sigma_i \quad (i,j = 1, 2, \ldots, n-1) \text{ if } |i - j| > 1$$

$$\sigma_i \sigma_{i+1} \sigma_i = \sigma_{i+1} \sigma_i \sigma_{i+1} \quad (i = 1, 2, \ldots, n-2). \tag{7.4}$$

This group acts naturally on topological sets of n disjoint strands with fixed endpoints – running downward and labeled from left to right – in the sense that each generator σ_i corresponds to a crossing of two contiguous strands labeled by i and $(i + 1)$, respectively: if σ_i stands for the crossing of the i–th strand over the $(i + 1)$–th one, then σ_i^{-1} represents the inverse operation with $\sigma_i \sigma_i^{-1} = \sigma_i^{-1} \sigma_i = e$, see Fig. 7.3 (top). An element of the braid group can be thought of as a 'word', such as for instance $\beta = \sigma_3^{-1} \sigma_2\, \sigma_3^{-1} \sigma_2 \sigma_1^3\, \sigma_2^{-1} \sigma_1 \sigma_2^{-2} \in \mathbf{B}_4$; the length $|\beta|$ of the word β is the number of its letters, where by a 'letter' we mean one of the generators or its inverse element. By a slight change of notation, denote by R_{ij} the over–crossing operation acting on two strands the endpoints of which are labeled by i and j. Then the second relation in (7.4) can be casted into the form of the algebraic *Yang–Baxter relation*

$$\mathsf{R}_{12}\, \mathsf{R}_{13}\, \mathsf{R}_{23} \;=\; \mathsf{R}_{23}\, \mathsf{R}_{13}\, \mathsf{R}_{12}, \tag{7.5}$$

represented pictorially in Fig. 7.3 (bottom).

Fig. 7.3 The generator σ_i and its inverse σ_i^{-1} (top). The algebraic Yang–Baxter equation (bottom)

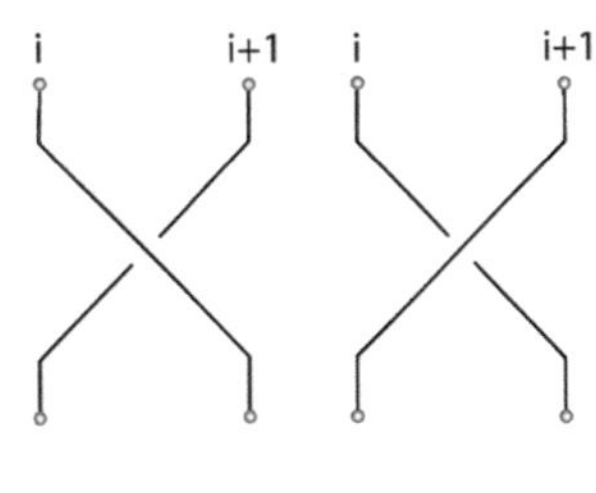

Fig. 7.4 The two types of
closures of a braid, namely
the plat closure (left) and the
standard closure (right), both
representing the trefoil knot

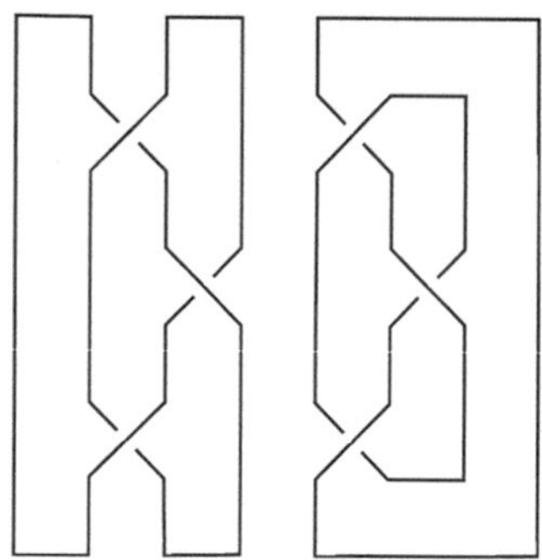

It is straightforward to get a link out of a braid: we have simply to 'close up' the
ends of an open braid β to get a *closed braid* $\hat{\beta}$ that reproduces the diagram of some
link L. Formally

$$\beta \xrightarrow{\text{closure}} \hat{\beta} \longleftrightarrow L. \qquad (7.6)$$

Notice however that this operation can be performed in two ways, denoted by $\hat{\beta}^{\,\text{st}}$
(the standard closure) and $\hat{\beta}^{\,\text{pl}}$ (the plat closure), respectively. In Fig. 7.4 the two
admissible closures of a same open braid are shown, where both closed braids can
be seen as deformations (by Reidemeister moves II and III) of the planar diagram of
the trefoil knot depicted in Fig. 7.1 (top).

As already pointed out, Alexander theorem does not establish a one–to–one
correspondence between links and braids. For instance, given a closed braid $\hat{\beta} = L$
with $\beta \in \mathbf{B}_n$, any other braid obtained from β by *conjugation*, namely $\beta' = \alpha\beta\alpha^{-1}$
(for some $\alpha \in \mathbf{B}_n$) has a closure $\hat{\beta}'$ which reproduces the same link L. Thus the
following question can be naturally raised.

*Problem 1A. Is it always possible to transform efficiently a given knot or link
into a closed braid?*

The answer is affirmative, since there exists a classical algorithm which performs
the reduction in a number of steps which is bounded from above by a polynomial
function of the braid index ([6], Sect. 2 and original references therein), where the
braid index of a (closed) braid is simply the number of its strands.

Taken for granted the above result about the efficiency of the reduction of any
link diagram to a closed braid, we can now exploit the algebraic properties of
braid groups. For what concerns in particular the issue of equivalence, Reidemeister
theorem can be recasted into Markov theorem. The following statement of this
theorem refers to the case of open braids, which captures the crucial features of the
construction, while the version involving closed braids can be found in [6], Sect. 2.

Equivalence of braids (Markov moves). Two braids are equivalent if they differ
by a finite sequence of Markov 'moves' of the following two types, together with
their inverse moves:

(i) change a braid $\beta \in \mathbf{B}_n$ to a conjugate element in the same group, $\beta \to \alpha\beta\alpha^{-1}$, with $\alpha \in \mathbf{B}_n$;

(ii) change $\beta \in \mathbf{B}_n$ to $i_n(\beta)\sigma_n^{\pm 1}$, where $i_n : \mathbf{B}_n \hookrightarrow \mathbf{B}_{n+1}$ is the natural inclusion obtained by disregarding the $(n+1)$–th strand and $\sigma_n, \sigma_n^{-1} \in \mathbf{B}_{n+1}$.

The next question arises in connection with the search for the most 'economical' representation of a knot diagram as a closed braid. The *minimum braid index* of a link L is the minimum number n for which there exists a braid $\beta \in \mathbf{B}_n$ whose closure $\hat{\beta}$ represents L.

> *Problem 1B. Does there exist an (efficient) algorithm to select, among the diagrams of a given link L, the diagram with the minimum braid index?*

At present no explicit algorithm for addressing this problem is known, so that its computational complexity class cannot be even evaluated (see [6], Sects. 2 and 4 for more details).

Coming to algorithmic problems characterized in purely algebraic terms, recall that braid groups belong to the class of *finitely presented groups*. Such groups are defined by means of a finite sets of *generators* together with *relations* among the generators and can have a finite order – as for the classical point groups of crystallography – or not – as happens for the braid group – (see [42] and older references therein). It was Max Dehn who stated the three 'fundamental problems' concerning a group G presented in terms of generators, denoted by $a, b, c, \ldots$, and relations $P, Q, R, \ldots$, namely

$$G \doteq \langle a, b, c, \ldots ; P, Q, R, \ldots \rangle . \tag{7.7}$$

Any element of G can be written (in multiplicative notation) as a *word* W in the alphabeth given by the generators and their inverse elements, as already done for the braid group. Note that the relations $P, Q, R, \ldots$ in (7.7) represent the minimal set of words in the generators, equivalent to the identity element e ('minimal' meaning that any other word equivalent to the identity can be reduced – by the use of the relations in the set – to the union of words in the same set).

Given the presentation (7.7) of the group G, Dehn problems are formulated as follows.

> *2A. The word problem. For an arbitrary word W in the generators, decide whether or not W defines the identity element in G.*
> *Equivalently: given two words W, W', decide whether $W = W'$.*

> *2B. The conjugacy problem. For two arbitrary words W_1 and W_2 in the generators, decide whether or not they are conjugate to each other. In a sharper form: find explicitly an element W' for which $W_2 = W'W_1(W')^{-1}$.*

> *2C. The isomorphism problem. For an arbitrary group G' defined by another presentation $G' \doteq \langle a', b', c', \ldots ; P', Q', R', \ldots \rangle$, decide whether or not G and G' are isomorphic.*

Except for the second issue in Problem 2B, we are in the presence of *decision problems*, namely problems that can be addressed by means of classical algorithms (running on a Turing machine) designed to answer 'yes' or 'no' to each of the above questions. Recall that the *time complexity function* $\mathfrak{f}_{\mathscr{A}}$ of an algorithm $\mathscr{A}$ is defined in terms of the size $\mathfrak{s}$ of the input. The size is in turn the length of an instance of the problem, such as the number of binary digits in a string representing a numerical instance or, in the case of finitely presented groups, the numbers of letters of the input words. An algorithm associated with a decision problem belongs to the complexity class **P** (**Polynomial**) if, for any instance of size $\mathfrak{s}$, the mapping

$$\mathfrak{s} \rightarrow \mathfrak{f}_{\mathscr{A}}(\mathfrak{s}) \tag{7.8}$$

is (bounded by) a polynomial function and to the class **NP** (**N**on deterministic **P**olynomial) if any guess on the answer can be checked in polynomial time [19]. Algorithmic problems endowed with complexity functions of exponential type have to be considered as intractable in the framework of classical information theory.

For what concerns the above list of algorithmic questions in the case of a generic group G, note preliminary that a solution of the complete conjugacy problem 2B (belonging to a certain complexity class) would imply a solution to Problem 2A (in the same class) since it would be sufficient to set $W' = e$ in the expression $W_2 = W'W_1(W')^{-1}$. It is also clear that the most difficult problem is the last one, which requires a 'global' inspection of the algebraic structures (generators plus relations) of the groups under examination. As for the braid group, we leave aside this problem and refer the reader to [6] (Sect. 1) for the definition of a presentation in terms of generators and relations alternative to the standard ones collected in (7.4).

The known results about the word and the conjugacy problems for the braid group $\mathbf{B}_n$ are briefly summarized below (see [6], Sect. 5 and original references therein).

2A. The solution to the word problem is polynomial, with a complexity function of the order

$$\mathscr{O}(|\beta|^2 n \ln n), \tag{7.9}$$

where $|\beta|$ is the length of the initial representative of the braid β and n is the braid index.

Surprisingly enough, the following problem, apparently very closely related to the word problem, turns out to belong to the class of **NP**–complete problems (recall that a particular **NP** problem is 'complete' if every other problem in the class can be polynomially reduced to it [19]).

Problem 2A'. Given a word β in the standard generators $\sigma_1, \sigma_2, \ldots, \sigma_{n-1}$ and their inverses, determine whether there is a shorter word β' which represents the same element in $\mathbf{B}_n$.

Finally, as for the second Dehn problem

2B. *The best known algorithm for the conjugacy problem is exponential in both $|\beta|$ and n.*

Remark It is worth noticing that the difficulty of solving the conjugacy problem in braid groups, as compared with the feasibility of the word problem, has been exploited for the construction of a public–key (classical) cryptosystem in [1]. As for cryptography, it has been recently proposed in [43] a classical protocol based on purely topological knot theory. However, such issues are not so close to the main topics of the present monography and thus we skip other details here.

7.4 Polynomial Invariants of Knots and Related Algorithmic Problems

Invariants of knots (links) of polynomial type arise (or can be reformulated) by resorting to *representations* of the braid group $\mathbf{B}_n$ in an algebra A, namely a vector space over some field (or ring) Λ endowed with a multiplication satisfying associative and distributive laws. The algebra must have a unit with respect to multiplication and for our purposes must be also finitely generated, namely its elements can be decomposed in terms of some finite 'basis' set, the number of elements of which equals the braid index n. The reason for considering an algebra should be clear if we recognize, on the one hand, that we can multiply braids $\in \mathbf{B}_n$ by simply composing their diagrams: given β_1 and $\beta_2 \in \mathbf{B}_n$ we get the product $\beta_1 \beta_2$ by placing the braid β_1 above β_2 and gluing the bottom free ends of β_1 with the top ends of β_2 (this operation was implicitly assumed in (7.4) and (7.5), see also Fig. 7.3). On the other hand, the operation associated with 'addition' of braids can be defined in terms of formal combinations of the type $a\beta_1 + b\beta_2$, for any $\beta_1, \beta_2 \in \mathbf{B}_n$ and $a, b \in \Lambda$.

A *representation of* $\mathbf{B}_n$ *inside the algebra* A is a map

$$\rho_\mathsf{A} : \mathbf{B}_n \longrightarrow \mathsf{A} \tag{7.10}$$

which satisfies

$$\rho_\mathsf{A}(\beta_1 \beta_2) = \rho_\mathsf{A}(\beta_1)\rho_\mathsf{A}(\beta_2) \quad \forall \beta_1, \beta_2 \in \mathbf{B}_n, \tag{7.11}$$

namely ρ_A is a group homomorphism from $\mathbf{B}_n$ to the multiplicative group G of the invertible elements of A (in particular: $\rho_\mathsf{A}(e) = 1$, where e is the identity element of $\mathbf{B}_n$ and 1 denotes the unit of A; $\rho_\mathsf{A}(\beta^{-1}) = [\rho_\mathsf{A}(\beta)]^{-1}, \forall \beta$). By using the standard generators of $\mathbf{B}_n$ defined in (7.4), it suffices to define the map (7.10) on the generators $\{\sigma_i\}$

$$\rho_\mathsf{A}(\sigma_i) := g_i \in \mathsf{G} \subset \mathsf{A}, \quad (i = 1, 2, \dots n - 1), \tag{7.12}$$

and extend linearly its action on products and sums of braids. Any pair of contiguous elements g_i and g_{i+1} must satisfy the *Yang–Baxter equation associated with the representation* ρ_{A}, namely

$$g_i\, g_{i+1}\, g_i \;=\; g_{i+1}\, g_i\, g_{i+1} \tag{7.13}$$

while $g_i g_j \;=\; g_j g_i$ for $|i-j| > 1$.

Once defined the representation ρ_{A} we may also introduce associated *matrix representations* of some fixed dimension N by representing A over the algebra of $(N \times N)$ matrices with entries in the field Λ

$$\mathsf{A} \;\longrightarrow\; \mathsf{M}(\Lambda, N). \tag{7.14}$$

If we restrict the domain of the above map to the group of invertible elements, the assignment (7.14) can be rephrased as the choice an N–dimensional vector space V over Λ, and thus we have the natural isomorphism

$$\mathsf{M}(\Lambda, N) \;\cong\; \mathsf{GL}_\Lambda\,(V,\,N), \tag{7.15}$$

where $\mathsf{GL}_\Lambda\,(V,\,N)$ is the general linear group of non–singular, Λ–linear maps $V \to V$. Loosely speaking, if we associate with a braid $\beta \in \mathbf{B}_n$ a matrix $M(\beta)$ obtained by means of a representation (7.14) of dimension $N = n$, then β can be characterized by the trace of $M(\beta)$ (the character of the representation). Such traces are candidates to be interpreted as invariants of links presented as closed braids, *cfr.* (7.6) in the previous section. A *trace function over the algebra* A is formally defined as a linear function over A and, by extension, over a matrix representation algebra (7.14)

$$\mathsf{A} \;\longrightarrow\; \mathsf{M}(\Lambda, N) \;\xrightarrow{\;\mathrm{Tr}\;}\; \Lambda \tag{7.16}$$

satisfying the property

$$\mathrm{Tr}\,(\,M(\beta)\,M'(\beta')\,) \;=\; \mathrm{Tr}\,(\,M'(\beta')\,M(\beta)\,). \tag{7.17}$$

for any $M(\beta), M'(\beta')$ which are the images under ρ_{A} of two braids $\beta, \beta' \in \mathbf{B}_n$. It can be shown that $\mathrm{Tr}(M(\beta))$ is a link invariant since it does not change under Markov move of type **(i)** defined in the previous section, namely

$$\mathrm{Tr}\,(M(\beta)) \;=\; \mathrm{Tr}\,(M'(\beta')) \quad \text{if } \beta \text{ and } \beta' \text{are conjugate.} \tag{7.18}$$

In other words, link invariants arising as Markov traces are *regular* isotopy invariants, as can be easily inferred comparing Reidemeister and Markov theorems.

The general algebraic setting outlined above is the framework underlying the original constructions of both the Jones [28, 29] and the HOMFLY [18, 40] polynomials for oriented links (note that these invariants take the same value in the orientations of the link components are reversed). In particular:

- the *Jones polynomial* of a link L, $J(L; t)$, is the Markov trace of the representation of $\mathbf{B}_n$ inside the Temperley–Lieb algebra $TL_n(t)$. It is a Laurent polynomial in one formal variable t with coefficients in $\mathbb{Z}$, namely it takes values in the ring $\Lambda \equiv \mathbb{Z}[t, t^{-1}]$;
- the *HOMFLY polynomial* $P(L; t, z)$ is obtained as a one–parameter family of Markov traces (parametrized by z) of the representation of $\mathbf{B}_n$ inside the Hecke algebra $H_n(t)$. It is a Laurent polynomial in two formal variables with coefficients in $\mathbb{Z}$, namely $\Lambda \equiv \mathbb{Z}[t^{\pm 1}, z^{\pm 1}]$.

As discussed at length below, the problem of evaluating (approximating) the Jones polynomial on a quantum computing machine has called many people's attention in the last decade. Purely algebraic definitions as outlined above seem however lacking in selecting *unitary* representation of the braid group $\mathbf{B}_n$ in a natural way. For instance, the approaches proposed in [2] and [56] provide two different types of Hilbert space structures and associated unitary representations of the braid group by resorting to clever, but 'ad hoc' constructions, on the one hand, and to standard quantum circuit processing, on the other. Contrarily, approaches inspired by Freedman's vision of the 'quantum field computer' [14] are ab initio framed within the physical background provided by CSW ($SU(2)$) topological quantum field theories, so that unitary representations of the braid group come out in a straightforward way. In particular, the formal variable of any polynomial to be associated with a unitary representation has to be a complex r–th root of unity

$$q := \exp(2\pi\, i/r) \ , r \in \mathbb{N}, \ r \geq 1 \tag{7.19}$$

and the idea is that, by letting r grow, the polynomial can be evaluated in more and more points lying on the unit circle in $\mathbb{C}$. The upgraded notation for the Jones polynomial is then

$$J(L; q) \ \in \ \mathbb{Z}[q, q^{-1}], \tag{7.20}$$

while the invariants we will consider in the following section are *colored* extensions of the Jones polynomial (7.20) denoted by

$$J(L; q; j_1, j_2, \ldots, j_M) \tag{7.21}$$

and parametrized by labels $\{j_1, j_2, \ldots, j_M\}$ (the 'colors') to be assigned to the S link components $\{L_i\}_{i=1,2,\ldots,S}$. >From the point of view of equivalence of links, $J(L; q; j_1, j_2, \ldots, j_S)$ turns out to be a 'regular isotopy' invariant (*cfr.* Markov theorem), but it can be shown that the quantity

$$\frac{q^{-3w(L)/4}}{q^{1/2} - q^{-1/2}} \ J(L; q; j_1, j_2, \ldots, j_S), \tag{7.22}$$

where $w(L)$ is the writhe of the link L defined in (7.3), is invariant under any type of ambient isotopies. The colored polynomials (7.22) reduce to Jones' when all the colors $j_1, j_2, \ldots, j_M$ are equal to a same j, with $j = 1/2$, but are genuine generalizations as far as they can distinguish knots with the same Jones polynomial, see e.g. [40].

Remark It is worth stressing that these invariants are 'universal', in the sense that they arise from a number of historically distinct approaches, ranging from the quasi tensor category approach by Drinfeld [11] and R–matrix representations obtained with the quantum group method [37, 50] to monodromy representations of the braid group in $2d$ conformal field theories [24, 39] (and references therein) up to $3d$ quantum Chern–Simons–Witten framework [25, 55] (and references therein) and Sect. 2.3 of the previous chapter. Referring to the representation ring of the quantum deformation of $SU(2)$ at a root of unity (reviewed in Sect. 4.1), the various unitary representations of the braid group exploited in these approaches turn out to be unitarily equivalent. ∎

Coming to algorithmic questions, we focus here on the Jones invariant (7.20), which is the simplest of the colored polynomials, on the one hand, and the prototype of invariants arising in a purely algebraic context, on the other. The reason why Jones' case is so crucial also in the computational context is actually due to the fact that a 'simpler' link invariant, the Alexander–Conway polynomial, can be computed efficiently, while the problem of computing 2–variable polynomials – such as the HOMFLY invariant – is **NP**–hard (see [6] for the definitions of the mentioned invariants and [27] for an account of computational questions). The issue of computational complexity of the Jones polynomial in classical information theory can be summarized as

> *Problem 3. How hard is it to determine the Jones polynomial of a link L?*

A quite exhaustive answer has been provided in [27], where the evaluation of the Jones polynomial of an alternating link $\tilde{L}$ at a root of unity q is shown to be #**P**–hard, namely computationally intractable in a very strong sense. Recall first that 'alternating' links are special instances of links, the planar diagrams of which exhibit over and under crossings, alternatively. Thus, the evaluation of the invariant of generic, non–alternating links is at least as hard. Secondly, the computation becomes feasible when the argument q of the polynomial is a 2nd, 3rd, 4th, 6th root of unity (refer to [27] or [43] for details on this technical issue). Finally, the #**P** complexity class can be defined as the class of enumeration problems in which structures or configurations to be counted are recognizable in polynomial time. More precisely, a problem π in #**P** is said #**P**–complete if, for any other problem π' in #**P**, π' is polynomial–time reducible to π; if a polynomial time algorithm were found for any such problem, it would follow that #**P** $\subseteq$ **P**. A problem is #**P**–hard if some #**P**–complete problem is polynomial–time reducible to it [19].

The intractability of Problem 3 relies on the fact that it is not possible to recognize in polynomial time all the equivalent configurations of a same link L, namely link diagrams $\{D(L), D'(L), D''(L), \ldots\}$, related to each other by (regular) isotopy. Coming back to the problems addressed in Sect. 2, since any link can be presented efficiently as a closed braid (Problem 1A), we are justified in switching to closed braids and implementing regular isotopy of diagrams by means of Markov move of type **(i)**. However, the intractability of Problem 1B (selecting the diagram with the minimum braid index) prevents us from selecting an 'optimal' representation of the isotopy class of diagrams that would provide, in turn, a unique standard configuration to be handled for computational purposes. Moreover, since Markov move **(i)** provides the practical implementation of the statement of the conjugacy problem in the braid group (Problem 2B), the whole matter could be reformulated within the group–theoretic setting as well (where the issue of the optimal presentation would be related with the **NP**–complete 'shorter word' problem stated in 2A' of Sect. 2).

In the discussion above, the relevant quantities encoding the 'size' of a typical instance of the computational problem – a link diagram L presented as a closed braid on n strands, $L = \hat{\beta}$ as in (7.6) – are of course the number of crossings κ and the braid index n. We might consider, instead of κ, the length $|\beta|$ of the open braid β associated with L, which equals the number of generators (and inverse generators) in the explicit expression of β as a word in $\mathbf{B}_n$. Finally, also the argument q of the Jones polynomial (7.20) is a relevant parameter since, when the integer r in (7.19) becomes $\gg 1$, we would reach more and more points on the unit circle in the complex plane, thus giving more and more accurate evaluations of the invariant. As for the algorithms to be addressed in the next section, their time complexity functions will be evaluated in terms of κ and n for each *fixed* value of the formal variable q.

The computational intractability of Problem 3 does not rules out the chance of an efficient 'approximate' calculation of the values of the Jones invariant.

> *Problem 4. How hard is it to approximate the Jones polynomial $\mathsf{J}(L, q)$ of a link L at a fixed root of unity q ($q \neq$ 2nd, 3rd, 4th, 6th root)?*

Loosely speaking, the approximation in question is given by a number Z such that, for any choice of a small $\delta > 0$, the numerical value of $\mathsf{J}(L, q)$ differs from Z by an amount ranging between $-\delta$ and $+\delta$. Then the value Z can be accepted as an approximation of the polynomial if Prob $\{| \mathsf{J}(L = \hat{\beta}, q) - Z| \leq \delta\} \geq \frac{3}{4}$ (see Sect. 4.3 below for the more details).

In the framework of classical complexity theory there do not exist algorithms to handle Problem 4 as it stands. Thus, at least at the time being, this problem is to be considered as intractable. The formal statement of the answer to Problem 4 in the quantum computational context was given in [7].

> *4.* The approximation of the Jones polynomial of a link presented as the closure of a braid at any fixed root of unity is **BPQ**–complete. Moreover, this problem is universal for quantum computation, namely is the 'prototype' of all problems efficiently solvable on a quantum computer.

Recall that **BQP** is the computational complexity class of problems which can be solved in polynomial time by a quantum computer with a probability of success at least $\frac{1}{2}$ for some fixed (bounded) error. In [7] it was actually proved that $\mathbf{P}^J = \mathbf{BQP}$, where $\mathbf{P}^J$ is defined as the class of languages accepted in polynomial time by a quantum Turing machine with an oracle for the language defined by Problem 4. This equality between computational classes implies that, if we find out an efficient quantum algorithm for Problem 4, then the problem itself is complete for the class **BQP**, namely each problem in this last class can be efficiently reduced to a proper approximate evaluation of the Jones polynomial of a suitable link (see [56] for a detailed discussion on this issue).

According to the above remarks, the search for efficient quantum algorithms for Problem 4 (not given explicitly in [7]) has represented a breakthrough in quantum information theory, and in the next section an account of the approach based on the spin network framework for quantum computation is addressed in some details.

7.5 Efficient Quantum Processing of Colored Jones Polynomials

The quantum algorithms – developed by one of the author in collaboration with Silvano Garnerone and Mario Rasetti – are designed to deal with both the colored link polynomials [20, 22] – viewed as vacuum expectation values of composite Wilson loop operators in Witten–Chern–Simons theory – and Witten–Turaev–Reshetikhin invariants of 3–manifolds presented as complements of a colored link in the 3–sphere [23] (see Sect. 2.3 of the previous chapter for the field–theoretic definitions). The goal is achieved through a two–level procedure summarized as follows.

- A specific unitary representation of the groupoid of colored oriented braids – associated with colored oriented links presented as plat closures of these braids – is processed on a q–spin network automaton (Sect. 4.1) in a number of steps that grows polynomially in the size of the input. Each elementary computational step is implemented in the automaton scheme by two types of *elementary* unitary transition matrices, braidings and changes of basis (fusion matrices), see Sect. 4.2 below.
- Such basic unitaries can be efficiently compiled and approximated by means of universal elementary gates acting on suitable qubit–registers (Sect. 4.3), thus providing efficiency of the whole algorithmic procedure within the *standard model* of the quantum circuit and a positive answer to a generalized version of Problem 4 discussed at the end of Sect. 3 above.

7.5.1   *Q-Spin Network Automata as Quantum Recognizers*

According to [53] a *quantum recognizer* is a particular type of finite–states quantum machine defined as a 5–tuple $\{Q, \mathcal{H}, X, Y, \mathbf{T}(Y|X)\}$, where

1. Q is a set of n basis states, the *internal states*;
2. $\mathcal{H}$ is an n–dimensional Hilbert space and we shall denote by $|\Psi_0\rangle \in \mathcal{H}$ a start state expressed in the given basis;
3. X and $Y = \{\text{accept, reject}, \epsilon\}$ are finite alphabets for input and output symbols respectively (ϵ denotes the null symbol);
4. $\mathbf{T}(Y|X)$ is the subset of $n \times n$ transition matrices of the form $\{T(y|x) = U(x)P(y); \ x \in X, y \in Y\}$, where $U(x)$ is a unitary matrix which determines the state vector evolution and $P(y)$ is a projection operator associated with the output measurement on (suitable complete sets of observables associated with) the upgraded state vector.

In this kind of machine the output alphabet is chosen in such a way that a word w written in the input alphabet X must be either accepted or rejected, while for the null symbol the requirement is $P(\epsilon) \equiv \mathbb{I}$ (the identity matrix). Thus the one–step transition matrices applied to the start state $|\Psi_0\rangle$ can in principle assume the forms

(a) $T(\epsilon|x) = U(x)P(\epsilon) \equiv U(x)\mathbb{I}$ $\forall x \in X$,
(b) $T(\text{accept}|x) = U(x)P(\text{accept})$ $\forall x \in X$ with $P(\text{accept}) \equiv |\Psi_0\rangle\langle\Psi_0|$,
(c) $T(\text{reject}|x) = U(x)P(\text{reject})$ $\forall x \in X$ with $P(\text{reject}) \equiv \mathbb{I} - |\Psi_0\rangle\langle\Psi_0|$,

according to whether no measure is performed (case a)), or the output is 'accept'/'reject', namely cases b)/c) respectively. In what follows this kind of *measure–once* automaton scheme will be adopted, so that there will be one automaton set up (initial state, unitary evolution and measuring) for each instance of the algorithmic problem under study.

The general axioms stated above can be suitably adapted to make this machine able to recognize a language $\mathscr{L}$ endowed with a word–probability distribution $\mathsf{p}(w)$ over the set of words $\{w\} \in \mathscr{L}$. In particular, for any word $w = x_1 x_2 \ldots x_l \in \mathscr{L}$ the recognizer one–step transition matrix elements are required to be of the form $T_{ij}(x_s) = U_{ij}(x_s)$ on reading each individual symbol $x_s \in w$, namely no measurement is performed at the intermediate steps (here i, j run from 1 to n, the dimension of the Hilbert space $\mathcal{H}$). Each $U_{ij}(x_s)$ must satisfy the condition

$$| U_{ij}(x_s) |^2 > 0, \tag{7.23}$$

and the recognizer upgrades the (normalized) initial state to

$$U(w) \, | \Psi_0\rangle \ \dot{=} \ U(x_l) \ldots U(x_1) \, | \Psi_0\rangle . \tag{7.24}$$

Then the machine assigns to the word w the number

$$\mathsf{p}(w) \ = \ | \langle\Psi_0| \, U(w) \, P(\text{accept}) \, U(w)|\Psi_0\rangle | \ \text{ with } \ 0 \le \mathsf{p}(w) \le 1 , \tag{7.25}$$

which corresponds to the probability of accepting the word w as a whole.

More generally, the machine accepts a word w according to an a priori probability distribution $Pr(w)$ with a word–probability threshold $0 \leq \delta \leq 1$ if

$$| Pr(w) - \mathfrak{p}(w) | \leq \delta, \quad \forall w \in \mathcal{L}. \tag{7.26}$$

In what follows the accuracy δ will be set to 0, so that the two probability distributions Pr and $\mathfrak{p}$ coincide.

The families of 'q–deformed' spin network automata (included in the upper box of the spin network simulator scheme in Sect. 1 of this chapter) are modeled on the tensor (monoidal) category associated with the representation ring of $\mathscr{U}_q(sl(2))$ and are defined for each fixed value of the root of unity q. The algebraic setting is actually the same we used to deal with Turaev–Viro quantum invariants in the previous chapter (*cfr.* in particular the quantum initial data of Sect. 4.1). For self consistency we remind here just the basic notions framed within a categorical background, referring for instance to [36] (Sect. 1) for a quite readable account.

The category in question is denoted here by $\mathfrak{R}(\mathscr{U}_q(sl(2)))$, and its basic objects are finite–dimensional $\mathscr{U}_q(sl(2))$–modules $\{V^j\}$ which are are irreducible if and only if the labels $\{j\}$ run over the finite set $\{0, \frac{1}{2}, 1, \frac{3}{2}, \ldots, (r-2)/2\}$. This distinguished family of irreps makes $\mathfrak{R}(\mathscr{U}_q(sl(2)))$ a finitely generated ring and V^j can be characterized by a scalar in the ground ring $\mathbb{C}$, the q–integer $[2j+1]_q$, $[n]_q = (q^{n/2} - q^{-n/2})/(q^{1/2} - q^{1/2})$. The ring structure is made explicit in terms of the direct sum $\oplus$ and tensor product $\otimes$ of irreps, namely

$$V^j \oplus V^k \in \mathfrak{R}(\mathscr{U}_q(sl(2)) \text{ if } j, k \leq (r-2)/2$$

$$V^j \otimes V^k \in \mathfrak{R}(\mathscr{U}_q(sl(2)) \text{ if } j + k \leq (r-2)/2, \tag{7.27}$$

where the ranges of the labels have to be suitably restricted. The analog of the Clebsch–Gordan series, providing the decomposition of the tensor product of two irreps into a (truncated) direct sum of irreps, reads

$$V^{j_1} \otimes V^{j_2} = \bigoplus_{j=|j_1-j_2|}^{\min\{j_1+j_2, \frac{(r-2)}{2}-j_1-j_2\}} V^j. \tag{7.28}$$

The ring $\mathfrak{R}(\mathscr{U}_q(sl(2)))$ is much richer than its 'classical' $SU(2)$–counterpart because it can be endowed with a *quasi–triangular Hopf algebra* structure. Roughly speaking, this means that, besides the standard operators $\oplus$ and $\otimes$ we can also introduce a comultiplication $\Delta : \mathscr{U}_q(sl(2)) \to \mathscr{U}_q(sl(2)) \otimes \mathscr{U}_q(sl(2))$, an antipode map $A : \mathscr{U}_q(sl(2)) \to \mathscr{U}_q(sl(2))$, a counit $\varepsilon : \mathscr{U}_q(sl(2)) \to \mathbb{C}$ and a distinguished invertible element

$$\mathsf{R} \in \mathscr{U}_q(sl(2)) \otimes \mathscr{U}_q(sl(2)) \tag{7.29}$$

called the *R–matrix*.

According to the categorical setting [57], $\mathfrak{R}(\mathcal{U}_q(sl(2)))$ is endowed with two morphisms, $\mathcal{F}$ (associator) and $\mathcal{R}$ (braiding)

$$(\mathfrak{R}(\mathcal{U}_q(sl(2)))\;;\;\mathcal{R}\;;\;\mathcal{F}). \tag{7.30}$$

The associator $\mathcal{F}$ relates different binary bracketing structures in the triple tensor product of modules V, U, W

$$\mathcal{F}\;:\;(V \otimes U) \otimes W \;\rightarrow\; V \otimes (U \otimes W), \tag{7.31}$$

and can be explicitly given in terms of the q–deformed counterpart of the $6j$–symbol (as a unitary matrix, it is also referred to as 'duality' or 'fusion' matrix borrowing the language of conformal field theories). The braiding morphism bears on the existence of the R–matrix (7.29) and is formally given by

$$\mathcal{R}_{V,W}\;:\;V \otimes W \;\rightarrow\; W \otimes V \tag{7.32}$$

$$\text{with}\quad \mathcal{R}_{W,V} \circ \mathcal{R}_{V,W} \neq \mathrm{Id}_{V \otimes W}.$$

Once suitable basis sets are chosen in the finite collection of irreducible spaces $\in \mathfrak{R}(\mathcal{U}_q(sl(2)))$, the unitary morphisms $\mathcal{R}$ and $\mathcal{F}$ can be made explicit (they satisfy the pentagon and two hexagon relations which are in correspondence with the Biedenharn–Elliott identity and the Racah identity for the q-$6j$ respectively, see their explicit expressions in Sect. 4 of the previous chapter).

Comparing the definition of the quantum recognizer, it should be clear that, once given a *unitary* representation of the braid group $\mathbf{B}_n$ complying with the algebraic framework outlined above, the data of the category (7.30) (for any fixed integer $r \geq 3$) are the ingredients of a recognizer designed to accept (reject) the language $\mathcal{L}(\mathbf{B}_n)$ of the braid group. More precisely, the input alphabet is represented by the set of generators $\{e, \sigma_i \sigma_i^{-1}\}$ $(i = 1, 2, \ldots, n - 1)$, a word w is an open braid written in terms of the generators and any $U(w)$ can be decomposed into a sequence of elementary unitaries, each satisfying (7.23), in agreement with the set up of the algebraic structure of the category (7.30). Then the probability of accepting the given word (7.25) is going to correspond to the (square of the modulus of) the colored Jones polynomial of the closure of the open braid associated with the word w. More details on the interplay between automaton calculations and the standard quantum circuit model – framed within an improved presentation of the main results proved in [20]–[23] – are addressed in the following two sections.

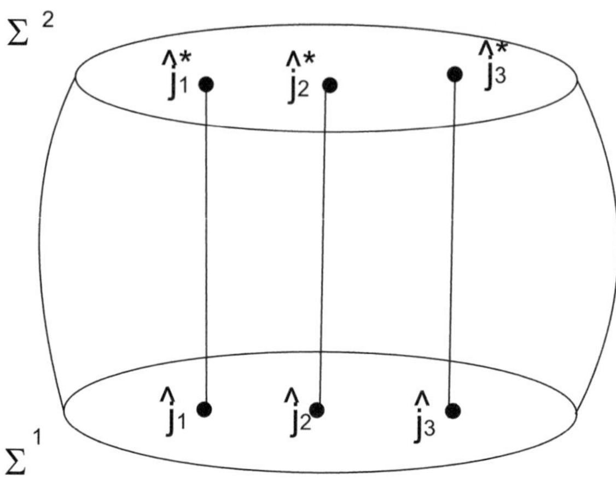

Fig. 7.5 Three Wilson lines intersecting Σ^1 and Σ^2

7.5.2 *Processing Colored Oriented Braids on Spin Network Q-Recognizers*

Recall from Sect. 2.3 of the previous chapter that observables in an $\mathcal{U}_q(sl(2))$ CSW (Chern–Simons–Witten) topological quantum field theory are vacuum expectation values of Wilson loop operators associated with knots (links) embedded in the ambient 3-manifold, *cfr.* the definition in (6.27). As pointed out in the remark at the end of that section the extension of the functionals to the case $\partial M^3 \neq \emptyset$ requires modifications of the classical CS action by suitable Wess–Zumino–Witten (WZW) terms which, once quantized, provide non–trivial quantum degrees of freedom on the boundary surfaces [9]. Consider in particular an oriented compact 3–manifold M^3 with, say, two boundaries Σ^1, Σ^2 (closed oriented surfaces). The field theoretic picture of a knot (link) smoothly embedded in $(M^3, \partial M^3)$ is represented (locally) by a number of Wilson lines carrying $\mathcal{U}_q(sl(2))$ irreps j_l^i intersecting the boundaries at some 'punctures' P_l^i as depicted schematically in Fig. 7.5.

Then the Wilson loop operator is to be associated with the knotted configuration arising after the identification of the two punctured boundary surfaces (endowed with the same topology and with opposite orientations) thus providing a 'vacuum' expectation value whose value is given by the colored Jones polynomial (7.21).

In [34] Kaul gave an explicit construction of these observable in the framework of CSW–WZW field theories (see also [49]). In view of the remarks in Sect. 4.1 above, it is the associated unitary representation of colored oriented braids which is going to be processed in the quantum–automaton calculation for colored Jones polynomials.

The basic ingredients of Kaul construction are oriented geometric braids as depicted in Fig. 7.6, the strands of which are endowed with colorings given by $\mathcal{U}_q(sl(2))$ irreps labels. An n–strand colored oriented braid is defined by two sets of assignments $\hat{j}_i = (j_i, \varepsilon_i)$ with $(i = 1, 2, \ldots, n)$, corresponding to the spin j_i labelling

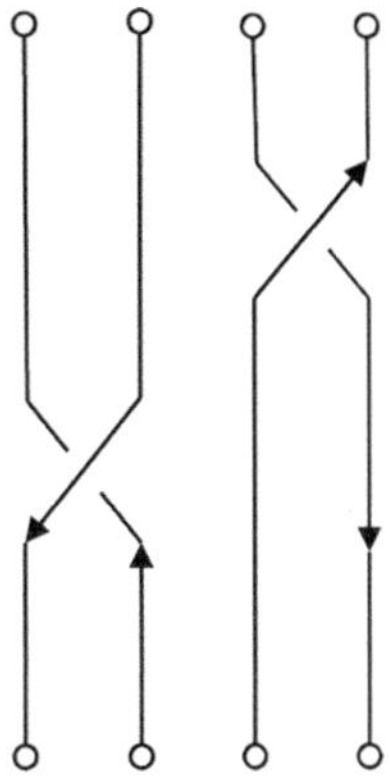

Fig. 7.6 An oriented braid
on four strands

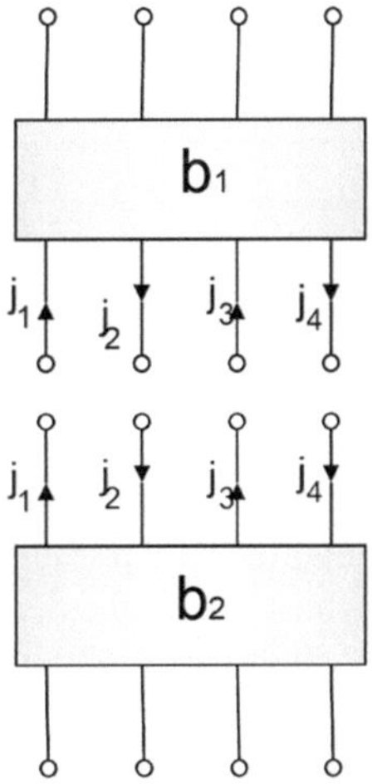

Fig. 7.7 The composition of
two colored oriented braids

the strand and to the orientation ε_i of the strand, with $\varepsilon_i = \pm 1$ (for the strand going into or away, respectively, from a horizontal rod from which the braid issues). The first set of assignments is associated to the upper rod, the second to the lower rod (we use the convention that two braids are composed in the downward direction). The conjugate of $\hat{j}_i$ is defined as $\hat{j}_i^* \equiv (j_i, -\varepsilon_i)$. It follows that the assignments on the lower rod are just a permutation of the conjugates of the assignments on the upper rod. A colored and oriented braid can thus be represented by the symbol

$$\sigma \begin{pmatrix} \hat{j}_1 \, \hat{j}_2 \, \cdots \, \hat{j}_n \\ \hat{l}_1 \, \hat{l}_2 \, \cdots \, \hat{l}_n \end{pmatrix}, \tag{7.33}$$

where $\hat{l}_j = \hat{j}_{\pi(i)}$ for some i and j and a permutation π of $\{1, 2, \ldots n\}$.

The composition of two colored oriented braids is well defined only if the orientations and the colors of the two braids match at the merging points, as shown in Fig. 7.7.

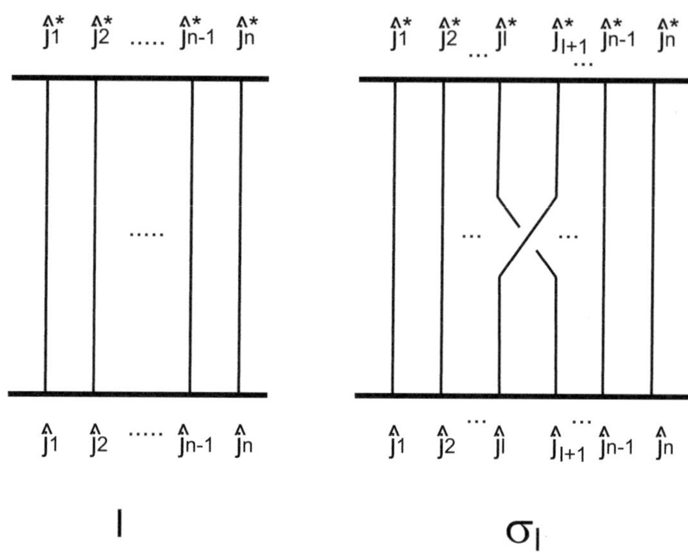

Fig. 7.8 One of the identities and the generator σ_l of the colored braid group

The group (actually a groupoid) of colored oriented braid is generated by a number of identities – one for each assignment of colors and orientations on a topologically trivial braid – and by the generators $\{\sigma_l\}$ $(l = 1, 2, \ldots n - 1)$ (Fig. 7.8) which satisfy the defining relations of $\mathbf{B}_n$ already stated in (7.4).

In order to obtain a link (namely a multicomponent knot) from a colored braid we need to 'close up' the open braid. According to [34] only the plat closure, or *platting*, of a colored braid is to be considered (recall from Fig. 7.4 that there exist two types of closures but in any case the procedure of transforming a knot diagram into a braid is efficiently implementable, *cfr.* Problem 1A of Sect. 2). The plat closure must be obviously performed on a braid with an even number of strands whose colors assignments match as follows (compare also Fig. 7.9)

$$\sigma \begin{pmatrix} \hat{j}_1 \ \hat{j}_1^* \ \hat{j}_2 \ \hat{j}_2^* \ \cdots \ \hat{j}_{2n} \ \hat{j}_{2n}^* \\ \hat{l}_1 \ \hat{l}_1^* \ \hat{l}_2 \ \hat{l}_2^* \ \cdots \ \hat{l}_{2n} \ \hat{l}_{2n}^* \end{pmatrix}. \tag{7.34}$$

The further step in Kaul's construction consists in looking at the embedding of the $2n$–strand braid into a 3–sphere S^3 with two 3–balls removed, giving rise to a 3–manifold with boundaries Σ^1, Σ^2 (topologically 2-spheres S^2 with opposite orientations). The intersections ('punctures') of the braid (7.34) with the boundaries inherit the colorings and orientations from the corresponding strands of the braid.

According to the general set up of topological quantum field theories [9, 55], finite dimensional Hilbert spaces $\mathscr{H}^1 \otimes \mathscr{H}^2$ are associated with the two boundaries Σ^1, Σ^2, and the basis sets in these spaces are the so–called conformal blocks of the boundary WZW conformal field theory at 'level' ℓ, with $2n$ external lines labeled by

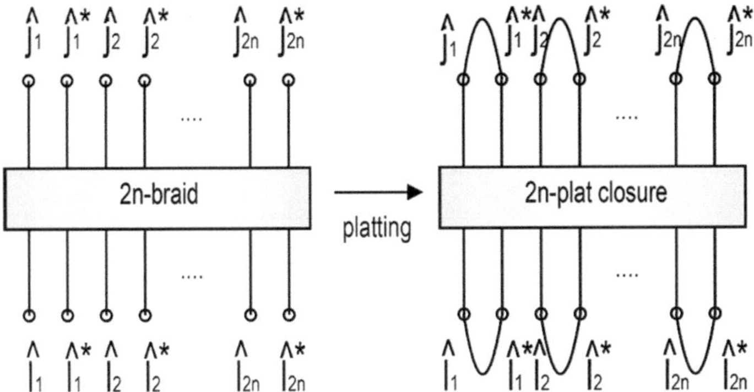

Fig. 7.9 The platting of a colored oriented braid σ on $2n$ strands

Fig. 7.10 A colored braid pattern B on $2m$ strands embedded into the 3-sphere bounded by Σ^1 and Σ^2, two copies of S^2 with opposite orientations

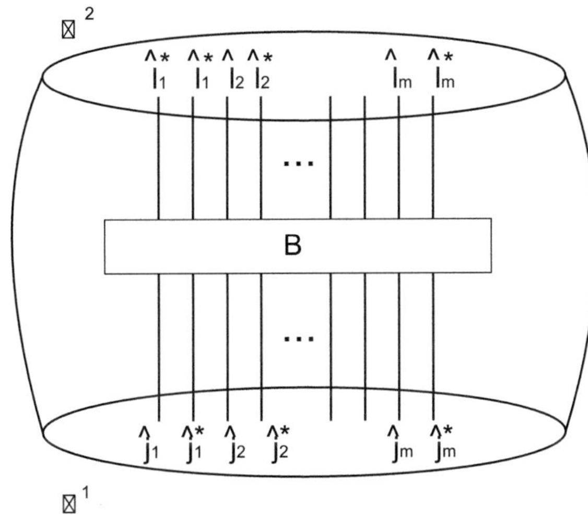

Fig. 7.11 The conformal block of type $\{\mathbf{p};\mathbf{r}\}$ (odd)

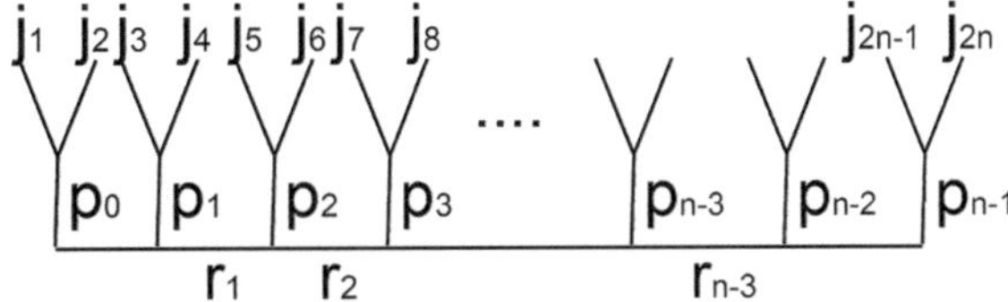

(in principle different) irreps of $SU(2)_q$ (ℓ is related to CSW coupling constant by $\ell = k + 2$, so that from now on we set $q = \exp\{2\pi i/\ell\}$) (Fig. 7.10). Two particular types of conformal block bases are needed to deal with braids the plat closures of which will give rise to colored oriented links, and their combinatorial patterns are shown in Figs. 7.11 and 7.12.

The (orthonormal) basis sets are constructed by taking particular binary coupling schemes of the $2n$ 'incoming' angular momentum variables $j_1, j_2, \ldots, j_{2n}$ which

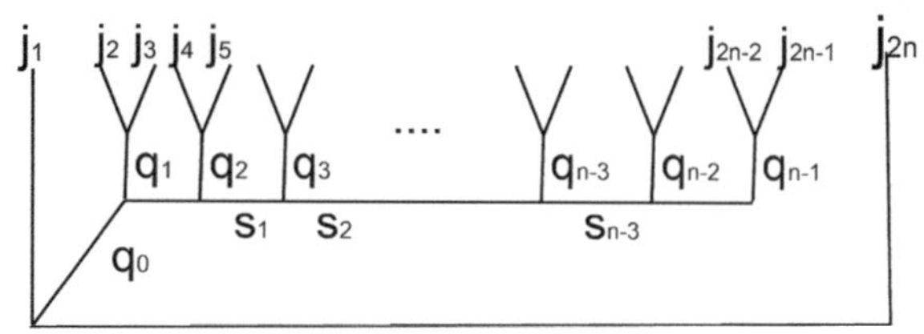

Fig. 7.12 The conformal block of type $\{\mathbf{q};\mathbf{s}\}$ (even)

must sum up to give a spin–0 total singlet state. The rest of the procedure can be carried out on explicitly applying the prescriptions outlined in Sect. 4.1 for the representation ring $\Re\left(\mathscr{U}_q(sl(2))\right)$ Looking first at the combinatorial structure of Fig. 7.11, the most general odd–coupled (orthonormal) basis is consistently labeled as

$$|\,\mathbf{p};\mathbf{r}\,\rangle^{\mathbf{j}}\,,\tag{7.35}$$

where $\mathbf{j}$ is a shorthand notation for the ordered string $j_1, j_2, \ldots j_{2n}$, $\mathbf{p}\equiv p_0, p_1,\ldots, p_{n-1}$ and $\mathbf{r}\equiv r_1, r_2,\ldots,r_{n-3}$. In the even–coupled case depicted in Fig. 7.12 the states of the basis are denoted by

$$|\,\mathbf{q};\mathbf{s}\,\rangle^{\mathbf{j}}\,,\tag{7.36}$$

where $\mathbf{j}$ is the same as before while $\mathbf{q}\equiv q_0, q_1,\ldots,q_{n-1}$ and $\mathbf{s}\equiv s_1, s_2,\ldots,s_{n-3}$.

The conformal blocks (7.35) [respectively (7.36)] are the eigenbasis of the odd [even] unitary braiding matrices $U(\sigma_{2l+1})$ $[U(\sigma_{2l})]$ (the matrix counterparts of the braiding morphism $\mathscr{R}$ in the categorical language)

$$U[\sigma_{2l+1}]|\,\mathbf{p};\mathbf{r}\,\rangle^{(\hat{j}_{2l+1}\hat{j}_{2l+2})}\rangle = \lambda_{p_l}\left(\hat{j}_{2l+1},\hat{j}_{2l+2}\right)|\,\mathbf{p};\mathbf{r}\,\rangle^{(\hat{j}_{2l+2}\hat{j}_{2l+1})}\rangle;\tag{7.37}$$

$$U[\sigma_{2l}]|\,\mathbf{q};\mathbf{s}\,\rangle^{(\hat{j}_{2l}\hat{j}_{2l+1})}\rangle = \lambda_{q_l}\left(\hat{j}_{2l},\hat{j}_{2l+1}\right)|\,\mathbf{q};\mathbf{s}\,\rangle^{(\hat{j}_{2l+1}\hat{j}_{2l})}\rangle\,.$$

The eigenvalues of the braiding matrices depend on the relative orientation of the strands. For right–handed half twists (i.e. over–crossings) their value is

$$\lambda_t\left(\hat{j},\hat{i}\right)\equiv (-)^{j+i-t}\,q^{(c_j+c_i)/2+c_{min(i,j)}-c_t/2},\tag{7.38}$$

for parallel oriented strands, while

$$\lambda_t\left(\hat{j},\hat{i}\right)\equiv (-)^{|j-i|-t}\,q^{-|c_j-c_i|/2+c_t/2},\tag{7.39}$$

if the orientation is anti–parallel. Here c_j is the quadratic Casimir operator $j(j+1)$ for the spin–j representation. Moreover the two basis sets are related to each other by

$$|\mathbf{p};\mathbf{r}\rangle^{\mathbf{j}} = \sum_{(\mathbf{q};\mathbf{s})} A^{(\mathbf{q};\mathbf{s})}_{(\mathbf{p};\mathbf{r})}\begin{bmatrix} j_1 & j_2 \\ j_3 & j_4 \\ \vdots & \vdots \\ j_{2n-1} & j_{2n} \end{bmatrix}|\mathbf{q};\mathbf{s}\rangle^{\mathbf{j}},\tag{7.40}$$

Fig. 7.13 Example of duality transformation (a 6-point correlator) between the two types of conformal blocks

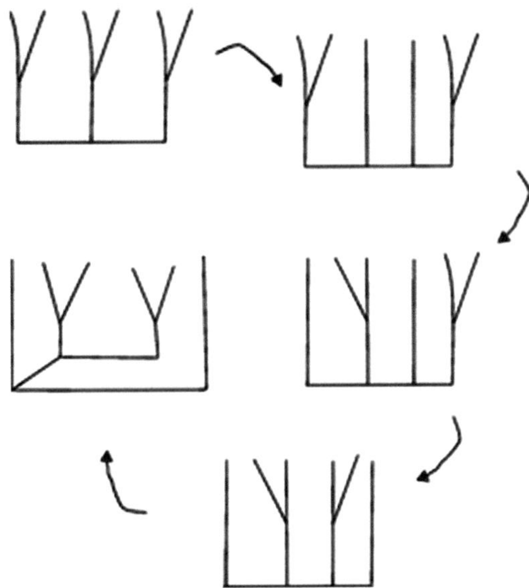

where the symbol $A^{(q;s)}_{(p;r)}[::]$ is a unitary *duality matrix* (or, borrowing the language from $SU(2)$ representation theory, a q-deformed $3nj$ recoupling coefficient). As pointed out several times, any such duality matrix can be decomposed into (sums of) products of a 'basic' duality matrix or q-$6j$ symbol corresponding to the categorical morphism $\mathscr{F}$. An explicit formula will be given in the following section, but a graphical illustration of the decomposition of (7.40) in the case of six primary fields is shown in Fig. 7.13.

As proved in [34], the colored polynomial of a link L, presented as the plat closure of a colored braid

$$\sigma \begin{pmatrix} \hat{j}_1 \ \hat{j}_1^* \ \cdots \ \hat{j}_n \ \hat{j}_n^* \\ \hat{l}_1 \ \hat{l}_1^* \ \cdots \ \hat{l}_n \ \hat{l}_n^* \end{pmatrix}$$

is given by

$$J[L;\mathbf{j};q] = \mathrm{Tr} \prod_{i=1}^{n} [2j_i + 1]_q \ \langle 0;0| \, U\left[\sigma\begin{pmatrix} \hat{j}_1 \ \hat{j}_1^* \ \cdots \ \hat{j}_n \ \hat{j}_n^* \\ \hat{l}_1 \ \hat{l}_1^* \ \cdots \ \hat{l}_n \ \hat{l}_n^* \end{pmatrix}\right] |0;0\rangle^{\mathbf{j}}, \qquad (7.41)$$

where $\mathbf{j} \equiv (j_1, j_2, \ldots, j_{2n})$ and $[2j_i + 1]_q$ is the quantum dimension of the irrep labeled by j_i. Thus $J[L;\mathbf{j};q]$ can be evaluated by taking the trace on the multi–indexes $\mathbf{j}, \mathbf{l}$ of the matrix elements of the above *composite braiding operator $U[\sigma]$* in the Kaul representation, e.g. with respect to the odd–coupled basis, where all the intermediate quantum numbers are constrained to give singlet eigenstates (a similar result would hold true for the even–coupled basis). Moreover, $U[\sigma]$ can be decomposed into a finite sequence of unitary matrices $U[\sigma_{2l+1}]$ (diagonal matrices in the odd–coupled basis adopted in (7.41)) and duality matrices of the type (7.40) to be applied whenever a switch to the even–coupled basis is needed, namely when an even $U[\sigma_{2l}]$ occurs in the decomposition (see [34] for the explicit expression of these matrices).

The construction outlined above can be cast into an effective process of calculation by resorting the concept of 'measure–once' quantum recognizer introduced in Sect. 4.1.

An $\mathscr{A}_q$ quantum recognizer is defined, for a fixed root of unity $q = \exp\{2\pi i/\ell\}$, by the 5–tuple $\{\mathscr{C}_{odd}, \mathscr{H}, B_{2n}, \{\text{accept, reject}, \epsilon\}, U(B_{2n})\}$, where

- $\mathscr{C}_{odd}$ is the odd–coupled conformal block basis of the boundary WZW theory (see (7.35) and Fig. 7.11).
- $\mathscr{H}$ is the ordered tensor product of $2n(2j_i + 1)$–dimensional Hilbert spaces supporting irreps labeled by a fixed set of j_i, with $j_i \leq (\ell-2)/2(i = 1, 2, \ldots, 2n)$.
- B_{2n} is the braid group on $2n$ strands whose generators $\mathfrak{g} \equiv \{\sigma_1, \sigma_2, \ldots, \sigma_{2n-1}\}$ and their inverses represent the input alphabet X (the identity element $e \in B_{2n}$ may play the role of the null symbol ϵ).
- $Y = \{$ accept, reject, ϵ $\}$ is the output alphabet.
- The transition matrices are expressed in terms of $U(B_{2n})$, denoting collectively the Kaul unitary representation matrices, while the projectors $P(y)$ $(y \in Y)$ are defined as in the general case given in Sect. 4.1.

According to the above identifications, we provide the automaton with an input word $w \in \mathbf{B}_{2n}$ of length κ (written in the alphabet $\mathfrak{g}$ by natural composition in $\mathbf{B}_{2n}$)

$$w = \sigma_{\alpha_1}^{\epsilon_1} \sigma_{\alpha_2}^{\epsilon_2} \cdots \sigma_{\alpha_\kappa}^{\epsilon_\kappa} ; \qquad \sigma_{\alpha_i} \in \mathfrak{g}, \epsilon_i = \pm 1 \tag{7.42}$$

and such that the (plat) closure of the $2n$–strand braid w gives the link L to be processed. Dropping for simplicity all the matrix indexes, the unitary evolution of the automaton is achieved on applying the sequence

$$U(w) = U(\sigma_{\alpha_\kappa}^{\epsilon_\kappa}) \, U(\sigma_{\alpha_{\kappa-1}}^{\epsilon_{\kappa-1}}) \cdots U(\sigma_{\alpha_1}^{\epsilon_1}) \tag{7.43}$$

to a start ket $|0; 0\rangle^{\mathbf{j}}$ expressed in the odd–coupled basis (7.35).

Whenever an odd–braiding $\sigma_\alpha = \sigma_{2i+1}$ (or $(\sigma_{2i+1})^{-1}$) occurs, the automaton one–step evolution upgrades the internal state by inserting the eigenvalue of the associated unitary $U(\sigma_{2i+1})$ and the two j–type labels of the internal state are switched , see (7.37). On the other hand, when an even–braiding $\sigma_\beta = \sigma_{2i}$ (or $(\sigma_{2i})^{-1}$) must be implemented, the automaton has to change the parity of the internal state by means of a duality matrix of the type (7.40), so that the effective transformation is given by a product $U(\sigma_\beta) A[::]$. Since any duality transformation can be split into a sequence of basic duality matrices as in Fig. 7.13, we may look at the single q-$6j$ symbol as an 'elementary', one–step evolution of the automaton. To assess this claim it is necessary to estimate the number of q-$6j$'s entering the generic q-$3nj$ recoupling coefficient and this can be done at once by resorting to an upper bound derived within the theory of twist–rotation graphs [12] (see also appendix A of [45]). In the present case this upper bound can be actually expressed in terms of $2n$, the braid index so that the time complexity function (the number of computational steps, recall its definition from (7.8)) for processing a braid–word in B_{2n} of length $|w| \equiv \kappa$ on a quantum recognizer $\mathscr{A}_q(w)$ grows at most as ($\tilde{N} := (2n - 1)$)

$$(2n, \kappa) \rightarrow \mathfrak{f}_{\mathscr{A}_q(w)}(2n, \kappa) \leq \kappa \cdot (\tilde{N} \ln \tilde{N}), \tag{7.44}$$

implying that the automaton model $\mathscr{A}_q$ processes efficiently any such braid w for fixed $q = \exp(2\pi i/(k + 2))$ (compare Sect. 4.3 below for a tighter bound).

Let us finally comment on the 'probability distribution' entering the definition of a quantum automaton that recognizes a language in a probabilistic sense (end of Sect. 4.1). On the basis of the expression of the colored link invariant given in (7.41) and by comparison with the word probability of a quantum recognizer defined in (7.25), it should be quite clear that the probability naturally associated with a link L processed on its own quantum automaton is the square modulus of its colored Jones polynomial (note that the positivity conditions in (7.23) are always satisfied). More details on the type of approximation are given at the end of the following section.

7.5.3 The Qubit Model and Approximate Evaluation of the Colored Jones Invariants

According to the second goal stated at the beginning of Sect. 4, we are going to show how to implement within a standard quantum circuit scheme the Kaul unitary representation of the (colored) braid group.

The starting point consists in providing a procedure for encoding the (automaton) basis states $\in \mathscr{C}_{odd}$ into a 'register' made of a suitable number of qubits, whereby showing how to realize efficiently in such a new representation space (actually $(\mathbb{C}^2)^{\otimes N}$ for a suitable N to be determined) the images of the braid group generators. Note that, according to (7.35), the quantum numbers which label each odd–vector ($\mathbf{j}, \mathbf{p}$ and $\mathbf{r}$) (each ranging over the finite set $\{0, \frac{1}{2}, \ldots, \frac{k}{2}\}$) are $(4n - 3)$, so that a register of $\lceil \log_2 (k + 1) \rceil$ qubits is sufficient to encode each one of them ($\lceil x \rceil$ denotes the smallest integer $\geq x$). The total number of qubits needed to encode for one basis vector is the product of the number of (mutually commuting) observables that specify a state times the number of qubits needed to specify the single quantum number, namely

$$(4n - 3) \times \lceil \log (k + 1) \rceil \tag{7.45}$$

which then grows linearly in the index of the braid group $2n$. An ordering on the quantum register must also be chosen, in order to associate with each block of $\lceil \log (k + 1) \rceil$ qubits a well defined quantum number of the system. The ordering adopted here is shown in Fig. 7.14.

Fig. 7.14 Register for the qubit–representation space

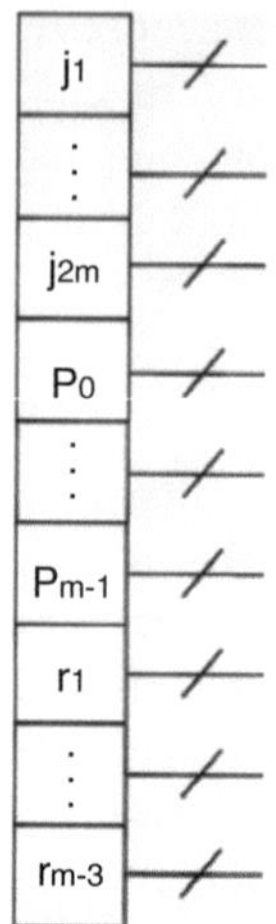

Recall from the previous section that in Kaul representation the odd generators of $\mathbf{B}_{2n}$ are mapped to unitary matrices which are diagonal in the representation space $\mathcal{H}$ of the automaton, while even elements of $\mathbf{B}_{2n}$ require a change of basis given by a duality matrix as formally given in (7.40). Each duality matrix, in turn, can be decomposed into a sequence of elementary duality matrices using the following explicit relation

$$
A_{(\mathbf{p};\mathbf{r})}^{(\mathbf{q};\mathbf{s})}
\begin{bmatrix}
j_1 & j_2 \\
\vdots & \vdots \\
j_{2n-1} & j_{2n}
\end{bmatrix}
\tag{7.46}
$$

$$
= \sum_{t_1,t_2,\ldots,t_{n-2}} \prod_{i=1}^{n-2} \left(A_{p_i}^{t_i}
\begin{bmatrix}
r_{i-1} & j_{2i+1} \\
j_{2i+2} & r_i
\end{bmatrix}
A_{t_i}^{s_{i-1}}
\begin{bmatrix}
t_{i-1} & q_i \\
s_i & j_{2n}
\end{bmatrix}
\right)
$$

$$
\times \prod_{l=0}^{n-2} A_{r_l}^{q_{l+1}}
\begin{bmatrix}
t_l & j_{2l+2} \\
j_{2l+3} & t_{l+1}
\end{bmatrix},
$$

where labels in multiple summations are constrained by triangular inequalities, compare also the (truncated) Clebsch–Gordan series (7.28). Here the q-$6j$ is looked at as a duality matrix $A[::]$, whereby the relation with the standard notation used in Sect. 2.2 of the previous chapter is

$$
A_{j_{23}}^{j_{12}}
\begin{bmatrix}
j_1 & j_2 \\
j_3 & j
\end{bmatrix}
:= (-)^{(j_1+j_2+j_3+j)} \, ([2j_{12}+1]_q [2_{23}+1]_q)^{1/2}
\begin{Bmatrix}
j_1 & j_2 & j_{12} \\
j_3 & j & j_{23}
\end{Bmatrix}_q .
\tag{7.47}
$$

Remark Recall from Sect. 1 of this chapter (in particular the final remark) the prominent role of the $6j$ as the elementary gate in the spin network framework of

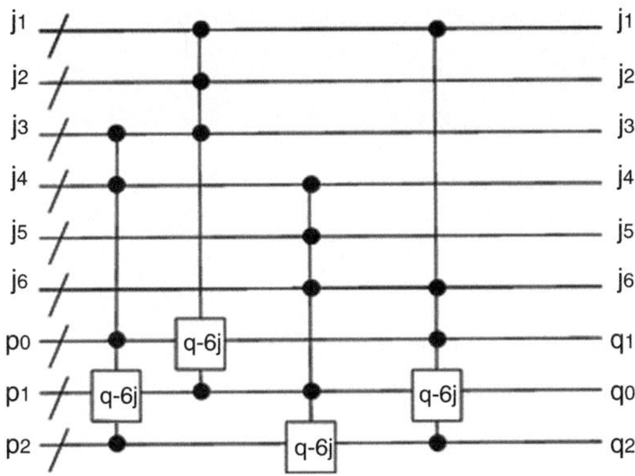

Fig. 7.15 The quantum circuit realization of the 6-point quantum correlator

quantum computation. Any *6j* symbol (either 'classical' or *q*-deformed) with all of its six entries fixed can be efficiently computed classically due to the finiteness of the Racah sum rule (see (6.8) in Sect. 1 of the previous chapter). On the other hand, the *6j* is, say, a $(2d + 1) \times (2d + 1)$ unitary matrix representing a change of basis, as given explicitly in (6.5) of Sect. 1 of the previous chapter, with the matrix indices j_{12}, j_{23} running over the interval of size $2d + 1$. Thus the assignment of a complexity class to the problem of calculating a *6j* consists in checking whether, as *d* increases, the calculation of the *6j* falls into the **BQP** class or not. The standard quantum circuit by means of which such a task can been achieved for a q-*6j* and for each $q =$ a root of unity is given below, while the analog problem involving a classical, $SU(2)$ *6j* seems still open. ∎

Since a q-*6j* is always applied on a $(5\lceil \log_2(k+1) \rceil)$-qubit register, it is a standard result that any such a gate can be efficiently compiled by resorting to a set of universal elementary gates (*cfr.* for instance [26]). Upon increasing the size of the input (by adding new crossings in the braid diagram) more and more q-*6j*'s may appear in the decomposition (7.46) of a *q-3nj*, but a single q-*6j* gate is still

$$2^{5\lceil \log(k+1) \rceil} \tag{7.48}$$

dimensional in the qubit–representation space. The counting of the number of *q-6j*–factors in (7.46) can be carried on by resorting to graphical representations, so that, for instance, the 6-point correlator in Fig. 7.13 of the previous section is associated with the quantum circuit depicted in Fig. 7.15. The dependence on *2n*, index of the braid group, of the number $\mathfrak{N}$ of q-*6j* 's in the decomposition of a given *q-3nj* amounts to

$$\mathfrak{N} = 3n - 5, \tag{7.49}$$

thus improving the rate of growth $\tilde{N} \ln \tilde{N}$ in the complexity function (7.44). Thus, on the basis of (7.45), (7.48), (7.44), and (7.49), Kaul unitary representation at a fixed root of unity *q* can be efficiently compiled within the standard quantum circuit model.

Let us state now the last algorithmic problem of this chapter, namely

Problem 5. How hard is it to approximate the colored Jones polynomial $J(L, \mathbf{j}; q)$ of a link L at a fixed root of unity q ($q \neq$ 2nd, 3rd, 4th, 6th root)?

which is a generalization of Problem 4 of Sect. 3.

The notion of approximation used in these contexts was formalized by Freedman et al. in [7]. Adopting here the terminology of [56], *additive approximation* means the following: given a normalized function $f(x)$, where x denotes an instance of a computational problem in the selected coding, one has an additive approximation of its value for each instance x if one can associate to it a random variable Z_x such that

$$\text{Prob } \{ |f(x) - Z_x| \leq \delta \} \geq \frac{3}{4} , \tag{7.50}$$

for arbitrary $\delta \geq 0$. The time needed to achieve the approximation is going to be polynomial in the size of the problem and in δ^{-1}.

Recall that, if $J(L, \mathbf{c}, q)$ denotes the colored Jones polynomial of a link L made of S components labeled by $\mathbf{c} = c_1, c_2, \ldots c_S$, the natural normalization is provided by the product of the quantum dimensions associated with $c_1, c_2, \ldots c_S$ (this is actually the value of the Jones polynomial for a collection of S unknots with the same labeling). Consider now a (Kaul–type) braid of length $\kappa \in \mathbf{B}_{2n}$ with a coloring $\mathbf{j}$ and fix a real $\delta > 0$. The first step of the procedure consists in generating an ensemble from which one can sample a random variable Z which is an additive approximation of the absolute value of the normalized colored Jones polynomial of the plat closure of the braid, in such a way that the following condition holds true

$$\text{Prob} \{ |J(L, \mathbf{j}, q) - Z| \leq \delta \} \geq \frac{3}{4} . \tag{7.51}$$

In order to check the efficiency of the above approximation, namely proving that it can be achieved in $\mathcal{O}(\text{Poly}(\kappa, 2n, \delta^{-1}))$ steps, two lemmas are needed (see [2, 56]).

Lemma 1 *Given a quantum circuit $\mathcal{U}$ of length $\mathcal{O}(\text{Poly}(n))$, acting on n qubits, and given a pure state $|\Phi\rangle$ which can be prepared in time $\mathcal{O}(\text{Poly}(n))$, then it is possible to make samplings in $\mathcal{O}(\text{Poly}(n))$–time from a set of two random variables σ and τ ($\sigma, \tau \in \{0, 1\}$), in such a way that $< \sigma + i\tau > = \langle \Phi | \mathcal{U} | \Phi \rangle$.*

Lemma 2 *Given a set of random variables $\{r_i\}$ such that*

$$r_i \geq 0 \quad , \quad \lim_{n \to \infty} \frac{1}{n} \sum_{i=1}^{n} r_i = m \quad , \quad \lim_{n \to \infty} \frac{1}{n} \sum_{i=1}^{n} \left(r_i^2 - \langle r_i^2 \rangle \right) = v^2 , \tag{7.52}$$

then $Prob(|n^{-1} \sum_i r_i - m| \geq \delta) \leq 2 \exp(-n\delta^2/(4v))$.

(The first lemma can be easily proved with linear algebra arguments while the second is a modified version of the well known Chernoff bound.)

Summing up, the positive answer to Problem 5 is achieved in two steps:

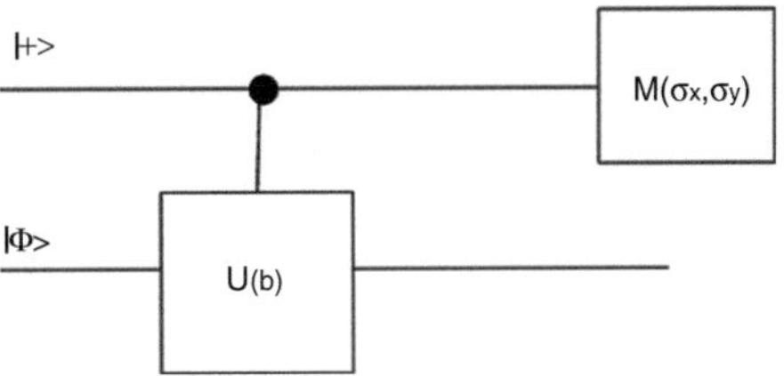

Fig. 7.16 The quantum circuit for the efficient approximation of colored Jones polynomials

- the qubit model for the Kaul representation can be efficiently compiled, as shown in the first part of this section;
- a sampling procedure can be implemented (efficiently) to extract the approximate value of the colored Jones polynomial.

The circuit that realizes the procedure as a whole is sketched in Fig. 7.16, where

$$|\Phi\rangle = |0; 0\rangle^{\mathbf{j}}$$

is the start state in the automaton calculation expressed in the qubit–representation while $U(b)$ stands for a suitable decomposition in elementary gates of Kaul unitary matrix associated with the colored braid b.

$$|+\rangle = \frac{1}{2}(|0\rangle + |1\rangle)$$

is a qubit state and the box in the upper right corner represents generically a measurment **M** on the Pauli matrices. Then the sampling lemma stated above ensures that a series of measurements of σ_x on the first qubit is going to provide the value for $Re(J(L, \mathbf{j}, q))$, while a series of measurements of σ_y provide the value for $Im(J(L, \mathbf{j}, q))$.

7.5.4   Extension to 3–Manifold Quantum Invariants

Recall that every closed, connected and orientable 3–manifold M^3 can be obtained by surgery on an unoriented framed link in the 3–sphere S^3 (*cfr.* Sect. 4.3 of the previous chapter where the Heegard splitting construction on a surgery link has been reviewed). The Reshetikhin–Turaev quantum invariants of 3-manifolds can be obtained as combinations of (colored) polynomial invariants of 'framed' unoriented links in S^3 [36, 40, 50]. (Within the CSW environment the necessity of introducing framings is physically motivated by the requirement of general covariance of the quantized field theory, see e.g. ch. 3 of [25].)

Loosely speaking, a framed oriented link $[L; \mathbf{f}]$ is obtained from a link L–thought of as made of knotted strings– by thickening its strings to get oriented 'ribbons'. If L has S knot components $K_1, K_2, \ldots, K_S$, for each K_s we introduce another closed path

K_s^f oriented in the same way as K_s and lying within an infinitesimal neighborhood of K_s in S^3. The overall topology of the link is not modified, but for each K_s we now have an extra variable $\tau(K_s)$ telling us how many times the oriented ribbon is 'twisted'. Denoting by $\mathbf{f} \doteq \{f_s = n(s), n(s) \in \mathbb{Z}\}$ ($s = 1, 2, \ldots, S$) the framing of the link L, f_s is the self–linking number of the band, or equivalently the linking number $lk(K_s, K_s^f) \equiv \chi(K_s, K_s^f)$ between the knot K_s and its framing curve K_s^f which winds $n(s)$ times in the right–handed direction. The *linking number* (for a link with more than one component knot) is defined for each pair of components (K_i, K_j) as $lk(K_i, K_j) = w(D_L) - w(D_{K_i}) - w(D_{K_j})$, where w represents the writhe of the associated knot diagram defined in (7.3). The twist of the band $\tau(K_s)$ is not independent from $lk(K_s, K_s^f)$, and the simplest choice we can made is to set $\tau(K_s) = w(K_s)$, where $w(K_s)$ is the writhe of the s–th component. The type of framing usually adopted within the CSW environment is the so–called 'vertical' framing.

Then, denoting by M_L^3 a closed, connected and oriented 3–manifold obtained by surgery in S^3 along an unoriented colored framed link $[L; \mathbf{f}, \mathbf{j}]$ with S link components, the

$$\mathscr{I}\,[M_L^3; \mathbf{f}; q] = \alpha^{-\sigma[L;\mathbf{f}]} \sum_{\{\mathbf{j}\}} \mu_{j_1}\, \mu_{j_2} \cdots \mu_{j_S}\, J\,[L; \mathbf{f}, \mathbf{j}; q] \tag{7.53}$$

is a topological invariant of the 3–manifold endowed with the framing assignment $\mathbf{f}$ for any fixed root of unity $q = \exp\{2\pi i/(k+2)\}$. In the above expression

$$\alpha \equiv \exp \frac{3\pi ik}{4(k+2)}; \quad \mu_j = \sqrt{\frac{2}{k+2}} \sin \frac{\pi(2j+1)}{k+2},$$

$\mathbf{j} \equiv (j_1, j_2, \ldots, j_S)$ run over $\{0, \frac{1}{2}, \ldots, \frac{k}{2}\}$ and the summation is performed over all admissible colorings. $J\,[L; \mathbf{f}, \mathbf{j}; q]$ is the 'unoriented' counterpart of the polynomial for the link L with coloring assignment $\mathbf{j}$ on its components and the summation is performed over all admissible colorings. Finally, $\sigma[L; \mathbf{f}]$ denotes the signature of the linking matrix $\chi\,[L; \mathbf{f}]$ namely the difference between the number of positive and negative eigenvalues of the symmetric matrix

$$\chi\,[L; \mathbf{f}] = \begin{pmatrix} n_1 & \chi(K_1,K_2) & \chi(K_1,K_3) & \cdots & \chi(K_1,K_S) \\ \chi(K_2,K_1) & n_2 & \chi(K_2,K_3) & \cdots & \chi(K_2,K_S) \\ \vdots & \cdots & \cdots & \cdots & \vdots \\ \chi(K_S,K_1) & \cdots & \cdots & \cdots & n_S \end{pmatrix} \tag{7.54}$$

where $\chi(K_i, K_j)$ is the linking number between the component knots K_i and K_j. (We refer the reader to, e.g, [36] for the proof of the invariance of (7.53) with respect to the proper set of Kirby moves.)

Note that the plat presentations of the colored links used to process Kaul representation in Sect. 4.2 (see (7.34) and Fig. 7.9) is different from the presentation given in the definition above, even though we keep on using the same notation $\mathbf{j}$

for the colorings. In the latter case a coloring is assigned to each of the S link components, while in the former we label the $2n$ strands of the associated braid with n colors. However, what really matters is the fact that both presentations of the link L give rise to the same invariant, as could be proved by exploiting the algebraic identities in the $\mathcal{U}_q(sl(2))$–representation ring. From the computational viewpoint this twofold choice does not matter as well because we might efficiently implement the transformation from a given link diagram to any associated closed braid on a classical computer (*cfr.* Problem 1A in section 2). Similarly, the overall dependence on the framing in the factor $\alpha^{-\sigma[L;\,\mathbf{f}]}$ of (7.53) can be easily taken into account.

In the paper [23] an efficient quantum algorithm for approximating the 3-manifold invariant $\mathcal{I}\,[M_L^3\,;\,\mathbf{f};\,q]$ has been provided for the first time. The starting point relies on a modification of the unitary representation of Sect. 4.2 provided by Kaul himself in [35], where in the CSW-WZW environment the vacuum expectation value of an unoriented $2n$–strands braid – the plat closure of which gives the surgery link – is worked out explicitly. In particular, a modification of the action of the braiding matrices on odd (even) conformal blocks (7.37) is needed. The analysis of the upgraded algorithmic procedure then follows the general, two–level approach stated at the beginning of Sect. 4 and carried out in Sects. 4.2 and 4.3 (automaton calculation plus quantum circuit–representation). Without entering into other details it suffices to conclude that the time complexity function for the evaluation of an additive approximation of the colored 3-manifold invariant (7.53) for a given framing $\mathbf{f}$ of the surgery link (with κ crossings) at a fixed root of unity grows as $\mathcal{O}(\mathrm{Poly}(\kappa, 2n, \delta^{-1}))$.

Remark In Sect. 4 of the previous chapter both state sum invariants of 3-manifolds (with or without boundary) and observables corresponding to colored fat graphs have been shown to rely on Turaev–Viro quantum initial data stated in Sect. 4.1. Actually it has been observed that invariants of colored (oriented) links are essentially the same for both presentations of the (oriented) closed ambient manifold M^3 (here we have been using Heegard splitting presentations along surgery links and there we resorted to colored triangulations). However, in Sect. 4.3 of the previous chapter the following particular form of the Turaev–Viro invariant for a handlebody N has been written as

$$< N, \varnothing > \equiv < N > = \omega^{-2} \sum_{\substack{J \in I^g \\ H \in I^g}} \prod_{i=1}^{g} \omega_{j_i}^2 \prod_{k=1}^{g} \omega_{h_k}^2 < \phi_H \,|\, \varnothing \,|\, \psi_J > . \qquad (7.55)$$

Without repeating here the details of this construction, $\{\phi_H\}$ and $\{\psi_J\}$ represent colored generating loops associated with the surgery link and its Heegard surface and ω's factors are quantum dimensions. The amplitudes $< \phi_H \,|\, \varnothing \,|\, \psi_J >$ can be evaluated on the basis of the rules given in Sect. 4.2.1 and contain (products of) q-$6j$ symbols, quantum dimensions and powers of the root of unity q. These combinations can be shown to amount to $3nj$ symbols – of definite types and for suitable choices of n – at least in the case of lens spaces (see the references given in the previous chapter). On the other hand, it is quite clear that the Turaev–Viro invariant for a closed 3-manifolds (which can be non orientable) is itself a combination of the same ingredients as before. Then the results of the present chapter on the efficiency of the approximate evaluation of (Jones and Reshetikhin–Turaev) quantum invariants are extended in a straightforward way to all of the state sum models of the Turaev–Viro type for each choice of the underlying admissible colorings. The goal is achieved with no need of selecting explicitly a representation of the (colored) braid group, but simply by resorting to the argument used above to decompose a general duality matrix $A[::]$ (Sect. 4.2) and to process in a qubit–representation a single q-$6j$ (Sect. 4.3). ∎

7.6 Quantum Computing and Quantized Geometries: An Outlook

Summarizing the content of this chapter, wit has been shown that all the significant quantities – partition functions and observables – in an $SU(2)$ quantum CSW theory can be efficiently approximated at finite values of the coupling constant k. The intrinsic field–theoretic solvability of CSW theory is thus reflected in its computability on a quantum computer.

Actually the crucial feature of possessing only global, purely topological degrees of freedom makes quantum CSW theory likely to be simulated within a computational scheme based on a discrete space of states [13] and able to implement efficiently polylocal braiding operations. The same conclusions hold true for all ab initio discretized colored state sums of the Turaev–Viro type encountered in the previous chapter, as noted in the remark above. Thus on the basis of Witten's celebrated results on relationships between gauge theories and gravity in $(2 + 1)$ dimensions [54] exploited many times through this book, we are actually computing (abstractly) features of quantized geometries but at the same time physical devices able to realize anyonic–type quantum computing [10, 38] should behave just like quantum $3d$ spacetimes.[2]

The idea that many (if not all) aspects of our reality may be thought of as 'outputs' of some kind of information processing is both appealing and intriguing [41]. In this connection the role of information theory and its tools are so enhanced that might become a unifying paradigm. Thus we may ask whether

---

[2]It has been recently proposed to describe topological phases of matter and anyonic–type vertex operators within a Turaev–Viro background [30, 31]. Such a proposal seems quite promising but at present only a few features of this approach have been worked out, so that we have not included it in this monography.

(i) an abstract universal model of computation, able to simulate any discrete quantum system including solvable topological field theories, must exists by its own; or

(ii) a (suitably chosen) quantum system is by itself a computing machine whose internal evolution can reproduce its own dynamics and, possibly, the proper dynamics of similar physical systems.

The classical version of hypothesis **(i)** is usually taken for granted as far as, on the one hand, a (probabilistic) Turing machine is capable of simulating the evolution of any classical system within a given accuracy, and, on the other, all concrete, finite–size realizations of the abstract machine obey the laws of classical physics. The feasibility of hypothesis **(ii)** depends heavily on which system is chosen as a simulator and which types of boundary or initial conditions must be imposed to reproduce the dynamical behavior of observed physical systems. Moreover, the concept of 'efficient' processing of information seems difficult to be handled without an abstract reference model of computation. With the previous remarks in mind, assumption **(i)** seems quite reliable when the reference model of computation is the spin network simulator (and associated q-automata), given its abstract definition as the discretized counterpart of the quantum field computer [14]. However spin networks – thought of as real ensembles of interacting angular momenta and spins of atomic or molecular systems – can play as well the role of the reference quantum system in statement **(ii)** (not to mention possible forthcoming experimental implementations of anyonic systems). Thus, much in the sense of **(ii)**, spin networks may act – under suitable constraints – as computing machines able to process information encoded into quantum spins to output 'quantized' 3-geometries obeying Einstein equations in the classical limit (compare the Ponzano–Regge model [47] in Sect. 2.1 of the previous chapter).

References

1. Anshel, I., Anshel, M., Goldfeld, D.: An algebraic method for public-key cryptography. Math. Res. Lett. **6**, 287–291 (1999)
2. Aharonov, D., Jones, V., Landau, Z.: A polynomial quantum algorithm for approximating the Jones polynomial. In: Kleinberg, J.M. (ed.) Proceedings of STOC 2006: 38th ACM Symposium on Theory of Computing, pp. 427–436 (2006)
3. Biedenharn, L.C., Louck, J.D.: Angular momentum in quantum physics: theory and applications. In: Rota G.-C. (ed.) Encyclopedia of Mathematics and Its Applications, vol. 8. Addison–Wesley Publications Co, Reading (1981)
4. Biedenharn, L.C., Louck, J.D.: The Racah–Wigner Algebra in Quantum Theory. In: Rota G.-C. (ed.) Encyclopedia of Mathematics and Its Applications, vol.9. Addison–Wesley Publications Co, Reading (1981)
5. Birman, J.S.: Braids, Links, and Mapping Class Groups. Princeton University Press, Princeton (1974)
6. Birman, J.S., Brendle, T.E.: Braids: a survey. In: Menasco, W., Thistlethwaite, M., (eds.) Handbook of Knot Theory. Elsevier, Amsterdam (2005) (eprintarXiv: math.GT/0409205)

7. Bordewich, M., Freedman, M., Lovász, L., Welsh, D.: Approximate counting and quantum computation. Combin. Probab. Comput. **14**, 737–754 (2005)
8. Carfora, M., Marzuoli, A., Rasetti, M.: Quantum tetrahedra. J. Phys. Chem. A **113**, 15376–15383 (2009)
9. Carlip, S.: Quantum Gravity in 2+1 Dimensions. Cambridge University Press, Cambridge (1998)
10. Das Sarma, S., Freedman, M., Nayak, C., Simon, S.H., Stern, A.: Non–Abelian anyons and topological quantum computation. Rev. Mod. Phys. **80**, 1083–1159 (2008)
11. Drinfel'd, V.G.: Quantum groups. In: Proceedings of ICM 1986 American Mathematical Society, Providence, pp. 798–820 (1987)
12. Fack, V., Lievens, S., Van der Jeugt, J.: On the diameter of the rotation graph of binary coupling trees. Discr. Math. **245**, 1–18 (2002)
13. Feynman, R.: Simulating physics with computers. Int. J. Theor. Phys. **21**, 467–488 (1982)
14. Freedman, M.H.: P/NP and the quantum field computer. Proc. Natl. Acad. Sci. U.S.A **95**, 98–101 (1998)
15. Freedman, M.H., Kitaev, A., Larsen, M., Wang, Z.: Topological quantum computation. Bull. Am. Math. Soc. **40**, 31–38 (2002)
16. Freedman, H.M., Kitaev, A., Wang, Z.: Simulation of topological field theories by quantum computer. Commun. Math. Phys. **227**, 587–603 (2002)
17. Freedman, H.M., Larsen, M., Wang, Z.: A modular functor which is universal for quantum computation. Commun. Math. Phys. **227**, 605–622 (2002)
18. Freyd, P., Yetter, D., Hoste, J., Lickorish, W., Millett, K., Ocneanu, A.: A new polynomial invariant of knots and links. Bull. Am. Math. Soc. **12**, 239–246 (1985)
19. Garey, M.R., Johnson, D.S.: Computers and intractability, a guide to the theory of NP–completeness. Freeman and Co., New York (1979)
20. Garnerone, S., Marzuoli, A., Rasetti, M.: Quantum computation of universal link invariants. Open Sys. Inf. Dyn. **13**, 373–382 (2006)
21. Garnerone, S., Marzuoli, A., Rasetti, M.: Quantum kinitting. Laser Phys. **16**, 1582–1594 (2006)
22. Garnerone, S., Marzuoli, A., Rasetti, M.: Quantum automata, braid group and link polynomials. Quant. Inform. Comput. **7**, 479–503 (2007)
23. Garnerone, S., Marzuoli, A., Rasetti, M.: Efficient quantum processing of 3–manifolds topological invariants. Adv. Theor. Math. Phys. **13**, 1–52 (2009)
24. Gomez, C., Ruiz–Altaba, M., Sierra, G.: Quantum Group in Two–Dimensional Physics. Cambridge University Press, Cambridge (1996)
25. Guadagnini, E.: The Link Invariants of the Chern–Simons Field Theory. W. de Gruyter, Berlin/Boston (1993)
26. Harrow, A.W., Recht, B., Chuang, I.L.: Efficient discrete approximations of quantum gates. J. Math. Phys. **43**, 4445–4451 (2002)
27. Jaeger, F., Vertigan, D.L., Welsh, D.J.A.: On the computational complexity of the Jones and Tutte polynomials. Math. Proc. Camb. Phil. Soc. **108**, 35–53 (1990)
28. Jones, V.F.R.: A polynomial invariant for knots via von Neumann algebras. Bull. Am. Math. Soc. **12**, 103–111 (1985)
29. Jones, V.F.R.: Hecke algebra representations of braid groups and link polynomials. Ann. Math. **126**, 335–388 (1987)
30. Kádár, Z., Marzuoli, A., Rasetti, M.: Braiding and entanglement in spin networks: a combinatorial approach to topological phases. Int. J. Quant. Inf. **7**(Suppl.), 195–203 (2009)
31. Kádár, Z., Marzuoli, A., Rasetti, M.: Microscopic description of 2D topological phases, duality and 3D state sums. Adv. Math. Phys. **2010**, 671039 (2010)
32. Kauffman, L.: Knots and Physics. World Scientific, Singapore (2001)
33. Kauffman, L.H., Lomonaco, J.S.: Topology and quantum computing. In: Entanglement and Decoherence. Lecture Notes in Physics, vol. 768, pp. 87–156. Springer, Berlin (2009)
34. Kaul, R.K.: Chern–Simons theory, colored oriented braids and link invariants. Commun. Math. Phys. **162**, 289–319 (1994)

35. Kaul, R.K.: Chern–Simons theory, knot invariants, vertex models and three–manifold invariants. In: Kaul, R.K. et al. (eds.) Horizons in World Physics, vol. 227. Nova Science Publications (1999)
36. Kirby, R., Melvin, P.: The 3-manifold invariant of Witten and Reshetikhin–Turaev for $sl(2, \mathbb{C})$. Invent. Math. **105**, 437–545 (1991)
37. Kirillov, A.N., Reshetikhin, N.Y.: In: Kac, V.G. (ed.) Infinite Dimensional Lie Algebras and Groups. Advanced Series in Mathematical Physics, vol. 7, pp. 285–339. World Scientific, Singapore (1988)
38. Kitaev, A.: Anyons in an exactly solved model and beyond. Ann. Phys. **321**, 2–111 (2006)
39. Kohno, T.: Hecke Algebra Representations of Braid Groups and Classical Yang-Baxter Equations. Advanced Studies in Pure Mathematics, vol. 16, pp. 255–269. Academic Press, Boston (1988)
40. Lickorish, W.B.R.: An Introduction to Knot Theory. Springer, New York (1997)
41. Lloyd, S.: A theory of quantum gravity based on quantum computation. eprint quant-ph/0501135v8 (2005)
42. Magnus, W., Karrass, A., Solitar, D.: Combinatorial Group Theory, 2nd edn. Dover Publications, New York (1976)
43. Marzuoli, A., Palumbo, G.: Post–quantum cryptography from mutant prime knots. arXiv: 1010.2055v1 [math-ph]
44. Marzuoli, A., Rasetti, M.: Spin network quantum simulator. Phys. Lett. A **306**, 79–87 (2002)
45. Marzuoli, A., Rasetti, M.: Computing spin networks. Ann. Phys. **318**, 345–407 (2005)
46. Nielsen, M.A., Chuang, I.L.: Quantum Computation and Quantum Information. Cambridge University Press, Cambridge (2000)
47. Ponzano, G., Regge, T.: Semiclassical limit of Racah coefficients. In: Bloch, F. et al. (eds.) Spectroscopic and Group Theoretical Methods in Physics, pp. 1–58. North–Holland, Amsterdam (1968)
48. Preskill, J.: Topological quantum computing for beginners. http://online.kitp.ucsb.edu.online/exotic_c04/preskill/ (2004)
49. Ramadevi, P., Govindarajan, T.R., Kaul, R.K.: Knot invariants from rational conformal field theories. Nucl. Phys. B **422**, 291–306 (1994)
50. Reshetikhin, N., Turaev, V.G.: Invariants of 3-manifolds via link polynomials and quantum groups. Invent. Math. **103**, 547–597 (1991)
51. Rolfsen, D.: Knots and Links. Publish or Perish Berkeley (1976)
52. Varshalovich, D.A., Moskalev, A.N., Khersonskii, V.K.: Quantum Theory of Angular Momentum. World Scientific, Singapore/Philadelphia (1988)
53. Wiesner, K., Crutchfield, J.P. Computation in finitary stocastic and quantum processes. Physica D **237**, 1175–1195 (2008)
54. Witten, E.: (2+1)-dimensional gravity as an exactly soluble system. Nucl. Phys. B **311**, 49–78 (1988/1989)
55. Witten, E.: Quantum field theory and the Jones polynomial. Commun. Math. Phys. **121**, 351–399 (1989)
56. Wocjan, P., Yard, J.: The Jones polynomial: quantum algorithms and applications in quantum complexity theory. Quant. Inform. Comput. **8**, 147–180 (2008)
57. Joyal, A., Street, R.: Braided tensor categories. Adv. Math. **102**, 20–78 (1993)
58. Yutsis, A.P., Levinson, I.B., Vanagas, V.V.: The Mathematical Apparatus of the Theory of Angular Momentum. Israel Program for Scientific Translations Ltd (1962)

Appendix A
Riemannian Geometry

In this appendix besides introducing notation and basic definitions we recollect some of the more technical results in Riemannian geometry we exploited in these lecture notes. We freely refer to the excellent exposition of the theory provided by [4, 8], and to [27] for further details.

A.1   Notation

In what follows, V^n will always denote a C^∞ compact n–dimensional manifold, $(n \geq 3)$. Unless otherwise stated, the manifold V^n will be supposed oriented, connected, and without boundary. Let (E, V^n, π) be a smooth vector bundle over V^n with projection map $\pi : E \longrightarrow V^n$. When no confusion arises the space of smooth sections $\{s \in C^\infty(V^n, E) \,|\, \pi \circ s = \mathrm{id}_V^n\}$ will simply be denoted by $C^\infty(V^n, E)$ instead of the standard $\Gamma(E)$ (to avoid confusion (Fig. A.1) with the already too numerous Γs appearing in Riemannian geometry: Christoffel symbols, harmonic coordinates, ...). If $U \subset V^n$ is an open set, we let $\{x^i\}_{i=1}^n$ be local coordinates for the points $p \in U$. The vector fields $\{\partial_i := \partial/\partial x^i\}_{i=1}^n \in C^\infty(U, TV^n)$ provide the local (positively oriented) coordinate basis for $T_p V^n$, $p \in U$. The corresponding dual basis for the cotangent spaces $T_p^* V^n$, is provided by the set of 1–forms $\{dx^i\}_{i=1}^n \in C^\infty(U, T^*V^n)$. In particular, the metric tensor $g \in C^\infty(V^n, \otimes_S^2 T^*V^n)$, where $C^\infty(V^n, \otimes_S^2 T^*V^n)$ denotes the set of smooth symmetric bilinear forms over V^n, has the local coordinates representation $g = g_{ik} dx^i \otimes dx^k$, where $g_{ik} := g(\partial_i, \partial_k)$, and the Einstein summation convention is in effect. The Riemannian volume form is $d\mu_g = \sqrt{\det g}\, dx^1 \wedge \ldots \wedge dx^n$, and we denote by $\mathrm{Vol}_g(V^n) := \int_{V^n} d\mu_g$ the volume of (the compact) (V^n, g). The distance between points $x, y \in V^n$, characterizing (V^n, g) as a metric space, is defined by setting

© Springer International Publishing AG 2017

M. Carfora, A. Marzuoli, *Quantum Triangulations*, Lecture Notes in Physics 942,
https://doi.org/10.1007/978-3-319-67937-2

Fig. A.1 …Riemannian geometry may have unexpected sides…and must be used responsibly!

$$d_g(x, y) := \inf_{\gamma \in \mathscr{C}(x,y)} \int_0^1 \sqrt{g(\dot{\gamma}(t), \dot{\gamma}(t))}\, dt, \tag{A.1}$$

where $\mathscr{C}(x, y)$ is the space of (piecewise C^1) curves $\gamma : [0, 1] \to V^n$ such that $\gamma(0) = x$ and $\gamma(1) = y$. If

$$|U|_g^2 := g^{i_1 h_1} \cdots g^{i_r h_r}\, U^{k_1 \ldots k_q}_{i_1 \ldots i_p}\, U^{j_1 \ldots j_q}_{h_i \ldots h_r}\, g_{k_1 j_1} \cdots g_{k_q j_q} \tag{A.2}$$

denotes the pointwise (squared) g–norm in the tensor bundle $T^{(r,q)} V^n := \otimes^r T^* V^n \otimes^q T V^n$, then we can endow the space of smooth (r, q)–tensor fields $C^\infty(V^n, T^{(r,q)} V^n)$ on V^n with the corresponding $L^p(d\mu_g)$ and $\mathscr{H}^{p,s}(d\mu_g)$ Sobolev norms

$$\|U\|_{L^p(d\mu_g)} \doteq \left[\int_{V^n} |U|_g^p \, d\mu_g \right]^{\frac{1}{p}}, \quad 1 \le p < \infty, \tag{A.3}$$

$$\|U\|_{\mathscr{H}^{p,s}(d\mu_g)} \doteq \sum_{i=0}^{s} \left[\int_{V^n} \left| \nabla^{(i)} U \right|_g^p \, d\mu_g \right]^{\frac{1}{p}}, \tag{A.4}$$

where s is an integer ≥ 0, and $\nabla^{(i)} U$ denotes the ith covariant derivative of the tensor field U. The completions of $C^\infty(V^n, T^{(r,q)} V^n)$ in these norms, define

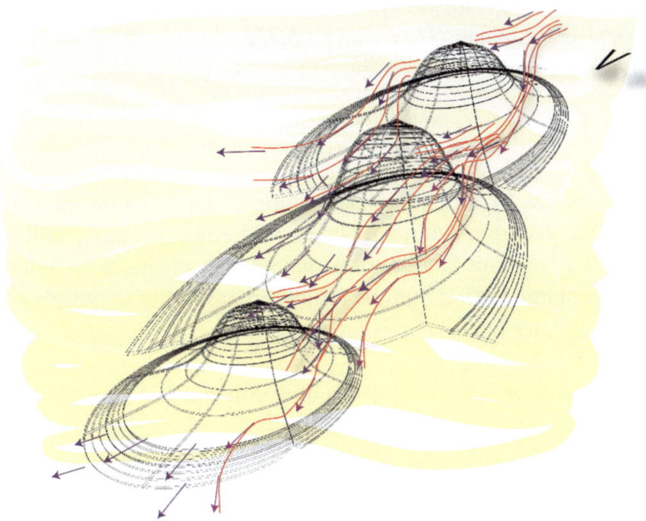

Fig. A.2 The group of diffeomorphisms

the Sobolev spaces of sections $\mathscr{H}^{p,s}(V^n, T^{(r,q)}V^n)$. In particular, for $p = 2$, we denote by $\mathscr{H}^s(V^n, T^{(r,q)}V^n)$ the Hilbert space $\mathscr{H}^{2,s}(V^n, T^{(r,q)}V^n)$. According to Sobolev embedding theorem, if $k \geq 0$ and $s > \frac{n}{2} + k$, then $\mathscr{H}^s(V^n, T^{(r,q)}V^n) \subset C^k(V^n, T^{(r,q)}V^n)$. Similarly, if we denote by $\mathscr{H}^s(V^n, V^n)$ the Sobolev space of maps $\phi : V^n \longrightarrow V^n$, then $\mathscr{H}^s(V^n, V^n) \subset C^k(V^n, V^n)$ is a dense continuous inclusion for $s > (n/2) + k$.

Let

$$\mathrm{Diff}^k(V^n) := \left\{ \phi \in C^k(V^n, V^n) \,\middle|\, \phi \text{ is a bijection and } \phi^{-1} \in C^k(V^n, V^n) \right\} \quad (A.5)$$

be the group of C^k diffeomorphisms of V^n, endowed with the compact–open C^k topology. The group of smooth diffeomorphisms $\mathrm{Diff}(V^n)$ is defined by the sequence of natural inclusions (Fig. A.2)

$$\mathrm{Diff}^1(V^n) \supset \mathrm{Diff}^2(V^n) \supset \ldots \mathrm{Diff}^k(V^n) \supset \ldots , \quad (A.6)$$

according to

$$\mathrm{Diff}(V^n) = \cap_{k=1}^{\infty} \mathrm{Diff}^k(V^n). \quad (A.7)$$

Recall that under a diffeomorphism $\phi : V^n \longrightarrow V^n$ the pull back from $(\phi(V^n), g)$ to (V^n, ϕ^*g) is given for $p \in V^n$ by

$$\left(\phi^*g\right)(V, W)\big|_p := g\left(\phi_*V, \phi_*W\right)\big|_{\phi(p)}, \quad (A.8)$$

where $\phi_* V$ denotes the push forward of V defined by the tangent map $\phi_* : T_p V^n \longrightarrow T_{\phi(p)} V^n$. In local coordinates (U, x^i) and $(\phi(U), y^a)$, $U \ni x^i \longmapsto \phi^a(x^i) := y^a(\phi(p)) \in \phi(U)$, we write

$$\left(\phi^* g\right)_{ik}(p) = \frac{\partial \phi^a}{\partial x^i}\frac{\partial \phi^b}{\partial x^k} g_{ab}(\phi(p)). \tag{A.9}$$

It is important to stress that the pull–back action of $\mathrm{Diff}(V^n)$ on the metric g is a right action since under composition of any two diffeomorphisms ϕ and ψ we have $(\psi \circ \phi)^* g = \phi^*(\psi^* g)$. This latter remark plays a role when, for technical reasons, it is necessary to enlarge $\mathrm{Diff}(V^n)$ to the (Hilbert manifold of the) diffeomorphisms of Sobolev class $\mathscr{H}^s$, for $s > (n/2) + 1$,

$$\mathscr{D}^s(V^n) := \left\{\phi \in \mathscr{H}^s(V^n, V^n) \mid \phi \text{ is a bijection and } \phi^{-1} \in \mathscr{H}^s(V^n, V^n)\right\}. \tag{A.10}$$

In such a framework the group of smooth diffeomorphisms can be recovered from $\mathscr{D}^s(V^n)$ by characterizing the smooth topology on $\mathrm{Diff}(V^n)$ as the limit of the topologies of $\mathscr{D}^s(V^n)$, by viewing $\mathrm{Diff}(V^n)$ as an Inverse Limit Hilbert group in the sense of Omori [26], i.e. $\mathrm{Diff}(V^n) = \cap_{s>n/2}\mathscr{D}^s(V^n)$. Since right multiplication is smooth, but left multiplication is not, $\mathscr{D}^s(V^n)$ is not "an infinite–dimensional" Lie group but only a topological group [11, 21].

A.2   Curvature and Scaling

Besides diffeomorphisms, the metric g is naturally acted upon also by overall rescalings according to for $\lambda \in \mathbb{R}_{>0}$ (Fig. A.3)

$$g \longmapsto \lambda g \tag{A.11}$$

Fig. A.3 Besides diffeomorphisms, the metric g is naturally acted upon also by overall rescalings

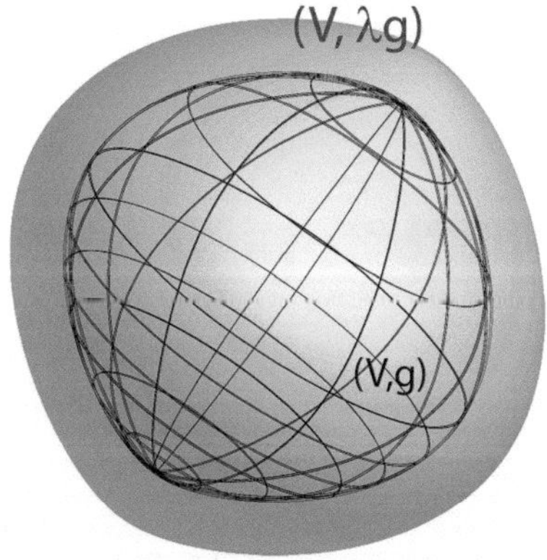

$$g^{-1} \longmapsto \lambda^{-1} g^{-1},$$

where $g^{-1} \in C^\infty(V^n, \otimes_S^2 TV^n)$ denotes the inverse metric providing the inner product in $T_p^* V^n$, $p \in V^n$. In local coordinates (U, x^i), we write $g_{ik} \longmapsto \lambda\, g_{ik}$ and $g^{ik} \longmapsto \lambda^{-1} g^{ik}$, respectively. Under the scaling (A.11) we immediately get an obvious dilation or contraction effect on distances and volume

$$d_g(p, q) \longmapsto d_{\lambda g}(p, q) = \lambda^{\frac{1}{2}} d_g(p, q)$$

$$\mathrm{Vol}_g(V^n) \longmapsto \mathrm{Vol}_{\lambda g}(V^n) = \lambda^{\frac{n}{2}} \mathrm{Vol}_g(V^n), \tag{A.12}$$

which is often useful to factor out in the study of Riemannian functional. In this connection it is worthwhile to recall how the basic Riemannian definitions interact with (A.11). We start with the Levi–Civita connection ∇^g of (V^n, g), ("∇" if there is no danger of confusion), defined by

$$2g\left(\nabla_X Y, Z\right) := X\left(g(Y, Z)\right) + Y\left(g(Z, X)\right) - Z\left(g(X, Y)\right) \tag{A.13}$$
$$+ g\left([X, Y], Z\right) - g\left([Y, Z], X\right) + g\left([Z, X], Y\right),$$

along with the torsion–free and metric compatibility conditions

$$\nabla_X Y - \nabla_Y X = [X, Y] \tag{A.14}$$

$$X\left(g(Y, Z)\right) = g\left(\nabla_X Y, Z\right) + g\left(Y, \nabla_X Z\right),$$

for all vector fields $X, Y, Z \in C^\infty(V^n, TV^n)$. In a local coordinate neighborhood $(U, \{x^i\})$ we write $\nabla_i \partial_j := \nabla_{\partial_i} \partial_j := \Gamma_{ij}^k(g)\, \partial_k$, where the Christoffel symbols $\Gamma_{ij}^k(g)$ are given by

$$\Gamma_{ij}^k(g) = \frac{1}{2} g^{k\ell} \left(\frac{\partial g_{j\ell}}{\partial x^i} + \frac{\partial g_{\ell i}}{\partial x^j} - \frac{\partial g_{ij}}{\partial x^\ell} \right). \tag{A.15}$$

If $(V^n, \phi^* g)$ is the pull–back of $(\phi(V^n), g)$ under a diffeomorphism $\phi : V^n \longrightarrow V^n$, then the Christoffel symbols $\Gamma_{ij}^k(g)$ and $\Gamma_{ij}^k(\phi^* g)$ are related by

$$\Gamma_{ij}^k(\phi^* g) \frac{\partial \phi^c}{\partial x^k} = \left(\Gamma_{ab}^c(g) \circ \phi\right) \frac{\partial \phi^a}{\partial x^i} \frac{\partial \phi^b}{\partial x^j} + \frac{\partial^2 \phi^c}{\partial x^i \partial x^j}, \tag{A.16}$$

whereas, under the metric rescaling (A.11), we easily compute from (A.15) $\Gamma_{ij}^k(\lambda g) = \Gamma_{ij}^k(g)$, hence

$$\nabla^{\lambda g} = \nabla^g. \tag{A.17}$$

Let us recall that if $\mathrm{Hess}_g f := \nabla df$ denotes the Hessian of f, $f \in C^\infty(V^n, \mathbb{R})$, then the Laplace–Beltrami operator $\triangle_g$ on (V^n, g) is defined by

$$\triangle_g f := \mathrm{tr}_g \left(Hess_g f\right) = g^{ik} \nabla_i \nabla_k f, \tag{A.18}$$

where we have used the geometer's sign convention according to which, on Euclidean space $(\mathbb{R}^n, \delta)$, $\triangle_\delta = \sum_{i=1}^{n} \frac{\partial^2}{\partial x_i^2}$. It is useful to have the expression of $\triangle_g$ available in local coordinates

$$\begin{aligned}
\triangle_g &= \frac{1}{\sqrt{\det g}} \frac{\partial}{\partial x^i} \left(\sqrt{\det g}\, g^{ij} \frac{\partial}{\partial x^j} \right) \\
&= g^{ij} \frac{\partial^2}{\partial x^i \partial x^j} + \frac{1}{\sqrt{\det g}} \frac{\partial}{\partial x^i} \left(\sqrt{\det g}\, g^{ij} \right) \frac{\partial}{\partial x^j} \\
&= g^{ij} \frac{\partial^2}{\partial x^i \partial x^j} - \Gamma^j \frac{\partial}{\partial x^j},
\end{aligned} \tag{A.19}$$

where we have defined

$$\Gamma^j := g^{hk} \Gamma_{hk}^j = -\frac{\partial}{\partial x^i} g^{ji} - g^{ij} \frac{\partial}{\partial x^i} \left(\log \sqrt{\det g} \right). \tag{A.20}$$

From the definitions of Hess_g and $\triangle_g$ it immediately follows that under the metric rescaling (A.11), we have

$$\mathrm{Hess}_{\lambda g} = \mathrm{Hess}_g,$$

$$\triangle_{\lambda g} = \lambda^{-1} \triangle_g. \tag{A.21}$$

The Riemann, the sectional, the Ricci and the scalar curvature operators (Fig. A.4) associated to (V^n, g) are respectively defined by

$$\mathrm{Rm}(g)(X, Y)Z := \left(\nabla_X \nabla_Y - \nabla_Y \nabla_X - \nabla_{[X,Y]} \right) Z, \tag{A.22}$$

$$\mathscr{S}ec(g)(X, Y) := \frac{g\left(\mathrm{Rm}(g)(X, Y)Y, X\right)}{g(X, X)g(Y, Y) - (g(X, Y))^2}, \tag{A.23}$$

$$\mathrm{Ric}(g)(Y, Z) := \mathrm{tr}_g \left(X \longmapsto \mathrm{Rm}(g)(X, Y)Z \right), \tag{A.24}$$

$$R(g) := trace\, \mathrm{Ric}(g), \tag{A.25}$$

for all vector fields $X, Y, Z \in C^\infty(V^n, TV^n)$. They are naturally equivariant under the action of $\mathrm{Diff}(V^n)$, whereas under the scaling action (A.11) we get

$$\mathrm{Rm}(\lambda g) = \mathrm{Rm}(g) \tag{A.26}$$

$$\mathrm{Sec}(\lambda g)(X, Y) = \lambda^{-1} \mathrm{Sec}(g)(X, Y)$$

$$\mathrm{Ric}(\lambda g) = \mathrm{Ric}(g)$$

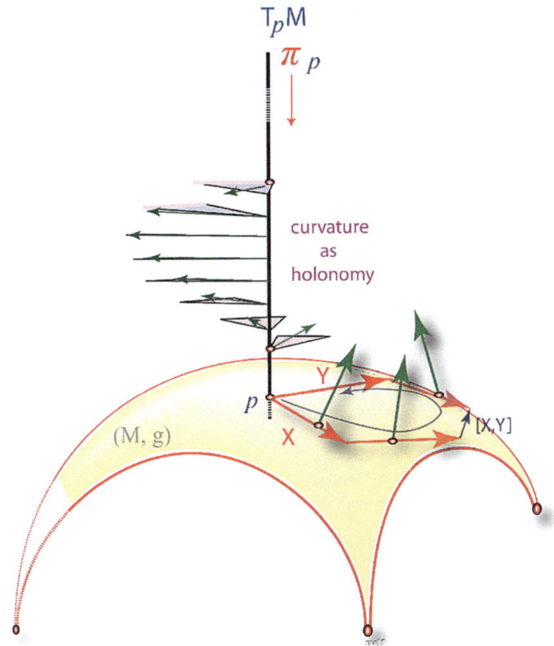

Fig. A.4 Curvature and parallel transport

$$R(\lambda g) = \lambda^{-1} R(g),$$

as easily follows from their definition and (A.17). Insight in the curvature operators is often provided by the expression of their components in a judiciously chosen coordinate system (U, x^i). To this end, let us state first our conventions for the components of $\mathrm{Rm}(g)$ and $\mathrm{Ric}(g)$, i.e.

$$\mathrm{Rm}(g)\left(\frac{\partial}{\partial x^i}, \frac{\partial}{\partial x^j}\right)\frac{\partial}{\partial x^k} := R^h_{\ ijk}\frac{\partial}{\partial x^h}, \tag{A.27}$$

$$R_{ik} := \sum_{h=1}^{n} R^h_{\ hik}, \tag{A.28}$$

where

$$R^h_{\ ijk} = \frac{\partial\, \Gamma^h_{jk}}{\partial x^i} - \frac{\partial\, \Gamma^h_{ki}}{\partial x^j} + \Gamma^h_{ir}\Gamma^r_{jk} - \Gamma^r_{ki}\Gamma^h_{jr}, \tag{A.29}$$

$$R_{ik} = \frac{\partial\, \Gamma^h_{ik}}{\partial x^h} - \frac{\partial\, \Gamma^h_{hk}}{\partial x^i} + \Gamma^h_{ik}\Gamma^r_{rh} - \Gamma^r_{hi}\Gamma^h_{kr}. \tag{A.30}$$

Notice that when considering the Riemann tensor components in covariant form, the upper index on the Riemann tensor is lowered into the 4–th position according to $R_{ijk\ell} := R^h_{\ ijk}\, g_{h\ell}$. We recall that, given a point $p \in (V^n, g)$, the exponential map $exp_p : T_p V^n \longrightarrow V^n$ is defined by (Fig. A.5)

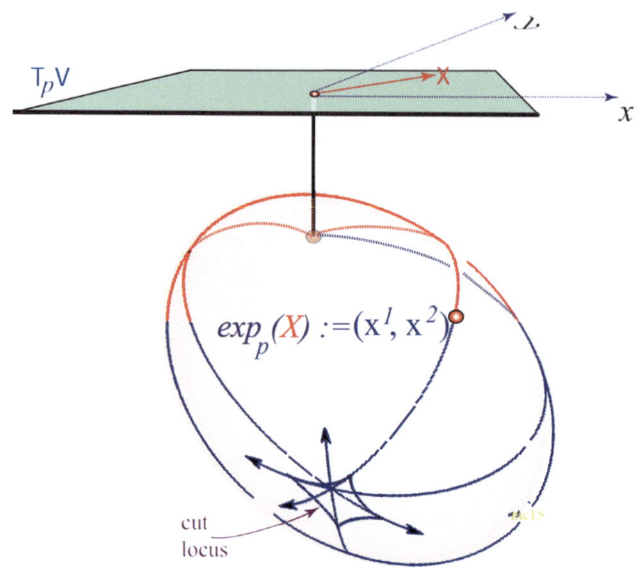

Fig. A.5 A two–dimensional rendering of the exponential map

$$X \longmapsto \exp_p(X) := \gamma_X(1),$$

where $\gamma_X : [0, \infty) \longrightarrow V^n$ is the constant speed geodesic issued from p with initial velocity $\dot{\gamma}_X(t = 0) = X$. If Cut$(p)$ denotes the cut locus of p in (V^n, g), namely the set of points $q \in V^n$ such that any minimal geodesic $\gamma(p, q)$ connecting p to q is not properly contained in another minimal geodesic issued from p, then normal geodesic coordinates $\{x^i\} : V^n - \mathrm{Cut}(p) \longrightarrow \mathbb{R}^n$ (based at the point $p \in V^n$) are defined by

$$\{x^i\}(q) := \left(\{X^i\} \circ \exp_p^{-1} \right)(q),$$

where $\{X^i\}$ are the components, in an orthonormal frame $\{e_{(i)}\}_{i=1}^n$ at $T_p V^n$, of the vector $\exp_p^{-1}(q) \in T_p V^n$. In normal coordinates one gets the classical Bertrand-Diguet-Puiseaux formulas providing a geometric interpretation, (see e.g. [4]), of the sectional, Ricci and scalar curvature, *viz.*

$$\ell(C(r)) = 2\pi r \left(1 - \frac{\mathscr{S}ec(X, Y)(p)}{6} r^2 + O\left(r^3\right) \right),$$

$$\sqrt{\det g(q)} = 1 - \frac{1}{6} R_{ik}(p) x^i(q) x^k(q) + O\left(r^3\right),$$

$$\mathrm{Vol}\,(B(p, r)) = \omega^n r^n \left(1 - \frac{R(p)}{6(n+2)} r^2 + O(r^3) \right),$$

(A.31)

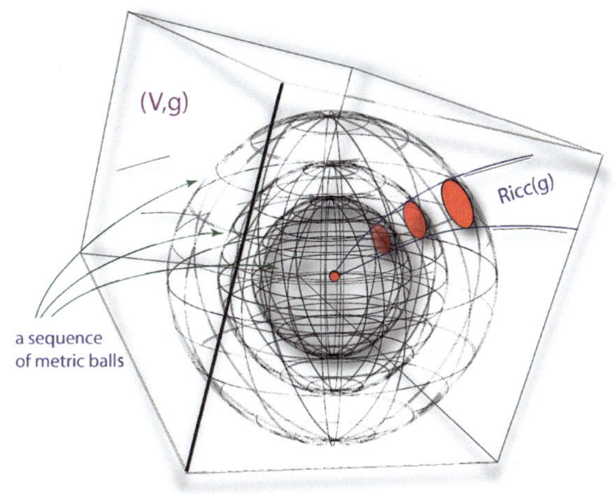

Fig. A.6 The geometrical meaning of Ricci curvature

where: (i) $r := d_{(g)}(q,p)$ is the Riemannian distance between q and p; (ii) $\ell(C(r))$ is the length of the circle of radius r in the geodesic surface generated by the unit vectors $X, Y \in T_p V^n$; (iii) $B(p, r)$ is the metric ball in (V^n, g) of radius r centered at p, and ω^n is the volume of the unit ball in Euclidean space. From these well–known expressions it follows that the sectional curvature measures the defect of the length of small circles $C(r)$ with respect to the Euclidean $2\pi r$. Similarly, the Ricci curvature measures the distortion of the solid angle, subtended by a small spherical sector in the direction $X = \exp_p^{-1}(q)$, with respect to the corresponding Euclidean value (Fig. A.6). Finally, the scalar curvature provides, as the radius varies, the distortion of the volume of small metric balls with respect to their Euclidean volume.

A.3   Some Properties of the Space of Riemannian Metrics

In the applications of Riemannian geometry discussed in these lecture notes we are not interested in a specific Riemannian metric g on V^n but rather in classes of metrics inducing on V^n the same geometric properties up to diffeomorphisms and rescalings [4]. This naturally calls into play the space of all smooth Riemannian metrics on V^n, the open convex cone (in the compact–open topology on $C^\infty(V^n, \otimes_S^2 T^* V^n)$) defined by (Fig. A.7)

$$\mathscr{M}et(V^n) := \left\{ g \in C^\infty(V^n, \otimes_S^2 T^* V^n) \mid g \text{ is positive definite} \right\}. \tag{A.32}$$

On $\mathscr{M}et(V^n)$ there is a natural action both of the group of diffeomorphisms $\mathrm{Diff}(V^n)$ and of the multiplicative group $\mathbb{R}^* := (\mathbb{R}_{>0}, \times)$ of positive real numbers. The former acting by pull–back

$$\mathrm{Diff}(V^n) \times \mathscr{M}et(V^n) \longrightarrow \mathscr{M}et(V^n) \tag{A.33}$$

$$(\phi, g) \longmapsto \phi^* g,$$

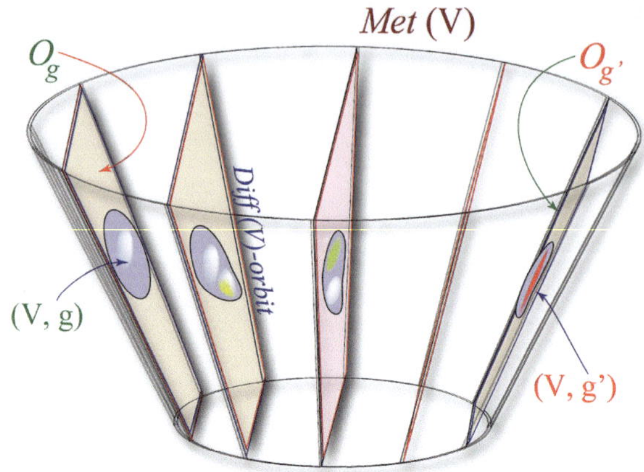

Fig. A.7 The ∞–dimensional cone of Riemannian metrics with its Diff(V^n) orbits $\mathscr{O}_g$

the latter acting by homotetic rescaling

$$\mathbb{R}^* \times \mathcal{M}et(V^n) \longrightarrow \mathcal{M}et(V^n) \tag{A.34}$$

$$(\lambda, g) \longmapsto \lambda\, g.$$

The quotient space $\mathcal{M}et(V^n)/\mathrm{Diff}(V^n)$ under (A.33) is the space of smooth Riemannian structures on V^n describing Riemannian manifolds with the same geometric properties, whereas the quotient $\mathcal{M}et(V^n)/(\mathbb{R}^* \times \mathrm{Diff}(V^n))$, associated with the composition of $\mathrm{Diff}(V^n)$ with the homotheties (A.34), describes Riemannian manifolds with the same metric properties up to an overall rescaling. Note that by normalizing volume we can write

$$\frac{\mathcal{M}et(V^n)}{\mathbb{R}^* \times \mathrm{Diff}(V^n)} \simeq \left\{ g \in \mathcal{M}et(V^n)\, \Big|\, \int_{V^n} d\mu_g = 1 \right\} \Big/ \mathrm{Diff}(V^n), \tag{A.35}$$

where $d\mu_g$ denotes the Riemannian measure on (V^n, g).

Most of our exploitation of the geometry of $\mathcal{M}et(V^n)$ is confined to its tangent space $\mathscr{T}_{(V^n,g)}\mathcal{M}et(V^n)$ at (V^n, g). Let us recall that an element of $\mathscr{T}_{(V^n,g)}\mathcal{M}et(V^n)$ is an equivalence class of curves of Riemannian metrics

$$(-\epsilon, \epsilon) \longrightarrow \mathcal{M}et(V^n) \tag{A.36}$$

$$\lambda \longmapsto g(\lambda)\,, \quad \text{with } g(0) = g,$$

where any two such a curve $\lambda \mapsto g_1(\lambda)$ and $\lambda \mapsto g_2(\lambda)$ are equivalent if they have the same velocity at $\lambda = 0$, i.e.

$$[g_1] \simeq [g_2] \tag{A.37}$$

$$\Leftrightarrow \left.\frac{dg_1(\lambda)}{d\lambda}\right|_{\lambda=0} = \left.\frac{dg_2(\lambda)}{d\lambda}\right|_{\lambda=0} = h \in C^\infty(V^n, \otimes_S^2 T^*V^n).$$

Hence we have the identification of the tangent space $\mathcal{T}_{(V^n,g)}\mathcal{M}et(V^n)$ with the space of sections $C^\infty(V^n, \otimes_S^2 T^*V^n)$,

$$\mathcal{T}_{(V^n,g)}\mathcal{M}et(V^n) \simeq C^\infty(V^n, \otimes_S^2 T^*V^n), \tag{A.38}$$

which serves as the local model for the infinite–dimensional (Frechét) manifold $\mathcal{M}et(V^n)$. To be somewhat informal, given $g \in \mathcal{M}et(V^n)$, the elements h of $\mathcal{T}_{(V^n,g)}\mathcal{M}et(V^n)$ can be interpreted as the infinitesimal variations of the given metric,

$$(-\epsilon, \epsilon) \longrightarrow \mathcal{M}et(V^n) \tag{A.39}$$

$$\lambda \longmapsto (g(\lambda))_{ik} = g_{ik} + \lambda\, h_{ik}\,, \quad 0 < \epsilon << 1.$$

Note that $\mathcal{M}et(V^n)$, being an open cone in $C^\infty(V^n, \otimes_S^2 T^*V^n)$, is contractible and the tangent bundle $\mathcal{T}\mathcal{M}et(V^n)$ of $\mathcal{M}et(V^n)$ is simply

$$\mathcal{T}\mathcal{M}et(V^n) = \mathcal{M}et(V^n) \times C^\infty(V^n, \otimes_S^2 T^*V^n). \tag{A.40}$$

Applications of Riemannian geometry to quantum field theory often exploit the Berger–Ebin decomposition [3] of $\mathcal{T}_{(V^n,g)}\mathcal{M}et(V^n)$. This decomposition allows us to discriminate between trivial (infinitesimal) variations of a metric, associated with the action of the diffeomorphisms group, and essential variations describing a change in the Riemannian structure of (V^n, g) (in this latter case one speaks of (infinitesimal) deformations of the given g: see [4] for the proper use of the term deformation *vs.* variation). The construction of this decomposition is somewhat delicate and owing to its importance in the applications discussed in these lecture notes, it is worthwhile to describe the various steps involved in its proof.

To begin with, let us recall that both $\mathcal{M}et(V^n)$ and $\mathrm{Diff}(V^n)$ are locally modeled on Frechét spaces of smooth sections and maps, where the necessary tools from analysis on Banach spaces, (basically a Picard iteration scheme), are not easily available. We need to enlarge $\mathcal{M}et(V^n)$ and $\mathrm{Diff}(V^n)$ to their Sobolev counterparts. To this end we consider the open subset of $\mathcal{H}^s(V^n, \otimes^2 T^*V^n)$ defining the Riemannian metrics $\mathcal{M}et^{\,s}(V^n)$ of Sobolev class $s > \frac{n}{2}$, and the (topological) group, $\mathcal{D}^{\,s'}(V^n)$, defined by the set of diffeomorphisms which, as maps $V^n \to V^n$ are an open subset of the Sobolev space of maps $\mathcal{H}^{s'}(V^n, V^n)$, with $s' \geq s + 1$, (see (A.10)). There is a natural projection map from $\mathcal{D}^{\,s'}(V^n)$ into $\mathcal{M}et^{\,s}(V^n)$

$$\pi : \mathcal{D}^{\,s'}(V^n) \longrightarrow \mathcal{O}_g^s \subset \mathcal{M}et^{\,s}(V^n) \tag{A.41}$$

$$\phi \longmapsto \pi(\phi) \doteq \phi^* g,$$

where

$$\mathcal{O}_g^s := \left\{ \tilde{g} \in \mathcal{M}et^{\,s}(V^n) \,|\, \tilde{g} = \phi^* g,\ \phi \in \mathcal{D}^{\,s'}(V^n) \right\} \tag{A.42}$$

is the $\mathscr{D}^{s'}(V^n)$–orbit of a given metric $g \in \mathscr{M}et^{\,s}(V^n)$, and $\phi^* g$ is the pull–back under $\phi \in \mathscr{D}^{s'}(V^n), s' \geq s + 1$. If $\mathscr{T}_{(V^n,g)}\mathscr{O}_g^s$ denotes the tangent space to any such an orbit, then $\mathscr{T}_{(V^n,g)}\mathscr{O}_g^s$ is the image of the operator

$$\delta_g^* : \mathscr{H}^{s+1}(V^n, TV^n) \to \mathscr{H}^s(V^n, \otimes^2 T^* V^n) \tag{A.43}$$

$$w \quad \mapsto \delta_g^*(w) \doteq \frac{1}{2}\,\mathscr{L}_w\, g,$$

where $\mathscr{L}_w\, g$ denotes the Lie derivative of the metric tensor g along the vector field w,

$$(\mathscr{L}_w\, g)_{ik} = \nabla_i w_k + \nabla_k w_i = w^h \partial_h g_{ik} + g_{hk}\partial_i w^h + g_{ih}\partial_k w^h. \tag{A.44}$$

Its (total) symbol in the (co)direction $\xi \in T_p^* V^n$ is provided by the injective bundle homomorphism

$$V_\xi^n[\delta_g^*] : T_p^* V^n \longrightarrow \otimes_S^2 T_p^* V^n \tag{A.45}$$

$$X_i\, dx^i \longmapsto V_\xi^n[\delta_g^*](X) = \frac{1}{2}\,(\xi_h X_k + \xi_k X_h)\, dx^h \otimes dx^k.$$

Note that the L^2 adjoint δ_g of δ_g^*, with respect to the inner product (A.3), is provided by (minus) the divergence operator

$$\delta_g : \mathscr{H}^s(V^n, \otimes^2 T^* V^n) \to \quad \mathscr{H}^{s-1}(V^n, T^* V^n) \tag{A.46}$$

$$h \quad \mapsto \delta_g\, h \doteq -g^{ij}\,\nabla_i h_{jk}\, dx^k,$$

with total symbol given by

$$V_\xi^n[\delta_g] : \otimes_S^2 T_p^* V^n \longrightarrow T_p^* V^n \tag{A.47}$$

$$w_{hk}\, dx^h \otimes dx^k \longmapsto V_\xi^n[\delta_g](w) = -g^{ih}\xi_i w_{hk} dx^k.$$

Since δ_g^* has injective symbol, Banach space theory implies that $\delta_g \circ \delta_g^*$ is an elliptic operator [3]. To expand in more detail, we compute the symbol of the composition $\delta_g \circ \delta_g^*$ according to

$$V_\xi^n[\delta_g \circ \delta_g^*] \; : T^* V^n \longrightarrow \otimes_S^2 T^* V^n \longrightarrow T^* V^n \tag{A.48}$$

$$X_i\, dx^i \longmapsto V_\xi^n[\delta_g^*](X) \longmapsto V_\xi^n[\delta_g] \circ V_\xi^n[\delta_g^*](X)$$

$$= V_\xi^n[\delta_g]\left(V_\xi^n[\delta_g^*](X)\right) = -\frac{1}{2}\,\left(|\xi|_g^2 \delta_k^i + \xi^i \xi_k\right) X_i\, dx^k,$$

and the ellipticity of the operator $\delta_g \circ \delta_g^*$ immediately follows by observing that the matrix $\left(|\xi|_g^2 \delta_k^i + \xi^i \xi_k\right)$ is non–singular with positive eigenvalues. By standard

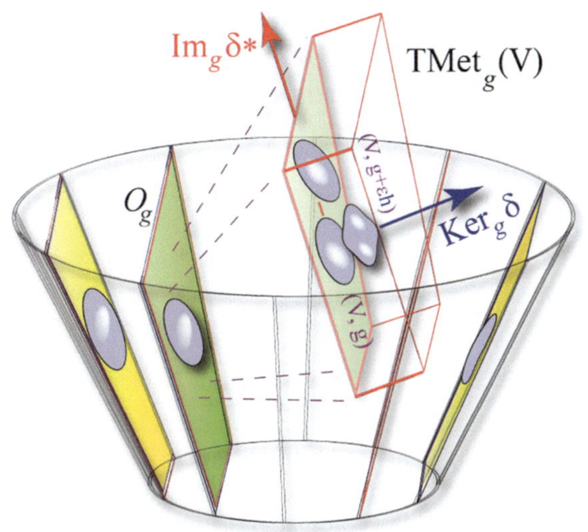

Fig. A.8 The geometry of the Berger–Ebin decomposition of $\mathscr{T}_{(V^n,g)}\mathscr{M}et(V^n)$, the tangent space to $\mathscr{M}et(V^n)$ at the metric g

elliptic theory the L^2–orthogonal subspace to Im δ_g^* in $\mathscr{T}_{(V^n,g)}\mathscr{M}et^s(V^n)$ is spanned by the (infinite–dimensional) kernel of δ_g. This entails the well–known Berger–Ebin $L^2(V^n, d\mu_g)$–orthogonal splitting [3, 8, 21] of the tangent space $\mathscr{T}_{(V^n,g)}\mathscr{M}et^s(V^n)$ (Fig. A.8),

$$\mathscr{T}_{(V^n,g)}\mathscr{M}et^s(V^n) \cong \left[\mathscr{T}_{(V^n,g)}\mathscr{M}et^s(V^n) \cap \ker \delta_g\right] \oplus \operatorname{Im}\delta_g^* \left[\mathscr{H}^{s+1}(V^n, TV^n)\right]$$

$$\text{(A.49)}$$

$$\cong \left[\mathscr{T}_{(V^n,g)}\mathscr{M}et^s(V^n) \cap \ker \delta_g\right] \oplus \mathscr{T}_{(V^n,g)}\mathcal{O}_g^s,$$

according to which, for any given tensor $h \in \mathscr{T}_{(V^n,g)}\mathscr{M}et^s(V^n)$, we can write

$$h_{ab} = h_{ab}^T + \mathscr{L}_w\, g_{ab}, \tag{A.50}$$

where h_{ab}^T denotes the div–free part of h, $(\nabla^a h_{ab}^T = 0)$, and where the vector field w is characterized as the solution, (unique up to the Killing vectors of (V^n, g)), of the elliptic PDE

$$\delta_g\, \delta_g^*\, w = \delta_g\, h. \tag{A.51}$$

We can also take into account the role of the scaling transformation (A.11) and refine (A.49) to the tangent space $\mathscr{T}_{(V^n,g)}\mathscr{M}et_1^s(V^n)$ to the space of metrics $\mathscr{M}et_1^s(V^n) := \left\{g \in \mathscr{M}et^s(V^n)\mid \int_V^n d\mu_g = 1\right\}$ of normalized total volume (cf. (A.35)). Since

$$\mathscr{T}_{(V^n,g)}\mathscr{M}et_1^s(V^n) = \left\{h \in \mathscr{H}^s(V^n, \otimes_S^2 T^*V^n)\mid \int_V^n \operatorname{tr}_g h\, d\mu_g = 0\right\}, \tag{A.52}$$

where $tr_g\, h := g^{ik}\, h_{ik}$, we have

$$\mathscr{T}_{(V^n,g)}\mathscr{M}et^s_1(V^n) \cong \left[\mathscr{T}_{(V^n,g)}\mathscr{M}et^s_1(V^n) \cap \ker \delta_g\right] \oplus \operatorname{Im} \delta^*_g \left[\mathscr{H}^{s+1}(V^n, TV^n)\right]$$

$$(A.53)$$

$$\cong \left[\mathscr{T}_{(V^n,g)}\mathscr{M}et^s_1(V^n) \cap \ker \delta_g\right] \oplus \mathscr{T}_{(V^n,g)}\mathscr{O}^s_g.$$

A.4   Ricci Curvature

In applications of Riemannian geometry to the quantum geometry of polyhedral surface (and its interplay with Ricci flow), Ricci curvature plays a central role. A deep insight in its geometric meaning is provided by the expression of its components in local harmonic coordinates $(U, \{x^i\})$. These latter are defined by requiring that each coordinate function $x^k : U \longrightarrow \mathbb{R}$ is harmonic. In particular, local solvability for elliptic PDE's implies, (compare with [7], Corollary 3.30), that for any given point $p \in (V^n, g)$ there always exists a neighborhood $U_p \subset V^n$ of p such that, (see (A.20)),

$$\triangle_g\, x^i = g^{jk}\left(\frac{\partial^2}{\partial x^j \partial x^k} - \Gamma^\ell_{jk}\frac{\partial}{\partial x^\ell}\right) x^i = -\Gamma^i = 0. \qquad (A.54)$$

When passing from normal geodesic coordinates to harmonic coordinates we gain control on the components of the metric tensor in terms of the Ricci curvature rather than of the full Riemann tensor (cf. [14] for a particularly clear comparison of these two coordinate systems from the point of view of geometric analysis). In particular, as first stressed by C. Lanczos [20], in harmonic coordinates the components of the Ricci tensor take the suggestive form, (see e.g. [27] Lemma 2.6 for an elegant computation),

$$R_{ik} \underset{(harmonic)}{=} -\frac{1}{2}\triangle_g\,(g_{ik}) + Q_{ik}\left(g^{-1}, \partial g\right), \qquad (A.55)$$

where for each fixed pair of indexes (i, k), the expression $\triangle_g\,(g_{ik})$ denotes the Laplace–Beltrami operator applied component wise to each g_{ik} as it were a *scalar function*, (in particular $\triangle_g g_{ik}$ is not the tensorial (rough) Laplacian $g^{ab}\nabla_a\nabla_b\, g_{ik}$ of the metric tensor g, a quantity that obviously vanishes identically for the Levi–Civita connection). In (A.55) $Q\left(g^{-1}, \partial g\right)$ denotes a sum of terms quadratic in the components of g, g^{-1} and their first derivatives, and whose explicit expression does not concern us here (compare with [27]). Hence in harmonic coordinates the Ricci curvature acts as a quasi–linear elliptic operator on the components of the metric, an observation that can be exploited to prove [9] that the metric tensor g, if not smooth, has maximal regularity in harmonic coordinates. The expression (A.55) also plays an inspiring role in Ricci flow theory, where the minus sign in front of

Fig. A.9 The Ricci curvature in harmonic coordinates as the generator of Brownian diffusion on the manifold (V^n, g)

the scalar Laplace–Beltrami $\triangle_g$ (g_{ik}) suggests the right PDE candidate for the Ricci flow among the many (apparently) possible ones. It is also worthwhile to observe that the factor $1/2$ appearing in (A.55), motivating the conventional "2" in the Ricci flow generator $-2\,\mathrm{Ric}(g)$, is not as incidental as is typically assumed. This is related to the fact that $\frac{1}{2}\triangle_g$ and not $\triangle_g$ is the generator of Brownian motion on (V^n, g) [15] (Fig. A.9).

Appendix B
A Capsule of Moduli Space Theory

In this Appendix we summarize notation and a few basic definitions of Riemann Moduli theory that we have been freely using in these lecture notes, (for details we mainly refer to [12, 13, 16]).

B.1 Riemann Surfaces with Marked Points and Divisors

We shall denote by $(M_g, \mathscr{C})$ a Riemann surface of genus g, (most often we simply write $(M, \mathscr{C})$ if the genus is clear from the context). Recall that $(M, \mathscr{C})$ is characterized by an atlas of local coordinate charts (U_k, φ_k) defined by maps $\varphi_k : U_k \to \mathbb{C}$ whose transition functions, $\varphi_h \circ \varphi_k^{-1} : \varphi_k(U_k \cap U_h) \to \varphi_h(U_h \cap U_k)$, are holomorphic maps between open subsets of $\mathbb{C}$. Any such a chart (U, φ) is said to provide a *local conformal parameter*, and for the generic $p \in U$ one sets $\varphi(p) := z = x + \sqrt{-1}\, y$. A N_0–*pointed surface* $((M; N_0), \mathscr{C})$, or surface with N_0 marked points, is an oriented closed, (connected), surface of genus g decorated with a distinguished set of N_0 pairwise distinct points $\{p_1, \ldots, p_{N_0}\}$. (Note that $((M; N_0), \mathscr{C})$ is distinct from the open Riemann surface $(M', \mathscr{C}) := (M, \mathscr{C}) \setminus \{p_1, \ldots, p_{N_0}\}$ obtained by removing from $(M, \mathscr{C})$ the points $\{p_1, \ldots, p_{N_0}\}$) (Fig. B.1).

The tangent and cotangent spaces at $p \in (M, \mathscr{C})$ are naturally obtained by tensoring with $\mathbb{C}$ the tangent $T_p M$ and cotangent space $T_p M^*$ of the underlying real surface M, i.e., $T_{\mathbb{C},p} M := T_p M \otimes \mathbb{C}$ and $T_{\mathbb{C},p} M^* := T_p M^* \otimes \mathbb{C}$. The respective basis induced by the local conformal parameter z are provided by the usual expressions

$$\frac{\partial}{\partial z} := \frac{1}{2}\left(\frac{\partial}{\partial x} - \sqrt{-1}\,\frac{\partial}{\partial y}\right), \quad \frac{\partial}{\partial \bar{z}} := \frac{1}{2}\left(\frac{\partial}{\partial x} + \sqrt{-1}\,\frac{\partial}{\partial y}\right), \tag{B.1}$$

$$dz := dx + \sqrt{-1}\, dy, \quad d\bar{z} := dx - \sqrt{-1}\, dy, \tag{B.2}$$

© Springer International Publishing AG 2017

M. Carfora, A. Marzuoli, *Quantum Triangulations*, Lecture Notes in Physics 942,
https://doi.org/10.1007/978-3-319-67937-2

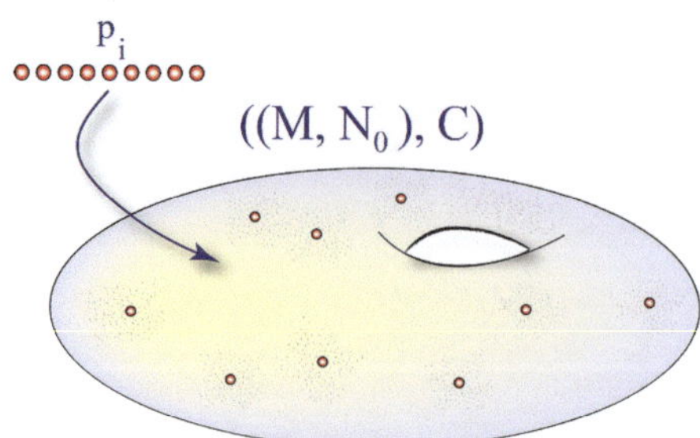

Fig. B.1 A N_0–pointed Riemann surface $((M, N_0), \mathscr{C})$ of genus g obtained by injecting a string of N_0 pairwise distinct points in $(M, \mathscr{C})$. Here $g = 1$ and $N_0 = 9$, we typically assume that $2g - 2 + N_0 > 0$, in such a case the automorphism group, $Aut\,(M, N_0)$, of the resulting pointed Riemann surface is finite. Recall that $Aut\,(M, N_0)$ is the largest group of conformal automorphisms that the Riemann surface $((M, N_0), \mathscr{C})$ can admit

$$dx \wedge dy = \frac{\sqrt{-1}}{2}\, dz \wedge d\bar{z}. \tag{B.3}$$

One can naturally split $T_{\mathbb{C},p}M$ into the holomorphic $T'_p M := \mathbb{C}\{\partial/\partial z\}$ and antiholomorphic $T''_p M := \mathbb{C}\{\partial/\partial \bar{z}\}$ tangent spaces at $p \in (M, \mathscr{C})$ according to $T_{\mathbb{C},p}M = T'_p M \oplus T''_p M$, with $T''_p M = \overline{T'_p M}$, (where the overline $\overline{\cdot}$ denotes complex conjugation). In this connection it is worthwhile recalling that a smooth map $f : M \to N$ between two surfaces M and N is holomorphic if and only if the corresponding tangent map $f_* : T_{\mathbb{C},p}M \longrightarrow T_{\mathbb{C},f(p)}N$ is such that $f_*\left(T'_p M\right) \subset T'_{f(p)}N$. This implies that there is a linear isomorphism between $T_p M$ and the holomorphic tangent space $T'_p M = \mathbb{C}\{\partial/\partial z\}$, an isomorphism, this latter, which provides a natural dictionary between geometrical quantities on the surface M and their corresponding realizations on $(M, \mathscr{C})$. Holomorphic maps also preserve the holomorphic–antiholomorphic decomposition of the cotangent space $T_{\mathbb{C},p}M^* = T'_p M^* \oplus T''_p M*$, and the corresponding splitting $\Omega(M) = \bigoplus_{p+q \leq 2} \Omega^{(p,q)}(M)$ of the space of differential forms on $(M, \mathscr{C})$. Explicitly, the spaces $\Omega^{(1,0)}(M)$, $\Omega^{(0,1)}(M)$, and $\Omega^{(1,1)}(M)$ of forms of type (p, q) are locally generated by the monomials $\varphi(z)\, dz$, $\varphi(z)\, d\bar{z}$, and $\varphi(z)\, dz \wedge d\bar{z}$, respectively. Corresponding to such a splitting we have the Dolbeault operators $\partial : \Omega^{(p,q)}(M) \to \Omega^{(p+1,q)}(M)$ and $\bar{\partial} : \Omega^{(p,q)}(M) \to \Omega^{(p,q+1)}(M)$, with $\partial\partial = 0$, $\bar{\partial}\,\bar{\partial} = 0$, $\partial\bar{\partial} = -\bar{\partial}\partial$, and $d = \partial + \bar{\partial}$, where d denotes the exterior derivative. Locally,

$$\partial := \frac{\partial}{\partial z}\, dz\,, \quad \bar{\partial} := \frac{\partial}{\partial \bar{z}}\, d\bar{z}, \tag{B.4}$$

In particular, a form $\psi \in \Omega^{(p,0)}$ is holomorphic if $\bar{\partial}\,\psi = 0$. A metric on a Riemann surface $(M, \mathscr{C})$ is conformal if locally it can be written as

$$ds^2 = h^2(z)\, dz \otimes d\bar{z}, \tag{B.5}$$

for some smooth function $h(z) > 0$. The corresponding Kähler form is given by

$$\varpi := \frac{\sqrt{-1}}{2} h^2(z)\, dz \wedge d\bar{z}. \tag{B.6}$$

The Gaussian curvature of the (smooth) metric $ds^2 = h^2(z)\, dz \otimes d\bar{z}$ is provided by

$$K := -\Delta_{ds^2} \ln h, \tag{B.7}$$

where Δ_{ds^2}, (Δ if is clear from the context which conformal metric we are using), is the Laplace–Beltrami operator with respect to the metric ds^2, *viz.*

$$\Delta := \frac{1}{h^2}\left(\frac{\partial^2}{\partial x^2} + \frac{\partial^2}{\partial y^2}\right) = \frac{4}{h^2(z)}\frac{\partial}{\partial z}\frac{\partial}{\partial \bar{z}}. \tag{B.8}$$

Let us recall, (see e.g. the very readable presentation in [13]), that a divisor D on a Riemann surface $(M, \mathscr{C})$ is a formal linear combination

$$D = \sum n_k\, z_k \tag{B.9}$$

where $n_k \in \mathbb{Z}$ and the z_k are points $\in M$. D is required to be locally finite, namely for any $q \in M$ there is a neighborhood $U_q \subset M$ of q such that $U_q \cap \{z_k\}$ contains only a finite number of the $\{z_k\}$ appearing in D. Since we typically deal with compact Riemann surfaces $(M, \mathscr{C})$, this implies that the formal sum defining D is finite. The set of divisors on $(M, \mathscr{C})$ naturally forms an additive (abelian) group $\mathscr{D}iv(M)$. The degree of a divisor $D = \sum n_k\, z_k$, (on a compact surface Riemann surface $(M, \mathscr{C})$), is defined according to

$$\deg D := \sum n_k. \tag{B.10}$$

Let f be a meromorphic function, i.e. the local quotient of two holomorphic functions on $(M, \mathscr{C})$. Let $k \in \mathbb{N}$, if f has a zero of order k at $z \in (M, \mathscr{C})$, define the order of f at z to be $ord_z f := k(> 0)$, conversely, if f has a pole of order k at $z \in (M, \mathscr{C})$ we define $ord_z f := -k$. The order of f at all other points is defined to be 0. The divisor of a meromorphic function f, $f \not\equiv 0$, is conventionally denoted by (f), and defined as

$$(f) := \sum (ord_{z_h} f)\, z_h, \tag{B.11}$$

where the sum is over all zeros and all poles of f. A divisor D is called a principal divisor if it is the divisor (f) of a meromorphic function $f \not\equiv 0$. Since a meromorphic function f on $(M, \mathscr{C})$ has the same number of zeros and poles (counted with multiplicity) it immediately follows that the degree of its divisor (f) satisfies

$$\deg (f) = 0. \tag{B.12}$$

More generally, if M^n is a (not necessarily compact) complex manifold of dimension n, a divisor can be naturally associated with hypersurfaces (i.e. $n-1$–dimensional submanifolds) of M^n. Explicitly, if V_i are irreducible analytic hypersurfaces of M^n, (i.e., for each point $p \in V_i \subset M^n$, V_i can be given in a neighborhood of p as the zeros of a single holomorphic function f, and it cannot be written as $V_i = \widetilde{V}_1 \cup \widetilde{V}_2$, where $\widetilde{V}_\alpha$, $\alpha = 1, 2$ are analytic hypersurfaces of M^n), then a divisor on M^n is a locally finite formal linear combination of irreducible analytic hypersurfaces of M^n

$$D = \sum_{V_k} n_k \, V_k. \tag{B.13}$$

Note that given any such a divisor, we can always find an open cover $\{U_\alpha\}$ of M^n such that in U_α, with $U_\alpha \cap V_i \neq \emptyset$, the hypersurface V_i has a local defining equation $V_i = \{g_{i\alpha}(z) = 0\}$ where $g_{i\alpha}(z)$ is a holomorphic function in U_α. In each U_α the set of $\{g_{i\alpha}(z)\}$, associated with all V_i for which $U_\alpha \cap V_i \neq \emptyset$, characterizes a meromorphic function f_α according to

$$f_\alpha := \prod_i g_{i\alpha}^{n_i}, \tag{B.14}$$

which are the local defining functions for the divisor D. The set of local defining functions $\{f_\alpha\}$ associated with an open covering $\{U_\alpha\}$ of M^n can be also used for characterizing the line bundle $[D]$ associated with the given divisor D, this is the bundle over M^n defined by the transition functions $\{\psi_{\alpha\beta} := f_\alpha / f_\beta\}$ for any $U_\alpha \cap U_\beta \neq \emptyset$. Note that the line bundle $[D]$ is trivial if and only if D is the divisor of a meromorphic function.

Recall that if $H_{DR}^{2k}(M^n, \mathbb{R})$ denotes the $2k$–th DeRham cohomology group of M^n, and $V \subset M^n$ is an analytic subvariety of (complex) dimension k, then the fundamental class (V) in the homology group $H_{2k}(M^n, \mathbb{R})$ is defined by the pairing

$$H_{DR}^{2k}(M^n, \mathbb{R}) \times H_{2k}(M^n, \mathbb{R}) \longrightarrow \mathbb{R} \tag{B.15}$$

$$\eta, \, (V) \qquad \qquad \longmapsto \int_V \eta.$$

By Poincaré duality this determines the fundamental class $\eta_V \in H_{DR}^{2n-2k}(M^n, \mathbb{R})$ of V. Thus, given a divisor $D = \sum_{V_k} n_k V_k$, we can introduce its fundamental class (D) and its Poincaré dual $\eta_D \in H_{DR}^2(M^n)$ according to

$$(D) := \sum n_k \, (V_k) \,, \tag{B.16}$$

$$\eta_D := \sum n_k \, \eta_{V_k}. \tag{B.17}$$

Given a line bundle L over M^n, locally defined by transition functions $\{\psi_{\alpha\beta}\}$ relative to a covering $\{U_\alpha\}$ of M^n, let

$$c_1(L) = \frac{\sqrt{-1}}{2\pi}\, \Xi, \tag{B.18}$$

be the (first) Chern class of L, where $\Xi|_{U_\alpha} = d\,\omega_\alpha$ denotes the curvature of L associated with any locally given connection 1–form ω_α. A basic result connecting line bundles, divisors and Chern classes is the observation that if L is the line bundle associated with a divisor D, then, (see [13] for a nicely commented proof),

$$c_1(L) = c_1([D]) = \eta_D \in H^2_{DR}(M^n). \tag{B.19}$$

B.2   The Teichmüller Space $\mathfrak{T}_{g,N_0}(M)$

In discussing the connection between polyhedral surfaces and Riemann surfaces we are naturally led to consider the relation between the space $POL_{g,N_0}(M)$ and $\mathfrak{M}_{g,N_0}$, the moduli space of N_0–pointed Riemann surfaces of genus g. This latter features as a basic object of study in a large variety of mathematical and physical applications of Riemann surface theory, and as such it is susceptible of many possible characterizations. From our perspective it is appropriate to adopt a Riemannian geometry viewpoint and define $\mathfrak{M}_{g,N_0}$ as a suitable quotient of the space of conformal classes of riemannian metrics the surface M can carry. For details and proofs we refer to [25].

Let us consider the set of all (smooth) Riemannian metrics on the genus–g surface M, (see Sect. 1.6),

$$\mathscr{M}et(M) \doteq \{g \in S_2(M)|\ g(x)(u,u) > 0 \ \text{if} \ u \neq 0\}, \tag{B.20}$$

where $S_2(M)$ is the space of symmetric bilinear forms on M, and let us denote by $C^\infty(M, \mathbb{R}_+)$ the group of positive smooth functions acting on metrics $g \in \mathscr{M}et(M)$ by pointwise multiplications. This action is free, smooth (and proper), and the quotient

$$\mathscr{C}onf(M) := \frac{\mathscr{M}et(M)}{C^\infty(M,\ \mathbb{R}_+)}, \tag{B.21}$$

is the Fréchet manifold of conformal structures. Note that $\mathscr{C}onf(M)$ naturally extends to the pointed surface $(M; N_0)$, obtained by decorating M with N_0 marked points $\{p_1, \ldots, N_0\}$, as long as the conformal class contains a smooth representative metric [29]. Let

$$\mathscr{M}et_{-1}(M; N_0) \hookrightarrow \mathscr{M}et\,(M; N_0) \tag{B.22}$$

be the set of metrics of constant curvature -1, describing the hyperbolic structures on $(M; N_0)$. If $\mathscr{D}iff_+(M)$ is the group of all orientation preserving diffeomorphisms then

$$\mathscr{D}iff_+(M; N_0) = \left\{ \psi \in Diff_+(M) : \psi \text{ preserves setwise } \{p_i\}_{i=1}^{N_0} \right\} \tag{B.23}$$

acts by pull-back on the metrics in $\mathscr{M}et_{-1}(M; N_0)$. Let $\mathscr{D}iff_0(M; N_0)$ be the subgroup consisting of diffeomorphisms which when restricted to $(M; N_0)$ are isotopic to the identity, then the Teichmüller space $\mathfrak{T}_{g,N_0}(M)$ associated with the genus g surface with N_0 marked points M is defined by

$$\mathfrak{T}_{g,N_0}(M) = \frac{\mathscr{M}et_{-1}(M; N_0)}{\mathscr{D}iff_0(M; N_0)}. \tag{B.24}$$

From a complex function theory perspective, $\mathfrak{T}_{g,N_0}(M)$ is characterized by fixing a reference complex structure $\mathscr{C}_0$ on $(M; N_0)$ (a marking) and considering the set of equivalence classes of complex structures $(\mathscr{C}, f)$ where $f : \mathscr{C}_0 \to \mathscr{C}$ is an orientation preserving quasi-conformal map, and where any two pairs of complex structures $(\mathscr{C}_1, f_1)$ and $(\mathscr{C}_2, f_2)$ are considered equivalent if $h \circ f_1$ is homotopic to f_2 via a conformal map $h : \mathscr{C}_1 \to \mathscr{C}_2$. Let us assume that the reference complex structure $\mathscr{C}_0$ admits an antiholomorophic reflection $j : \mathscr{C}_0 \to \mathscr{C}_0$. Since any orientation reversing diffeomorphism $\widetilde{\varphi}$ can be written as $\varphi \circ j$ for some orientation preserving diffeomorphism φ, the (extended) mapping class group

$$\mathfrak{Map}(M; N_0) \doteq \mathscr{D}iff(M; N_0) / \mathscr{D}iff_0(M; N_0) \tag{B.25}$$

acts naturally on $\mathfrak{T}_{g,N_0}(M)$ according to

$$\mathfrak{Map}(M; N_0) \times \mathfrak{T}_{g,N_0}(M) \longrightarrow \mathfrak{T}_{g,N_0}(M) \tag{B.26}$$

$$\begin{cases} (\varphi, (\mathscr{C}, f)) \longmapsto (\mathscr{C}, f \circ \varphi^{-1}), & \varphi \in \mathscr{D}iff_+(M; N_0) \\ (\widetilde{\varphi}, (\mathscr{C}, f)) \longmapsto (\mathscr{C}^*, f \circ j \circ \varphi^{-1}) & \widetilde{\varphi} \in \mathscr{D}iff(M; N_0) - \mathscr{D}iff_+(M; N_0) \end{cases},$$

where the conjugate surface $\mathscr{C}^*$ is the Riemann surface locally described by the complex conjugate coordinate charts associated with $\mathscr{C}$. It follows that the Teichmüller space $\mathfrak{T}_{g,N_0}(M)$ can be also seen as the universal cover of the moduli space $\mathfrak{M}_{g,N_0}$ of genus g Riemann surfaces with $N_0(T)$ marked points defined by

$$\mathfrak{M}_{g,N_0} = \frac{\mathfrak{T}_{g,N_0}(M)}{\mathfrak{Map}(M; N_0)} \tag{B.27}$$

This characterization represents the moduli associated to $((M; N_0), \mathscr{C})$ by an equivalence class of Riemannian metrics $[ds^2]$ on M, where two metrics $ds^2_{(1)}$ and $ds^2_{(2)}$ define the same point $[ds^2]$ of $\mathfrak{M}_{g,N_0}$ if and only if there exists $f \in C^\infty(M, \mathbb{R}_+)$

and an orientation preserving diffeomorphism $\psi \in \mathscr{D}iff_{g,N_0}$ fixing each marked point p_k individually such that $ds^2_{(2)} = f\,\Psi^*\left(ds^2_{(1)}\right)$. According to the remark above, the class $[ds^2]$ may contain singular metrics provided that they are conformal to a smooth one. As emphasized by M. Troyanov, this implies that we can represent $[ds^2] \in \mathfrak{M}_{g,N_0}$ by conical metrics, as well.

For a genus g Riemann surfaces with $N_0(T) > 3$ marked points the complex vector space $Q_{N_0}(M)$ of (holomorphic) quadratic differentials is defined by tensor fields ψ described, in a locally uniformizing complex coordinate chart (U, ζ), by a holomorphic function $\mu : U \to \mathbb{C}$ such that $\psi = \mu(\zeta)d\zeta \otimes d\zeta$. Away from the discrete set of the zeros of ψ, we can locally choose a canonical conformal coordinate $z_{(\psi)}$ (unique up to $z_{(\psi)} \mapsto \pm z_{(\psi)} + const$) by integrating the holomorphic 1-form $\sqrt{\psi}$, i.e.,

$$z_{(\psi)} = \int^{\zeta} \sqrt{\mu(\zeta')d\zeta' \otimes d\zeta'}, \tag{B.28}$$

so that $\psi = dz_{(\psi)} \otimes dz_{(\psi)}$. If we endow $Q_{N_0}(M)$ with the L^1-(Teichmüller) norm

$$\|\psi\| \doteq \int_M |\psi|, \tag{B.29}$$

then the Banach space of integrable quadratic differentials on M,

$$Q_{N_0}(M) \doteq \{\psi|,\ \|\psi\| < +\infty\}, \tag{B.30}$$

is non-empty and consists of meromorphic quadratic differentials whose only singularities are (at worst) simple poles at the N_0 distinguished points of (M, N_0). $Q_{N_0}(M)$ is finite dimensional and, (according to the Riemann-Roch theorem), it has complex dimension

$$\dim_{\mathbb{C}} Q_{N_0}(M) = 3g - 3 + N_0(T), \tag{B.31}$$

From the viewpoint of Riemannian geometry, a quadratic differential is basically a transverse-traceless two tensor deforming a Riemannian structure to a nearby inequivalent Riemannian structure. Thus a quadratic differential $\psi = \mu(\zeta)d\zeta \otimes d\zeta$ also encodes information on possible deformations of the given complex structure. Explicitly, by performing an affine transformation with constant dilatation $K > 1$, one defines a new uniformizing variable $z'_{(\psi)}$ associated with $\psi = \mu(\zeta)d\zeta \otimes d\zeta$ by deforming the variable $z_{(\psi)}$ defined by (B.28) according to

$$z'_{(\psi)} = KRe(z_{(\psi)}) + \sqrt{-1}Im(z_{(\psi)}). \tag{B.32}$$

The new metric associated with such a deformation is provided by

$$\left|dz'_{(\psi)}\right|^2 = \frac{(K+1)^2}{4}\left|dz_{(\psi)} + \frac{K-1}{K+1}d\bar{z}_{(\psi)}\right|^2. \tag{B.33}$$

Since $dz^2_{(\psi)}$ is the given quadratic differential $\psi = \mu(\zeta)d\zeta \otimes d\zeta$, we can equivalently write $\left|dz'_{(\psi)}\right|^2$ as

$$\left|dz'_{(\psi)}\right|^2 = \frac{(K+1)^2}{4}\,|\mu|\left|d\zeta + \frac{K-1}{K+1}\left(\frac{\overline{\mu}}{|\mu|^{1/2}}\right)d\overline{\zeta}\right|^2, \tag{B.34}$$

where $(\overline{\mu}/|\mu|)\,d\overline{\zeta} \otimes d\zeta^{-1}$ is the (Teichmüller-) Beltrami form associated with the quadratic differential ψ. If we consider quadratic differentials $\psi = \mu(\zeta)d\zeta \otimes d\zeta$ in the open unit ball $Q^{(1)}_{N_0}(M) \doteq \{\psi|,\ \|\mu(\zeta)\| < 1\}$ in the Teichmüller norm(B.29), then there is a natural choice for the constant K provided by

$$K = \frac{1 + \|\mu(\zeta)\|}{1 - \|\mu(\zeta)\|}. \tag{B.35}$$

In this latter case we get

$$\left|dz'_{(\psi)}\right|^2 = \frac{\|\mu(\zeta)\|^2}{(\|\mu(\zeta)\| - 1)^2}\,|\mu|\left|d\zeta + \frac{1}{\|\mu(\zeta)\|}\left(\frac{\overline{\mu}}{|\mu|^{1/2}}\right)d\overline{\zeta}\right|^2. \tag{B.36}$$

According to Teichmüller's existence theorem any complex structure on can be parametrized by the metrics (B.36) as $\psi = \mu(\zeta)d\zeta \otimes d\zeta$ varies in $Q^{(1)}_{N_0}(M)$. This is equivalent to saying that for any given (M, g), (with $(M, N_0; [g])$ defining a reference complex structure $\mathscr{C}_0$ on (M, N_0)), and any diffeomorphism $f \in \mathscr{D}iff_0(M; N_0)$ mapping (M, g) into[1] (M, g_1), there is a quadratic differential $\psi \in Q^{(1)}_{N_0}(M)$ and a biholomorphic map $F \in \mathscr{D}iff_0(M; N_0)$, homotopic to f such that $[F^* g_1]$ is given by the conformal class associated with (B.36). This is the familiar point of view which allows to identify Teichmüller space with the open unit ball $Q^{(1)}_{N_0}(M)$ in the space of quadratic differentials $Q_{N_0}(M)$. It is also worthwhile noticing that (B.36) allows us to consider the open unit ball $Q^{(1)}_{N_0}(M)$ in the space of quadratic differential as providing a slice for the combined action of $\mathscr{D}iff_0(M; N_0)$ and of the conformal group $\mathscr{C}onf^s(M; N_0)$ on the space of Riemannian metrics $\mathscr{M}et(M; N_0)$, i.e.

$$Q^{(1)}_{N_0}(M) \hookrightarrow \frac{\mathscr{M}et(M; N_0)}{\mathscr{D}iff_0(M; N_0)} \simeq \mathscr{C}onf^s(M; N_0) \times \mathfrak{T}_{g,N_0}(M) \tag{B.37}$$

$$\left[\left|dz'_\psi\right|^2\right] \longmapsto \frac{\|\mu\|^2}{(\|\mu\| - 1)^2}\,|\mu| \cdot \left|d\zeta + \frac{1}{\|\mu\|}\left(\frac{\overline{\mu}}{|\mu|^{1/2}}\right)d\overline{\zeta}\right|^2,$$

where

$$\mathscr{C}onf^s(M; N_0) \doteq \left\{f : M \to \mathbb{R}^+ | f \in H^s(M, \mathbb{R})\right\} \tag{B.38}$$

---

[1] With $(M, N_0; [g_1])$ a complex structure distinct from $(M, N_0; [g])$.

denotes the (Weyl) space of conformal factors defined by of all positive (real valued) functions on M, $f \in H^s(M, \mathbb{R})$, whose derivatives up to the order s exist in the sense of distributions and are represented by square integrable functions.

In line with the above remarks, one introduces also the space $B_{N_0}(M)$ of (L^∞ measurable) Beltrami differentials, i.e. of tensor fields $\varpi = v(\zeta)d\bar{\zeta}\otimes d\zeta^{-1}$, sections of $k^{-1}\otimes\bar{k}$, (k being the holomorphic cotangent bundle to M), with $\sup_M |v(\zeta)| < \infty$. The space of Beltrami differentials is naturally identified with the tangent space to $\mathfrak{T}_{g,N_0}(M)$, i.e.,

$$\varpi = v(\zeta)d\bar{\zeta} \otimes d\zeta^{-1} \in T_{\mathscr{C}}\mathfrak{T}_{g,N_0}(M), \tag{B.39}$$

(with $\mathscr{C}$ a complex structure in $\mathfrak{T}_{g,N_0}(M)$). The two spaces $Q_{N_0}(M)$ and $B_{N_0}(M)$ can be naturally paired according to

$$\langle\psi|\varpi\rangle = \int_M \mu(\zeta)v(\zeta)d\zeta d\bar{\zeta}. \tag{B.40}$$

In such a sense $Q_{N_0}(M)$ is $\mathbb{C}$-anti-linear isomorphic to $T_{\mathscr{C}}\mathfrak{T}_{g,N_0}(M)$, and can be canonically identified with the cotangent space $T^*_{\mathscr{C}}\mathfrak{T}_{g,N_0}(M)$ to $\mathfrak{T}_{g,N_0}(M)$. On the cotangent bundle $T^*_{\mathscr{C}}\mathfrak{T}_{g,N_0}(M)$ we can define the Weil-Petersson metric as the inner product between quadratic differentials corresponding to the L^2- norm provided by

$$\|\psi\|^2_{WP} \doteq \int_M h^{-2}(\zeta)\,|\psi(\zeta)|^2\,|d\zeta|^2\,, \tag{B.41}$$

where $\psi \in Q_{N_0}(M)$ and $h(\zeta)\,|d\zeta|^2$ is the hyperbolic metric on M. Note that $\frac{\psi}{h}$ is a Beltrami differential on M, thus if we introduce a basis $\{\frac{\partial}{\partial\mu_\alpha}\}_{\alpha=1}^{3g-3+N_0}$ of the vector space of harmonic Beltrami differentials on (M, N_0), we can write

$$G_{\alpha\bar{\beta}} = \int_M \frac{\partial}{\partial\mu_\alpha}\,\frac{\partial}{\partial\bar{\mu}_\beta}\,h(\zeta)\,|d\zeta|^2 \tag{B.42}$$

for the components of the Weil-Petersson metric on the tangent space to $\mathfrak{T}_{g,N_0}(M)$. We can introduce the corresponding Weil-Petersson Kähler form according to

$$\omega_{WP} := \sqrt{-1}G_{\alpha\bar{\beta}}dZ^\alpha \wedge d\bar{Z}^\beta, \tag{B.43}$$

where $\{dZ^\alpha\}$ are the basis, in $Q_{N_0}(M)$, dual to $\{\frac{\partial}{\partial\mu_\alpha}\}$ under the pairing (B.40). Such Kähler potential is invariant under the mapping class group $\mathfrak{Map}(M; N_0)$ to the effect that the Weil-Petersson volume 2-form ω_{WP} on $\mathfrak{T}_{g,N_0}(M)$ descends on $\mathfrak{M}_{g,N_0}$, and it has a (differentiable) extension, in the sense of orbifold, to $\overline{\mathfrak{M}}_{g,N_0}$.

B.3   Some Properties of the Moduli Spaces $\mathfrak{M}_{g,N_0}$

It is well-known that the moduli space $\mathfrak{M}_{g,N_0}$ is a connected orbifold space of complex dimension $3g - 3 + N_0(T)$ and that, although in general non complete, it admits a stable compactification (Deligne-Mumford) $\overline{\mathfrak{M}}_{g,N_0}$. This latter is, by definition, the moduli space of stable N_0-pointed surfaces of genus g, where a stable surface is a compact Riemann surface with at most ordinary double points such that its parts are hyperbolic. Topologically, a stable pointed surface (or, equivalently a stable curve, in the complex sense), is obtained by considering a finite collection of embedded circles in $M' := M/\{p_1, \ldots p_{N_0}\}$, each in a distinct isotopy class relative to M', and such that none of these circles bound a disk in M containing at most one p_k. By contracting each such a circle, and keeping track of the marked points in such a way that any component of the resulting surface can support a hyperbolic metric, we get the stable pointed surface of genus g. Thus, the closure $\partial\mathfrak{M}_{g,N_0}$ of $\mathfrak{M}_{g,N_0}$ in $\overline{\mathfrak{M}}_{g,N_0}$ consists of stable surfaces with double points, and gives rise to a stratification decomposing $\overline{\mathfrak{M}}_{g,N_0}$ into subvarieties. By definition, a stratum of codimension k is the component of $\overline{\mathfrak{M}}_{g,N_0}$ parametrizing stable surfaces (of fixed topological type) with k double points. The orbifold $\overline{\mathfrak{M}}_{g,N_0}$ is endowed with $N_0(T)$ natural line bundles $\mathscr{L}_i$ defined by the cotangent space to M at the i-th marked point.

A basic observation in moduli space theory is the fact that any point p on a stable curve $((M; N_0), \mathscr{C}) \in \overline{\mathfrak{M}}_{g,N_0}$ defines a natural mapping

$$((M; N_0), \mathscr{C}) \longrightarrow \overline{\mathfrak{M}}_{g,N_0+1} \tag{B.44}$$

that determines a stable curve $((M; N_0 + 1), \mathscr{C}') \in \overline{\mathfrak{M}}_{g,N_0+1}$. Explicitly, as long as the point p is disjoint from the set of marked points $\{p_k\}_{k=1}^{N_0}$ one simply defines $((M; N_0+1), \mathscr{C}')$ to be $((M; N_0, \{p\}), \mathscr{C})$. If the point $p = p_h$ for some $p_h \in \{p_k\}_{k=1}^{N_0}$, then: (i) for any $1 \leq i \leq N_0$, with $i \neq h$, identify $p_i' \in ((M; N_0 + 1), \mathscr{C}')$ with the corresponding p_i; (ii) take a thrice–pointed sphere $\mathbb{CP}^1_{(0,1,\infty)}$, label with a sub-index h one of its marked points $(0, 1, \infty)$, say ∞_h, and attach it to the given $p_h \in [((M; N_0), \mathscr{C})]$; (iii) relabel the remaining two marked points $(0, 1) \in \mathbb{CP}^1_{(0,1,\infty)}$ as p_h' and p_{N_0+1}'. In this way, we get a genus g noded surface

$$s_h\left[((M; N_0), \mathscr{C})\right] = \left[((M; N_0 + 1), \mathscr{C}')\right] \doteq ((M; N_0), \mathscr{C})_h \cup \mathbb{CP}^1_{(0,1,\infty_h)} \tag{B.45}$$

with a rational tail and with a double point corresponding to the original marked point p_h. Finally if p happens to coincide with a node, then $[((M; N_0 + 1), \mathscr{C}')]$ results from setting $p_j' \doteq p_j$ for any $1 \leq i \leq N_0$ and by: (i) normalizing $[((M; N_0), \mathscr{C})]$ at the node (i.e., by separating the branches of $[((M; N_0), \mathscr{C})]$ at p); (ii) inserting a copy of $\mathbb{CP}^1_{(0,1,\infty)}$ with $\{0, \infty\}$ identified with the preimage of p and with $p_{N_0+1}' \doteq 1 \in \mathbb{CP}^1_{(0,1,\infty)}$. Conversely, let

$$\pi : \overline{\mathfrak{M}}_{g,N_0+1} \longrightarrow \overline{\mathfrak{M}}_{g,N_0} \tag{B.46}$$

$$[((M; N_0 + 1), \mathscr{C})] \vdash \quad \overset{\textit{forget}}{\underset{\textit{collapse}}{\&}} \quad \longrightarrow \left[((M; N_0), \mathscr{C}') \right]$$

the projection which forgets the $(N_0 + 1)$st marked point and collapse to a point any irreducible unstable component of the resulting curve. The fiber of π over $((M; N_0), \mathscr{C})$ is parametrized by the map (B.44), and if $((M; N_0), \mathscr{C})$ has a trivial automorphism group $Aut[((M; N_0), \mathscr{C})]$ then $\pi^{-1}((M; N_0), \mathscr{C})$ is by definition the surface $((M; N_0), \mathscr{C})$, otherwise it is identified with the quotient

$$((M; N_0), \mathscr{C})/Aut[((M; N_0), \mathscr{C})]. \tag{B.47}$$

Thus, under the action of π, we can consider $\overline{\mathfrak{M}}_{g,N_0+1}$ as a family (in the orbifold sense) of Riemann surfaces over $\overline{\mathfrak{M}}_{g,N_0}$ and we can identify $\overline{\mathfrak{M}}_{g,N_0+1}$ with the universal curve $\overline{\mathscr{C}}_{g,N_0}$,

$$\pi : \overline{\mathscr{C}}_{g,N_0} \longrightarrow \overline{\mathfrak{M}}_{g,N_0} . \tag{B.48}$$

Note that, by construction, $\overline{\mathscr{C}}_{g,N_0}$ (but for our purposes is more profitable to think in terms of $\overline{\mathfrak{M}}_{g,N_0+1}$) comes endowed with the N_0 natural sections $s_1, \ldots, s_{N_0}$

$$s_h : \overline{\mathfrak{M}}_{g,N_0} \longrightarrow \overline{\mathscr{C}}_{g,N_0} \tag{B.49}$$

$$[((M; N_0), \mathscr{C})] \longmapsto s_h \left[((M; N_0), \mathscr{C}) \right] \doteq ((M; N_0), \mathscr{C})_h \cup \mathbb{CP}^1_{(0,1,\infty_h)},$$

defined by (B.45).

The images of the sections s_i characterize a divisor $\{D_i\}_{i=1}^{N_0}$ in $\overline{\mathscr{C}}_{g,N_0}$ which has a great geometric relevance in discussing the topology of $\overline{\mathfrak{M}}_{g,N_0}$. Such a study exploits the properties of the tautological classes over $\overline{\mathscr{C}}_{g,N_0}$ generated by the sections $\{s_i\}_{i=1}^{N_0}$ and by the corresponding divisors $\{D_i\}_{i=1}^{N_0}$. To define such classes, recall that the cotangent bundle (in the orbifold sense) to the fibers of the universal curve $\pi : \overline{\mathscr{C}}_{g,N_0} \longrightarrow \overline{\mathfrak{M}}_{g,N_0}$ gives rise to a holomorphic line bundle $\omega_{g,N_0} \doteq \omega_{\overline{\mathscr{C}}_{g,N_0}/\overline{\mathfrak{M}}_{g,N_0}}$ over $\overline{\mathscr{C}}_{g,N_0}$ (the relative dualizing sheaf of $\pi : \overline{\mathscr{C}}_{g,N_0} \longrightarrow \overline{\mathfrak{M}}_{g,N_0}$), this is essentially the sheaf of 1-forms with a natural polar behavior along the possible nodes of the Riemann surface describing the fiber of π. It is worthwhile to discuss the behavior of the relative dualizing sheaf ω_{g,N_0} restricted to the generic divisor D_h generated by the section s_h. To this end, let $z_1(h)$ and $z_2(\infty_h)$ denote local coordinates defined in the disks $\Delta_{p_h} \doteq \{|z_1(h)| < 1\}$ and $\Delta_{\infty_h} \doteq \{|z_2(\infty_h)| < 1\}$ respectively centered around the marked points $p_h \in ((M; N_0), \mathscr{C})$, and $\infty \in \mathbb{CP}^1_{(0,1,\infty)}$. Let $\Delta_{t_h} = \{t_h \in \mathbb{C} : |t_h| < 1\}$. Consider the analytic family $s_h(t_h)$ of surfaces of genus g defined over Δ_{t_h} and obtained by removing the disks $|z_1(h)| < |t_h|$ and $|z_2(\infty_h)| < |t_h|$ from $((M; N_0), \mathscr{C})$ and $\mathbb{CP}^1_{(0,1,\infty)}$ and gluing the resulting surfaces through the annulus $\{(z_1(h), z_2(\infty_h))| \ z_1(h)z_2(\infty_h) = t_h, \ t_h \in \Delta_{t_h}\}$ by identifying the points of coordinate $z_1(h)$ with the points of coordinates $z_2(\infty_h) = t_h/z_1(h)$. The family $s_h(t_h) \to \Delta_{t_h}$ opens the node $z_1(h)z_2(\infty_h) = 0$ of the section $s_h|_{((M;N_0),\mathscr{C})}$. Note that in such a way we can independently and holomorphically open the distinct nodes of the various sections $\{s_k\}_{k=1}^{N_0}$. More generally, while opening the node we can also vary the complex structure of $((M;N_0),\mathscr{C})$ by introducing local complex coordinates $(\tau_\alpha)_{\alpha=1}^{3g-3+N_0}$ for $\mathfrak{M}_{g,N_0}$ around $((M;N_0),\mathscr{C})$. If

$$s_h(\tau_\alpha, t_h) \to \mathfrak{M}_{g,N_0} \times \Delta_{t_h} \tag{B.50}$$

denotes the family of surfaces opening of the node, then in the corresponding coordinates (τ_α, t_h) the divisor D_h, image of the section s_h, is locally defined by the equation $t_h = 0$. Similarly, the divisor $D \doteq \sum_{h=1}^{N_0} D_h$ is characterized by the locus of equation $\prod_{h=1}^{N_0} t_h = 0$.

The elements of the dualizing sheaf $\omega_{g,N_0}|_{s_h(t_h)} \doteq \omega_{g,N_0}(D_h)$ are differential forms $u(h) = u_1 dz_1(h) + u_2 dz_2(\infty_h)$ such that $u(h) \wedge dt_h = f dz_1(h) \wedge dz_2(\infty_h)$, where f is a holomorphic function of $z_1(h)$ and $z_2(\infty_h)$. By differentiating $z_1(h)z_2(\infty_h) = t_h$, one gets $f = u_1 z_1(h) - u_2 z_2(\infty_h)$ which is the defining relation for the forms in $\omega_{g,N_0}(D_h)$. In particular, by choosing $u_1 = f/2z_1(h)$, and $u_2 = f/2z_2(\infty_h)$ we get the local isomorphism between the sheaf of holomorphic functions $\mathscr{O}_{s_h(t_h)}$ over $s_h(t_h)$ and $\omega_{g,N_0}(D_h)$

$$f \longmapsto u(h) = f\left(\frac{1}{2} \frac{dz_1(h)}{z_1(h)} - \frac{1}{2} \frac{dz_2(\infty_h)}{z_2(\infty_h)} \right). \tag{B.51}$$

If we set $f = f_0 + f_1(z_1(h)) + f_2(z_2(\infty_h))$, where f_0 is a constant and $f_1(0) = 0 = f_2(0)$, then on the noded surface s_h, $(t_h = 0)$, we get from the relation $z_2(\infty_h) dz_1(h) + z_1(h) dz_2(\infty_h) = 0$,

$$u_h|_{z_2(\infty_h)=0} = \frac{f_0 + f_1(z_1(h))}{z_1(h)} dz_1(h), \tag{B.52}$$

$$u_h|_{z_1(h)=0} = -\frac{f_0 + f_2(z_2(\infty_h))}{z_2(\infty_h)} dz_2(\infty_h),$$

on the two branches $\Delta_{p_h} \cap ((M;N_0),\mathscr{C})$ and $\Delta_{\infty_h} \cap \mathbb{CP}^1_{(0,1,\infty)}$ of the node where $z_1(h)$ and $z_2(\infty_h)$ are a local coordinate (i.e. $z_2(\infty_h) = 0$ and $z_1(h) = 0$, respectively). Thus, near the node of s_h, $\omega_{g,N_0}(D_h)$ is generated by $\frac{dz_1(h)}{z_1(h)}$ and $\frac{dz_2(\infty_h)}{z_2(\infty_h)}$ subjected to the relation $\frac{dz_1(h)}{z_1(h)} + \frac{dz_2(\infty_h)}{z_2(\infty_h)} = 0$. Stated differently, a section of the sheaf $\omega_{g,N_0}(D_h)$ pulled back to the smooth normalization $((M;N_0),\mathscr{C})_{p_h} \bigsqcup \mathbb{CP}^1_{(0,1,\infty_h)}$ of s_h can be identified with a meromorphic 1-form with at most simple poles at the marked points p_h and ∞_h which are identified under the normalization map, and with opposite residues at such marked points. By extending such a construction to all N_0 sections $\{s_h\}_{h=1}^{N_0}$, we can define the line bundle

$$\omega_{g,N_0}(D) \doteq \omega_{g,N_0}\left(\sum_{i=1}^{N_0} D_i \right) \longrightarrow \overline{\mathscr{C}}_{g,N_0} \tag{B.53}$$

as ω_{g,N_0} twisted by the divisor $D \doteq \sum_{h=1}^{N_0} D_h$, viz. the line bundle locally generated by the differentials $\frac{dz_1(h)}{z_1(h)}$ for $z_1(h) \neq 0$ and $-\frac{dz_2(\infty_h)}{z_2(\infty_h)}$ for $z_2(\infty_h) \neq 0$, with $z_1(h)z_2(\infty_h) = 0$, and $h = 1,\ldots,N_0$. As above, $\{z_1(h)\}_{h=1}^{N_0}$ are local variables at the marked points $\{p_h\}_{i=1}^{N_0} \in ((M;N_0),\mathscr{C})$, whereas $z_2(\infty_h)$ is the corresponding variable in the thrice–pointed sphere $\mathbb{CP}^1_{(0,1,\infty)}$.

If, roughly speaking, we interpret $\overline{\mathfrak{M}}_{g,N_0+1}$, as a bundle of surfaces over $\overline{\mathfrak{M}}_{g,N_0}$, then the stratification of $\overline{\mathfrak{M}}_{g,N_0+1}$ into subvarieties $\{V_k\}$ provides an intuitive motivation for the existence of characteristic classes ($\mathbb{Q}$-Poincaré duals of the $\{V_k\}$) that describe the topological properties of $\overline{\mathfrak{M}}_{g,N_0}$. Two such families of classes have proven to be particularly relevant. The first is obtained by pulling back ω_{g,N_0} to $\overline{\mathfrak{M}}_{g,N_0}$ by means of the sections s_k. In this way one gets the line bundle

$$s_k^* \omega_{g,N_0} \doteq \mathscr{L}_k \to \overline{\mathfrak{M}}_{g,N_0} \tag{B.54}$$

whose fiber at the moduli point $((M;N_0),\mathscr{C})$ is defined by the cotangent space $T^*_{(M,p_k)}$ to $((M;N_0),\mathscr{C})$ at the marked point p_k. By taking the first Chern class $c_1(\mathscr{L}_k)$ of the resulting bundles $\{\mathscr{L}_k\}_{k=1}^{N_0}$ one gets the Witten classes $\psi_{(g,N_0),k} \in H^2(\overline{\mathfrak{M}}_{g,N_0};\mathbb{Q})$

$$\psi_{(g,N_0),k} \doteq c_1(\mathscr{L}_k). \tag{B.55}$$

Conversely, if we take the Chern class $c_1\left(\omega_{g,N_0}(D)\right)$ of the line bundle $\omega_{g,N_0}(D)$ and intersect it with itself $j \geq 0$ times, then one can define the Mumford $k_{(g,N_0)j}$ classes $\in H^{2j}(\overline{\mathfrak{M}}_{g,N_0};\mathbb{Q})$ according to [1]

$$k_{(g,N_0)j} \doteq \pi_* \left(\left(c_1\left(\omega_{g,N_0}(D)\right)\right)^{j+1}\right), \tag{B.56}$$

where π_* denotes fiber integration. If $\widehat{\psi}_{N_0+1} \doteq \psi_{(g,N_0+1),N_0+1}$ denotes the Witten class in $H^2(\overline{\mathfrak{M}}_{g,N_0+1};\mathbb{Q})$ associated with the last marked point, then $k_{(g,N_0)j}$ can be also defined as

$$k_{(g,N_0)j} \doteq \pi_* \left(\left(\widehat{\psi}_{N_0+1}\right)^{j+1}\right) \tag{B.57}$$

It is worthwhile recalling that $k_{(g,N_0),1}$ is the class of the Weil-Petersson Kähler form ω_{W-P} on $\overline{\mathfrak{M}}_{g,N_0}$, [1],

$$k_{(g,N_0),1} = \pi_* \left(\left(c_1\left(\omega_{g,N_0}(D)\right)\right)^2\right) = \frac{1}{2\pi^2}[\omega_{W-P}]. \tag{B.58}$$

B.4   Strebel Theorem

A basic result in phrasing the correspondence between Riemann surfaces and combinatorial structures is provided by Strebel's theory of holomorphic (and meromorphic) quadratic differentials [28]. Recall that these objects are the holomorphic (meromorphic) sections of $T'M \otimes T'M$, i.e. the tensor fields on $(M, \mathscr{C})$ that can be locally written as $\Phi = \phi(z)\, dz \otimes dz$, for some holomorphic (meromorphic) $\phi(z)$. Geometrically the role of holomorphic quadratic differential stems from the basic observation that for a given point $p \in U$, the function $\zeta(q) := \int_p^q \sqrt{\phi(z)\, dz^2}$ provides a local conformal parameter in a neighborhood $U' \subset U$ of p if $\phi(p) \neq 0$. In terms of the coordinate ζ we can write $\Phi = d\zeta \otimes d\zeta$, and the sets $\zeta^{-1}\{z\,|\, \Im z = const\}$, and $\zeta^{-1}\{z\,|\, \Re z = const\}$ foliate U' in the standard $\zeta = X + \sqrt{-1}\, Y$ orthogonal way. In general, the structure of the foliation induced by Φ around its zeros and poles is quite more sophisticated, however the case relevant to ribbon graphs and polyhedral surfaces can be fully described by the Strebel theorem. For future reference, it is worthwhile having it handy in the elegant formulation provided by Mulase and Penkawa [24]:

Theorem B.1 *Let $((M; N_0), \mathscr{C})$ be a closed Riemann surface of genus $g \geq 0$ with $N_0 \geq 1$ marked points $\{p_k\}_{k=1}^{N_0}$, where $2-2g-N_0 < 0$. Let us denote by $(L_1, \ldots, L_{N_0})$ an ordered N_0–tuple of positive real numbers. Then there is a unique (Jenkins–Strebel) meromorphic quadratic differential Φ on $((M; N_0), \mathscr{C})$ such that:*

 (i) Φ is holomorphic on $M \setminus \{p_1, \ldots, p_{N_0}\}$;
(ii) Φ has a double pole at each p_k, $k = 1, \ldots, N_0$;
(iii) The union of all non–compact horizontal leaves $\zeta^{-1}\{z\,|\, \Im z = const\}$ form a closed subset $\subset ((M; N_0), \mathscr{C})$ of measure zero;
(iv) Every compact horizontal leaf λ is a simple loop circling around one of the poles. In particular, if λ_k is the loop around the pole p_k, then $L_k = \oint_{\lambda_k} \sqrt{\Phi}$, (the branch of $\sqrt{\Phi}$ is chosen so that the integral is positive when the circuitation along λ is along the positive orientation induced by $((M; N_0), \mathscr{C}))$;
 (v) Every non–compact horizontal leaf of Φ is bounded by zeros of Φ. Conversely, every zero of degree m of Φ bounds $m + 2$ horizontal leaves;
(vi) If N_2 denotes the number of zeros of the quadratic differential Φ, then Φ induces a unique cell–decomposition of $((M; N_0), \mathscr{C})$ with N_0 2–cells, N_1 1–cells, and N_2 0–cells where $N_1 = N_0 + N_2 - 2 + 2g$. The 1–skeleton of this cell decomposition is a metric ribbon graph with vertex–valency ≥ 3.

It is straightforward to check that Strebel's theorem characterizes a map $\mathscr{S}$: $\mathfrak{M}_{g,N_0} \times \mathbb{R}_+^{N_0} \to RG_{g,N_0}^{met}$ which, given a sequence of N_0 positive real numbers $\{L(k)\}$, associates to a pointed Riemann surface $((M; N_0), \mathscr{C}) \in \mathfrak{M}_{g,N_0}$ a metric ribbon graph $\Gamma \in RG_{g,N_0}^{met}$ with N_0 labelled boundary components $\{\partial\Gamma(k)\}$ of perimeters $\{L(k)\}$. This map is actually a bijection since, given a metric ribbon graph $\Gamma \in \mathbb{R}_+^{|e(\Gamma)|}$ with N_0 labelled boundary components $\{\partial\Gamma(k)\}$ of perimeters $\{L(k)\}$, we can, out of such combinatorial data, construct a decorated Riemannian surface

$((M; N_0), \mathscr{C}; \{L(K)\}) \in \mathfrak{M}_{g,N_0} \times \mathbb{R}_+^{N_0}$. The characterization of the correspondance $\mathscr{S}^{-1} : \Gamma \mapsto ((M; N_0), \mathscr{C}; \{L(K)\})$ is described in a cristal clear way in [24], and one eventually establishes [19] the

Theorem B.2 *Strebel theory defines a natural bijection*

$$\mathfrak{M}_{g,N_0} \times \mathbb{R}_+^{N_0} \simeq RG_{g,N_0}^{met}, \tag{B.59}$$

between the decorated moduli space $\mathfrak{M}_{g,N_0} \times \mathbb{R}_+^{N_0}$ *and the space of all metric ribbon graphs* RG_{g,N_0}^{met} *with* N_0 *labelled boundary components.*

B.5   The Teichmüller Space of Surfaces with Boundaries

This is also the appropriate place for a few remarks on moduli space theory for surfaces with boundaries. The elements of the Teichmüller space $\mathfrak{T}_{g,N_0}^{\partial}$ of hyperbolic surfaces Ω with N_0 geodesic boundary components are marked Riemann surface modelled on a surface S_{g,N_0} of genus g with complete finite-area metric of constant Gaussian curvature -1, (and with N_0 geodesic boundary components $\partial S = \sqcup \partial S_j$), i.e., a triple (S_{g,N_0}, f, Ω) where $f : S_{g,N_0} \to \Omega$ is a quasiconformal homeomorphism, (the marking map), which extends uniquely to a homeomorphism from $S_{g,N_0} \cup \partial S$ onto $\Omega \cup \partial \Omega$. Any two such a triple $(S_{g,N_0}, f_1, \Omega_{(1)})$ and $(S_{g,N_0}, f_2, \Omega_{(2)})$ are considered equivalent iff there is a biholomorphism $h : \Omega_{(1)} \to \Omega_{(2)}$ such that $f_2^{-1} \circ h \circ f_1 : S_{g,N_0} \cup \partial S \to S_{g,N_0} \cup \partial S$ is homotopic to the identity via continuous mappings pointwise fixing ∂S.

For a given string $L = (L_1, \ldots, L_{N_0}) \in \mathbb{R}_{\geq 0}^{N_0}$, we let $\mathfrak{T}_{g,N_0}^{\partial}(L)$ denote the Teichmüller space of hyperbolic surfaces Ω with geodesic boundary components of length

$$(|\partial \Omega_1|, \ldots, |\partial \Omega_{N_0}|) = (L_1, \ldots, L_{N_0}) \doteq L \in \mathbb{R}_{\geq 0}^{N_0}. \tag{B.60}$$

This characterizes the boundary length map

$$\mathfrak{L} : \mathfrak{T}_{g,N_0}^{\partial} \longrightarrow \mathbb{R}_{\geq 0}^{N_0} \tag{B.61}$$

$$\Omega \longmapsto \mathfrak{L}(\Omega) = (|\partial \Omega_1|, \ldots, |\partial \Omega_{N_0}|) ,$$

and we can write $\mathfrak{T}_{g,N_0}^{\partial}(L) := \mathfrak{L}^{-1}(L)$. Note that, by convention, a boundary component such that $|\partial \Omega_j| = 0$ is a cusp and moreover $\mathfrak{T}_{g,N_0}(L = 0) = \mathfrak{T}_{g,N_0}$, where $\mathfrak{T}_{g,N_0}$ is the Teichmüller space of hyperbolic surfaces with N_0 punctures, (with $6g - 6 + 2N_0 \geq 0$). For each given string $L = (L_1, \ldots, L_{N_0})$ there is a natural action on $\mathfrak{T}_{g,N_0}(L)$ of the mapping class group $\mathfrak{Map}_{g,N_0}^{\partial}$ defined by the group of all the isotopy classes of orientation preserving homeomorphisms of Ω which leave each boundary component $\partial \Omega_j$ pointwise (and isotopy-wise) fixed. This action changes the marking f of S_{g,N_0} on Ω, and characterizes the quotient space

$$\mathfrak{M}_{g,N_0}(L) \doteq \frac{\mathfrak{T}_{g,N_0}(L)}{\mathfrak{Map}^{\partial}_{g,N_0}} \tag{B.62}$$

as the moduli space of Riemann surfaces (homeomorphic to S_{g,N_0}) with N_0 boundary components of length $|\partial\Omega_k| = L_k$. Note again that when $\{L_k \to 0\}$, $\mathfrak{M}_{g,N_0}(L)$ reduces to the usual moduli space $\mathfrak{M}_{g,N_0}$ of Riemann surfaces of genus g with N_0 marked points. It is worthwhile noticing that, since the boundary components of a surface $\Omega_{g,N_0} \in \mathfrak{T}^{\partial}_{g,N_0}$ are left pointwise fixed, any surface Ω_{g,N_0} in $\mathfrak{T}^{\partial}_{g,N_0}$, for $N_0 > 1$, can be embedded into a surface $\Omega'_{g+1,N_0-1} \in \mathfrak{T}^{\partial}_{g+1,N_0-1}$ by glueing the two legs of a pair of pants onto two of the boundary components of Ω_{g,N_0}. The attachment of a torus with two boundary components allows also to include $\Omega_{g,1}$ in $\Omega_{g+1,1}$. Such a chain of embeddings induces a corresponding chain of embeddings of the mapping class group $\mathfrak{Map}^{\partial}_{g,N_0}$ into $\mathfrak{Map}^{\partial}_{g+1,N_0-1}$. Under direct limit, this gives rise to a notion of stable mapping class group playing a basic role in the study of the cohomology of the moduli space.

We conclude this notational capsule by specializing to the boundary case the characterization of the Weil–Petersson inner product. The reader will find many more details in the remarkable and very informative papers by G. Mondello, (in particular [22]). As in Sect. B.2, we introduce the real vector space of holomorphic quadratic differentials $Q^{\partial}_{N_0}(\Omega)$ whose restrictions to $\partial\Omega$ is real. The corresponding space of Beltrami differentials, identified with the tangent space $T_{\mathscr{C}}\,\mathfrak{T}^{\partial}_{g,\,N_0}$, will be denoted by $B_{N_0}(\Omega)$ and the they are paired according to

$$(\mu\,d\zeta \otimes d\zeta, \nu\,d\bar{\zeta} \otimes d\zeta^{-1}) \longmapsto \int_{\Omega} \mu\,\nu\,d\zeta\,d\bar{\zeta}, \tag{B.63}$$

where $\mu\,d\zeta \otimes d\zeta \in Q^{\partial}_{N_0}(\Omega)$ and $\nu\,d\bar{\zeta} \otimes d\zeta^{-1}) \in B_{N_0}(\Omega)$. As in the boundaryless case, by noticing that the ratio between a quadratic differential and the hyperbolic metric $h\,d\zeta \otimes d\bar{\zeta}$ on Ω is a Beltrami differential, we can define the Weil–Petersson inner product on $T_{\mathscr{C}}\,\mathfrak{T}^{\partial}_{g,\,N_0}$ according to

$$G^{\partial}_{\alpha\bar{\beta}} = \int_M \frac{\partial}{\partial\mu_\alpha}\,\frac{\partial}{\partial\bar{\mu}_\beta}\,h(\zeta)\,|d\zeta|^2 \tag{B.64}$$

where $\{\frac{\partial}{\partial\mu_\alpha}\}_{\alpha=1}^{3g-3+2N_0}$ is a basis of the vector space of harmonic Beltrami differentials on Ω. The corresponding Weil-Petersson form is provided by

$$\omega_{WP} := \sqrt{-1}\,G^{\partial}_{\alpha\bar{\beta}}\,dZ^\alpha \wedge d\bar{Z}^\beta, \tag{B.65}$$

where $\{dZ^\alpha\}$ are the basis, in $Q^\partial_{N_0}(M)$, dual to $\{\frac{}{\mu_\alpha}\}$ under the pairing (B.63). Note in particular that

$$\eta^{\alpha\bar{\beta}} := \sqrt{-1} \int_\Omega h^{-2}\, dZ^\alpha d\bar{Z}^\beta\, |d\zeta|^2, \tag{B.66}$$

defines the Weil–Petersson Poisson tensor associated with ω_{WP}. It is important to stress that, (because of the presence of the boundaries), the Weil–Petersson form is degenerate (in particular it is not Kähler) on $\mathfrak{T}^\partial_{g,N_0}$. However, the Poisson structure defined by $\eta^{\alpha\bar{\beta}}$ induces a foliation in $\mathfrak{T}^\partial_{g,N_0}$ whose symplectic leaves are the spaces $\mathfrak{T}^\partial_{g,N_0}(L)$, of Riemann surfaces with given boundary length vector $L \in \mathbb{R}^{N_0}_{\geq 0}$, endowed with ω_{WP}. G. Mondello [22] has provided a nice geometrical characterization of the Poisson structure $\eta^{\alpha\bar{\beta}}$ in terms of ideal hyperbolic triangulations of the surfaces $\Omega \in \mathfrak{T}^\partial_{g,N_0}$. As we have seen, this characterization plays an important role in Chap. 3, and it hints to even deeper connections between the geometry of the space of polyhedral surfaces and hyperbolic geometry.

Appendix C
Spectral Theory on Polyhedral Surfaces

In this appendix we briefly discuss the basic facts of spectral theory of Laplace type operators on polyhedral surfaces that we have exploited in these lecture notes.

C.1 Kokotov's Spectral Theory on Polyhedral Surfaces

Spectral theory for cone manifolds has a long standing tradition, (a fine sample of classical works is provided by [2, 5, 6, 10, 18]). Here we shall mainly refer to the elegant results recently obtained by A. Kokotov [17]. They provide a rather complete analysis of the determinant of the Laplacian on polyhedral surfaces. It must be stressed that whereas the study of the determinant of the Laplacian in the smooth setting is a well–developed subject, results in the polyhedral case have been sparse and often subjected to quite restrictive hypotheses [10, 21].

As a consequence of the presence of the conical points $\{p_1, \ldots, N_0\}$, the Laplacian Δ on the Riemann surface $((M, N_0), \mathscr{C}_{sg})$ is not an essentially self-adjoint operator. There are (infinitely) many possible self–adjoint extensions of Δ, with domains typically determined by the behavior of functions (formally) harmonic at the conical points. To take care of this extension problem in a natural way, let us recall that the minimal domain $\mathscr{D}_{min}$ of the Laplacian on $C_0^\infty(M', \mathbb{R})$ consists of the graph closure on the set $C_0^\infty(M', \mathbb{R})$, where a function $u \in \mathscr{D}_{min}$ if there is a sequence $\{u_k\} \in C_0^\infty(M', \mathbb{R})$ and a function $w \in L^2(M)$ such that $u_k \to u$ and $\Delta u_k \to w$ in $L^2(M)$, where $L^2(M)$ denotes the space of square summable functions on $((M, N_0), \mathscr{C}_{sg})$. Similarly, the maximal domain $\mathscr{D}_{max}$ for Δ, corresponding to the domain of the adjoint operator to Δ on $\mathscr{D}_{min}$, is the subspace of functions $v \in L^2(M)$ such that for all $u \in \mathscr{D}_{min}$ there is a function $f \in L^2(M)$ with $\langle v, \Delta u \rangle = \langle f, u \rangle$, where $\langle \cdot, \cdot \rangle$ denotes the $L^2(M)$ pairing on $((M, N_0), \mathscr{C}_{sg})$. All possible distinct self–adjoint extensions of Δ on $((M, N_0), \mathscr{C}_{sg})$ are parametrized by a domain $\mathscr{D}_\Delta$, with $\mathscr{D}_{min} \subseteq \mathscr{D}_\Delta \subseteq \mathscr{D}_{max}$ [23]. Explicitly, we can consider without loss of generality the

© Springer International Publishing AG 2017

M. Carfora, A. Marzuoli, *Quantum Triangulations*, Lecture Notes in Physics 942,

https://doi.org/10.1007/978-3-319-67937-2

case in which we have just one conical point $\{p\}$ with conical angle θ, and denote by z the local conformal parameter in a neighborhood of p. One introduces[17] the functions (formally harmonic on $((M,p),\,\mathscr{C}_{sg}))$, defined by

$$V^k_\pm(z) := |z|^{\pm\frac{2\pi k}{\theta}} \exp\left\{\sqrt{-1}\,\frac{2\pi\,k}{\theta}\,\arg z\right\}, \qquad k > 0, \tag{C.1}$$

$$V^0_+ := 1\,, \qquad\qquad V^0_- := \ln|z|. \tag{C.2}$$

These functions are in $L^2(M)$ as long as $k < \frac{\theta}{2\pi}$, and we can consider the linear subspaces $\mathscr{E}$ of $L^2(M)$ generated by the functions $\varphi\,V^k_\pm(z)$, with $0 \le k < \frac{\theta}{2\pi}$, where φ is a C^∞–mollifier of the characteristic function of the cone with vertex at p, e.g., $\varphi(z) := c\,\exp(|z|^2 - 1)^{-1}$, $|z| < 1$, and $\varphi(z) = 0$, for $|z| \ge 1$, with $\varphi(z = 0) = 1$, (assuming that the region isometric to the cone corresponds to $|z| < 1$). The self–adjoint extension of Δ are then parametrized by the subspaces of functions $\mathscr{E}$ such that[17]

$$\langle\Delta u, v\rangle - \langle u, \Delta v\rangle = \lim_{\varepsilon\searrow 0+}\oint\left(u\,\frac{\partial v}{\partial r} - v\,\frac{\partial u}{\partial r}\right) = 0, \tag{C.3}$$

for all $u,\,v \in \mathscr{E}$, and where $r := |z|$, (note that in the above characterization one exploits the delicate fact that any $u \in \mathscr{D}_{min}$ is such that $u(z) = O(r)$ as $r \searrow 0$–see section 3.2.1 of [17] for details). The Friedrics extension, which is the relevant one for the results to follow, is associated with the subspace $\mathscr{E}$ generated by the functions $\varphi\,V^k_+(z), 0 \le k < \frac{\theta}{2\pi}$. The associated domain $\mathscr{D}_\Delta := \mathscr{D}_{min} + \mathscr{E}$ comprises functions which are bounded near the apex of the cone.

Denote by $\mathscr{H}(z,z';\eta)$ the heat kernel for the Friedrichs extension of the Laplacian Δ. This is a distribution on $M \times M \times [0,\infty)$ such that, for all $u \in \mathscr{D}_\Delta := \mathscr{D}_{min} + \mathscr{E}$ away from the conical points, the convolution $\int_M \mathscr{H}(z,z';\eta)\,u(z')\,dA_{z'}$ is smooth for all $\eta > 0$, and

$$\left(\frac{\partial}{\partial\eta} + \Delta^{(z)}\right)\mathscr{H}(z,z';\eta) = 0, \tag{C.4}$$

$$\lim_{\eta\searrow 0+}\mathscr{H}(z,z';\eta) = \delta(z,z'),$$

where $(z,z';\eta) \in (M \times M\backslash Diag\,(M \times M)) \times [0,\infty)$, and $\Delta^{(z)}$ denotes the Laplacian with respect to the variable z. The Dirac initial condition is understood in the distributional sense, i.e., for any smooth $u \in C^\infty_0(M',\mathbb{R}) \cap \mathscr{D}_\Delta$, $\int_M \mathscr{H}(z,z';\eta)\,u(z')\,dA_{z'} \to u(z)$, as $\eta \searrow 0^+$, where the limit is meant in the uniform norm on $C^\infty_0(M',\mathbb{R})$. We are now ready to state Kokotov's main results

Theorem C.1 (Kokotov[17], Th.1) *Let $((M,N_0),\,\mathscr{C}_{sg})$ be the Riemann surface, with conical singularities $Div(T) := \sum_{k=1}^{N_0}\left(\frac{\Theta(k)}{2\pi} - 1\right)$, associated with the polyhe-*

dral manifold (P_T, M), and let Δ be (the Friedrichs extension) of the corresponding Laplace operator. Then

(i) Δ has a discrete spectral resolution, the eigenvalues $0 = \lambda_0 < \lambda_1 \leq \lambda_2 \leq \ldots \to \infty$ have finite multiplicities, and the associated spectral counting function $N(\lambda) := Card\,[k \geq 1 \,:\, \lambda_k \leq \lambda] = O(\lambda)$, as $\lambda \to \infty$.

(ii) If $Tr\, e^{\eta \Delta}$ denotes the heat trace of the heat kernel $\mathscr{H}(z, z'; \eta)$ associated with $(\Delta, \mathscr{D}_\Delta)$, then, for some $\varepsilon > 0$, the asymptotics

$$Tr\, e^{\eta \Delta} := \int_M \mathscr{H}(z, z; \eta)\, dA = \frac{Area\,((M, N_0), \mathscr{C}_{sg})}{4\pi\, \eta} \tag{C.5}$$

$$+ \frac{1}{12} \sum_{k=1}^{N_0} \left\{ \frac{2\pi}{\Theta(k)} - \frac{\Theta(k)}{2\pi} \right\} + O(e^{-\varepsilon/\eta}),$$

holds in the uniform norm on $C^\infty(M', \mathbb{R})$.

Let

$$\zeta_\Delta(s) := \sum_{\lambda_k > 0} \frac{1}{\lambda_k^s} \tag{C.6}$$

denote the ζ–function associated with the positive part of the spectrum of the operator $(\Delta, \mathscr{D}_\Delta)$, then we have

Theorem C.2 (Kokotov[17]) *The function $\zeta_\Delta(s)$ is holomorphic in the half–plane $\{\Re\, s > 1\}$, and there is an entire function $e(s)$ such that*

$$\zeta_\Delta(s) = \frac{1}{\Gamma(s)} \left\{ \frac{Area\,((M, N_0), \mathscr{C}_{sg})}{4\pi\,(s-1)} \right. \tag{C.7}$$

$$\left. + \left[\frac{1}{12} \sum_{k=1}^{N_0} \left\{ \frac{2\pi}{\Theta(k)} - \frac{\Theta(k)}{2\pi} \right\} - 1 \right] \frac{1}{s} + e(s) \right\},$$

where $\Gamma(s)$ denotes the Gamma function. $\zeta_\Delta(s)$ is a regular at $s = 0$, and one can define the Ray–Singer ζ–regularized determinate of $(\Delta, \mathscr{D}_\Delta)$ according to

$$\overset{'}{\det}\, \Delta := \exp\left\{ -\zeta'_\Delta(s = 0) \right\}. \tag{C.8}$$

As a direct consequence of this representation, one has[17] the

Corollary C.1 *Let $((M, N_0), \mathscr{C}_{sg})$ and $((\widetilde{M, N_0}), \mathscr{C}_{sg})$ two homothetic Riemann surfaces, with (the same) conical singularities $Div(T) := \sum_{k=1}^{N_0} \left(\frac{\Theta(k)}{2\pi} - 1 \right) \sigma^0(k)$, and let ds^2 and $\widetilde{ds}^2 = \kappa\, ds^2$, with $k > 0$ a positive constant, be the respective* *conical metrics, (see Theorem 2.1). If we denote by* $\det' \Delta$ *and* $\det' \widetilde{\Delta}$ *the associated* ζ*–regularized determinants, then one has the rescaling*

$$\det{}' \widetilde{\Delta} \;=\; \kappa^{-\left(\frac{\chi(M)}{6}-1\right)-\frac{1}{12}\sum_{k=1}^{N_0}\left\{\frac{2\pi}{\Theta(k)}+\frac{\Theta(k)}{2\pi}-2\right\}}\; \det{}' \Delta. \tag{C.9}$$

For $\eta \in [0, 1]$, let us consider two distinct families of polyhedral surfaces

$$\eta \mapsto (T_{(1)}, M)_\eta \in POL_{g, N_0}(M) \tag{C.10}$$

$$\eta \mapsto (T_{(2)}, M)_\eta \in POL_{g, \widehat{N}_0}(M), \tag{C.11}$$

(note that generally $N_0 \neq \widehat{N}_0$, and that η may be allowed to vary on a smooth parameter manifold [17]). We assume that the corresponding vertex sets $\{\sigma_{(1)}^0(k, ; \eta)\}_{k=1}^{N_0}$ and $\{\sigma_{(2)}^0(h, \eta)\}_{h=1}^{\widehat{N}_0}$ are disjoint for all $\eta \in [0, 1]$, and that they support distinct η–independent conical singularities $\{\Theta_{(1)}(k)\}$ and $\{\Theta_{(2)}(h)\}$. We also assume that $(T_{(1)}, M)_\eta$, and $(T_{(2)}, M)_\eta$, $\eta \in [0, 1]$, define the same (η–independent) conformal structure $((M, N_0), \mathscr{C}_{sg}^{(1)}) \simeq ((M, \widehat{N}_0), \mathscr{C}_{sg}^{(2)})$. Let $\{p_k(\eta)\}_{k=1}^{N_0} \in M$ and $\{q_h(\eta)\}_{h=1}^{\widehat{N}_0} \in M$ be the disjoint sets of points associated with the divisors

$$Div(T_{(1)}, \ \eta) := \sum_{k=1}^{N_0} \left(\frac{\Theta_{(1)}(k)}{2\pi} - 1 \right) p_k(\eta), \tag{C.12}$$

and

$$Div(T_{(2)}, \ \eta) := \sum_{h=1}^{\widehat{N}_0} \left(\frac{\Theta_{(2)}(h)}{2\pi} - 1 \right) q_h(\eta). \tag{C.13}$$

According to (2.70) the conical metric $ds_{T_{(1)}}^2$ of $((M, N_0), \mathscr{C}_{sg}^{(1)})$ around the generic conical point $p_k(\eta)$ is given, in term of a local conformal parameter $t(k, \ \eta)$, by

$$ds_{T_{(1)}, \ (k)}^2 := \frac{[L(k)]^2}{4\pi^2 \ |t(k, \ \eta)|^2} \ |t(k, \ \eta)|^{2\left(\frac{\Theta_{(1)}(k)}{2\pi}\right)} \ |d\,t(k, \ \eta)|^2 \,, \tag{C.14}$$

whereas the conical metric $ds_{T_{(2)}}^2$ of $((M, \widehat{N}_0), \mathscr{C}_{sg}^{(2)})$ around the generic conical point $q_h(\eta)$ is given, in term of a local conformal parameter $z(h, \ \eta)$, by

$$ds_{T_{(2)}, \ (k)}^2 := \frac{[L'(h)]^2}{4\pi^2 \ |z(h, \ \eta)|^2} \ |z(h, \ \eta)|^{2\left(\frac{\Theta_{(2)}(h)}{2\pi}\right)} \ |d\,z(h, \ \eta)|^2 \,. \tag{C.15}$$

Since in the metric $ds^2_{T_{(1)}}$ the points $\{q_h(\eta)\}_{h=1}^{\widehat{N_0}} \in M$, supporting the conical singularities of $ds^2_{T_{(2)}}$, are regular points, we can assume that there are smooth functions $g_{(2,\,h)}(\eta)$ of $z(h,\,\eta)$ such that in a neighborhood of $q_h(\eta)$ the metric $ds^2_{T_{(1)}}$ takes the form

$$ds^2_{T_{(1)}},\Big|_{q_h(\eta)} := \left| g_{(2,\,h)}(z(h,\,\eta)) \right|^2 \left| d\,z(h,\,\eta) \right|^2 . \tag{C.16}$$

Similarly, we can assume that there are smooth functions $f_{(1,\,k)}(\eta)$ such that in a neighborhood of $p_k(\eta)$ the metric $ds^2_{T_{(2)}}$ takes the form

$$ds^2_{T_{(2)}},\Big|_{p_k(\eta)} := \left| f_{(1,\,k)}(t(k,\,\eta)) \right|^2 \left| d\,t(k,\,\eta) \right|^2 . \tag{C.17}$$

With these notational remarks along the way, we have the following

Theorem C.3 (Kokotov [17]) *If* $\det \Delta^{(1)}$ *and* $\det \Delta^{(2)}$ *respectively denote the* ζ–*regularized determinants of the (Friedrichs extension of the) Laplacian associated with the conical metrics* $ds^2_{T_{(1)}}$ *and* $ds^2_{T_{(2)}}$, *then there is a constant C independent of* $\eta \in [0,\,1]$ *such that*

$$\frac{\det' \Delta^{(1)}}{\det' \Delta^{(2)}} = C \; \frac{Area\left((M, N_0),\, \mathscr{C}_{sg}^{(1)}\right)}{Area\left((M, \widehat{N}_0),\, \mathscr{C}_{sg}^{(2)}\right)} \; \frac{\prod_{h=1}^{\widehat{N_0}} \left| \mathfrak{g}_{(2,\,h)} \right|^{\frac{1}{6}\left(\frac{\Theta^{(2)}(h)}{2\pi}-1\right)}}{\prod_{k=1}^{N_0} \left| \mathfrak{f}_{(1,\,k)} \right|^{\frac{1}{6}\left(\frac{\Theta^{(1)}(k)}{2\pi}-1\right)}}, \tag{C.18}$$

where $Area\left((M, N_0),\, \mathscr{C}_{sg}^{(j)}\right)$, $j = 1, 2$, *denotes the area of the Riemann surface* $\left((M, N_0),\, \mathscr{C}_{sg}^{(1)}\right)$ *in the corresponding conical metric* $ds^2_{T_{(j)}}$, *and where we have set* $\mathfrak{f}_{(1,\,k)} := f_{(1,\,k)}(t(k,\,\eta) = 0)$ *and* $\mathfrak{g}_{(2,\,h)} := g_{(2,\,h)}(z(h,\,\eta) = 0)$.

As emphasized by Kokotov, this results extends to polyhedral surfaces Polyakov's formula describing the scaling, under a conformal transformation, of the determinant of the Laplacian on a smooth Riemann surface.

References

1. Arbarello, E., Cornalba, M.: Combinatorial and algebro-geometrical classes on the moduli spaces of curves. J. Algebraic Geom. **5**, 705–749 (1996)
2. Aurell, E., Salomonson, P.: Further results on functional determinants of Laplacians in simplicial complexes. hep-th/9405140
3. Berger, M., Ebin, D.: Some decomposition of the space of symmetric tensors on a Riemannian manifold. J. Differ. Geom. **3**, 379–392 (1969)
4. Besse, A.L.: Einstein Manifolds. Ergebnisse der Mathematik und ihrer Grenzgebiete, vol. 10. Springer, Berlin/Heidelberg (1987)

5. Carslaw, H.S.: The Green's function for a wedge of any angle, and other problems in the conduction of heat. Proc. Lond. Math. Soc. **8**, 365–374 (1910)
6. Cheeger, J.: Spectral geometry of singular Riemannian spaces. J. Differ. Geom. **18**, 575–657 (1983)
7. Chow, B., Knopf, D.: The Ricci Flow: An Introduction. Mathematical Surveys and Monographs, vol. 110. American Mathematical Society, Providence (2004)
8. Chow, B., Lu, P., Ni, L.: Hamilton's Ricci Flow. Graduate Studies in Mathematics, Vol. 77. Providence, American Mathematical Society (2007)
9. DeTurck, D.M., Kazdan, J.L.: Some regularity theorems in Riemannian geometry. Ann. Sci. École Norm. Sup. **14**, 249–260 (1981)
10. D'Hoker, E., Phong, D.H.: Functional determinants on Mandelstam diagrams. Commun. Math. Phys. **124**, 629–645 (1989)
11. Ebin, D., Marsden, J.: Groups of diffeomorphisms and the motion of an incompressible fluid. Ann. Math. **92**, 102–163 (1970)
12. Farkas, H.M., Kra, I.: Riemann Surfaces, 2nd edn. Springer, New York (1992)
13. Griffiths, P., Harris, J.: Principles of Algebraic Geometry. Wiley, New York (1978)
14. Hebey, E.: Nonlinear Analysis on Manifolds: Sobolev Spaces and Inequalities. Courant Institute Lecture Notes, vol. 5. American Mathematical Society, Providence (2000)
15. Hsu, E.P.: Stochastic Analysis on Manifolds. AMS Graduate Series in Mathematics, vol. 38. American Mathematical Society, Providence (2002)
16. Jost, J.: Compact Riemann Surfaces, 2nd end. Springer, Berlin/Heidelberg (2002)
17. Kokotov, A.: Compact polyhedral surfaces of an arbitrary genus and determinant of Laplacian. Proc. Am. Math. Soc. **141**, 725–735 (2013)
18. Kondratjev, V.: Boundary value problems for elliptic equations in domains with conical points. Proc. Moscow Math. Soc. **16**, 219–292 (1967)
19. Kontsevitch, M.: Intersection theory on the moduli space of curves and the matrix Airy functions. Commun. Math. Phys. **147**, 1–23 (1992)
20. Lanczos, C.: Ein Vereinfachendes Koordinatensystem für die Einsteinschen Gravitationsgleichungen. Phys. Zeits. **23**, 537–539 (1922)
21. Menotti, P., Peirano, P.P.: Functional integration on two dimensional Regge geometries. Nucl.Phys. **B473**, 426–454 (1996). hep-th/9602002
22. Mondello, G.: Triangulated Riemann surfaces with boundary and the Weil-Petersson Poisson structure. J. Differ. Geom. **81**, 391–436 (2009)
23. Mooers, E.A.: Heat kernel asymptotics on manifolds with conical singularities. Journal D'Analyse Mathematique **78**, 1–36 (1999)
24. Mulase, M., Penkava, M., Ribbon graphs, quadratic differentials on Riemann surfaces, and algebraic curves defined over $\overline{\mathbb{Q}}$. Asian J. Math. **2**(4), 875–920 (1998). math-ph/9811024
25. Nag, S.: The Complex Analytic Theory of Teichmüller Spaces. Canadian Mathematical Society Series of Monographs. Wiley, New York (1988)
26. Omori, H.: On the group of diffeomorphisms of a compact manifold. Glob. Anal. (Proc. Symp. Pure Math.) **15**, 167–183 (1968)
27. Petersen, P.: Riemannian Geometry. Graduate Text in Mathematics, vol. 171. Springer, New York (1998)
28. Strebel, K.: Quadratic Differentials. Springer, Berlin (1984)
29. Troyanov, M.: On the moduli space of singular Euclidean surfaces. arXiv:math/0702666v2 [math.DG]

Index

© Springer International Publishing AG 2017

M. Carfora, A. Marzuoli, *Quantum Triangulations*, Lecture Notes in Physics 942,

https://doi.org/10.1007/978-3-319-67937-2